STUDENT'S SOLUTIONS MANUAL

to accompany

Calculus

AND ITS APPLICATIONS

SEVENTH EDITION

STUDENT'S
SOLUTIONS MANUAL

JUDITH A. PENNA

to accompany

Calculus

AND ITS APPLICATIONS

SEVENTH EDITION

Marvin L. Bittinger

Indiana University—Purdue University at Indianapolis

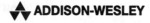 **ADDISON-WESLEY**

An imprint of Addison Wesley Longman, Inc.

Reading, Massachusetts • Menlo Park, California • New York • Harlow, England
Don Mills, Ontario • Sydney • Mexico City • Madrid • Amsterdam

Reproduced by Addison Wesley Longman from camera-ready copy supplied by the author.

Copyright © 2000 Addison Wesley Longman.

ISBN   0-201-33865-3

1 2 3 4 5 6 7 8 9 10 VG 0302010099

Table of Contents

Chapter 1

Functions, Graphs, and Models

Exercise Set 1.1

1. Graph $y = x - 1$.

We choose some x-values and calculate the corresponding y-values to find some ordered pairs that are solutions of the equation. Then we plot the points and draw the graph.

x	y	(x, y)
-4	-5	$(-4, -5)$
0	-1	$(0, -1)$
3	2	$(3, 2)$

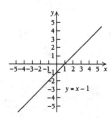

3. Graph $y = -3x$.

We choose some x-values and calculate the corresponding y-values to find some ordered pairs that are solutions of the equation. Then we plot the points and draw the graph.

x	y	(x, y)
-1	3	$(-1, 3)$
0	0	$(0, 0)$
2	-6	$(2, -6)$

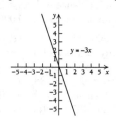

5. Graph $y = \frac{2}{3}x - 4$.

We choose some x-values and calculate the corresponding y-values to find some ordered pairs that are solutions of the equation. We use multiples of 3 for x to avoid fractional values for y. Then we plot the points and draw the graph.

x	y	(x, y)
-3	-6	$(-3, -6)$
0	-4	$(0, -4)$
3	-2	$(3, -2)$

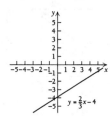

7. Graph $x + y = -4$.

We will solve for y first.

$$x + y = -4$$
$$y = -x - 4$$

Now we choose some x-values and calculate the corresponding y-values to find some ordered pairs that are solutions of the equation. Then we plot the points and draw the graph.

x	y	(x, y)
-4	0	$(-4, 0)$
-2	-2	$(-2, -2)$
1	-5	$(1, -5)$

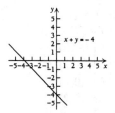

9. Graph $8y - 2x = 4$.

We will solve for y first.

$$8y - 2x = 4$$
$$8y = 2x + 4$$
$$y = \frac{1}{8}(2x + 4)$$
$$y = \frac{1}{4}x + \frac{1}{2}$$

Now we choose some x-values and calculate the corresponding y-values to find some ordered pairs that are solutions of the equation. Then we plot the points and draw the graph.

x	y	(x, y)
-2	0	$(-2, 0)$
2	1	$(2, 1)$
6	2	$(6, 2)$

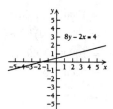

11. Graph $y = 3 - x^2$.

x	y	(x, y)
-2	-1	$(-2, -1)$
-1	2	$(-1, 2)$
0	3	$(0, 3)$
1	2	$(1, 2)$
2	-1	$(2, -1)$

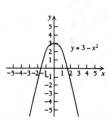

13. Graph $x = 2 - y^2$.

Since x is expressed in terms of y we first choose values for y and then compute x. Then plot the points that are found and draw the graph.

x	y	(x, y)
-2	-2	$(-2, -2)$
1	-1	$(1, -1)$
2	0	$(2, 0)$
1	1	$(1, 1)$
-2	2	$(-2, 2)$

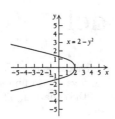

15. Graph $y = |x|$.

x	y	(x, y)
-5	5	$(-5, 5)$
-3	3	$(-3, 3)$
0	0	$(0, 0)$
2	2	$(2, 2)$
4	4	$(4, 4)$

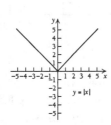

17. a) $A = P(1 + i)^t$
$$= 1000(1 + 0.06)^1$$
$$= 1000(1.06)$$
$$= \$1060$$

b) $A = P\left(1 + \dfrac{i}{n}\right)^{nt}$
$$= 1000\left(1 + \dfrac{0.06}{2}\right)^{2 \cdot 1}$$
$$= 1000(1 + 0.03)^2$$
$$= 1000(1.03)^2$$
$$= 1000(1.0609)$$
$$= \$1060.90$$

c) $A = P\left(1 + \dfrac{i}{n}\right)^{nt}$
$$= 1000\left(1 + \dfrac{0.06}{4}\right)^{4 \cdot 1}$$
$$= 1000(1 + 0.015)^4$$
$$= 1000(1.015)^4$$
$$= 1000(1.061363551)$$
$$= 1061.363551$$
$$\approx \$1061.36$$

d) $A = P\left(1 + \dfrac{i}{n}\right)^{nt}$
$$= 1000\left(1 + \dfrac{0.06}{365}\right)^{365 \cdot 1}$$
$$= 1000(1 + 0.000164383)^{365}$$
$$= 1000(1.000164383)^{365}$$
$$= 1000(1.06183131)$$
$$= 1061.83131$$
$$\approx \$1061.83$$

e) There are $24 \cdot 365$, or 8760, hours in one year.
$$A = P\left(1 + \dfrac{i}{n}\right)^{nt}$$
$$= 1000\left(1 + \dfrac{0.06}{8760}\right)^{8760 \cdot 1}$$
$$= 1000(1 + 0.000006849)^{8760}$$
$$= 1000(1.000006849)^{8760}$$
$$= 1000(1.061836374)$$
$$= 1061.836374$$
$$\approx \$1061.84$$

19. $M = P\left[\dfrac{\frac{i}{12}\left(1 + \frac{i}{12}\right)^n}{\left(1 + \frac{i}{12}\right)^n - 1}\right]$

We substitute \$18,000 for P, $9\frac{3}{4}$% (or 0.0975) for i, and 36 for n ($3 \times 12 = 36$). Then we use a calculator to perform the computation.
$$M = 18,000\left[\dfrac{\frac{0.0975}{12}\left(1 + \frac{0.0975}{12}\right)^{36}}{\left(1 + \frac{0.0975}{12}\right)^{36} - 1}\right]$$
$$\approx \$578.70$$

21. a) Locate 60 on the horizontal axis and go directly up to the graph. Then move left to the vertical axis and read the value there. We estimate that the incidence of breast cancer in women of age 60 is about 325 per 100,000 women.

b) Locate 400 on the vertical axis and move horizontally across to the graph. There are two x-values that correspond to the y-value of 400. They are about 67 and 88, so for women age 67 and for women age 88 the incidence of breast cancer is about 400 per 100,000 women.

c) The highest point on the graph corresponds to the x-value of about 79, so the largest incidence of breast cancer occurs in women of age 79.

d) $\boxed{tw}$

23.

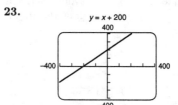

25.

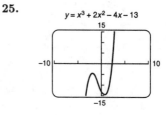

27. First we solve for y.

$$9.6x + 4.2y = -100$$
$$4.2y = -9.6x - 100$$
$$y = \frac{-9.6x - 100}{4.2}$$

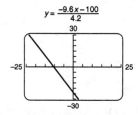

29. First we solve for y.

$$x = 4 + y^2$$
$$x - 4 = y^2$$
$$\pm\sqrt{x - 4} = y$$

Then graph $y_1 = \sqrt{x - 4}$ and $y_2 = -\sqrt{x - 4}$ in the same window.

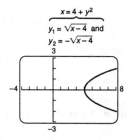

Exercise Set 1.2

1. The correspondence is a function, because each member of the domain corresponds to exactly one number of the range.

3. The correspondence is not a function, because one member of the domain, 6, corresponds to more than one member of the range.

5. The correspondence is a function, because each member of the domain corresponds to exactly one number of the range.

7. This correspondence is a function, because each member of the class has exactly one seat number.

9. This correspondence is a function, because each shape has exactly one positive number for its perimeter.

11. This correspondence is not a function, because a textbook has more than one even-numbered page.

13. a) $f(x) = 2x + 3$

$$f(4.1) = 2(4.1) + 3 = 8.2 + 3 = 11.2$$
$$f(4.01) = 2(4.01) + 3 = 8.02 + 3 = 11.02$$
$$f(4.001) = 2(4.001) + 3 = 8.002 + 3 = 11.002$$
$$f(4) = 2(4) + 3 = 8 + 3 = 11$$

Input	Output
4.1	11.2
4.01	11.02
4.001	11.002
4	11

b) $f(x) = 2x + 3$

$$f(5) = 2(5) + 3 = 10 + 3 = 13$$
$$f(-1) = 2(-1) + 3 = -2 + 3 = 1$$
$$f(k) = 2(k) + 3 = 2k + 3$$
$$f(1 + t) = 2(1 + t) + 3 = 2 + 2t + 3 = 2t + 5$$
$$f(x + h) = 2(x + h) + 3 = 2x + 2h + 3$$

15. $g(x) = x^2 - 3$

$$g(-1) = (-1)^2 - 3 = 1 - 3 = -2$$
$$g(0) = 0^2 - 3 = 0 - 3 = -3$$
$$g(1) = 1^2 - 3 = 1 - 3 = -2$$
$$g(5) = 5^2 - 3 = 25 - 3 = 22$$
$$g(u) = u^2 - 3$$
$$g(a + h) = (a + h)^2 - 3 = a^2 + 2ah + h^2 - 3$$
$$g(1 - h) = (1 - h)^2 - 3 = 1 - 2h + h^2 - 3$$
$$= h^2 - 2h - 2$$

17. $f(x) = \dfrac{1}{(x + 3)^2}$

a) $f(4) = \dfrac{1}{(4 + 3)^2} = \dfrac{1}{7^2} = \dfrac{1}{49}$

$f(-3) = \dfrac{1}{(-3 + 3)^2} = \dfrac{1}{0^2} = \dfrac{1}{0}$; since division by 0 is undefined, $f(-3)$ does not exist

$f(0) = \dfrac{1}{(0 + 3)^2} = \dfrac{1}{3^2} = \dfrac{1}{9}$

$f(a) = \dfrac{1}{(a + 3)^2}$

$f(t + 1) = \dfrac{1}{(t + 1 + 3)^2} = \dfrac{1}{(t + 4)^2}$

$f(t + 3) = \dfrac{1}{(t + 3 + 3)^2} = \dfrac{1}{(t + 6)^2}$

$f(x + h) = \dfrac{1}{(x + h + 3)^2}$

b) $f(x) = \dfrac{1}{x^2 + 6x + 9}$

This function takes an input, squares it, adds six times the input, adds 9, and then takes the reciprocal of the result.

19. Graph $f(x) = 2x + 3$.

We first choose any number for x and then determine $f(x)$, or y.

$$f(-2) = 2(-2) + 3 = -4 + 3 = -1$$
$$f(-1) = 2(-1) + 3 = -2 + 3 = 1$$
$$f(0) = 2 \cdot 0 + 3 = 0 + 3 = 3$$
$$f(1) = 2 \cdot 1 + 3 = 2 + 3 = 5$$

x	y	(x,y)
-2	-1	$(-2,-1)$
-1	1	$(-1,1)$
0	3	$(0,3)$
1	5	$(1,5)$

Next we plot the input-output pairs from the table and draw the graph.

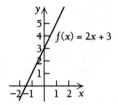

21. Graph $g(x) = -4x$.

We first choose any number for x and then determine $g(x)$, or y.

$g(-1) = -4(-1) = 4$

$g(0) = -4 \cdot 0 = 0$

$g(1) = -4 \cdot 1 = -4$

x	y	(x,y)
-1	4	$(-1,4)$
0	0	$(0,0)$
1	-4	$(1,-4)$

Next we plot the input-output pairs from the table and draw the graph.

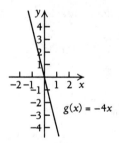

23. Graph $f(x) = x^2 - 1$.

We first choose any number for x and then determine $f(x)$, or y.

$f(-2) = (-2)^2 - 1 = 4 - 1 = 3$

$f(-1) = (-1)^2 - 1 = 1 - 1 = 0$

$f(0) = 0^2 - 1 = 0 - 1 = -1$

$f(1) = 1^2 - 1 = 1 - 1 = 0$

$f(2) = 2^2 - 1 = 4 - 1 = 3$

x	y	(x,y)
-2	3	$(-2,3)$
-1	0	$(-1,0)$
0	-1	$(0,-1)$
1	0	$(1,0)$
2	3	$(2,3)$

Next we plot the input-output pairs from the table and draw the graph.

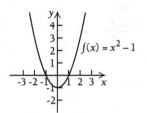

25. Graph $g(x) = x^3$.

We first choose any number for x and then determine $g(x)$, or y.

$g(-2) = (-2)^3 = -8$

$g(-1) = (-1)^3 = -1$

$g(0) = 0^3 = 0$

$g(1) = 1^3 = 1$

$g(2) = 2^3 = 8$

x	y	(x,y)
-2	-8	$(-2,-8)$
-1	-1	$(-1,-1)$
0	0	$(0,0)$
1	1	$(1,1)$
2	8	$(2,8)$

Next we plot the input-output pairs from the table and draw the graph.

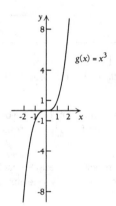

27.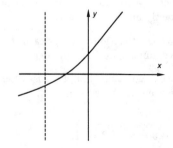

No vertical line meets the graph more than once. Thus, the graph is a graph of a function.

29.

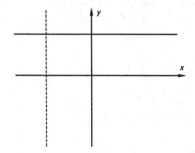

No vertical line meets the graph more than once. Thus, the graph is a graph of a function.

31.

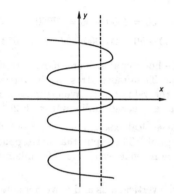

A vertical line (in fact, many) meets the graph more than once. Therefore, the graph is not the graph of a function.

33.

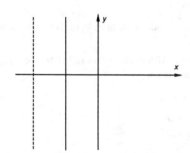

The dashed vertical line can be moved to the right to meet the graph more than once. Thus, the graph is not the graph of a function.

35.

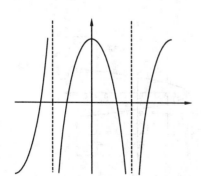

No vertical line meets the graph more than once. Thus, this is the graph of a function.

37.

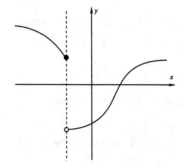

No vertical line meets the graph more than once. Thus, this is the graph of a function.

39. a) Graph $x = y^2 - 1$.

We first choose any number for y (since x is expressed in terms of y) and then determine x.

For $y = -2$, $x = (-2)^2 - 1 = 4 - 1 = 3$.

For $y = -1$, $x = (-1)^2 - 1 = 1 - 1 = 0$.

For $y = 0$, $x = 0^2 - 1 = 0 - 1 = -1$.

For $y = 1$, $x = 1^2 - 1 = 1 - 1 = 0$.

For $y = 2$, $x = 2^2 - 1 = 4 - 1 = 3$.

x	y	(x, y)
3	-2	$(3, -2)$
0	-1	$(0, -1)$
-1	0	$(-1, 0)$
0	1	$(0, 1)$
3	2	$(3, 2)$

Next we plot the ordered pairs and draw the graph.

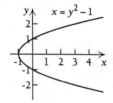

b)

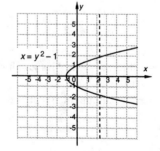

A vertical line (in fact, many) meets the graph more than once. Therefore the graph of $x = y^2 - 1$ is not the graph of a function.

41. $f(x) = x^2 - 3x$

$f(x + h) = (x + h)^2 - 3(x + h)$ Substituting

$\qquad\qquad = x^2 + 2xh + h^2 - 3x - 3h$

43. Graph: $f(x) = \begin{cases} 1 \text{ for } x < 0, \\ -1 \text{ for } x \geq 0 \end{cases}$

First graph $f(x) = 1$ for inputs less than 0.

x	y	(x, y)
$-\dfrac{1}{2}$	1	$\left(-\dfrac{1}{2}, 1\right)$
-1	1	$(-1, 1)$
-2	1	$(-2, 1)$
-3.21	1	$(-3.21, 1)$
-4	1	$(-4, 1)$

For any input less than 0, the output is 1.

Plot the input-output pairs from the table and draw this part of the graph. Since the number 0 is not an input, the point $(0, 1)$ is not part of the graph. Therefore an open circle is used at the point $(0, 1)$.

Next graph $f(x) = -1$ for inputs greater than or equal to 0.

x	y	(x, y)
0	-1	$(0, -1)$
1	-1	$(1, -1)$
$1\dfrac{3}{4}$	-1	$\left(1\dfrac{3}{4}, -1\right)$
2.7	-1	$(2.7, -1)$
3	-1	$(3, -1)$

For any input greater than or equal to 0, the output is -1.

Plot the input-output pairs from the table and draw this part of the graph. Since the number 0 is an input, the point $(0, -1)$ is part of the graph. Therefore a solid circle is used at the point $(0, -1)$.

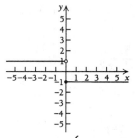

45. Graph: $f(x) = \begin{cases} -3 \text{ for } x = -2, \\ x^2 \text{ for } x \neq -2 \end{cases}$

First graph $f(x) = -3$ for $x = -2$. This graph consists of only one point, $(-2, -3)$.

Next graph $f(x) = x^2$ for all inputs except -2.

x	y	(x, y)
-3	9	$(-3, 9)$
-1	1	$(-1, 1)$
0	0	$(0, 0)$
1	1	$(1, 1)$
2	4	$(2, 4)$

Plot the input-output pairs from the table and draw this part of the graph. Since the number -2 is not an input,

the point $(-2, 4)$ is not part of the graph. Therefore an open circle is used at the point $(-2, 4)$.

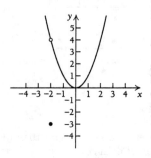

47. $R(x) = 200x + 50$

$R(10) = 200 \cdot 10 + 50 = 2000 + 50 = \$2050$

$R(100) = 200 \cdot 100 + 50 = 20,000 + 50 = \$20,050$

49. a) Locate 2 on the horizontal axis and move directly up to the graph. Then move across to the vertical axis and read the value there. We estimate that the revenue was about \$40 million for week 2.

b) Locate 9 on the vertical axis and move horizontally across to the graph. The input that corresponds to 9 is 6, so the revenue was about \$9 million in week 6.

c) Locate 30 on the vertical axis and move horizontally across to the graph. The input that corresponds to 30 is 3, so the revenue was about \$30 million in week 3.

51. a) Yes; a unique "scale of impact" number is assigned to each event.

b) The inputs are the events; the outputs are the scale of impact numbers.

53. $2y^2 + 3x = 4x + 5$

$\qquad 2y^2 = x + 5$

$\qquad\quad y^2 = \dfrac{x + 5}{2}$

$\qquad\quad\ y = \pm\sqrt{\dfrac{x + 5}{2}}$

We sketch the graph:

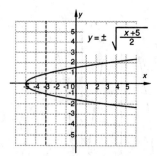

Since a vertical line (in fact, many) meets the graph more than once, this is not a function.

55. $(3y^{3/2})^2 = 72x$

$9y^3 = 72x$

$y^3 = 8x$

$y = 2\sqrt[3]{x}$

We sketch the graph:

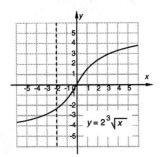

$y = 2\sqrt[3]{x}$

Since no vertical line meets the graph more than once, this is a function.

57. $\boxed{tw}$

59.

X	Y1	
-3	.6	
-2	ERROR	
-1	-1	
0	-.75	
1	-1	
2	ERROR	
3	.6	

X = -3

61. This is left to the student. Compare your graphs to the hand-drawn graphs in Exercises 43-46.

Exercise Set 1.3

1. $(0, 5)$

3. $[-9, -4]$

5. $[x, x + h]$

7. (p, ∞)

9. $[-3, 3]$

11. $[-14, -11)$

13. $(-\infty, -4]$

15. a) Locate 1 on the horizontal axis and then find the point on the graph for which 1 is the first coordinate. From that point, look to the vertical axis to find the corresponding y-coordinate, 1. Thus, $f(1) = 1$.

b) The domain is the set of all x-values in the graph. It is $\{-3, -1, 1, 3, 5\}$.

c) To determine which member(s) of the domain are paired with 2, locate 2 on the vertical axis. From there look left and right to the graph to find any points for which 2 is the second coordinate. One such point exists, $(3, 2)$. Thus, the x-value for which $f(x) = 2$ is 3.

d) The range is the set of all y-values in the graph. It is $\{-1, 0, 1, 2, 3\}$.

17. a) Locate 1 on the horizontal axis and then find the point on the graph for which 1 is the first coordinate. From that point, look to the vertical axis to find the corresponding y-coordinate, 4. Thus, $f(1) = 4$.

b) The domain is the set of all x-values in the graph. It is $\{-5, -3, 1, 2, 3, 4, 5\}$.

c) To determine which member(s) of the domain are paired with 2, locate 2 on the vertical axis. From there look left and right to the graph to find any points for which 2 is the second coordinate. Three such points exist. They are $(-5, 2)$, $(-3, 2)$, and $(4, 2)$. Thus, the x-values for which $f(x) = 2$ are -5, -3, and 4.

d) The range is the set of all y-values in the graph. It is $\{-3, 2, 4, 5\}$.

19. a) Locate 1 on the horizontal axis and then find the point on the graph for which 1 is the first coordinate. From that point, look to the vertical axis to find the corresponding y-coordinate, about $2\frac{2}{3}$. Thus, $f(1) \approx 2\frac{2}{3}$.

b) The set of all x-values in the graph extends from -2 to 3, so the domain is $\{x| -2 \le x \le 3\}$, or $[-2, 3]$.

c) To determine which member(s) of the domain are paired with 2, locate 2 on the vertical axis. From there look left and right to the graph to find any points for which 2 is the second coordinate. One such point exists. Its first coordinate appears to be about $1\frac{3}{4}$. Thus, the x-value for which $f(x) = 2$ is about $1\frac{3}{4}$.

d) The set of all y-values in the graph extends from 1 to 5, so the range is $\{y| 1 \le y \le 5\}$, or $[1, 5]$.

21. a) Locate 1 on the horizontal axis and the find the point on the graph for which 1 is the first coordinate. From that point, look to the vertical axis to find the corresponding y-coordinate, -2. Thus $f(1) = -2$

b) The set of all x-values in the graph extends from -4 to 2, so the domain is $\{x| -4 \le x \le 2\}$, or $[-4, 2]$.

c) To determine which member(s) of the domain are paired with 2, locate 2 on the vertical axis. From there look left and right to the graph to find any points for which 2 is the second coordinate. One such point exists, $(-2, 2)$. Thus, the x-value for which $f(x) = 2$ is -2.

d) The set of all y-values in the graph extends from -3 to 3, so the range is $\{y| -3 \le y \le 3\}$, or $[-3, 3]$.

23. a) Locate 1 on the horizontal axis and then find the point on the graph for which 1 is the first coordinate. From that point, look to the vertical axis to find the corresponding y-coordinate, 3. Thus, $f(1) = 3$.

b) The set of all x-values in the graph extends from -3 to 3, so the domain is $\{x| -3 \leq x \leq 3\}$, or $[-3,3]$.

c) To determine which member(s) of the domain are paired with 2, locate 2 on the vertical axis. From there look left and right to the graph to find any points for which 2 is the second coordinate. Two such points exist. They are about $\left(-1\frac{2}{5}, 2\right)$ and $\left(1\frac{2}{5}, 2\right)$. Thus, the x-values for which $f(x) = 2$ are about $-1\frac{2}{5}$ and $1\frac{2}{5}$.

d) The set of all y-values in the graph extends from -5 to 4, so the range is $\{y|-5 \leq y \leq 4\}$, or $[-5,4]$.

25. a) Locate 1 on the horizontal axis and then find the point on the graph for which 1 is the first coordinate. From that point, look to the vertical axis to find the corresponding y-coordinate, -1. Thus, $f(1) = -1$.

b) The set of all x-values in the graph extends from -6 to 5, so the domain is $\{x| -6 \leq x \leq 5\}$, or $[-6,5]$.

c) To determine which member(s) of the domain are paired with 2, locate 2 on the vertical axis. From there look left and right to the graph to find any points for which 2 is the second coordinate. Three such points exist, $(-4,2)$, $(0,2)$ and $(3,2)$. Thus, the x-values for which $f(x) = 2$ are -4, 0, and 3.

d) The set of all y-values in the graph extends from -2 to 2, so the range is $\{y|-2 \leq y \leq 2\}$, or $[-2,2]$.

27. $f(x) = \dfrac{7}{5-x}$

Since $\dfrac{7}{5-x}$ cannot be calculated when the denominator is 0, we find the x-value that causes $5 - x$ to be 0:

$5 - x = 0$

$5 = x$ Adding x on both sides

Thus, 5 is not in the domain of f, while all other real numbers are. The domain of f is

$\{x|x \text{ is a real number and } x \neq 5\}$, or

$(-\infty, 5) \cup (5, \infty)$.

29. $f(x) = 4 - 5x$

We can calculate $4 - 5x$ for any value of x, so the domain is the set of all real numbers.

31. $f(x) = x^2 - 2x + 3$

We can calculate $x^2 - 2x + 3$ for any value of x, so the domain is the set of all real numbers.

33. $f(x) = \dfrac{x-2}{3x+4}$

Since $\dfrac{x-2}{3x+4}$ cannot be calculated when the denominator is 0, we find the x-value that causes $3x + 4$ to be 0:

$3x + 4 = 0$

$3x = -4$

$x = -\dfrac{4}{3}$

Thus, $-\dfrac{4}{3}$ is not in the domain of f, while all other real numbers are. The domain of f is

$\left\{x\middle|x \text{ is a real number and } x \neq -\dfrac{4}{3}\right\}$, or

$\left(\infty, -\dfrac{4}{3}\right) \cup \left(-\dfrac{4}{3}, \infty\right)$.

35. $f(x) = |x-4|$

We can calculate $|x-4|$ for any value of x, so the domain is the set of all real numbers.

37. $f(x) = \dfrac{x^2 - 3x}{|4x-7|}$

Since $\dfrac{x^2 - 3x}{|4x-7|}$ cannot be calculated when the denominator is 0, we find the x-values that causes $|4x - 7|$ to be 0:

$|4x - 7| = 0$

$4x - 7 = 0$

$4x = 7$

$x = \dfrac{7}{4}$

Thus, $\dfrac{7}{4}$ is not in the domain of f, while all other real numbers are. The domain of f is

$\left\{x\middle|x \text{ is a real number and } x \neq \dfrac{7}{4}\right\}$, or

$\left(-\infty, \dfrac{7}{4}\right) \cup \left(\dfrac{7}{4}, \infty\right)$.

39. $g(x) = \dfrac{-11}{4+x}$

Since $\dfrac{-11}{4+x}$ cannot be calculated when the denominator is 0, we find the x-value that causes $4 + x$ to be 0:

$4 + x = 0$

$x = -4$

Thus, -4 is not in the domain of g, while all other real numbers are. The domain of g is

$\{x|x \text{ is a real number and } x \neq -4\}$, or

$(-\infty, -4) \cup (-4, \infty)$.

41. $g(x) = 8 - x^2$

We can calculate $8 - x^2$ for any value of x, so the domain is the set of all real numbers.

43. $g(x) = 4x^3 + 5x^2 - 2x$

We can calculate $4x^3 + 5x^2 - 2x$ for any value of x, so the domain is the set of all real numbers.

45. $g(x) = \dfrac{2x-3}{6x-12}$

Since $\dfrac{2x-3}{6x-12}$ cannot be calculated when the denominator is 0, we find the x-values that cause $6x - 12$ to be 0:

$$6x - 12 = 0$$
$$6x = 12$$
$$x = 2$$

Thus, 2 is not in the domain of g, while all other real numbers are. The domain of g is

$\{x | x \text{ is a real number and } x \neq 2\}$, or

$(-\infty, 2) \cup (2, \infty)$.

47. $g(x) = |x| + 1$

We can calculate $|x| + 1$ for any value of x, so the domain is the set of all real numbers.

49. $g(x) = \dfrac{x^2 + 2x}{|10x - 20|}$

Since $\dfrac{x^2 + 2x}{|10x - 20|}$ cannot be calculated when the denominator is 0, we find the x-value that causes $|10x - 20|$ to be 0:

$$|10x - 20| = 0$$
$$10x - 20 = 0$$
$$10x = 20$$
$$x = 2$$

Thus, 2 is not in the domain of g, while all other real numbers are. The domain of g is

$\{x | x \text{ is a real number and } x \neq 2\}$, or

$(-\infty, 2) \cup (2, \infty)$.

51. The input -1 has the output -8, so $f(-1) = -8$;

the input 0 has the output 0, so $f(0) = 0$;

the input 1 has the output -2, so $f(1) = -2$.

53. a) We use the compound interest formula from Theorem 2 in Section 1.1 and substitute 10,000 for P, 2 for n, and 8 for t.

$$A = P\left(1 + \frac{i}{n}\right)^{nt}$$

$$A = 10,000\left(1 + \frac{i}{2}\right)^{2 \cdot 8}$$

$$A = 10,000\left(1 + \frac{i}{2}\right)^{16}$$

b) The interest rate must be a positive number, so the domain is the set of all positive real numbers.

55. $\boxed{tw}$

57. $\boxed{tw}$

59. Exercise 29: all real numbers; exercise 32: $[3, \infty)$; exercise 44: all real numbers; exercise 47: $[1, \infty)$; exercise 48: $[0, \infty)$

---

1. Graph $y = -4$.

The graph consists of all ordered pairs whose second coordinate is -4. Thus, y must be -4, but x can be any number.

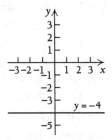

3. Graph $x = 4.5$.

The graph consists of all ordered pairs whose first coordinate is 4.5. Thus, x must be 4.5, but y can be any number.

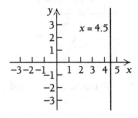

5. Graph $y = -3x$.

We first make a table of values. We choose any number for x and then determine y by substitution.

When $x = -2$, $y = -3(-2) = 6$.

When $x = 0$, $y = -3 \cdot 0 = 0$.

When $x = 1$, $y = -3 \cdot 1 = -3$.

Plot these ordered pairs and draw the graph.

x	y
-2	6
0	0
1	-3

The function $y = -3x$, or $y = -3x + 0$, has slope -3 and y-intercept $(0, 0)$.

7. Graph $y = 0.5x$.

We first make a table of values. We choose any number for x and then determine y by substitution.

When $x = -4$, $y = 0.5(-4) = -2$.

When $x = 0$, $y = 0.5(0) = 0$.

When $x = 2$, $y = 0.5(2) = 1$.

Plot these ordered pairs and draw the graph.

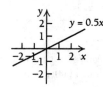

x	y
-4	-2
0	0
2	1

The function $y = 0.5x$, or $y = 0.5x + 0$, has slope 0.5 and y-intercept $(0, 0)$.

9. Graph $y = -2x + 3$.

We first make a table of values. We choose any number for x and then determine y by substitution.

When $x = -1$, $y = -2(-1) + 3 = 2 + 3 = 5$.

When $x = 0$, $y = -2 \cdot 0 + 3 = 0 + 3 = 3$.

When $x = 3$, $y = -2 \cdot 3 + 3 = -6 + 3 = -3$.

Plot these ordered pairs and draw the graph.

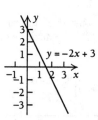

x	y
-1	5
0	3
3	-3

The function $y = -2x + 3$ has slope -2 and y-intercept $(0, 3)$.

11. Graph $y = -x - 2$.

We first make a table of values. We choose any number for x and then determine y by substitution.

When $x = -5$, $y = -(-5) - 2 = 5 - 2 = 3$.

When $x = 0$, $y = -0 - 2 = -2$.

When $x = 2$, $y = -2 - 2 = -4$.

Plot these ordered pairs and draw the graph.

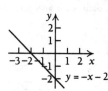

x	y
-5	3
0	-2
2	-4

The function $y = -x - 2$, or $y = -1 \cdot x - 2$, has slope -1 and y-intercept $(0, -2)$.

13. Solve the equation for y.

$$2x + y - 2 = 0$$

$$y = -2x + 2 \quad \text{Adding } -2x + 2$$

The slope is -2.
The y-intercept is $(0, 2)$.

15. Solve the equation for y.

$$2x + 2y + 5 = 0$$

$$2y = -2x - 5 \quad \text{Adding } -2x - 5$$

$$y = -x - \frac{5}{2} \quad \text{Multiplying by } \frac{1}{2}$$

The slope is -1.

The y-intercept is $\left(0, -\frac{5}{2}\right)$.

17.
$$y - y_1 = m(x - x_1)$$

$$y - (-5) = -5(x - 1) \quad \text{Substituting}$$

$$y + 5 = -5x + 5$$

$$y = -5x$$

19.
$$y - y_1 = m(x - x_1)$$

$$y - 3 = -2(x - 2) \quad \text{Substituting}$$

$$y - 3 = -2x + 4$$

$$y = -2x + 7$$

21.
$$y = mx + b$$

$$y = \frac{1}{2}x + (-6) \quad \text{Substituting}$$

$$y = \frac{1}{2}x - 6$$

23.
$$y - y_1 = m(x - x_1)$$

$$y - 3 = 0(x - 2) \quad \text{Substituting}$$

$$y - 3 = 0$$

$$y = 3$$

25.
$$m = \frac{y_2 - y_1}{x_2 - x_1}$$

$$m = \frac{1 - (-2)}{-2 - (-4)} \quad \begin{array}{l}\text{Substituting 1 for } y_2, -2 \text{ for} \\ y_1, -2 \text{ for } x_2, \text{ and } -4 \text{ for } x_1\end{array}$$

$$= \frac{1 + 2}{-2 + 4}$$

$$= \frac{3}{2}$$

It does not matter which point is taken first, as long as we subtract coordinates in the same order. We could also find m as follows.

$$m = \frac{-2 - 1}{-4 - (-2)} \quad \begin{array}{l}\text{Substituting } -2 \text{ for } y_2, 1 \text{ for} \\ y_1, -4 \text{ for } x_2, \text{ and } -2 \text{ for } x_1\end{array}$$

$$= \frac{-2 - 1}{-4 + 2}$$

$$= \frac{-3}{-2}$$

$$= \frac{3}{2}$$

27. $m = \dfrac{y_2 - y_1}{x_2 - x_1}$

$m = \dfrac{\dfrac{4}{5} - \dfrac{1}{2}}{-3 - \dfrac{2}{5}}$ Substituting

$= \dfrac{\dfrac{8}{10} - \dfrac{5}{10}}{-\dfrac{15}{5} - \dfrac{2}{5}}$

$= \dfrac{\dfrac{3}{10}}{-\dfrac{17}{5}}$

$= \dfrac{3}{10} \cdot \left(-\dfrac{5}{17}\right)$

$= -\dfrac{15}{170}$

$= -\dfrac{3}{34}$

29. $m = \dfrac{y_2 - y_1}{x_2 - x_1}$

$= \dfrac{-9 - (-7)}{3 - 3}$ Substituting

$= \dfrac{-9 + 7}{3 - 3}$

$= \dfrac{-2}{0}$

Since we cannot divide by 0, the slope of the line through $(3, -7)$ and $(3, -9)$ is undefined. The line has no slope.

31. $m = \dfrac{y_2 - y_1}{x_2 - x_1}$

$m = \dfrac{3 - 3}{-1 - 2}$ Substituting

$= \dfrac{0}{-3}$

$= 0$

33. $m = \dfrac{y_2 - y_1}{x_2 - x_1}$

$m = \dfrac{3(x + h) - 3x}{x + h - x}$ Substituting

$= \dfrac{3x + 3h - 3x}{x + h - x}$

$= \dfrac{3h}{h}$

$= 3$

35. $m = \dfrac{y_2 - y_1}{x_2 - x_1}$

$m = \dfrac{[2(x + h) + 3] - (2x + 3)}{(x + h) - x}$ Substituting

$= \dfrac{2x + 2h + 3 - 2x - 3}{x + h - x}$

$= \dfrac{2h}{h}$

$= 2$

37. From Exercise 25, we know that the slope of the line is $\dfrac{3}{2}$. Using the point $(-4, -2)$, we substitute in the point-slope equation.

$$y - (-2) = \frac{3}{2}[x - (-4)]$$

$$y + 2 = \frac{3}{2}(x + 4)$$

$$y + 2 = \frac{3}{2}x + 6$$

$$y = \frac{3}{2}x + 4$$

We could also use the point $(-2, 1)$:

$$y - 1 = \frac{3}{2}[x - (-2)]$$

$$y - 1 = \frac{3}{2}(x + 2)$$

Simplifying this, we also get $y = \dfrac{3}{2}x + 4$.

39. From Exercise 27, we know that the slope is $-\dfrac{3}{34}$. Using the point $\left(\dfrac{2}{5}, \dfrac{1}{2}\right)$, we substitute in the point-slope equation.

$$y - \frac{1}{2} = -\frac{3}{34}\left(x - \frac{2}{5}\right)$$

$$y - \frac{1}{2} = -\frac{3}{34}x + \frac{6}{170}$$

$$y = -\frac{3}{34}x + \frac{91}{170}$$

We could also use $\left(-3, \dfrac{4}{5}\right)$:

$$y - \frac{4}{5} = -\frac{3}{34}[x - (-3)]$$

$$y - \frac{4}{5} = -\frac{3}{34}(x + 3)$$

Simplifying this, we also get $y = -\dfrac{3}{34}x + \dfrac{91}{170}$.

41. From Exercise 29, we know that the line containing $(3, -7)$ and $(3, -9)$ has no slope. The graph is a line which contains all ordered pairs whose first coordinate is 3. The second coordinate can be any number. The line is vertical. The equation is $x = 3$.

43. From Exercise 31, we know that the slope of the line containing $(2, 3)$ and $(-1, 3)$ is 0. The graph consists of all ordered pairs whose second coordinate is 3. The first coordinate can be any number. The line is horizontal. The equation is $y = 3$.

45. From Exercise 33, we know that the slope is 3. Using the point $(x, 3x)$, we substitute in the point-slope equation:

$$y - 3x = 3(x - x)$$

$$y - 3x = 3 \cdot 0$$

$$y - 3x = 0$$

$$y = 3x$$

We could also use $(x + h, \; 3(x + h))$:

$$y - 3(x + h) = 3[x - (x - h)]$$

Simplifying this, we also get $y = 3x$.

47. From Exercise 35, we know that the slope is 2. Using the point $(x, 2x+3)$, we substitute in the point-slope equation:

$$y - (2x + 3) = 2(x - x)$$
$$y - 2x - 3 = 2 \cdot 0$$
$$y - 2x - 3 = 0$$
$$y = 2x + 3$$

We could also use $(x + h, \; 2(x + h) + 3)$:

$$y - [2(x + h) + 3] = 2[x - (x + h)]$$

Simplifying this, we also get $y = 2x + 3$.

49. Slope $= \dfrac{0.4}{5} = \dfrac{4}{50} = \dfrac{2}{25} = 0.08 = 8\%$

51. Slope $= \dfrac{43.33}{1238} = 0.035 = 3.5\%$

53. The rate of change can be found using the coordinates of any two points on the line. We use $(1991, 800)$ and $(1995, 1200)$.

$$\text{Rate} = \frac{\text{Change in tuition and fees}}{\text{corresponding change in time}}$$
$$= \frac{1200 - 800}{1995 - 1991}$$
$$= \frac{400}{4}$$
$$= \$100 \text{ per year}$$

55. a) A is directly proportional to P if there is some positive constant m such that $A = mP$.

We let P represent the investment and $8\%P$ represent the interest earned on the investment in one year. Then

$$A = P + 8\%P$$
$$A = 1 \cdot P + 0.08P$$
$$A = 1.08P$$

Since A can be expressed as a positive constant times P, we say A is directly proportional to P.

b) $A = 1.08P$

$A = 1.08(\$100)$ Substituting $100 for P

$A = \$108$

c) $\qquad A = 1.08P$

$\$259.20 = 1.08P$ Substituting $259.20 for A

$\dfrac{\$259.20}{1.08} = P$

$\$240 = P$

57. a) Total costs = Variable costs + Fixed costs

To produce x calculators it costs $20 per calculator in addition to the fixed costs of $100,000. That is, the variable costs are $20x$ dollars. The total cost is

$$C(x) = 20x + 100,000$$

b) The total revenue from the sale of x calculators is $45 per calculator. That is, the total revenue is given by the function

$$R(x) = 45x$$

c) Total profit = Total revenue − Total costs

$$P(x) = R(x) - C(x)$$
$$P(x) = 45x - (20x + 100,000)$$
$$P(x) = 45x - 20x - 100,000$$
$$P(x) = 25x - 100,000$$

d) $\qquad P(x) = 25x - 100,000$

$$P(150,000) = 25(150,000) - 100,000$$
$$= 3,750,000 - 100,000$$
$$= 3,650,000$$

A profit of $3,650,000 will be realized if the expected sales of 150,000 calculators occur.

e) The profit will be $0 if the firm breaks even. We set the profit function equal to 0 and solve for x.

$$P(x) = 25x - 100,000$$
$$0 = 25x - 100,000$$
$$100,000 = 25x$$
$$4000 = x$$

Thus, to break even the firm must sell 4000 calculators. (Note that we could also have solved the equation $R(x) = C(x)$, or $45x = 20x + 100,000$, to find the break-even value of x.)

59. a) Total costs = Variable costs + Fixed costs

In addition to the fixed cost of the lawnmower, $250, mowing costs $1 per lawn. Then for x lawns, the variable cost is $1 \cdot x$, or x, dollars. Thus,

$$C(x) = x + 250$$

b) $\qquad P(x) = R(x) - C(x)$

$$P(x) + C(x) = R(x)$$
$$(9x - 250) + x + 250 = R(x)$$
$$10x = R(x)$$

The total revenue from mowing x lawns is $10x$, so the student charges $10 per lawn.

c) We solve $P(x) = 0$ to find the break-even values of x:

$$9x - 250 = 0$$
$$9x = 250$$
$$x = 27\frac{7}{9}$$
$$x \approx 28 \qquad \text{Rounding up}$$

The student must mow 28 lawns before making a profit.

61. a) If R is directly proportional to T, then there is some positive constant m such that $R = mT$.

To find m substitute 12.51 for R and 3 for T in the equation $R = mT$ and solve for m.

$$R = mT$$
$$12.51 = m \cdot 3$$
$$\frac{12.51}{3} = m$$
$$4.17 = m$$

The variation constant is 4.17. The equation of variation is $R = 4.17T$.

b) $R = 4.17T$ Equation of variation

$R = 4.17(6)$ Substituting 6 for T

$= 25.02$

The R-factor for insulation that is 6 inches thick is 25.02.

63. a) If B is directly proportional to W, then there is some positive constant m such that $B = mW$.

To find m substitute 200 for W and 5 for B in the equation $B = mW$ and solve for m.

$$B = mW$$
$$5 = m \cdot 200$$
$$\frac{5}{200} = m$$
$$0.025 = m$$

The variation constant is 0.025. The equation of variation is $B = 0.025W$.

b) 0.025 is equivalent to 2.5%, so the resulting equation is

$$B = 2.5\%W.$$

The weight, B, of a human's brain is 2.5% of the body weight, W.

c) $B = 0.025W$ Equation of variation

$B = 0.025(120)$ Substituting 120 for W

$B = 3$

The weight of the brain of a person who weighs 120 lb is 3 lb.

65. a) $D(F) = 2F + 115$

$$D(0°) = 2 \cdot 0 + 115 = 0 + 115 = 115 \text{ ft}$$
$$D(-20°) = 2(-20) + 115 = -40 + 115 = 75 \text{ ft}$$
$$D(10°) = 2 \cdot 10 + 115 = 20 + 115 = 135 \text{ ft}$$
$$D(32°) = 2 \cdot 32 + 115 = 64 + 115 = 179 \text{ ft}$$

b) $\boxed{tw}$

67. a) $M(x) = 2.89x + 70.64$

$M(26) = 2.89(26) + 70.64$ Substituting

$= 75.14 + 70.64$

$= 145.78$

The male was 145.78 cm tall.

b) $F(x) = 2.75x + 71.48$

$F(26) = 2.75(26) + 71.48$ Substituting

$= 71.5 + 71.48$

$= 142.98$

The female was 142.98 cm tall.

69. a) $A(t) = 0.08t + 19.7$

$$A(0) = 0.08(0) + 19.7 = 0 + 19.7 = 19.7$$
$$A(1) = 0.08(1) + 19.7 = 0.08 + 19.7 = 19.78$$
$$A(10) = 0.08(10) + 19.7 = 0.8 + 19.7 = 20.5$$
$$A(30) = 0.08(30) + 19.7 = 2.4 + 19.7 = 22.1$$
$$A(50) = 0.08(50) + 19.7 = 4 + 19.7 = 23.7$$

b) $1998 - 1950 = 48$, so we find $A(48)$:

$A(48) = 0.08(48) + 19.7 = 3.84 + 19.7 = 23.54$

The median age of women at first marriage in 1998 is 23.54.

c) Plot the points found in parts (a) and (b) and draw the graph.

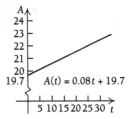

71. $\boxed{tw}$

73. Answers may vary.

Exercise Set 1.5

1. Graph $y = \frac{1}{2}x^2$ and $y = -\frac{1}{2}x^2$ on the same set of axes.

Find some ordered pairs that are solutions of $y = \frac{1}{2}x^2$, keeping the results in a table. Since the domain of $y = \frac{1}{2}x^2$ consists of all real numbers, we choose any number for x and then find y.

When $x = 0$, $y = \frac{1}{2} \cdot 0^2 = \frac{1}{2} \cdot 0 = 0$.

When $x = 2$, $y = \frac{1}{2} \cdot 2^2 = \frac{1}{2} \cdot 4 = 2$.

When $x = -2$, $y = \frac{1}{2} \cdot (-2)^2 = \frac{1}{2} \cdot 4 = 2$.

When $x = 3$, $y = \frac{1}{2} \cdot 3^2 = \frac{1}{2} \cdot 9 = \frac{9}{2}$.

When $x = -3$, $y = \frac{1}{2} \cdot (-3)^2 = \frac{1}{2} \cdot 9 = \frac{9}{2}$.

x	y
0	0
2	2
-2	2
3	$\frac{9}{2}$
-3	$\frac{9}{2}$

Find some ordered pairs that are solutions of $y = -\frac{1}{2}x^2$, keeping the results in a table. Since the domain of $y = -\frac{1}{2}x^2$ consists of all real numbers, we choose any number for x and then find y.

When $x = $ 0, $y = -\frac{1}{2} \cdot 0^2 = -\frac{1}{2} \cdot 0 = 0$.

When $x = $ 2, $y = -\frac{1}{2} \cdot 2^2 = -\frac{1}{2} \cdot 4 = -2$.

When $x = -2$, $y = -\frac{1}{2} \cdot (-2)^2 = -\frac{1}{2} \cdot 4 = -2$.

When $x = $ 3, $y = -\frac{1}{2} \cdot 3^2 = -\frac{1}{2} \cdot 9 = -\frac{9}{2}$.

When $x = -3$, $y = -\frac{1}{2} \cdot (-3)^2 = -\frac{1}{2} \cdot 9 = -\frac{9}{2}$.

x	y
0	0
2	-2
-2	-2
3	$-\frac{9}{2}$
-3	$-\frac{9}{2}$

Plot the ordered pairs for $y = \frac{1}{2}x^2$ and draw the graph with a dashed line. Plot the ordered pairs for $y = -\frac{1}{2}x^2$ and draw the graph with a solid line. For $y = \frac{1}{2}x^2$ the coefficient of x^2 is positive, so the graph opens upward. For $y = -\frac{1}{2}x^2$ the coefficient of x^2 is negative, so the graph opens downward.

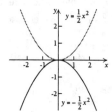

3. Graph $y = x^2$ and $y = (x-1)^2$ on the same set of axes.

Find some ordered pairs that are solutions of $y = x^2$, keeping the results in a table. Since the domain of $y = x^2$ consists of all real numbers, we choose any number for x and then find y.

When $x = $ 0, $y = 0^2 = 0$.

When $x = $ 1, $y = 1^2 = 1$.

When $x = -1$, $y = (-1)^2 = 1$.

When $x = $ 2, $y = 2^2 = 4$.

When $x = -2$, $y = (-2)^2 = 4$.

x	y
0	0
1	1
-1	1
2	4
-2	4

Find some ordered pairs that are solutions of $y = (x-1)^2$, keeping the results in a table. Since the domain of $y = (x-1)^2$ consists of all real numbers, we choose any number for x and then find y.

When $x = $ 1, $y = (1-1)^2 = 0^2 = 0$.

When $x = $ 0, $y = (0-1)^2 = (-1)^2 = 1$.

When $x = $ 2, $y = (2-1)^2 = 1^2 = 1$.

When $x = -1$, $y = (-1-1)^2 = (-2)^2 = 4$.

When $x = $ 3, $y = (3-1)^2 = 2^2 = 4$.

x	y
1	0
0	1
2	1
-1	4
3	4

Plot the ordered pairs for $y = x^2$ and draw the graph with a dashed line. Plot the ordered pairs for $y = (x-1)^2$, or $y = x^2 - 2x + 1$, and draw the graph with a solid line.

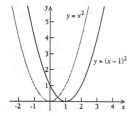

5. Graph $y = x^2$ and $y = (x+1)^2$ on the same set of axes.

We found some ordered pairs that are solutions of $y = x^2$, in Exercise 3.

Find some ordered pairs that are solutions of $y = (x+1)^2$, keeping the results in a table. Since the domain of $y = (x+1)^2$ consists of all real numbers, we choose any number for x and then find y.

When $x = -1$, $y = (-1+1)^2 = 0^2 = 0$.

When $x = -2$, $y = (-2+1)^2 = (-1)^2 = 1$.

When $x = $ 0, $y = (0+1)^2 = 1^2 = 1$.

When $x = -3$, $y = (-3+1)^2 = (-2)^2 = 4$.

When $x = $ 1, $y = (1+1)^2 = 2^2 = 4$.

x	y
-1	0
-2	1
0	1
-3	4
1	4

Plot the ordered pairs for $y = x^2$ and draw the graph with a dashed line. Plot the ordered pairs for $y = (x+1)^2$, or $y = x^2 + 2x + 1$, and draw the graph with a solid line.

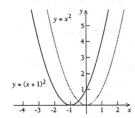

7. Graph $y = |x|$ and $y = |x+3|$ on the same set of axes.

Find some ordered pairs that are solutions of $y = |x|$. The domain of $y = |x|$ is the set of all real numbers. We choose any number for x and then find y.

When $x = 0$, $y = |0| = 0$.

When $x = 2$, $y = |2| = 2$.

When $x = -2$, $y = |-2| = 2$.

When $x = 5$, $y = |5| = 5$.

When $x = -5$, $y = |-5| = 5$.

x	y
0	0
2	2
-2	2
5	5
-5	5

Find some ordered pairs that are solutions of $y = |x+3|$. The domain of $y = |x+3|$ is the set of all real numbers. We choose any number for x and then find y.

When $x = -3$, $y = |-3+3| = |0| = 0$.

When $x = -4$, $y = |-4+3| = |-1| = 1$.

When $x = -2$, $y = |-2+3| = |1| = 1$.

When $x = -6$, $y = |-6+3| = |-3| = 3$.

When $x = 0$, $y = |0+3| = |3| = 3$.

x	y
-3	0
-4	1
-2	1
-6	3
0	3

Plot the ordered pairs for $y = |x|$ and draw the graph with a dashed line. Plot the ordered pairs for $y = |x+3|$ and draw the graph with a solid line.

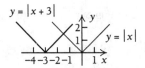

9. Graph $y = x^3$ and $y = x^3 + 1$ on the same set of axes.

Find some ordered pairs that are solutions of $y = x^3$. The domain of $y = x^3$ is the set of all real numbers. We choose any number for x and then find y.

When $x = 0$, $y = 0^3 = 0$.

When $x = -1$, $y = (-1)^3 = -1$.

When $x = 1$, $y = 1^3 = 1$.

When $x = -2$, $y = (-2)^3 = -8$.

When $x = 2$, $y = 2^3 = 8$.

x	y
0	0
-1	-1
1	1
-2	-8
2	8

Find some ordered pairs that are solutions of $y = x^3 + 1$. The domain of $y = x^3 + 1$ is the set of all real numbers. We choose any number for x and then find y.

When $x = 0$, $y = 0^3 + 1 = 1$.

When $x = -1$, $y = (-1)^3 + 1 = -1 + 1 = 0$.

When $x = 1$, $y = 1^3 + 1 = 1 + 1 = 2$.

When $x = -2$, $y = (-2)^3 + 1 = -8 + 1 = -7$.

When $x = 2$, $y = 2^3 + 1 = 8 + 1 = 9$.

x	y
0	1
-1	0
1	2
-2	-7
2	9

Plot the ordered pairs for $y = x^3$ and draw the graph with a dashed line. Plot the ordered pairs for $y = x^3 + 1$ and draw the graph with a solid line.

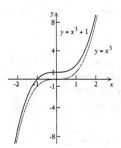

11. Graph $y = \sqrt{x}$ and $y = \sqrt{x+1}$ on the same set of axes.

Find some ordered pairs that are solutions of $y = \sqrt{x}$. The domain consists of only the nonnegative real numbers. We choose for x any number in the interval $[0, \infty)$ and then find y.

When $x = 0$, $y = \sqrt{0} = 0$.

When $x = 1$, $y = \sqrt{1} = 1$.

When $x = 4$, $y = \sqrt{4} = 2$.

When $x = 9$, $y = \sqrt{9} = 3$.

x	y
0	1
1	1
4	2
9	3

Find some ordered pairs that are solutions of $y = \sqrt{x + 1}$. The domain of this function is restricted to those input values that result in the value of the radicand, $x + 1$, being greater than or equal to 0. To determine the domain, we solve the inequality $x + 1 \geq 0$.

$$x + 1 \geq 0$$

$$x \geq -1 \qquad \text{Adding } -1$$

The domain consists of all real numbers greater than or equal to -1. We choose any real number greater than or equal to -1 and then find y.

When $x = -1$, $y = \sqrt{-1 + 1} = \sqrt{0} = 0$.

When $x = \;\;\; 0$, $y = \sqrt{0 + 1} = \sqrt{1} = 1$.

When $x = \;\;\; 3$, $y = \sqrt{3 + 1} = \sqrt{4} = 2$.

When $x = \;\;\; 8$, $y = \sqrt{8 + 1} = \sqrt{9} = 3$.

x	y
-1	0
0	1
3	2
8	3

Plot the ordered pairs for $y = \sqrt{x}$ and draw the graph with a dashed line. Plot the ordered pairs for $y = \sqrt{x + 1}$ and draw the graph with a solid line.

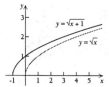

13. Graph $y = x^2 - 4x + 3$.

Before making a table of values we should recognize that $y = x^2 - 4x + 3$ is a quadratic function whose cup-shaped graph opens upward (the coefficient of x^2, 1, is positive). We first find the vertex, or turning point. The x-coordinate of the vertex is

$$x = -\frac{b}{2a} = -\frac{-4}{2 \cdot 1} = 2.$$

Substituting 2 for x in the equation, we find the second coordinate of the vertex:

$$y = x^2 - 4x + 3 = 2^2 - 4 \cdot 2 + 3 =$$

$$4 - 8 + 3 = -1.$$

The vertex is $(2, -1)$.

We can find the first coordinates of points where the graph of the function intersects the x-axis, if they exist, by solving the quadratic equation $x^2 - 4x + 3 = 0$.

$$x^2 - 4x + 3 = 0$$

$$(x - 3)(x - 1) = 0$$

$$x - 3 = 0 \;\;\text{ or }\;\; x - 1 = 0 \;\;\text{ Principle of Zero Products}$$
$$x = 3 \;\;\text{ or }\;\;\;\;\;\; x = 1$$

The graph of $y = x^2 - 4x + 3$ has x-intercepts $(3, 0)$ and $(1, 0)$.

We choose some x-values on each side of the vertex and compute y-values.

When $x = 0$, $y = 0^2 - 4 \cdot 0 + 3 = 3$.

When $x = 4$, $y = 4^2 - 4 \cdot 4 + 3 = 3$.

x	y
3	0
1	0
2	-1
0	3
4	3

Plot these ordered pairs and draw the graph.

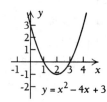

15. Graph $y = -x^2 + 2x - 1$.

We should recognize that $y = -x^2 + 2x - 1$ is a quadratic function whose graph opens downward (the coefficient of x^2, -1, is negative). We first find the vertex:

The x-coordinate is

$$x = -\frac{b}{2a} = -\frac{2}{2(-1)} = 1.$$

We substitute 1 for x in the equation to find the second coordinate of the vertex:

$$y = -1^2 + 2 \cdot 1 - 1 = -1 + 2 - 1 = 0.$$

The vertex is $(1, 0)$. Note that this is also the x-intercept. (There cannot be another x-intercept since the vertex is the highest point on the graph.) We choose some x-values on each side of the vertex and compute y-values.

When $x = \;\;\; 0$, $y = -0^2 + 2 \cdot 0 - 1 = -1$.

When $x = \;\;\; 2$, $y = -2^2 + 2 \cdot 2 - 1 = -1$.

When $x = -1$, $y = -(-1)^2 + 2(-1) - 1 = -4$.

When $x = \;\;\; 3$, $y = -3^2 + 2 \cdot 3 - 1 = -4$.

x	y
1	0
0	-1
2	-1
-1	-4
3	-4

Plot these ordered pairs and draw the graph.

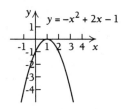

$$y = -x^2 + 2x - 1$$

17. Graph $y = \dfrac{2}{x}$.

Note that 0 is not in the domain of this function since it would yield a 0 denominator.

We find some ordered pairs that are solutions of $y = \dfrac{2}{x}$ by choosing any real number except 0 for x and finding y.

Here we list only a few substitutions.

When $x = \dfrac{1}{2}$, $y = \dfrac{2}{\frac{1}{2}} = 2 \cdot \dfrac{2}{1} = 4$.

When $x = -2$, $y = \dfrac{2}{-2} = -1$.

When $x = 4$, $y = \dfrac{2}{4} = \dfrac{1}{2}$.

x	y
$\dfrac{1}{2}$	4
1	2
2	1
4	$\dfrac{1}{2}$
6	$\dfrac{1}{3}$

x	y
$-\dfrac{1}{2}$	-4
-1	-2
-2	-1
-4	$-\dfrac{1}{2}$
-6	$-\dfrac{1}{3}$

Plot these ordered pairs and draw the graph.

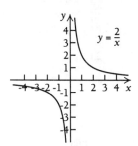

$$y = \frac{2}{x}$$

19. Graph $y = \dfrac{-2}{x}$.

Note that 0 is not in the domain of this function since it would yield a 0 denominator.

We find some ordered pairs that are solutions of $y = \dfrac{-2}{x}$ by choosing any real number except 0 for x and finding y.

Here we list only a few substitutions.

When $x = \dfrac{1}{2}$, $y = \dfrac{-2}{\frac{1}{2}} = -2 \cdot \dfrac{2}{1} = -4$.

When $x = 4$, $y = \dfrac{-2}{4} = -\dfrac{1}{2}$.

When $x = -\dfrac{1}{3}$, $y = \dfrac{-2}{-\frac{1}{3}} = -2 \cdot \dfrac{-3}{1} = 6$.

When $x = -2$, $y = \dfrac{-2}{-2} = 1$.

x	y
$\dfrac{1}{2}$	-4
1	-2
2	-1
4	$-\dfrac{1}{2}$
6	$-\dfrac{1}{3}$

x	y	
$-\dfrac{1}{2}$	4	Table of values
-1	2	for $y = \dfrac{-2}{x}$
-2	1	
-4	$\dfrac{1}{2}$	
-6	$\dfrac{1}{3}$	

Plot these ordered pairs and draw the graph.

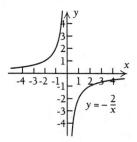

$$y = -\frac{2}{x}$$

21. Graph $y = \dfrac{1}{x^2}$.

Note that 0 is not in the domain of this function since it would yield a 0 denominator.

We find some ordered pairs that are solutions of $y = \dfrac{1}{x^2}$ by choosing any real number except 0 for x and finding y.

When $x = -2$, $y = \dfrac{1}{(-2)^2} = \dfrac{1}{4}$.

When $x = -1$, $y = \dfrac{1}{(-1)^2} = \dfrac{1}{1} = 1$.

When $x = -\dfrac{1}{2}$, $y = \dfrac{1}{\left(-\frac{1}{2}\right)^2} = \dfrac{1}{\frac{1}{4}} = 1 \cdot \dfrac{4}{1} = 4$.

When $x = \dfrac{1}{2}$, $y = \dfrac{1}{\left(\frac{1}{2}\right)^2} = \dfrac{1}{\frac{1}{4}} = 1 \cdot \dfrac{4}{1} = 4$.

When $x = 1$, $y = \dfrac{1}{1^2} = \dfrac{1}{1} = 1$.

When $x = 2$, $y = \dfrac{1}{2^2} = \dfrac{1}{4}$.

x	y
-2	$\frac{1}{4}$
-1	1
$-\frac{1}{2}$	4
$\frac{1}{2}$	4
1	1
2	$\frac{1}{4}$

Plot these ordered pairs and draw the graph.

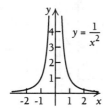

23. Graph $y = \sqrt[3]{x}$.

Find some ordered pairs that are solutions of $y = \sqrt[3]{x}$ keeping the results in a table. Since the domain of $y = \sqrt[3]{x}$ consists of all real numbers, we choose any number for x and then find y.

When $x = -6$, $y = \sqrt[3]{-6} \approx -1.817$.

When $x = -4$, $y = \sqrt[3]{-4} \approx -1.587$.

When $x = -1$, $y = \sqrt[3]{-1} \approx -1$.

When $x = -0.5$, $y = \sqrt[3]{-0.5} \approx -0.794$.

When $x = 0$, $y = \sqrt[3]{0} = 0$.

When $x = 0.5$, $y = \sqrt[3]{0.5} \approx 0.794$.

When $x = 1$, $y = \sqrt[3]{1} = 1$.

When $x = 4$, $y = \sqrt[3]{4} \approx 1.587$.

When $x = 6$, $y = \sqrt[3]{6} \approx 1.817$.

x	y
-6	-1.8
-4	-1.6
-1	-1
-0.5	-0.8
0	0
0.5	0.8
1	1
4	1.6
6	1.8

Plot these ordered pairs and draw the graph.

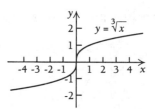

25. $f(x) = \dfrac{x^2 - 9}{x + 3}$

First we simplify the function.

$$f(x) = \frac{x^2 - 9}{x + 3} = \frac{(x + 3)(x - 3)}{x + 3} = \frac{x + 3}{x + 3} \cdot \frac{x - 3}{1} = x - 3, \ x \neq -3$$

The number -3 is not in the domain of the function, because it would result in division by 0. We can express the function as

$$y = f(x) = x - 3, \ x \neq -3.$$

We substitute any value for x other than -3 and compute the corresponding y-value.

x	y
-5	-8
-4	-7
-2	-5
0	-3
1	-2
3	0
5	2

We plot these points and draw the graph. The open circle at the point $(-3, -6)$ indicates that it is not part of the graph.

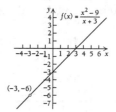

27. $f(x) = \dfrac{x^2 - 1}{x - 1}$

First we simplify the function.

$$f(x) = \frac{x^2 - 1}{x - 1} = \frac{(x + 1)(x - 1)}{x - 1} = \frac{x - 1}{x - 1} \cdot \frac{x + 1}{1} = x + 1, \ x \neq 1$$

The number 1 is not in the domain of the function, because it would result in division by 0. We can express the function as

$$y = f(x) = x + 1, \ x \neq 1.$$

We substitute any value for x other than 1 and compute the corresponding y-value.

x	y
-4	-3
-2	-1
-1	0
0	1
2	3
3	4

We plot these points and draw the graph. The open circle at the point $(1, 2)$ indicates that it is not part of the graph.

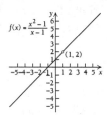

29. $\qquad x^2 - 2x = 2$

$x^2 - 2x - 2 = 0 \qquad$ Standard form

$a = 1,\, b = -2,\, \text{and } c = -2$

Then use the quadratic formula.

$x = \dfrac{-b \pm \sqrt{b^2 - 4ac}}{2a}$

$x = \dfrac{-(-2) \pm \sqrt{(-2)^2 - 4(1)(-2)}}{2 \cdot 1}$

$\qquad\qquad$ Substituting 1 for a, -2 for b, and -2 for c

$= \dfrac{2 \pm \sqrt{4 + 8}}{2} = \dfrac{2 \pm \sqrt{12}}{2}$

$= \dfrac{2 \pm 2\sqrt{3}}{2} \qquad (\sqrt{12} = \sqrt{4 \cdot 3} = 2\sqrt{3})$

$= \dfrac{2(1 \pm \sqrt{3})}{2 \cdot 1} = 1 \pm \sqrt{3}$

The solutions are $1 + \sqrt{3}$ and $1 - \sqrt{3}$.

31. $\qquad x^2 + 6x = 1$

$x^2 + 6x - 1 = 0 \qquad$ Standard notation

$a = 1,\, b = 6,\, \text{and } c = -1$

Then use the quadratic formula.

$x = \dfrac{-b \pm \sqrt{b^2 - 4ac}}{2a}$

$x = \dfrac{-6 \pm \sqrt{6^2 - 4(1)(-1)}}{2 \cdot 1} \qquad$ Substituting

$= \dfrac{-6 \pm \sqrt{36 + 4}}{2} = \dfrac{-6 \pm \sqrt{40}}{2}$

$= \dfrac{-6 \pm 2\sqrt{10}}{2} \qquad (\sqrt{40} = \sqrt{4 \cdot 10} = 2\sqrt{10})$

$= \dfrac{2(-3 \pm \sqrt{10})}{2 \cdot 1} = -3 \pm \sqrt{10} \quad$ Simplifying

The solutions are $-3 + \sqrt{10}$ and $-3 - \sqrt{10}$.

33. $\qquad 4x^2 = 4x + 1$

$4x^2 - 4x - 1 = 0 \qquad$ Standard notation

$a = 4,\, b = -4,\, \text{and } c = -1$

Then use the quadratic formula.

$x = \dfrac{-b \pm \sqrt{b^2 - 4ac}}{2a}$

$x = \dfrac{-(-4) \pm \sqrt{(-4)^2 - 4(4)(-1)}}{2 \cdot 4} \qquad$ Substituting

$= \dfrac{4 \pm \sqrt{16 + 16}}{8} = \dfrac{4 \pm \sqrt{32}}{8}$

$= \dfrac{4 \pm 4\sqrt{2}}{8} \qquad (\sqrt{32} = \sqrt{16 \cdot 2} = 4\sqrt{2})$

$= \dfrac{4(1 \pm \sqrt{2})}{4 \cdot 2} = \dfrac{1 \pm \sqrt{2}}{2}$

The solutions are $\dfrac{1 + \sqrt{2}}{2}$ and $\dfrac{1 - \sqrt{2}}{2}$.

35. $3y^2 + 8y + 2 = 0 \quad$ Standard form

$a = 3,\, b = 8,\, \text{and } c = 2$

Then use the quadratic formula.

$y = \dfrac{-b \pm \sqrt{b^2 - 4ac}}{2a}$

$y = \dfrac{-8 \pm \sqrt{8^2 - 4 \cdot 3 \cdot 2}}{2 \cdot 3} \qquad$ Substituting

$= \dfrac{-8 \pm \sqrt{64 - 24}}{6} = \dfrac{-8 \pm \sqrt{40}}{6}$

$= \dfrac{-8 \pm 2\sqrt{10}}{6} \qquad (\sqrt{40} = \sqrt{4 \cdot 10} = 2\sqrt{10})$

$= \dfrac{2(-4 \pm \sqrt{10})}{2 \cdot 3} = \dfrac{-4 \pm \sqrt{10}}{3}$

The solutions are $\dfrac{-4 + \sqrt{10}}{3}$ and $\dfrac{-4 - \sqrt{10}}{3}$.

37. $\sqrt{x^3} = x^{3/2} \quad$ (The index is 2; $\sqrt[n]{a^m} = a^{m/n}$.)

39. $\sqrt[5]{a^3} = a^{3/5} \quad (\sqrt[n]{a^m} = a^{m/n})$

41. $\sqrt[7]{t} = \sqrt[7]{t^1} \quad (t = t^1)$

$\qquad = t^{1/7} \quad (\sqrt[n]{a^m} = a^{m/n})$

43. $\dfrac{1}{\sqrt[3]{t^4}} = \dfrac{1}{t^{4/3}} \quad (\sqrt[n]{a^m} = a^{m/n})$

$\qquad = t^{-4/3} \quad \left(\dfrac{1}{a^n} = a^{-n}\right)$

45. $\dfrac{1}{\sqrt{t}} = \dfrac{1}{t^{1/2}} \quad$ (The index is 2; $\sqrt[n]{a^m} = a^{m/n}$)

$\qquad = t^{-1/2} \quad \left(\dfrac{1}{a^n} = a^{-n}\right)$

47. $\dfrac{1}{\sqrt{x^2 + 7}} = \dfrac{1}{(x^2 + 7)^{1/2}} \quad$ (The index is 2;

$\qquad\qquad\qquad\qquad\qquad\qquad \sqrt[n]{a^m} = a^{m/n})$

$\qquad = (x^2 + 7)^{-1/2} \quad \left(\dfrac{1}{a^n} = a^{-n}\right)$

49. $x^{1/5} = \sqrt[5]{x^1} \quad (a^{m/n} = \sqrt[n]{a^m})$

$\qquad = \sqrt[5]{x} \quad (x^1 = x)$

51. $y^{2/3} = \sqrt[3]{y^2}$ $\quad (a^{m/n} = \sqrt[n]{a^m})$

53. $t^{-2/5} = \dfrac{1}{t^{2/5}}$ $\quad \left(a^{-n} = \dfrac{1}{a^n}\right)$

$\qquad = \dfrac{1}{\sqrt[5]{t^2}}$ $\quad (a^{m/n} = \sqrt[n]{a^m})$

55. $b^{-1/3} = \dfrac{1}{b^{1/3}}$ $\quad \left(a^{-n} = \dfrac{1}{a^n}\right)$

$\qquad = \dfrac{1}{\sqrt[3]{b^1}}$ $\quad (a^{m/n} = \sqrt[n]{a^m})$

$\qquad = \dfrac{1}{\sqrt[3]{b}}$ $\quad (b^1 = b)$

57. $e^{-17/6} = \dfrac{1}{e^{17/6}}$ $\quad \left(a^{-n} = \dfrac{1}{a^n}\right)$

$\qquad = \dfrac{1}{\sqrt[6]{e^{17}}}$ $\quad (a^{m/n} = \sqrt[n]{a^m})$

59. $(x^2 - 3)^{-1/2} = \dfrac{1}{(x^2 - 3)^{1/2}}$ $\quad \left(a^{-n} = \dfrac{1}{a^n}\right)$

$\qquad = \dfrac{1}{\sqrt{x^2 - 3}}$ $\quad (a^{m/n} = \sqrt[n]{a^m})$

61. $\quad 9^{3/2}$

$\qquad = (9^{1/2})^3$ $\quad \left(\dfrac{3}{2} = \dfrac{1}{2} \cdot 3\right)$

$\qquad = (\sqrt{9})^3$ $\quad (a^{1/n} = \sqrt[n]{a})$

$\qquad = 3^3$ $\quad (\sqrt{9} = 3)$

$\qquad = 27$

63. $\quad 64^{2/3}$

$\qquad = (64^{1/3})^2$ $\quad \left(\dfrac{2}{3} = \dfrac{1}{3} \cdot 2\right)$

$\qquad = (\sqrt[3]{64})^2$ $\quad (a^{1/n} = \sqrt[n]{a})$

$\qquad = 4^2$ $\quad (\sqrt[3]{64} = 4)$

$\qquad = 16$

65. $\quad 16^{3/4}$

$\qquad = (16^{1/4})^3$ $\quad \left(\dfrac{3}{4} = \dfrac{1}{4} \cdot 3\right)$

$\qquad = (\sqrt[4]{16})^3$ $\quad (a^{1/n} = \sqrt[n]{a})$

$\qquad = 2^3$ $\quad (\sqrt[4]{16} = 2)$

$\qquad = 8$

67. The domain of a rational function is restricted to those input values that do not result in division by 0.

To determine the domain of

$f(x) = \dfrac{x^2 - 25}{x - 5}$

we set the denominator equal to 0 and solve.

$x - 5 = 0$

$\qquad x = 5$

Thus 5 is not in the domain. The domain consists of all real numbers except 5.

69. The domain of a rational function is restricted to those input values that do not result in division by 0.

To determine the domain of

$f(x) = \dfrac{x^3}{x^2 - 5x + 6}$

we set the denominator equal to 0 and solve.

$x^2 - 5x + 6 = 0$

$(x - 3)(x - 2) = 0$

$x - 3 = 0 \text{ or } x - 2 = 0 \quad \text{Principle of Zero Products}$

$\quad x = 3 \text{ or } \qquad x = 2$

Thus 3 and 2 are not in the domain. The domain consists of all real numbers except 3 and 2.

71. The domain of $f(x) = \sqrt{5x + 4}$ is restricted to those input values that result in the value of the radicand, $5x + 4$, being greater than or equal to 0. To determine the domain we solve the inequality $5x + 4 \geq 0$.

$5x + 4 \geq 0$

$\quad 5x \geq -4$

$\qquad x \geq -\dfrac{4}{5}$

The domain consists of all real numbers greater than or equal to $-\dfrac{4}{5}$, or $\left[-\dfrac{4}{5}, \infty\right)$.

73. We set $D(p) = S(p)$ and solve:

$1000 - 10p = 250 + 5p$

$\qquad 750 = 15p$

$\qquad 50 = p$

Thus, $p_E = \$50$. To find x_E we substitute p_E into either $D(p)$ or $S(p)$. We use $D(p)$:

$x_E = D(p_E) = D(50) = 1000 - 10 \cdot 50 =$

$1000 - 500 = 500$

The equilibrium quantity is 500 units and the equilibrium point is $(\$50, 500)$.

75. We set $D(p) = S(p)$ and solve:

$\dfrac{5}{p} = \dfrac{p}{5}$

$25 = p^2$

$0 = p^2 - 25$

$0 = (p + 5)(p - 5)$

$p + 5 = 0 \quad \text{or} \quad p - 5 = 0$

$\quad p = -5 \text{ or } \qquad p = 5$

Only 5 makes sense in this context. Thus, $p_E = \$5$. To find x_E we substitute p_E into either $D(p)$ or $S(p)$. We use $S(p)$:

$x_E = S(p_E) = S(5) = \dfrac{5}{5} = 1$

The equilibrium quantity is 1 unit and the equilibrium point is $(\$5, 1)$.

77. We set $D(p) = S(p)$ and solve:

$(p - 3)^2 = p^2 + 2p + 1$

$p^2 - 6p + 9 = p^2 + 2p + 1$

$-6p + 9 = 2p + 1$

$8 = 8p$

$1 = p$

Thus $p_E = \$1$. To find x_E we substitute p_E into either $D(p)$ or $S(p)$. We use $S(p)$.

$$x_E = S(p_E) = S(1) = 1^2 + 2 \cdot 1 + 1 = 1 + 2 + 1 = 4$$

Thus the equilibrium quantity is 4 units, and the equilibrium point is $(\$1, 4)$.

79. We set $D(p) = S(p)$ and solve:

$$5 - p = \sqrt{p + 7}, \quad 0 \le p \le 5$$
$$(5 - p)^2 = (\sqrt{p + 7})^2 \quad \text{Squaring both sides}$$
$$25 - 10p + p^2 = p + 7$$
$$18 - 11p + p^2 = 0$$
$$(9 - p)(2 - p) = 0$$
$$9 - p = 0 \quad \text{or} \quad 2 - p = 0$$
$$9 = p \quad \text{or} \quad 2 = p$$

Since 9 is not in the domain of $D(p)$, $0 \le p \le 5$, the equilibrium price $p_E = \$2$. To find x_E we substitute p_E into either $D(p)$ or $S(p)$. We use $D(p)$:

$$x_E = D(p_E) = D(2) = 5 - 2 = 3$$

The equilibrium quantity is 3 units, and the equilibrium point is $(\$2, 3)$.

81. Dividends D are inversely proportional to prime rate R.

$$D = \frac{k}{R}$$
$$13\frac{7}{8} = \frac{k}{8\frac{1}{4}\%} \quad \text{Substituting}$$
$$13.875 = \frac{k}{0.0825}$$
$$1.1446875 = k$$

The equation of variation is $D = \dfrac{1.1446875}{R}$.

We find D when R is $9\frac{1}{2}\%$.

$$D = \frac{1.1446875}{9\frac{1}{2}\%} \quad \text{Substituting}$$
$$D = \frac{1.1446875}{0.095}$$
$$D \approx 12.05$$

The dividends would be about $12.05 per share.

83. $f(x) = \dfrac{1}{6}x^3 + \dfrac{1}{2}x^2 + \dfrac{1}{3}x$

$$f(7) = \frac{1}{6} \cdot 7^3 + \frac{1}{2} \cdot 7^2 + \frac{1}{3} \cdot 7$$
$$= \frac{343}{6} + \frac{49}{2} + \frac{7}{3}$$
$$= \frac{343}{6} + \frac{147}{6} + \frac{14}{6}$$
$$= \frac{504}{6}$$
$$= 84$$

There are 84 oranges when there are 7 layers.

$$f(10) = \frac{1}{6} \cdot 10^3 + \frac{1}{2} \cdot 10^2 + \frac{1}{3} \cdot 10$$
$$= \frac{1000}{6} + \frac{100}{2} + \frac{10}{3}$$
$$= \frac{1000}{6} + \frac{300}{6} + \frac{20}{6}$$
$$= \frac{1320}{6}$$
$$= 220$$

There are 220 oranges when there are 10 layers.

$$f(12) = \frac{1}{6} \cdot 12^3 + \frac{1}{2} \cdot 12^2 + \frac{1}{3} \cdot 12$$
$$= \frac{1728}{6} + \frac{144}{2} + \frac{12}{3}$$
$$= \frac{1728}{6} + \frac{432}{6} + \frac{24}{6}$$
$$= \frac{2184}{6}$$
$$= 364$$

There are 364 oranges when there are 12 layers.

85. a) In 2002, $t = 32$.

$P = 1000(32)^{5/4} + 14,000 \approx 90,109$ particles per cubic centimeter.

In 2005, $t = 2005 - 1970 = 35$.

$P = 1000(35)^{5/4} + 14,000 \approx 99,130$ particles per cubic centimeter.

In 2010, $t = 2010 - 1970 = 40$.

$P = 1000(40)^{5/4} + 14,000 \approx 114,595$ particles per cubic centimeter.

b) Plot the points found above and others, if necessary, and draw the graph.

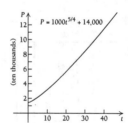

87. $\boxed{tw}$

89.

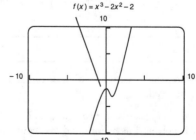

Zero: 2.359

91.

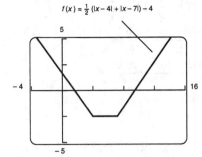

$f(x) = \frac{1}{2}(|x-4| + |x-7|) - 4$

Zeros: 1.5, 9.5

93.

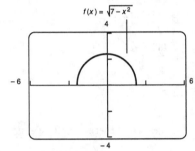

$f(x) = \sqrt{7-x^2}$

Zeros: -2.646, 2.646

95.

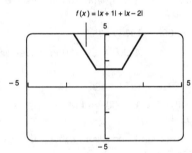

$f(x) = |x+1| + |x-2|$

Zeros: none

97.

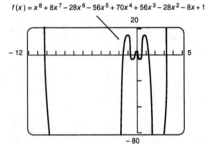

$f(x) = x^8 + 8x^7 - 28x^6 - 56x^5 + 70x^4 + 56x^3 - 28x^2 - 8x + 1$

Zeros: -10.153, -1.871, -0.821, -0.303, 0.099, 0.535, 1.219, 3.297

Exercise Set 1.6

---

1. The points rise at a steady rate, so a linear function $f(x) = mx + b$ could be used.

3. The points rise, then fall, then rise again, so a polynomial function that is neither linear nor quadratic could be used.

5. The points fall and then rise in a curved manner that could be modeled with a quadratic function $f(x) = ax^2 + bx + c$, $a > 0$.

7. a)

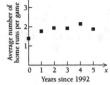

The data could be modeled by a linear function.

b) Consider a linear function $H(x) = mx + b$, where H is the average number of home runs per game in the National League x years after 1992. We use the points $(1, 1.72)$ and $(5, 1.83)$ to find m and b. We substitute and obtain a system of equations.

$$1.72 = m \cdot 1 + b, \text{ or } 1.72 = m + b, \quad (1)$$
$$1.83 = m \cdot 5 + b, \text{ or } 1.83 = 5m + b \quad (2)$$

Subtracting each side of Equation (1) from the corresponding side of Equation (2), we get

$$0.11 = 4m$$
$$0.0275 = m.$$

Now substitute 0.0275 for m in either Equation (1) or (2) and solve for b. We use Equation (1).

$$1.72 = 0.0275 + b$$
$$1.6925 = b$$

Then we have $H(x) = 0.0275x + 1.6925$.

c) In 2000, $x = 2000 - 1992 = 8$.

$$H(8) = 0.0275(8) + 1.6925 \approx 1.91$$

In 2000, the average number of home runs per game is about 1.91.

In 2005, $x = 2005 - 1992 = 13$.

$$H(13) = 0.0275(13) + 1.6925 = 2.05$$

In 2005, the average number of home runs per game will be 2.05.

9. a) Answers will vary depending on the points that are used. We use $(2, 3.3)$ and $(6, 6.5)$. Consider a linear function $S(x) = mx + b$, where S is the net sales of The Gap, in billions of dollars, x years after 1992. Substitute to obtain a system of equations.

$$3.3 = m \cdot 2 + b, \text{ or } 3.3 = 2m + b, \quad (1)$$
$$6.5 = m \cdot 6 + b, \text{ or } 6.5 = 6m + b \quad (2)$$

Subtracting each side of Equation (1) from the corresponding side of Equation (2) we get

$$3.2 = 4m$$
$$0.8 = m$$

Now substitute 0.8 for m in Equation (1) and solve for b.

$$3.3 = 2(0.8) + b$$
$$3.3 = 1.6 + b$$
$$1.7 = b$$

Then we have $S(x) = 0.8x + 1.7$.

b)

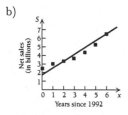

Years since 1992

c) In 2000, $x = 2000 - 1992 = 8$.

$S(8) = 0.8(8) + 1.7 = 8.1$

The net sales will be $8.1 billion in 2000.

In 2005, $x = 2005 - 1992 = 13$.

$S(13) = 0.8(13) + 1.7 = 12.1$

The net sales will be $12.1 billion in 2005.

11. a) Consider a quadratic function $N(x) = ax^2 + bx + c$, where N is the net earnings of Payless ShoeSource, in millions of dollars, x years after 1994. Substitute the coordinates of the given points in the function.

$$131.5 = a \cdot 0^2 + b \cdot 0 + c,$$
$$54.0 = a \cdot 1^2 + b \cdot 1 + c,$$
$$107.7 = a \cdot 2^2 + b \cdot 2 + c, \text{ or}$$

$$131.5 = c,$$
$$54.0 = a + b + c,$$
$$107.7 = 4a + 2b + c$$

Solving this system of equations, we get $a = 65.6$, $b = -143.1$, and $c = 131.5$. Thus, we have $N(x) = 65.6x^2 - 143.1x + 131.5$.

b) In 2002, $x = 2002 - 1994 = 8$.

$S(8) = 65.6(8)^2 - 143.1(8) + 131.5 = 3185.1$

Net earnings will be about $3185 million in 2002.

In 2010, $x = 2010 - 1994 = 16$.

$S(16) = 65.6(16)^2 - 143.1(16) + 131.5 = 14,635.5$

Net earnings will be about $14,636 million in 2010.

c) $\boxed{tw}$

13. a) We look for a function of the form $N(x) = ax^2 + bx + c$, where $N(x)$ represents the number of nighttime accidents (for every 200 million km) and s represents the travel speed (in km/h). We substitute the given values of x and $N(x)$.

$$400 = a(60)^2 + b(60) + c,$$
$$250 = a(80)^2 + b(80) + c,$$
$$250 = a(100)^2 + b(100) + c,$$
$$\text{or}$$
$$400 = 3600a + 60b + c,$$
$$250 = 6400a + 80b + c,$$
$$250 = 10,000a + 100b + c.$$

Solving the system of equations, we get $a = 0.1875$, $b = -33.75$, $c = 1750$.

Thus, the function $N(x) = 0.1875x^2 - 33.75x + 1750$ fits the data.

b) Find $N(50)$.

$N(50) = 0.1875(50)^2 - 33.75(50) + 1750 = 531.25$

At 50 km/h, 531 accidents per 200,000,000 km driven occur.

15. $\boxed{tw}$

17. a) $H(x) = 0.114x + 1.52$

b) $H(8) \approx 2.43$; $H(13) \approx 3.00$

c) $\boxed{tw}$

19. a) $N(x) = 24.675x^2 - 69.435x + 123.315$

b) $N(8) \approx \$1147$ million; $N(16) \approx \$5329$ million

c) $\boxed{tw}$

Chapter 2

Differentiation

1. Since $\lim\limits_{x \to 1^-} f(x) = 2$ and $\lim\limits_{x \to 1^+} f(x) = -1$, then $\lim\limits_{x \to 1} f(x)$ does not exist. The function is not continuous at $x = 1$ and is therefore not continuous over the whole real line.

3. The function is continuous over the whole real line since it is continuous at each point in $(-\infty, \infty)$. Note that the graph can be traced without lifting the pencil from the paper.

5. a) As inputs x approach 1 from the right, outputs $f(x)$ approach -1. Thus, the limit from the right is -1.

$$\lim_{x \to 1^+} f(x) = -1$$

As inputs x approach 1 from the left, outputs $f(x)$ approach 2. Thus, the limit from the left is 2.

$$\lim_{x \to 1^-} f(x) = 2$$

Since the limit from the left, 2, is not the same as the limit from the right, -1, we say $\lim\limits_{x \to 1} f(x)$ does not exist.

b) When the input is 1, the output, $f(1)$, is -1.

$$f(1) = -1$$

c) Since the limit at $x = 1$ does not exist, the function is not continuous at $x = 1$.

d) As inputs x approach -2 from the left, outputs $f(x)$ approach 3. Thus, the limit from the left is 3.

$$\lim_{x \to -2^-} f(x) = 3$$

As inputs x approach -2 from the right, outputs $f(x)$ approach 3. Thus, the limit from the right is 3.

$$\lim_{x \to -2^+} f(x) = 3$$

Since the limit from the left, 3, is the same as the limit from the right, 3, we have

$$\lim_{x \to -2} f(x) = 3.$$

e) When the input is -2, the output, $f(-2)$, is 3.

$$f(-2) = 3$$

f) The function $f(x)$ is continuous at $x = -2$, because

1) $f(-2)$ exists, $f(-2) = 3$,

2) $\lim\limits_{x \to -2} f(x)$ exists, $\lim\limits_{x \to -2} f(x) = 3$,
and

3) $\lim\limits_{x \to -2} f(x) = f(-2) = 3$.

7. a) As inputs x approach 1 from the left, outputs $h(x)$ approach 2. Thus, the limit from the left is 2.

$$\lim_{x \to 1^-} h(x) = 2$$

As inputs x approach 1 from the right, outputs $h(x)$ approach 2. Thus, the limit from the right is 2.

$$\lim_{x \to 1^+} h(x) = 2$$

Since the limit from the left, 2, is the same as the limit from the right, 2, we have $\lim\limits_{x \to 1} h(x) = 2$.

b) When the input is 1, the output, $h(1)$, is 2.

$$h(1) = 2$$

c) The function $h(x)$ is continuous at $x = 1$, because

1) $h(1)$ exists, $h(1) = 2$,

2) $\lim\limits_{x \to 1} h(x)$ exists, $\lim\limits_{x \to 1} h(x) = 2$,
and

3) $\lim\limits_{x \to 1} h(x) = h(1) = 2$.

d) As inputs x approach -2 from the left, outputs $h(x)$ approach 0. Thus, the limit from the left is 0.

$$\lim_{x \to -2^-} h(x) = 0$$

As inputs x approach -2 from the right, outputs $h(x)$ approach 0. Thus, the limit from the right is 0.

$$\lim_{x \to -2^+} h(x) = 0$$

Since the limit from the left, 0, is the same as the limit from the right, 0, we have

$$\lim_{x \to -2} h(x) = 0.$$

e) When the input is -2, the output, $h(-2)$, is 0.

$$h(-2) = 0$$

f) The function $h(x)$ is continuous at $x = -2$, because

1) $h(-2)$ exists, $h(-2) = 0$,

2) $\lim\limits_{x \to -2} h(x)$ exists, $\lim\limits_{x \to -2} h(x) = 0$,
and

3) $\lim\limits_{x \to -2} h(x) = h(-2) = 0$.

9. a) As inputs x approach 1 from the right, outputs $f(x)$ approach 3.

$$\lim_{x \to 1^+} f(x) = 3$$

b) As inputs x approach 1 from the left, outputs $f(x)$ approach 3.

$$\lim_{x \to 1^-} f(x) = 3$$

c) Since the limit from the right, 3, is the same as the limit from the left, 3, we have $\lim\limits_{x \to 1} f(x) = 3$.

d) When the input is 1, the output is 2.

$$f(1) = 2$$

e) The function is not continuous at $x = 1$, because $f(1) = 2$, but $\lim\limits_{x \to 1} f(x) = 3$.

f) $f(2)$ exists. $(f(2) = 4 - 2 = 2)$

   $\lim\limits_{x \to 2^-} f(x) = 2$ and $\lim\limits_{x \to 2^+} f(x) = 2$, so

   $\lim\limits_{x \to 2} f(x)$ exists. $(\lim\limits_{x \to 2} f(x) = 2)$

   $\lim\limits_{x \to 2} f(x) = f(2) = 2$

   Thus, f is continuous at $x = 2$.

11. a) As inputs x approach -2 from the right, outputs $f(x)$ approach 2. Thus, the statement $\lim\limits_{x \to -2^+} f(x) = 1$ is false.

b) As inputs x approach -2 from the left, outputs $f(x)$ approach 0. Thus, the statement $\lim\limits_{x \to -2^-} f(x) = 0$ is true.

c) $\lim\limits_{x \to -2^-} f(x) = 0$ but $\lim\limits_{x \to -2^+} f(x) = 2$, so the statement $\lim\limits_{x \to -2^-} f(x) = \lim\limits_{x \to -2^+} f(x)$ is false.

d) Since the limit from the left, 0, is not the same as the limit from the right, 2, $\lim\limits_{x \to -2} f(x)$ does not exist. Thus, the given statement is false.

e) $\lim\limits_{x \to -2} f(x)$ does not exist (see part (d)), so the statement $\lim\limits_{x \to -2} f(x) = 2$ is false.

f) $\lim\limits_{x \to 0^-} f(x) = 0$ and $\lim\limits_{x \to 0^+} f(x) = 0$, so $\lim\limits_{x \to 0} f(x) = 0$ and the given statement is true.

g) From the graph we see that $f(0) = 2$, so the given statement is true.

h) Since $\lim\limits_{x \to -2} f(x)$ does not exist (see part (d)), the statement that f is continuous at $x = -2$ is false.

i) $\lim\limits_{x \to 0} f(x) = 0$ but $f(0) = 2$, so the statement that f is continuous at $x = 0$ is false.

j) $f(-1)$ exists. $(f(-1) = 1)$

   $\lim\limits_{x \to -1^-} f(x) = 1$ and $\lim\limits_{x \to -1^+} f(x) = 1$, so

   $\lim\limits_{x \to -1} f(x)$ exists. $(\lim\limits_{x \to -1} f(x) = 1)$

   $\lim\limits_{x \to -1} f(x) = f(-1) = 1$

   Thus, the statement that f is continuous at $x = -1$ is true.

13. The function p is not continuous at 1 since $\lim\limits_{x \to 1} p(x)$ does not exist.

The function p is continuous at $1\frac{1}{2}$, because

1) $p\left(1\frac{1}{2}\right)$ exists, $p\left(1\frac{1}{2}\right) = 55\cancel{c}$,

2) $\lim\limits_{x \to 1\frac{1}{2}} p(x)$ exists, $\lim\limits_{x \to 1\frac{1}{2}} p(x) = 55\cancel{c}$, and

3) $\lim\limits_{x \to 1\frac{1}{2}} p(x) = p\left(1\frac{1}{2}\right) = 55\cancel{c}$.

The function p is not continuous at $x = 2$ since $\lim\limits_{x \to 2} p(x)$ does not exist.

The function p is continuous at 2.53, because

1) $p(2.53)$ exists, $p(2.53) = 77\cancel{c}$,

2) $\lim\limits_{x \to 2.53} p(x)$ exists, $\lim\limits_{x \to 2.53} p(x) = 77\cancel{c}$, and

3) $\lim\limits_{x \to 2.53} p(x) = p(2.53) = 77\cancel{c}$.

15. As inputs x approach 1 from the left, outputs $p(x)$ approach $33\cancel{c}$. Thus the limit from the left is $33\cancel{c}$.

   $\lim\limits_{x \to 1^-} p(x) = 33\cancel{c}$

As inputs x approach 1 from the right, outputs $p(x)$ approach $55\cancel{c}$. Thus the limit from the right is $55\cancel{c}$.

   $\lim\limits_{x \to 1^+} p(x) = 55\cancel{c}$

Since the limit from the left, $33\cancel{c}$, is not the same as the limit from the right, $55\cancel{c}$, we have

   $\lim\limits_{x \to 1} p(x)$ does not exist.

17. As inputs x approach 2.3 from the left, outputs $p(x)$ approach $77\cancel{c}$. Thus the limit from the left is $77\cancel{c}$.

   $\lim\limits_{x \to 2.3^-} p(x) = 77\cancel{c}$

As inputs x approach 2.3 from the right, outputs $p(x)$ approach $77\cancel{c}$. Thus the limit from the right is $77\cancel{c}$.

   $\lim\limits_{x \to 2.3^+} p(x) = 77\cancel{c}$

Since the limit from the left, $77\cancel{c}$, is the same as the limit from the right, $77\cancel{c}$, we have

   $\lim\limits_{x \to 2.3} p(x) = 77\cancel{c}$.

19. As inputs x approach 12 from the left, outputs $C(x)$ approach \$20,000. Thus, $\lim\limits_{x \to 12^-} C(x) = \$20,000$.

As inputs x approach 12 from the right, outputs $C(x)$ approach \$28,000. Thus, $\lim\limits_{x \to 12^+} C(x) = \$28,000$.

Since the limit from the left is not the same as the limit from the right, $\lim\limits_{x \to 12} C(x)$ does not exist.

21. $C(x)$ is not continuous at $x = 12$, because $\lim\limits_{x \to 12} C(x)$ does not exist.

$C(x)$ is continuous at $x = 20$, because

1) $C(20)$ exists,

2) $\lim\limits_{x \to 20} C(x)$ exists, and

3) $\lim\limits_{x \to 20} C(x) = C(20) = \$32,000$.

23. $\boxed{tw}$

25. As inputs t approach 20 from the right, outputs $N(t)$ approach 100. Thus, $\lim\limits_{t \to 20^+} N(t) = 100$.

As inputs t approach 20 from the left, outputs $N(t)$ approach 30. Thus, $\lim\limits_{t \to 20^-} N(t) = 30$.

Since the limits from the right, 100, is not the same an the limit from the left, 30, $\lim\limits_{t \to 20} N(t)$ does not exist.

27. Since $\lim\limits_{t \to 20} N(t)$ does not exist, N is not continuous at $t = 20$.

N is continuous at $t = 30$, because

1) $N(30)$ exists,

2) $\lim\limits_{t \to 30} N(t)$ exists, and

3) $\lim\limits_{t \to 30} N(t) = N(30) = 100.$

29. $\boxed{tw}$

31. a) $\lim\limits_{x \to 0^-} f(x) = -2$ but $\lim\limits_{x \to 0^+} f(x) = 2$, so $\lim\limits_{x \to 0} f(x)$ does not exist.

b) $\lim\limits_{x \to -2^-} f(x) = 2$ and $\lim\limits_{x \to -2^+} f(x) = 2$, so $\lim\limits_{x \to -2} f(x) = 2.$

33. $\boxed{tw}$

Exercise Set 2.2

1. Find $\lim\limits_{x \to 1}(x^2 - 3)$.

From the continuity principles it follows that $x^2 - 3$ is continuous. Thus the limit can be found by substitution.

$$\lim_{x \to 1}(x^2 - 3) = 1^2 - 3 \quad \text{Substituting}$$
$$= 1 - 3$$
$$= -2$$

3. Find $\lim\limits_{x \to 0} \dfrac{3}{x}$.

The function $\dfrac{3}{x}$ is not continuous at $x = 0$, and there is no algebraic simplification for $\dfrac{3}{x}$. Thus, direct substitution is not possible. We can use input-output tables or a graph. Here we use a graph.

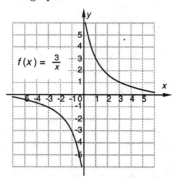

$$f(x) = \frac{3}{x}$$

As x approaches 0 from the left, the outputs become more and more negative without bound. These numbers do not approach any real number, though it might be said that "the limit from the left is $-\infty$ (negative infinity)." As x approaches 0 from the right, the outputs increase without bound. These numbers do not approach any real number, though it might be said that "the limit from the right is ∞ (infinity)." For a limit to exist, the limits from the left and right must both exist and be the same.

Thus, $\lim\limits_{x \to 0} \dfrac{3}{x}$ does not exist.

5. Find $\lim\limits_{x \to 3}(2x + 5)$.

From the continuity principles it follows that $2x + 5$ is continuous. Thus the limit can be found by substitution.

$$\lim_{x \to 3}(2x + 5) = 2 \cdot 3 + 5 \quad \text{Substituting}$$
$$= 6 + 5$$
$$= 11$$

7. Find $\lim\limits_{x \to -5} \dfrac{x^2 - 25}{x + 5}$.

The function $\dfrac{x^2 - 25}{x + 5}$ is not continuous at $x = -5$. We use algebraic simplification and then some limit principles.

$$\lim_{x \to -5} \frac{x^2 - 25}{x + 5}$$
$$= \lim_{x \to -5} \frac{(x + 5)(x - 5)}{x + 5} \quad \text{Factoring the numerator}$$
$$= \lim_{x \to -5}(x - 5) \quad \text{Simplifying, assuming } x \neq -5$$
$$= \lim_{x \to -5} x - \lim_{x \to -5} 5 \quad \text{By L3}$$
$$= -5 - 5 \quad \text{By L2 and L1}$$
$$= -10$$

9. Find $\lim\limits_{x \to -2} \dfrac{5}{x}$.

The function $\dfrac{5}{x}$ is continuous at all real numbers except 0. Since $\dfrac{5}{x}$ is continuous at $x = -2$, we can substitute to find the limit.

$$\lim_{x \to -2} \frac{5}{x} = \frac{5}{-2} \quad \text{Substituting}$$
$$= -\frac{5}{2}$$

11. Find $\lim\limits_{x \to 2} \dfrac{x^2 + x - 6}{x - 2}$.

The function $\dfrac{x^2 + x - 6}{x - 2}$ is not continuous at $x = 2$. We use algebraic simplification and then some limit principles.

$$\lim_{x \to 2} \frac{x^2 + x - 6}{x - 2}$$
$$= \lim_{x \to 2} \frac{(x - 2)(x + 3)}{x - 2} \quad \text{Factoring the numerator}$$
$$= \lim_{x \to 2}(x + 3) \quad \text{Simplifying, assuming } x \neq 2$$
$$= \lim_{x \to 2} x + \lim_{x \to 2} 3 \quad \text{By L3}$$
$$= 2 + 3 \quad \text{By L2 and L1}$$
$$= 5$$

13. Find $\lim\limits_{x \to 5} \sqrt[3]{x^2 - 17}$.

From the continuity principles it follows that $\sqrt[3]{x^2 - 17}$ is continuous. Thus the limit can be found by substitution.

$$\lim_{x \to 5} \sqrt[3]{x^2 - 17} = \sqrt[3]{5^2 - 17} \qquad \text{Substituting}$$

$$= \sqrt[3]{25 - 17}$$

$$= \sqrt[3]{8}$$

$$= 2$$

15. Find $\lim_{x \to 1}(x^4 - x^3 + x^2 + x + 1)$.

From the continuity principles it follows that $x^4 - x^3 + x^2 + x + 1$ is continuous. Thus the limit can be found by substitution.

$$\lim_{x \to 1}(x^4 - x^3 + x^2 + x + 1)$$

$$= 1^4 - 1^3 + 1^2 + 1 + 1 \qquad \text{Substituting}$$

$$= 1 - 1 + 1 + 1 + 1$$

$$= 3$$

17. Find $\lim_{x \to 2} \dfrac{1}{x - 2}$.

The function $\dfrac{1}{x - 2}$ is not continuous at $x = 2$, and there is no algebraic simplification for $\dfrac{1}{x - 2}$. Thus, direct substitution is not possible. We can use input-output tables or a graph. Here we use a graph.

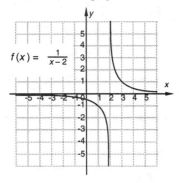

As x approaches 2 from the left, the outputs get more and more negative without bound. Thus, $\lim_{x \to 2^-} \dfrac{1}{x - 2} = -\infty$. As x approaches 2 from the right, the outputs increase without bound. Thus, $\lim_{x \to 2^+} \dfrac{1}{x - 2} = \infty$. We say that $\lim_{x \to 2} \dfrac{1}{x - 2}$ does not exist.

> Remember: For a limit to exist, the limits from the left and right must both exist and be the same.

19. Find $\lim_{x \to 2} \dfrac{3x^2 - 4x + 2}{7x^2 - 5x + 3}$.

The rational function $\dfrac{3x^2 - 4x + 2}{7x^2 - 5x + 3}$ is continuous at all real numbers except those for which $7x^2 - 5x + 3 = 0$. Since $7x^2 - 5x + 3 \neq 0$ when $x = 2$, the function is continuous at $x = 2$. Thus we can substitute to find the limit.

$$\lim_{x \to 2} \frac{3x^2 - 4x + 2}{7x^2 - 5x + 3}$$

$$= \frac{3 \cdot 2^2 - 4 \cdot 2 + 2}{7 \cdot 2^2 - 5 \cdot 2 + 3} \qquad \text{Substituting}$$

$$= \frac{12 - 8 + 2}{28 - 10 + 3}$$

$$= \frac{6}{21}$$

$$= \frac{2}{7} \qquad \text{Simplifying}$$

21. Find $\lim_{x \to 2} \dfrac{x^2 + x - 6}{x^2 - 4}$.

The function $\dfrac{x^2 + x - 6}{x^2 - 4}$ is not continuous at $x = 2$ ($x^2 - 4 = 0$ when $x = 2$). We first use algebraic simplification.

$$\lim_{x \to 2} \frac{x^2 + x - 6}{x^2 - 4}$$

$$= \lim_{x \to 2} \frac{(x - 2)(x + 3)}{(x - 2)(x + 2)} \quad \begin{array}{l}\text{Factoring numerator and}\\\text{denominator}\end{array}$$

$$= \lim_{x \to 2} \frac{(x + 3)}{(x + 2)} \quad \text{Simplifying, assuming } x \neq 2$$

Using input-output tables, we see that as inputs x approach 2 from the left, outputs $\dfrac{x + 3}{x + 2}$ approach $\dfrac{5}{4}$. As inputs x approach 2 from the right, outputs $\dfrac{x + 3}{x + 2}$ approach $\dfrac{5}{4}$. Since the limit from the left, $\dfrac{5}{4}$, is the same as the limit from the right, $\dfrac{5}{4}$, we have

$$\lim_{x \to 2} \frac{x + 3}{x + 2} = \frac{5}{4}.$$

Thus, $\lim_{x \to 2} \dfrac{x^2 + x - 6}{x^2 - 4} = \dfrac{5}{4}$.

Note that limit principles could have been used as an alternative to input-output tables in finding the limit.

23. Find $\lim_{h \to 0}(6x^2 + 6xh + 2h^2)$.

We treat x as a constant since we are interested only in the way in which the expression varies when $h \to 0$. Since $6x^2 + 6xh + 2h^2$ is continuous at $h = 0$, we can substitute to find the limit.

$$\lim_{h \to 0}(6x^2 + 6xh + 2h^2)$$

$$= 6x^2 + 6x \cdot 0 + 2 \cdot 0^2 \qquad \text{Substituting}$$

$$= 6x^2 + 0 + 0$$

$$= 6x^2$$

25. Find $\lim_{h \to 0} \dfrac{-2x - h}{x^2(x + h)^2}$.

We treat x as a constant since we are interested only in the way in which the expression varies when $h \to 0$. Since $\dfrac{-2x - h}{x^2(x + h)^2}$ is continuous at $h = 0$, we can substitute to find the limit.

$$\lim_{h\to 0}\frac{-2x-h}{x^2(x+h)^2}=\frac{-2x-0}{x^2(x+0)^2} \quad \text{Substituting 0 for } h$$

$$=\frac{-2x}{x^2\cdot x^2}$$

$$=\frac{-2x}{x^4}$$

$$=\frac{-2}{x^3} \qquad \text{Simplifying}$$

27. It is helpful to graph the function.

$$f(x)=\begin{cases} 1, & \text{for } x\neq 2 \\ -1, & \text{for } x=2 \end{cases}$$

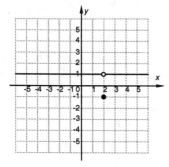

a) Find $\lim_{x\to 0} f(x)$.

As inputs x approach 0 from the left, outputs $f(x)$ approach 1. Thus, the limit from the left is 1. As inputs x approach 0 from the right, outputs $f(x)$ approach 1, so the limit from the right is 1. Since the limit from the left, 1, is the same as the limit from the right, 1, we have

$$\lim_{x\to 0} f(x)=1.$$

b) Find $\lim_{x\to 2^-} f(x)$.

As inputs x approach 2 from the left, outputs $f(x)$ approach 1. Thus, the limit from the left is 1.

$$\lim_{x\to 2^-} f(x)=1$$

c) Find $\lim_{x\to 2^+} f(x)$.

As inputs x approach 2 from the right, outputs $f(x)$ approach 1. Thus, the limit from the right is 1.

$$\lim_{x\to 2^+} f(x)=1$$

d) Since the limit from the left, 1, is the same as the limit from the right, 1, we have

$$\lim_{x\to 2} f(x)=1.$$

e) The function $f(x)$ is continuous at $x=0$ because

1) $f(0)$ exists, $f(0)=1$,

2) $\lim_{x\to 0} f(x)$ exists, $\lim_{x\to 0} f(x)=1$, and

3) $\lim_{x\to 0} f(x)=f(0)=1$.

The function $f(x)$ is not continuous at $x=2$ because the $\lim_{x\to 2} f(x)$ is not the same as the function value at $x=2$.

$$\lim_{x\to 2} f(x)=1, \text{ but } f(2)=-1.$$

$$\lim_{x\to 2} f(x)\neq f(2)$$

29.

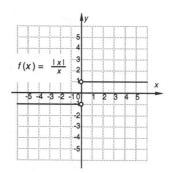

$$f(x)=\frac{|x|}{x}$$

$$\lim_{x\to 0^-}\frac{|x|}{x}=-1, \quad \lim_{x\to 0^+}\frac{|x|}{x}=1, \text{ so } \lim_{x\to 0}\frac{|x|}{x} \text{ does not exist.}$$

31. 6

33. -0.2887, or $-\dfrac{1}{2\sqrt{3}}$

35. 0.75

37. 0.25

Exercise Set 2.3

1. a) $f(x)=7x^2$

so

$$f(x+h)=7(x+h)^2 \qquad \text{Substituting } x+h \text{ for } x$$

$$=7(x^2+2xh+h^2)$$

$$=7x^2+14xh+7h^2$$

Then

$$\frac{f(x+h)-f(x)}{h} \qquad \text{Difference quotient}$$

$$=\frac{(7x^2+14xh+7h^2)-7x^2}{h} \qquad \text{Substituting}$$

$$=\frac{14xh+7h^2}{h}$$

$$=\frac{h(14x+7h)}{h} \qquad \text{Factoring the numerator}$$

$$=\frac{h}{h}\cdot(14x+7h)$$

$$=14x+7h, \quad \text{Simplified difference quotient}$$

or $7(2x+h)$

b) The difference quotient column in the table can be easily completed using the simplified difference quotient.

$14x+7h$ Simplified difference quotient

$=14(4)+7(2)=56+14=70$
Substituting 4 for x and 2 for h;

$=14(4)+7(1)=56+7=63$
Substituting 4 for x and 1 for h;

$=14(4)+7(0.1)=56+0.7=56.7$
Substituting 4 for x and 0.1 for h;

$= 14(4) + 7(0.1) = 56 + 0.07 = 56.07$
$\quad$ Substituting 4 for x and 0.01 for h.

The values in the table are 70, 63, 56.7, and 56.07.

3. a) $f(x) = -7x^2$

so

$f(x + h) = -7(x + h)^2$ $\quad$ Substituting $x + h$ for x
$\qquad = -7(x^2 + 2xh + h^2)$
$\qquad = -7x^2 - 14xh - 7h^2$

Then
$$\frac{f(x+h) - f(x)}{h} \qquad \text{Difference quotient}$$

$$= \frac{(-7x^2 - 14xh - 7h^2) - (-7x^2)}{h} \qquad \text{Substituting}$$

$$= \frac{-7x^2 - 14xh - 7h^2 + 7x^2}{h}$$

$$= \frac{-14xh - 7h^2}{h}$$

$$= \frac{h(-14x - 7h)}{h} \qquad \text{Factoring the numerator}$$

$$= \frac{h}{h} \cdot (-14x - 7h)$$

$$= -14x - 7h, \qquad \text{Simplified difference quotient}$$

or $-7(2x + h)$

b) The difference quotient column in the table can be easily completed using the simplified difference quotient.

$\qquad -14x - 7h$ $\quad$ Simplified difference quotient

$= -14(4) - 7(2) = -56 - 14 = -70$
$\qquad$ Substituting 4 for x and 2 for h;
$= -14(4) - 7(1) = -56 - 7 = -63$
$\qquad$ Substituting 4 for x and 1 for h;
$= -14(4) - 7(0.1) = -56 - 0.7 = -56.7$
$\qquad$ Substituting 4 for x and 0.1 for h;
$= -14(4) - 7(0.1) = -56 - 0.07 = -56.07$
$\qquad$ Substituting 4 for x and 0.01 for h.

The values in the table are -70, -63, -56.7, and -56.07.

5. a) $f(x) = 7x^3$

so

$f(x + h) = 7(x + h)^3$ $\quad$ Substituting $x + h$ for x
$\qquad = 7(x^3 + 3x^2h + 3xh^2 + h^3)$
$\qquad = 7x^3 + 21x^2h + 21xh^2 + 7h^3$

Then

$$\frac{f(x+h) - f(x)}{h} \qquad \text{Difference quotient}$$

$$= \frac{(7x^3 + 21x^2h + 21xh^2 + 7h^3) - 7x^3}{h}$$
$$\qquad\qquad\qquad \text{Substituting}$$

$$= \frac{21x^2h + 21xh^2 + 7h^3}{h}$$

$$= \frac{h(21x^2 + 21xh + 7h^2)}{h} \qquad \text{Factoring the numerator}$$

$$= \frac{h}{h} \cdot (21x^2 + 21xh + 7h^2)$$

$$= 21x^2 + 21xh + 7h^2 \qquad \text{Simplified difference quotient}$$

or $7(3x^2 + 3xh + h^2)$

b) The difference quotient column in the table can be easily completed using the simplified difference quotient.

$\qquad 21x^2 + 21xh + 7h^2$
$\qquad$ Simplified difference quotient
$= 21 \cdot 4^2 + 21 \cdot 4 \cdot 2 + 7 \cdot 2^2$

$\qquad$ Substituting 4 for x and 2 for h
$= 21 \cdot 16 + 21 \cdot 8 + 7 \cdot 4$
$= 336 + 168 + 28$
$= 532;$

$= 21 \cdot 4^2 + 21 \cdot 4 \cdot 1 + 7 \cdot 1^2$
$\qquad$ Substituting 4 for x and 1 for h
$= 21 \cdot 16 + 21 \cdot 4 + 7 \cdot 1$
$= 336 + 84 + 7$
$= 427;$

$= 21(4)^2 + 21(4)(0.1) + 7(0.1)^2$
$\qquad$ Substituting 4 for x and 0.1 for h
$= 21(16) + 21(0.4) + 7(0.01)$
$= 336 + 8.4 + 0.07$
$= 344.47;$

$= 21(4)^2 + 21(4)(0.01) + 7(0.01)^2$
$\qquad$ Substituting 4 for x and 0.01 for h
$= 21(16) + 21(0.04) + 7(0.0001)$
$= 336 + 0.84 + 0.0007$
$= 336.8407$

The values in the table are 532, 427, 344.47, and 336.8407.

7. a) $f(x) = \dfrac{5}{x}$

so

$f(x+h) = \dfrac{5}{x+h}$ Substituting $x+h$ for x

Then

$\dfrac{f(x+h) - f(x)}{h}$ Difference quotient

$= \dfrac{\dfrac{5}{x+h} - \dfrac{5}{x}}{h}$ Substituting

$= \dfrac{\dfrac{5}{x+h} \cdot \dfrac{x}{x} - \dfrac{5}{x} \cdot \dfrac{x+h}{x+h}}{h}$ Multiplying by 1

$= \dfrac{\dfrac{5x}{x(x+h)} - \dfrac{5(x+h)}{x(x+h)}}{h}$

$= \dfrac{\dfrac{5x - 5x - 5h}{x(x+h)}}{h}$

$= \dfrac{\dfrac{-5h}{x(x+h)}}{\dfrac{h}{1}}$ $h = \dfrac{h}{1}$

$= \dfrac{-5h}{x(x+h)} \cdot \dfrac{1}{h}$ Multiplying by the reciprocal

$= \dfrac{h}{h} \cdot \dfrac{-5}{x(x+h)}$

$= \dfrac{-5}{x(x+h)}$ Simplified difference quotient

b) The difference quotient column in the table can be easily completed using the simplified difference quotient.

$\dfrac{-5}{x(x+h)}$ Simplified difference quotient

$= \dfrac{-5}{4(4+2)} = \dfrac{-5}{24} \approx -0.2083$
Substituting 4 for x and 2 for h;

$= \dfrac{-5}{4(4+1)} = \dfrac{-5}{20} = -0.25$
Substituting 4 for x and 1 for h;

$= \dfrac{-5}{4(4+0.1)} = \dfrac{-5}{16.4} \approx -0.3049$
Substituting 4 for x and 0.1 for h;

$= \dfrac{-5}{4(4+0.01)} = \dfrac{-5}{16.04} \approx -0.3117$
Substituting 4 for x and 0.01 for h.

The values in the table are -0.2083, -0.25, -0.3049, and -0.3117.

9. a) $f(x) = -2x + 5$

so

$f(x+h) = -2(x+h) + 5$ Substituting $x+h$ for x
$= -2x - 2h + 5$

Then

$\dfrac{f(x+h) - f(x)}{h}$ Difference quotient

$= \dfrac{(-2x-2h+5) - (-2x+5)}{h}$ Substituting

$= \dfrac{-2x - 2h + 5 + 2x - 5}{h}$

$= \dfrac{-2h}{h}$

$= -2$ Simplified difference quotient

b) The difference quotient is -2 for all values of x and h, so all values in the difference quotient column of the table are -2.

11. a) $f(x) = x^2 - x$

so

$f(x+h) = (x+h)^2 - (x+h)$
Substituting $x+h$ for x
$= x^2 + 2xh + h^2 - x - h$

Then

$\dfrac{f(x+h) - f(x)}{h}$ Difference quotient

$= \dfrac{(x^2+2xh+h^2-x-h) - (x^2-x)}{h}$
Substituting

$= \dfrac{x^2 + 2xh + h^2 - x - h - x^2 + x}{h}$

$= \dfrac{2xh + h^2 - h}{h}$

$= \dfrac{h(2x + h - 1)}{h}$

$= 2x + h - 1$ Simplified difference quotient

b) The difference quotient column in the table can be easily completed using the simplified difference quotient.

$2x + h - 1$ Simplified difference quotient

$= 2(4) + 2 - 1 = 9$
Substituting 4 for x and 2 for h;

$= 2(4) + 1 - 1 = 8$
Substituting 4 for x and 1 for h;

$= 2(4) + 0.1 - 1 = 7.1$
Substituting 4 for x and 0.1 for h;

$= 2(4) + 0.01 - 1 = 7.01$
Substituting 4 for x and 0.01 for h.

The values in the table are 9, 8, 7.1, and 7.01.

13. a) When the input is 0, the output, $U(0)$, is 0.
When the input is 1, the output, $U(1)$, is 70.

Let $(x_1, y_1) = (0, 0)$ and $(x_2, y_2) = (1, 70)$.

The average rate of change is
$$\frac{y_2 - y_1}{x_2 - x_1} = \frac{70 - 0}{1 - 0} = \frac{70}{1} =$$
70 pleasure units/unit of product.

When the input is 1, the output, $U(1)$, is 70.
When the input is 2, the output, $U(2)$, is 109.

Let $(x_1, y_1) = (1, 70)$ and $(x_2, y_2) = (2, 109)$.

The average rate of change is
$$\frac{y_2 - y_1}{x_2 - x_1} = \frac{109 - 70}{2 - 1} = \frac{39}{1} =$$
39 pleasure units/unit of product.

When the input is 2, the output, $U(2)$, is 109.
When the input is 3, the output, $U(3)$, is 138.

Let $(x_1, y_1) = (2, 109)$ and $(x_2, y_2) = (3, 138)$.

The average rate of change is
$$\frac{y_2 - y_1}{x_2 - x_1} = \frac{138 - 109}{3 - 2} = \frac{29}{1} =$$
29 pleasure units/unit of product.

When the input is 3, the output, $U(3)$, is 138.
When the input is 4, the output, $U(4)$, is 161.

Let $(x_1, y_1) = (3, 138)$ and $(x_2, y_2) = (4, 161)$.

The average rate of change is
$$\frac{y_2 - y_1}{x_2 - x_1} = \frac{161 - 138}{4 - 3} = \frac{23}{1} =$$
23 pleasure units/unit of product.

b) $\boxed{tw}$

15. a) $R(x) = -0.01x^2 + 1000x$ Total revenue

$R(301) = -0.01(301)^2 + 1000(301)$

Substituting 301 for x

$= -0.01(90,601) + 1000(301)$

$= -906.01 + 301,000$

$= 300,093.99$

The total revenue from the sale of 301 units is $300,093.99.

b) $R(x) = -0.01x^2 + 1000x$ Total revenue

$R(300) = -0.01(300)^2 + 1000(300)$

Substituting 300 for x

$= -0.01(90,000) + 1000(300)$

$= -900 + 300,000$

$= 299,100$

The total revenue from the sale of 300 units is $299,100.

c) $R(301) = 300,093.99$ Part (a)

$R(300) = 299,100$ Part (b)

$R(301) - R(300) = 300,093.99 - 299,100$

$= 993.99$

The change in total revenue from the sale of 300 units to 301 units is $993.99.

d) The average rate of change of the total revenue with respect to the number of units sold, as the number of units sold changes from 300 to 301, is
$$\frac{R(301) - R(300)}{301 - 300}$$
$$= \frac{300,093.99 - 299,100}{301 - 300}$$
$$= \frac{993.99}{1}$$
$$= \$993.99$$

e) $\boxed{tw}$

17. a) We will find the average rate of change in millions of dollars per year. Let $(x_1, y_1) = (1993, \ 2.960)$ and $(x_2, y_2) = (1997, \ 5.284)$.
$$\frac{y_2 - y_1}{x_2 - x_1} = \frac{5.284 - 2.960}{1997 - 1993} = \frac{2.324}{4} = 0.581$$
The average rate of change is $0.581 million per year.

b) We will find the average rate of change in millions of dollars per year. Let $(x_1, y_1) = (1995, 3.723)$ and $(x_2, y_2) = (1997, \ 5.284)$.
$$\frac{y_2 - y_1}{x_2 - x_1} = \frac{5.284 - 3.723}{1997 - 1995} = \frac{1.561}{2} = 0.7805$$
The average rate of change is about $0.781 million per year.

c) $\boxed{tw}$

19. a) As t changes from 0 to 8, $M(t)$ changes from 0 to 10. The average rate of change is $\frac{10 - 0}{8 - 0} = \frac{10}{8} = \frac{5}{4}$, or 1.25 words per minute.

As t changes from 8 to 16, $M(t)$ changes from 10 to 20. The average rate of change is $\frac{20 - 10}{16 - 8} = \frac{10}{8} = \frac{5}{4}$, or 1.25 words per minute.

As t changes from 16 to 24, $M(t)$ changes from 20 to 25. The average rate of change is $\frac{25 - 20}{24 - 16} = \frac{5}{8}$, or 0.625 words per minute.

As t changes from 24 to 32, $M(t)$ changes from 25 to 25. The average rate of change is $\frac{25 - 25}{32 - 24} = \frac{0}{8}$, or 0 words per minute.

As t changes from 32 to 36, $M(t)$ changes from 25 to 25. The average rate of change is $\frac{25 - 25}{36 - 32} = \frac{0}{4}$, or 0 words per minute.

b) $\boxed{tw}$

21. a) $s(t) = 16t^2$ s is distance in feet;
                              t is time in seconds

$s(3) = 16 \cdot 3^2$ Substituting 3 for t

$= 16 \cdot 9$

$= 144$

The object will fall 144 feet in 3 seconds.

b) $s(t) = 16t^2$

$s(5) = 16 \cdot 5^2$ Substituting 5 for t

$\qquad = 16 \cdot 25$

$\qquad = 400$

The object will fall 400 feet in 5 seconds.

c) The average rate of change of distance with respect to time during the time from 3 to 5 seconds is

$$\frac{s(5) - s(3)}{5 - 3}$$

$$= \frac{400 \text{ feet} - 144 \text{ feet}}{5 \text{ seconds} - 3 \text{ seconds}}$$

$$= \frac{256 \text{ feet}}{2 \text{ seconds}}$$

$$= 128 \frac{\text{ft}}{\text{sec}}$$

The average velocity is 128 ft/sec.

23. a) For each curve, as t changes from 0 to 4, $P(t)$ changes from 0 to 500. Thus, the average growth rate for each is

$$\frac{500 - 0}{4 - 0} = \frac{500}{4} = 125 \text{ million people/yr.}$$

b) $\boxed{tw}$

c) Country A:

As t changes from 0 to 1, $P(t)$ changes from 0 to 290. The average growth rate is

$$\frac{290 - 0}{1 - 0} = \frac{290}{1} = 290 \text{ million people/yr.}$$

As t changes from 1 to 2, $P(t)$ changes from 290 to 250. The average growth rate is

$$\frac{250 - 290}{2 - 1} = \frac{-40}{1} = -40 \text{ million people/yr.}$$

As t changes from 2 to 3, $P(t)$ changes from 250 to 200. The average growth rate is

$$\frac{200 - 250}{3 - 2} = \frac{-50}{1} = -50 \text{ million people/yr.}$$

As t changes from 3 to 4, $P(t)$ changes from 200 to 500. The average growth rate is

$$\frac{500 - 200}{4 - 3} = \frac{300}{1} = 300 \text{ million people/yr.}$$

Country B:

As t changes from 0 to 1, $P(t)$ changes from 0 to 125. The average growth rate is

$$\frac{125 - 0}{1 - 0} = \frac{125}{1} = 125 \text{ million people/yr.}$$

As t changes from 1 to 2, $P(t)$ changes from 125 to 250. The average growth rate is

$$\frac{250 - 125}{2 - 1} = \frac{125}{1} = 125 \text{ million people/yr.}$$

As t changes from 2 to 3, $P(t)$ changes from 250 to 375. The average growth rate is

$$\frac{375 - 250}{3 - 2} = \frac{125}{1} = 125 \text{ million people/yr.}$$

As t changes from 3 to 4, $P(t)$ changes from 375 to 500. The average growth rate is

$$\frac{500 - 375}{4 - 3} = \frac{125}{1} = 125 \text{ million people/yr.}$$

d) $\boxed{tw}$

25. $\boxed{tw}$

27. $f(x) = ax^2 + bx + c$

so

$f(x + h) = a(x + h)^2 + b(x + h) + c$

$\qquad\qquad$ Substituting $x + h$ for x

$\qquad = a(x^2 + 2xh + h^2) + b(x + h) + c$

$\qquad = ax^2 + 2axh + ah^2 + bx + bh + c$

Then

$$\frac{f(x + h) - f(x)}{h} \quad \text{Difference quotient}$$

$$= \frac{(ax^2 + 2axh + ah^2 + bx + bh + c) - (ax^2 + bx + c)}{h}$$

$$\qquad\qquad\qquad\qquad\qquad \text{Substituting}$$

$$= \frac{2axh + ah^2 + bh}{h}$$

$$= \frac{h(2ax + ah + b)}{h} \quad \text{Factoring the numerator}$$

$$= 2ax + ah + b \quad \text{Simplified difference quotient}$$

29. $f(x) = \sqrt{x}$

so

$f(x + h) = \sqrt{x + h}$ Substituting $x + h$ for x

Then

$$\frac{f(x + h) - f(x)}{h} \quad \text{Difference quotient}$$

$$= \frac{\sqrt{x + h} - \sqrt{x}}{h} \quad \text{Substituting}$$

$$= \frac{\sqrt{x + h} - \sqrt{x}}{h} \cdot \frac{\sqrt{x + h} + \sqrt{x}}{\sqrt{x + h} + \sqrt{x}}$$

$$\qquad\qquad\qquad\qquad \text{Multiplying by 1}$$

$$= \frac{(x + h) - x}{h(\sqrt{x + h} + \sqrt{x})}$$

$$= \frac{h}{h(\sqrt{x + h} + \sqrt{x})}$$

$$= \frac{1}{\sqrt{x + h} + \sqrt{x}} \quad \text{Simplified difference quotient}$$

31. $f(x) = \dfrac{1}{x^2}$

so

$$f(x + h) = \frac{1}{(x + h)^2} \quad \text{Substituting } x + h \text{ for } x$$

Then

$$\frac{f(x+h) - f(x)}{h} \quad \text{Difference quotient}$$

$$= \frac{\dfrac{1}{(x+h)^2} - \dfrac{1}{x^2}}{h} \quad \text{Substituting}$$

$$= \frac{\dfrac{1}{(x+h)^2} \cdot \dfrac{x^2}{x^2} - \dfrac{1}{x^2} \cdot \dfrac{(x+h)^2}{(x+h)^2}}{h}$$

$$= \frac{\dfrac{x^2 - (x+h)^2}{x^2(x+h)^2}}{\dfrac{h}{1}}$$

$$= \frac{x^2 - x^2 - 2xh - h^2}{x^2(x+h)^2} \cdot \frac{1}{h}$$

$$= \frac{-2xh - h^2}{x^2(x+h)^2} \cdot \frac{1}{h}$$

$$= \frac{h(-2x - h)}{x^2(x+h)^2} \cdot \frac{1}{h}$$

$$= \frac{-2x - h}{x^2(x+h)^2} \quad \text{Simplified difference quotient}$$

33. $f(x) = \dfrac{x}{1+x}$

so

$$f(x+h) = \frac{x+h}{1+x+h} \quad \text{Substituting } x+h \text{ for } x$$

Then

$$\frac{f(x+h) - f(x)}{h} \quad \text{Difference quotient}$$

$$= \frac{\dfrac{x+h}{1+x+h} - \dfrac{x}{1+x}}{h} \quad \text{Substituting}$$

$$= \frac{\dfrac{x+h}{1+x+h} \cdot \dfrac{1+x}{1+x} - \dfrac{x}{1+x} \cdot \dfrac{1+x+h}{1+x+h}}{\dfrac{h}{1}}$$

$$= \frac{(x^2 + xh + x + h) - (x + x^2 + xh)}{(1+x)(1+x+h)} \cdot \frac{1}{h}$$

$$= \frac{h}{(1+x)(1+x+h)} \cdot \frac{1}{h}$$

$$= \frac{1}{(1+x)(1+x+h)} \quad \text{Simplified difference quotient}$$

Exercise Set 2.4

1. a), b)

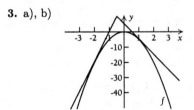

c)

$$\frac{f(x+h) - f(x)}{h}$$

$$= \frac{5(x+h)^2 - 5x^2}{h}$$

$$= \frac{5(x^2 + 2xh + h^2) - 5x^2}{h}$$

$$= \frac{5x^2 + 10xh + 5h^2 - 5x^2}{h}$$

$$= \frac{10xh + 5h^2}{h}$$

$$= \frac{h(10x + 5h)}{h}$$

$$= 10x + 5h \quad \text{Simplified difference quotient}$$

Find the limit of the difference quotient as $h \to 0$.

$$\lim_{h \to 0} \frac{f(x+h) - f(x)}{h}$$

$$= \lim_{h \to 0} (10x + 5h)$$

$$= 10x$$

Thus, $f'(x) = 10x$.

d) $f'(-2) = 10(-2) = -20$ Substituting -2 for x

$f'(0) = 10 \cdot 0 = 0$ Substituting 0 for x

$f'(1) = 10 \cdot 1 = 10$ Substituting 1 for x

3. a), b)

c)

$$\frac{f(x+h) - f(x)}{h}$$

$$= \frac{-5(x+h)^2 - (-5x^2)}{h}$$

$$= \frac{-5(x^2 + 2xh + h^2) - (-5x^2)}{h}$$

$$= \frac{-5x^2 - 10xh - 5h^2 + 5x^2}{h}$$

$$= \frac{-10xh - 5h^2}{h}$$

$$= \frac{h(-10x - 5h)}{h}$$

$$= -10x - 5h \quad \text{Simplified difference quotient}$$

Find the limit of the difference quotient as $h \to 0$.

$$\lim_{h \to 0} \frac{f(x+h) - f(x)}{h}$$

$$= \lim_{h \to 0} (-10x - 5h)$$

$$= -10x$$

Thus, $f'(x) = -10x$.

d) $f'(-2) = -10(-2) = 20$

$\quad f'(0) = -10 \cdot 0 = 0$

$\quad f'(1) = -10 \cdot 1 = -10$

5. a), b)

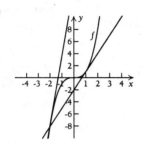

c) $\dfrac{f(x+h) - f(x)}{h}$

$= \dfrac{(x+h)^3 - x^3}{h}$

$= \dfrac{x^3 + 3x^2h + 3xh^2 + h^3 - x^3}{h}$

$= \dfrac{3x^2h + 3xh^2 + h^3}{h}$

$= \dfrac{h(3x^2 + 3xh + h^2)}{h}$

$= 3x^2 + 3xh + h^2 \quad$ Simplified difference quotient

Find the limit of the difference quotient as $h \to 0$.

$\qquad \lim\limits_{h \to 0} \dfrac{f(x+h) - f(x)}{h}$

$= \lim\limits_{h \to 0} (3x^2 + 3xh + h^2)$

$= 3x^2$

Thus, $f'(x) = 3x^2$.

d) $f'(-2) = 3(-2)^2 = 3 \cdot 4 = 12$

$\quad f'(0) = 3 \cdot 0^2 = 3 \cdot 0 = 0$

$\quad f'(1) = 3 \cdot 1^2 = 3 \cdot 1 = 3$

7. a), b)

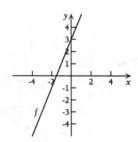

All the tangent lines are identical to the graph of the original function.

c) $\dfrac{f(x+h) - f(x)}{h}$

$= \dfrac{2(x+h) + 3 - (2x + 3)}{h}$

$= \dfrac{2x + 2h + 3 - 2x - 3}{h}$

$= \dfrac{2h}{h}$

$= 2 \qquad$ Simplified difference quotient

Find the limit of the difference quotient as $h \to 0$.

$\qquad \lim\limits_{h \to 0} \dfrac{f(x+h) - f(x)}{h}$

$= \lim\limits_{h \to 0} 2$

$= 2$

Thus, $f'(x) = 2$.

d) $f'(-2) = 2$

$\quad f'(0) = 2$

$\quad f'(1) = 2$

9. a), b)

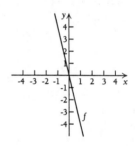

All the tangent lines are identical to the graph of the original function.

c) $\dfrac{f(x+h) - f(x)}{h}$

$= \dfrac{-4(x+h) - (-4x)}{h}$

$= \dfrac{-4x - 4h + 4x}{h}$

$= \dfrac{-4h}{h}$

$= -4 \qquad$ Simplified difference quotient

Find the limit of the difference quotient as $h \to 0$.

$\qquad \lim\limits_{h \to 0} \dfrac{f(x+h) - f(x)}{h}$

$= \lim\limits_{h \to 0} -4$

$= -4$

Thus, $f'(x) = -4$.

d) $f'(-2) = -4$

$\quad f'(0) = -4$

$\quad f'(1) = -4$

11. a), b)

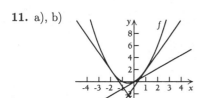

c)
$$\frac{f(x+h) - f(x)}{h}$$

$$= \frac{(x+h)^2 + (x+h) - (x^2 + x)}{h}$$

$$= \frac{x^2 + 2xh + h^2 + x + h - x^2 - x}{h}$$

$$= \frac{2xh + h^2 + h}{h}$$

$$= \frac{h(2x + h + 1)}{h}$$

$$= 2x + h + 1 \quad \text{Simplified difference quotient}$$

Find the limit of the difference quotient as $h \to 0$.

$$\lim_{h \to 0} \frac{f(x+h) - f(x)}{h}$$

$$= \lim_{h \to 0} (2x + h + 1)$$

$$= 2x + 1$$

Thus, $f'(x) = 2x + 1$.

d) $f'(-2) = 2(-2) + 1 = -3$

$$f'(0) = 2 \cdot 0 + 1 = 1$$

$$f'(1) = 2 \cdot 1 + 1 = 3$$

13. a), b)

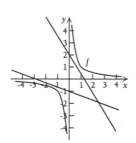

There is no tangent line for $x = 0$.

c)
$$\frac{f(x+h) - f(x)}{h}$$

$$= \frac{\dfrac{1}{x+h} - \dfrac{1}{x}}{h}$$

$$= \frac{\dfrac{1}{x+h} \cdot \dfrac{x}{x} - \dfrac{1}{x} \cdot \dfrac{x+h}{x+h}}{h}$$

$$= \frac{\dfrac{x - x - h}{x(x+h)}}{h}$$

$$= \frac{\dfrac{-h}{x(x+h)}}{h}$$

$$= \frac{-h}{x(x+h)} \cdot \frac{1}{h}$$

$$= \frac{-1}{x(x+h)} \quad \text{Simplified difference quotient}$$

Find the limit of the difference quotient as $h \to 0$.

$$\lim_{h \to 0} \frac{f(x+h) - f(x)}{h}$$

$$= \lim_{h \to 0} \frac{-1}{x(x+h)}$$

$$= \frac{-1}{x \cdot x}$$

$$= \frac{-1}{x^2}$$

Thus, $f'(x) = \dfrac{-1}{x^2}$.

d) $f'(-2) = \dfrac{-1}{(-2)^2} = \dfrac{-1}{4} = -\dfrac{1}{4}$

$f'(0)$ does not exist because $f(0)$ does not exist. Thus, f is not differentiable at 0. This is confirmed when we substitute to find $f'(0)$: $f'(0) = \dfrac{-1}{0^2}$ is undefined.

$$f'(1) = \frac{-1}{1^2} = \frac{-1}{1} = -1$$

15.
$$\frac{f(x+h) - f(x)}{h}$$

$$= \frac{m(x+h) - mx}{h}$$

$$= \frac{mx + mh - mx}{h}$$

$$= \frac{mh}{h}$$

$$= m \quad \text{Simplified difference quotient}$$

Find the limit of the difference quotient as $h \to 0$.

$$\lim_{h \to 0} \frac{f(x+h) - f(x)}{h}$$

$$= \lim_{h \to 0} m$$

$$= m$$

Thus, $f'(x) = m$.

17. From Example 3 we know that $f'(x) = 2x$.

$f'(3) = 2 \cdot 3 = 6$, so the slope of the line tangent to the curve at $(3, 9)$ is 6. We substitute the point $(3, 9)$ and the slope 6 in the point-slope equation to find the equation of the tangent line:

$$y - y_1 = m(x - x_1)$$

$$y - 9 = 6(x - 3)$$

$$y - 9 = 6x - 18$$

$$y = 6x - 9$$

$f'(-1) = 2(-1) = -2$, so the slope of the line tangent to the curve at $(-1, 1)$ is -2. We substitute the point $(-1, 1)$ and the slope -2 in the point-slope equation to find the equation of the tangent line:

$$y - y_1 = m(x - x_1)$$
$$y - 1 = -2[x - (-1)]$$
$$y - 1 = -2(x + 1)$$
$$y - 1 = -2x - 2$$
$$y = -2x - 1$$

$f'(10) = 2 \cdot 10 = 20$, so the slope of the line tangent to the curve at $(10, 100)$ is 20. We substitute the point $(10, 100)$ and the slope 20 in the point-slope equation to find the equation of the tangent line:

$$y - y_1 = m(x - x_1)$$
$$y - 100 = 20(x - 10)$$
$$y - 100 = 20x - 200$$
$$y = 20x - 100$$

19. From Exercise 14, we know that the simplified difference quotient is $\dfrac{-5}{x(x + h)}$.

Find the limit of the difference quotient as $h \to 0$.

$$\lim_{h \to 0} \frac{f(x + h) - f(x)}{h} = \lim_{h \to 0} \frac{-5}{x(x + h)}$$
$$= \frac{-5}{x \cdot x}$$
$$= \frac{-5}{x^2}$$

Thus, $f'(x) = \dfrac{-5}{x^2}$.

$f'(1) = \dfrac{-5}{1^2} = -5$, so the slope of the line tangent to the curve at $(1, 5)$ is -5. We substitute the point $(1, 5)$ and the slope -5 in the point-slope equation to find the equation of the tangent line:

$$y - y_1 = m(x - x_1)$$
$$y - 5 = -5(x - 1)$$
$$y - 5 = -5x + 5$$
$$y = -5x + 10$$

$f'(-1) = \dfrac{-5}{(-1)^2} = -5$, so the slope of the line tangent to the curve at $(-1, -5)$ is -5. We substitute the point $(-1, -5)$ and the slope -5 in the point-slope equation to find the equation of the tangent line:

$$y - y_1 = m(x - x_1)$$
$$y - (-5) = -5[x - (-1)]$$
$$y + 5 = -5(x + 1)$$
$$y + 5 = -5x - 5$$
$$y = -5x - 10$$

$f'(100) = \dfrac{-5}{100^2} = -0.0005$, so the slope of the line tangent to the curve at $(100, 0.05)$ is -0.0005. We substitute the point $(100, 0.05)$ and the slope -0.0005 in the point-slope equation to find the equation of the tangent line:

$$y - 0.05 = -0.0005(x - 100)$$
$$y - 0.05 = -0.0005x + 0.05$$
$$y = -0.0005x + 0.1$$

21. First we find $f'(x)$:

$$\frac{f(x + h) - f(x)}{h} = \frac{4 - (x + h)^2 - (4 - x^2)}{h}$$
$$= \frac{4 - (x^2 + 2xh + h^2) - 4 + x^2}{h}$$
$$= \frac{4 - x^2 - 2xh - h^2 - 4 + x^2}{h}$$
$$= \frac{-2xh - h^2}{h}$$
$$= \frac{h(-2x - h)}{h}$$
$$= -2x - h$$

$$f'(x) = \lim_{h \to 0}(-2x - h) = -2x$$

$f'(-1) = -2(-1) = 2$, so the slope of the line tangent to the curve at $(-1, 3)$ is 2. We substitute the point $(-1, 3)$ and the slope 2 in the point-slope equation to find the equation of the tangent line:

$$y - y_1 = m(x - x_1)$$
$$y - 3 = 2[x - (-1)]$$
$$y - 3 = 2(x + 1)$$
$$y - 3 = 2x + 2$$
$$y = 2x + 5$$

$f'(0) = -2 \cdot 0 = 0$, so the slope of the line tangent to the curve at $(0, 4)$ is 0. We substitute the point $(0, 4)$ and the slope 0 in the point-slope equation to find the equation of the tangent line:

$$y - y_1 = m(x - x_1)$$
$$y - 4 = 0(x - 0)$$
$$y - 4 = 0$$
$$y = 4$$

$f'(5) = -2 \cdot 5 = -10$, so the slope of the line tangent to the curve at $(5, -21)$ is -10. We substitute the point $(5, -21)$ and the slope -10 in the point-slope equation to find the equation of the tangent line:

$$y - y_1 = m(x - x_1)$$
$$y - (-21) = -10(x - 5)$$
$$y + 21 = -10x + 50$$
$$y = -10x + 29$$

23. If a function has a "sharp point" or "corner," it will not have a derivative at that point. Thus, the function is not differentiable at x_3, x_4, and x_6. The function has a vertical tangent at x_{12}. Vertical lines have no slope, hence there is no derivative at the point x_{12}. Also, if a function is discontinuous at some point a, then it is not differentiable at a. The function is discontinuous at point x_0. Thus, it is also not differentiable at x_0.

25. The postage function does not have any "sharp points" or "corners" nor does it have any vertical tangents. However, it is discontinuous at all natural numbers. Therefore, the postage function is not differentiable for 1, 2, 3, 4, and so on.

27. $\boxed{tw}$

29. We found the simplified difference quotient in Exercise 31, Exercise Set 2.3. We now find the limit of the difference quotient as $h \to 0$.

$$\lim_{h \to 0} \frac{-2x - h}{x^2 (x + h)^2}$$

$$= \frac{-2x}{x^2 \cdot x^2} \qquad \begin{array}{l}[\text{As } h \to 0, \, (-2x - h) \to -2x \\ \text{and } (x + h) \to x]\end{array}$$

$$= \frac{-2x}{x^4}$$

$$= \frac{-2}{x^3}$$

Thus, $f'(x) = \dfrac{-2}{x^3}$.

31. We found the simplified difference quotient in Exercise 33, Exercise Set 2.3. We now find the limit of the difference quotient as $h \to 0$.

$$\lim_{h \to 0} \frac{1}{(1 + x)(1 + x + h)}$$

$$= \frac{1}{(1 + x)(1 + x)} \qquad [\text{As } h \to 0, \, (1 + x + h) \to (1 + x).]$$

$$= \frac{1}{(1 + x)^2}, \text{ or } \frac{1}{1 + 2x + x^2}$$

Thus, $f'(x) = \dfrac{1}{(1 + x)^2}$.

33. The domain of a rational function is restricted to those input values that do not result in division by 0. The domain for $f(x) = \dfrac{x^2 - 9}{x + 3}$ consists of all real numbers except -3. Since $f(-3)$ does not exist, the function is not continuous at -3. Thus, the function is not differentiable at $x = -3$.

35. - 39.

Exercise Set 2.5

1. $\quad y = x^7$

$$\frac{dy}{dx} = \frac{d}{dx} x^7$$

$$= 7x^{7-1} \qquad \text{Theorem 1}$$

$$= 7x^6$$

3. $\quad y = 15 \qquad \text{Constant function}$

$$\frac{dy}{dx} = \frac{d}{dx} 15$$

$$= 0 \qquad \text{Theorem 2}$$

5. $\quad y = 4x^{150}$

$$\frac{dy}{dx} = \frac{d}{dx} 4x^{150}$$

$$= 4 \cdot \frac{d}{dx} x^{150} \qquad \text{Theorem 3}$$

$$= 4 \cdot 150 x^{150-1} \qquad \text{Theorem 1}$$

$$= 600 x^{149}$$

7. $\quad y = x^3 + 3x^2$

$$\frac{dy}{dx} = \frac{d}{dx} (x^3 + 3x^2)$$

$$= \frac{d}{dx} x^3 + \frac{d}{dx} 3x^2 \qquad \text{Theorem 4}$$

$$= \frac{d}{dx} x^3 + 3 \cdot \frac{d}{dx} x^2 \qquad \text{Theorem 3}$$

$$= 3x^{3-1} + 3 \cdot 2x^{2-1} \qquad \text{Theorem 1}$$

$$= 3x^2 + 6x$$

9. $\quad y = 8\sqrt{x}$

$$y = 8x^{1/2}$$

$$\frac{dy}{dx} = \frac{d}{dx} 8x^{1/2}$$

$$= 8 \cdot \frac{d}{dx} x^{1/2} \qquad \text{Theorem 3}$$

$$= 8 \cdot \frac{1}{2} x^{1/2-1} \qquad \text{Theorem 1}$$

$$= 4x^{-1/2}, \text{ or } \frac{4}{x^{1/2}}, \text{ or } \frac{4}{\sqrt{x}}$$

11. $\quad y = x^{0.07}$

$$\frac{dy}{dx} = \frac{d}{dx} x^{0.07}$$

$$= 0.07 x^{0.07-1} \qquad \text{Theorem 1}$$

$$= 0.07 x^{-0.93}, \text{ or } \frac{0.07}{x^{0.93}}$$

13. $\quad y = \frac{1}{2} x^{4/5}$

$$\frac{dy}{dx} = \frac{d}{dx} \frac{1}{2} x^{4/5}$$

$$= \frac{1}{2} \cdot \frac{d}{dx} x^{4/5} \qquad \text{Theorem 3}$$

$$= \frac{1}{2} \cdot \frac{4}{5} x^{4/5-1} \qquad \text{Theorem 1}$$

$$= \frac{2}{5} x^{-1/5}, \text{ or } \frac{2}{5x^{1/5}}, \text{ or } \frac{2}{5\sqrt[5]{x}}$$

15. $\quad y = x^{-3}$

$$\frac{dy}{dx} = \frac{d}{dx} x^{-3}$$

$$= -3x^{-3-1} \qquad \text{Theorem 1}$$

$$= -3x^{-4}, \text{ or } \frac{-3}{x^4}$$

17. $\quad y = 3x^2 - 8x + 7$

$$\frac{dy}{dx} = \frac{d}{dx} (3x^2 - 8x + 7)$$

$$= \frac{d}{dx} 3x^2 - \frac{d}{dx} 8x + \frac{d}{dx} 7 \qquad \text{Theorem 4}$$

$$= 3 \cdot \frac{d}{dx} x^2 - 8 \cdot \frac{d}{dx} x + \frac{d}{dx} 7 \qquad \text{Theorem 3}$$

$$= 3 \cdot 2x^{2-1} - 8 \cdot 1x^{1-1} + 0 \qquad \begin{array}{l}\text{Theorem 1,} \\ \text{Theorem 2}\end{array}$$

$$= 6x - 8 \qquad\qquad x^{1-1} = x^0 = 1$$

19. $y = \sqrt[4]{x} - \dfrac{1}{x}$

$\quad\quad y = x^{1/4} - x^{-1}$

$\quad\quad \dfrac{dy}{dx} = \dfrac{d}{dx}(x^{1/4} - x^{-1})$

$\quad\quad\quad\quad = \dfrac{d}{dx}x^{1/4} - \dfrac{d}{dx}x^{-1} \quad\quad$ Theorem 4

$\quad\quad\quad\quad = \dfrac{1}{4}x^{1/4-1} - (-1)x^{-1-1} \quad$ Theorem 1

$\quad\quad\quad\quad = \dfrac{1}{4}x^{-3/4} + x^{-2}$, or $\dfrac{1}{4\sqrt[4]{x^3}} + \dfrac{1}{x^2}$

21. $f(x) = 0.64x^{2.5}$

$\quad\quad \dfrac{d}{dx}f(x) = \dfrac{d}{dx}0.64x^{2.5}$

$\quad\quad\quad\quad = 0.64 \cdot \dfrac{d}{dx}x^{2.5} \quad$ Theorem 3

$\quad\quad\quad\quad = 0.64(2.5x^{2.5-1}) \quad$ Theorem 1

$\quad\quad\quad\quad = 1.6x^{1.5})$

$\quad\quad f'(x) = 1.6x^{1.5}$

23. $f(x) = \dfrac{5}{x} - x$

$\quad\quad \dfrac{d}{dx}f(x) = \dfrac{d}{dx}(5x^{-1} - x)$

$\quad\quad\quad\quad = \dfrac{d}{dx}5x^{-1} - \dfrac{d}{dx}x \quad\quad$ Theorem 4

$\quad\quad\quad\quad = 5 \cdot \dfrac{d}{dx}x^{-1} - \dfrac{d}{dx}x \quad\quad$ Theorem 3

$\quad\quad\quad\quad = 5 \cdot (-1)x^{-1-1} - 1x^{1-1} \quad$ Theorem 1

$\quad\quad\quad\quad = -5x^{-2} - 1 \quad\quad\quad\quad x^{1-1} = x^0 = 1$

$\quad\quad f'(x) = -5x^{-2} - 1$, or $\dfrac{-5}{x^2} - 1$

25. $f(x) = 4x - 7$

$\quad\quad \dfrac{d}{dx}f(x) = \dfrac{d}{dx}(4x - 7)$

$\quad\quad\quad\quad = \dfrac{d}{dx}4x - \dfrac{d}{dx}7 \quad\quad$ Theorem 4

$\quad\quad\quad\quad = 4 \cdot \dfrac{d}{dx}x - \dfrac{d}{dx}7 \quad\quad$ Theorem 3

$\quad\quad\quad\quad = 4 \cdot 1 - 0 \quad$ Theorem 1, $\dfrac{d}{dx}x = 1 \cdot x^{1-1} = $

$\quad\quad\quad\quad\quad\quad\quad\quad\quad x^0 = 1$

$\quad\quad\quad\quad\quad\quad\quad\quad$ Theorem 2, $\dfrac{d}{dx}7 = 0$

$\quad\quad\quad\quad = 4$

$\quad\quad f'(x) = 4$

27. $f(x) = 4x + 9$

$\quad\quad \dfrac{d}{dx}f(x) = \dfrac{d}{dx}(4x + 9)$

$\quad\quad\quad\quad = \dfrac{d}{dx}4x + \dfrac{d}{dx}9 \quad\quad$ Theorem 4

$\quad\quad\quad\quad = 4 \cdot \dfrac{d}{dx}x + \dfrac{d}{dx}9 \quad\quad$ Theorem 3

$\quad\quad\quad\quad = 4 \cdot 1 + 0 \quad\quad$ Theorems 1 and 2

$\quad\quad\quad\quad = 4$

$\quad\quad f'(x) = 4$

29. $f(x) = \dfrac{x^4}{4} = \dfrac{1}{4}x^4$

$\quad\quad \dfrac{d}{dx}f(x) = \dfrac{d}{dx}\dfrac{1}{4}x^4$

$\quad\quad\quad\quad = \dfrac{1}{4} \cdot \dfrac{d}{dx}x^4 \quad\quad$ Theorem 3

$\quad\quad\quad\quad = \dfrac{1}{4} \cdot 4x^{4-1} \quad\quad$ Theorem 1

$\quad\quad\quad\quad = x^3$

$\quad\quad f'(x) = x^3$

31. $f(x) = -0.01x^2 - 0.5x + 70$

$\quad\quad \dfrac{d}{dx}f(x) = \dfrac{d}{dx}(-0.01x^2 - 0.5x + 70)$

$\quad\quad\quad\quad = \dfrac{d}{dx}(-0.01x^2) - \dfrac{d}{dx}0.5x + \dfrac{d}{dx}70$

$\quad\quad\quad\quad\quad\quad\quad\quad$ Theorem 4

$\quad\quad\quad\quad = -0.01 \cdot \dfrac{d}{dx}x^2 - 0.5 \cdot \dfrac{d}{dx}x + \dfrac{d}{dx}70$

$\quad\quad\quad\quad\quad\quad\quad\quad$ Theorem 3

$\quad\quad\quad\quad = -0.01 \cdot 2x - 0.5 \cdot 1 + 0$

$\quad\quad\quad\quad\quad\quad\quad\quad$ Theorems 1 and 2

$\quad\quad\quad\quad = -0.02x - 0.5$

$\quad\quad f'(x) = -0.02x - 0.5$

33. $f(x) = 3x^{-2/3} + x^{3/4} + x^{6/5} + \dfrac{8}{x^3}$

$\quad\quad\quad\quad = 3x^{-2/3} + x^{3/4} + x^{6/5} + 8x^{-3}$

$\quad\quad f'(x) = 3\left(-\dfrac{2}{3}\right)x^{-2/3-1} + \dfrac{3}{4} \cdot x^{3/4-1} +$

$\quad\quad\quad\quad\quad \dfrac{6}{5} \cdot x^{6/5-1} + 8(-3)x^{-3-1}$

$\quad\quad\quad\quad\quad\quad\quad\quad$ Theorems 1 and 3

$\quad\quad\quad\quad = -2x^{-5/3} + \dfrac{3}{4}x^{-1/4} + \dfrac{6}{5}x^{1/5} - 24x^{-4}$

35. $f(x) = \dfrac{2}{x} - \dfrac{x}{2} = 2x^{-1} - \dfrac{1}{2}x$

$\quad\quad f'(x) = 2(-1)x^{-1-1} - \dfrac{1}{2} \cdot 1$

$\quad\quad\quad\quad = -2x^{-2} - \dfrac{1}{2}$, or $-\dfrac{2}{x^2} - \dfrac{1}{2}$

37. $f(x) = \dfrac{16}{x} - \dfrac{8}{x^3} + \dfrac{1}{x^4}$

$\quad = 16x^{-1} - 8x^{-3} + x^{-4}$

$\quad f'(x) = 16(-1)x^{-1-1} - 8(-3)x^{-3-1} + (-4)x^{-4-1}$

$\quad = -16x^{-2} + 24x^{-4} - 4x^{-5}$

39. $f(x) = \sqrt{x} + \sqrt[3]{x} - \sqrt[4]{x} + \sqrt[5]{x}$

$\quad = x^{1/2} + x^{1/3} - x^{1/4} + x^{1/5}$

$\quad f'(x) = \dfrac{1}{2}x^{1/2-1} + \dfrac{1}{3}x^{1/3-1} - \dfrac{1}{4}x^{1/4-1} + \dfrac{1}{5}x^{1/5-1}$

$\quad = \dfrac{1}{2}x^{-1/2} + \dfrac{1}{3}x^{-2/3} - \dfrac{1}{4}x^{-3/4} + \dfrac{1}{5}x^{-4/5}$

41. $f(x) = x^3 - 2x + 1$

$\quad f'(x) = 3x^2 - 2 \cdot 1 + 0 = 3x^2 - 2$

At $(2, 5)$, $f'(x) = f'(2) = 3 \cdot 2^2 - 2 = 3 \cdot 4 - 2 = 12 - 2 = 10$, so we have

$$y - y_1 = m(x - x_1)$$
$$y - 5 = 10(x - 2)$$
$$y - 5 = 10x - 20$$
$$y = 10x - 15$$

At $(-1, 2)$, $f'(x) = f'(-1) = 3(-1)^2 - 2 = 3 \cdot 1 - 2 = 3 - 2 = 1$, so we have

$$y - y_1 = m(x - x_1)$$
$$y - 2 = 1[x - (-1)]$$
$$y - 2 = x + 1$$
$$y = x + 3$$

At $(0, 1)$, $f'(x) = f'(0) = 3 \cdot 0^2 - 2 = 0 - 2 = -2$, so we have

$$y - y_1 = m(x - x_1)$$
$$y - 1 = -2(x - 0)$$
$$y - 1 = -2x$$
$$y = -2x + 1$$

43. $y = x^2$ so $\dfrac{dy}{dx} = 2x$

A horizontal line has slope 0. We first find the values of x for which $dy/dx = 0$. That is, we want to find x such that $2x = 0$.

Solve: $2x = 0$

$\quad x = 0$ Multiplying by $\dfrac{1}{2}$

Next we find the point on the graph. We determine the second coordinate from the original equation $y = x^2$.

For $x = 0$, $y = 0^2 = 0$.

At the point $(0, 0)$, there is a horizontal tangent.

45. $y = -x^3$ so $\dfrac{dy}{dx} = -3x^2$

A horizontal tangent has slope 0. We first find the values of x for which $dy/dx = 0$. That is, we want to find x such that $-3x^2 = 0$.

Solve: $-3x^2 = 0$

$\quad x^2 = 0$ Multiplying by $-\dfrac{1}{3}$

$\quad x = 0$ Taking the square root

Next we find the point on the graph. We determine the second coordinate from the original equation $y = -x^3$.

For $x = 0$, $y = -0^3 = 0$.

At the point $(0, 0)$, there is a horizontal tangent.

47. $y = 3x^2 - 5x + 4$

$\quad \dfrac{dy}{dx} = 6x - 5$

A horizontal tangent has slope 0. We first find the values of x for which $dy/dx = 0$. That is, we want to find x such that $6x - 5 = 0$.

Solve: $6x - 5 = 0$

$\quad 6x = 5$

$\quad x = \dfrac{5}{6}$

Next we find the point on the graph. We determine the second coordinate from the original equation $y = 3x^2 - 5x + 4$.

For $x = \dfrac{5}{6}$, $y = 3\left(\dfrac{5}{6}\right)^2 - 5\left(\dfrac{5}{6}\right) + 4$

$\quad = 3\left(\dfrac{25}{36}\right) - \left(\dfrac{25}{6}\right) + 4$

$\quad = \dfrac{75}{36} - \dfrac{25}{6} \cdot \dfrac{6}{6} + 4 \cdot \dfrac{36}{36}$

$\quad = \dfrac{75}{36} - \dfrac{150}{36} + \dfrac{144}{36}$

$\quad = \dfrac{69}{36}$

$\quad = \dfrac{23}{12}$

At the point $\left(\dfrac{5}{6}, \dfrac{23}{12}\right)$, there is a horizontal tangent.

49. $y = -0.01x^2 - 0.5x + 70$

$\quad \dfrac{dy}{dx} = -0.02x - 0.5$

First find the values of x for which $dy/dx = 0$. That is, we want to find x such that $-0.02x - 0.5 = 0$.

Solve: $-0.02x - 0.5 = 0$

$\quad -0.02x = 0.5$

$\quad x = \dfrac{0.5}{-0.02}$

$\quad x = -25$

Next we find the point on the graph. We determine the second coordinate from the original equation $y = -0.01x^2 - 0.5x + 70$.

For $x = -25$, $y = -0.01(-25)^2 - 0.5(-25) + 70$

$\quad = -0.01(625) + 12.5 + 70$

$\quad = -6.25 + 12.5 + 70$

$\quad = 76.25$

At the point $(-25, 76.25)$, there is a horizontal tangent.

51. $y = 2x + 4$ Linear function

$\dfrac{dy}{dx} = 2$ Slope is 2

For all values of x, $dy/dx = 2$. There are no values of x for which $dy/dx = 0$. Thus, there are no points on the graph at which there is a horizontal tangent.

53. $y = 4$ Horizontal line

$\dfrac{dy}{dx} = 0$ Slope is 0

For all values of x, $dy/dx = 0$. The tangent line is horizontal at all points on the graph.

55. $y = -x^3 + x^2 + 5x - 1$

$\dfrac{dy}{dx} = -3x^2 + 2x + 5$

First find the values of x for which $dy/dx = 0$. That is, we want to find x such that $-3x^2 + 2x + 5 = 0$. We can simplify our work by multiplying by -1. We get $3x^2 - 2x - 5 = 0$.

We factor and solve:

$3x^2 - 2x - 5 = 0$

$(3x - 5)(x + 1) = 0$ Factoring

$3x - 5 = 0$ or $x + 1 = 0$ Principle of zero products

$3x = 5$ or $x = -1$

$x = \dfrac{5}{3}$ or $x = -1$

Next we find the points on the graph. We determine the second coordinate from the original equation $y = -x^3 + x^2 + 5x - 1$.

For $x = \dfrac{5}{3}$, $y = -\left(\dfrac{5}{3}\right)^3 + \left(\dfrac{5}{3}\right)^2 + 5 \cdot \dfrac{5}{3} - 1$

$= -\dfrac{125}{27} + \dfrac{25}{9} + \dfrac{25}{3} - 1$

$= -\dfrac{125}{27} + \dfrac{75}{27} + \dfrac{225}{27} - \dfrac{27}{27}$

$= \dfrac{148}{27}$

For $x = -1$, $y = -(-1)^3 + (-1)^2 + 5(-1) - 1$

$= 1 + 1 - 5 - 1$

$= -4$

At the points $\left(\dfrac{5}{3}, \dfrac{148}{27}\right)$ and $(-1, -4)$, there are horizontal tangents.

57. $y = \dfrac{1}{3}x^3 - 3x + 2$

$\dfrac{dy}{dx} = x^2 - 3$

First find the values of x for which $dy/dx = 0$. That is, we want to find x such that $x^2 - 3 = 0$.

Solve: $x^2 - 3 = 0$

$x^2 = 3$

$x = \pm\sqrt{3}$

Next we find the point on the graph. We determine the second coordinate from the original equation

$y = \dfrac{1}{3}x^3 - 3x + 2.$

For $x = \sqrt{3}$, $y = \dfrac{1}{3}(\sqrt{3})^3 - 3\sqrt{3} + 2$

$= \dfrac{1}{3}(3\sqrt{3}) - 3\sqrt{3} + 2$

$= \sqrt{3} - 3\sqrt{3} + 2$

$= 2 - 2\sqrt{3}$

For $x = -\sqrt{3}$, $y = \dfrac{1}{3}(-\sqrt{3})^3 - 3(-\sqrt{3}) + 2$

$= \dfrac{1}{3}(-3\sqrt{3}) - 3(-\sqrt{3}) + 2$

$= -\sqrt{3} + 3\sqrt{3} + 2$

$= 2 + 2\sqrt{3}$

At the points $(\sqrt{3}, 2 - 2\sqrt{3})$ and $(-\sqrt{3}, 2 + 2\sqrt{3})$, there are horizontal tangents.

59. $y = 20x - x^2$

$\dfrac{dy}{dx} = 20 - 2x$

First find the values of x for which $dy/dx = 1$. That is, we want to find x such that $20 - 2x = 1$.

Solve: $20 - 2x = 1$

$-2x = -19$

$x = \dfrac{19}{2}$

Next we find the point on the graph. We determine the second coordinate from the original equation $y = 20x - x^2$.

For $x = \dfrac{19}{2}$, $y = 20\left(\dfrac{19}{2}\right) - \left(\dfrac{19}{2}\right)^2$

$= 190 - \dfrac{361}{4}$

$= \dfrac{399}{4}$

At the point $\left(\dfrac{19}{2}, \dfrac{399}{4}\right)$, there is a tangent line which has slope 1.

61. $y = -0.025x^2 + 4x$

$\dfrac{dy}{dx} = -0.05x + 4$

First find the values of x for which $dy/dx = 1$. That is, we want to find x such that $-0.05x + 4 = 1$.

Solve: $-0.05x + 4 = 1$

$-0.05x = -3$

$x = 60$

Next we find the point on the graph. We determine the second coordinate from the original equation

$y = -0.025x^2 + 4x$.

For $x = 60$, $y = -0.025 \cdot 60^2 + 4 \cdot 60$

$= -0.025 \cdot 3600 + 240$

$= -90 + 240$

$= 150$

At the point $(60, 150)$, there is a tangent line which has slope 1.

63. $y = \dfrac{1}{3}x^3 + 2x^2 + 2x$

$$\frac{dy}{dx} = x^2 + 4x + 2$$

First find the values of x for which $dy/dx = 1$. That is, we want to find x such that $x^2 + 4x + 2 = 1$.

Solve: $x^2 + 4x + 2 = 1$

$\qquad\quad x^2 + 4x + 1 = 0$ $\quad$ Adding -1

This is a quadratic equation, not readily factorable, so we use the quadratic formula where $a = 1$, $b = 4$, and $c = 1$.

$$x = \frac{-b \pm \sqrt{b^2 - 4ac}}{2a}$$

$$x = \frac{-4 \pm \sqrt{4^2 - 4 \cdot 1 \cdot 1}}{2 \cdot 1} \quad \begin{array}{l}\text{Substituting 1 for } a, \\ \text{4 for } b\text{, 1 for } c\end{array}$$

$$= \frac{-4 \pm \sqrt{16 - 4}}{2}$$

$$= \frac{-4 \pm \sqrt{12}}{2}$$

$$= \frac{-4 \pm 2\sqrt{3}}{2} \qquad\qquad \sqrt{12} = \sqrt{4 \cdot 3} = 2\sqrt{3}$$

$$= \frac{2(-2 \pm \sqrt{3})}{2}$$

$$= -2 \pm \sqrt{3}$$

Next we find the points on the graph. We determine the second coordinate from the original equation $y = \dfrac{1}{3}x^3 + 2x^2 + 2x$.

For $x = -2 + \sqrt{3}$,

$$y = \frac{1}{3}(-2 + \sqrt{3})^3 + 2(-2 + \sqrt{3})^2 + 2(-2 + \sqrt{3})$$

$$= \frac{1}{3}(-26 + 15\sqrt{3}) + 2(7 - 4\sqrt{3}) - 4 + 2\sqrt{3}$$

$$= -\frac{26}{3} + 5\sqrt{3} + 14 - 8\sqrt{3} - 4 + 2\sqrt{3}$$

$$= \frac{4}{3} - \sqrt{3}$$

For $x = -2 - \sqrt{3}$,

$$y = \frac{1}{3}(-2 - \sqrt{3})^3 + 2(-2 - \sqrt{3})^2 + 2(-2 - \sqrt{3})$$

$$= \frac{1}{3}(-26 - 15\sqrt{3}) + 2(7 + 4\sqrt{3}) - 4 - 2\sqrt{3}$$

$$= -\frac{26}{3} - 5\sqrt{3} + 14 + 8\sqrt{3} - 4 - 2\sqrt{3}$$

$$= \frac{4}{3} + \sqrt{3}$$

At the points $\left(-2 + \sqrt{3},\ \dfrac{4}{3} - \sqrt{3}\right)$ and $\left(-2 - \sqrt{3},\ \dfrac{4}{3} + \sqrt{3}\right)$, there are tangent lines which have slope 1.

65. $y = x^4 - \dfrac{4}{3}x^2 - 4$

$$\frac{dy}{dx} = 4x^3 - \frac{8}{3}x$$

First find the values of x for which $dy/dx = 0$. That is, we want to find x such that $4x^3 - \dfrac{8}{3}x = 0$.

Solve: $\qquad 4x^3 - \dfrac{8}{3}x = 0$

$$12x^3 - 8x = 0$$

$$4x(3x^2 - 2) = 0$$

$4x = 0 \ $ or $ \ 3x^2 - 2 = 0$

$x = 0 \ $ or $\qquad x^2 = \dfrac{2}{3}$

$x = 0 \ $ or $\qquad x = \pm\sqrt{\dfrac{2}{3}}$

Next we find the points on the graph. We determine the second coordinate from the original equation

$$y = x^4 - \frac{4}{3}x^2 - 4.$$

For $x = 0$, $y = 0^4 - \dfrac{4}{3} \cdot 0^2 - 4 = -4$

For $x = \pm\sqrt{\dfrac{2}{3}}$, $y = \left(\pm\sqrt{\dfrac{2}{3}}\right)^4 - \dfrac{4}{3}\left(\pm\sqrt{\dfrac{2}{3}}\right)^2 - 4$

$$= \frac{4}{9} - \frac{4}{3} \cdot \frac{2}{3} - 4$$

$$= \frac{4}{9} - \frac{8}{9} - \frac{36}{9}$$

$$= -\frac{40}{9}$$

At the points $(0, -4)$, $\left(\sqrt{\dfrac{2}{3}}, -\dfrac{40}{9}\right)$, and $\left(-\sqrt{\dfrac{2}{3}}, -\dfrac{40}{9}\right)$, there are horizontal tangents.

67. $y = x(x - 1) = x^2 - x$

$$\frac{dy}{dx} = 2x - 1$$

69. $y = (x - 2)(x + 3) = x^2 + x - 6$

$$\frac{dy}{dx} = 2x + 1$$

71. $y = \dfrac{x^5 + x}{x^2} = \dfrac{x^5}{x^2} + \dfrac{x}{x^2} = x^3 + x^{-1}$

$$\frac{dy}{dx} = 3x^2 - x^{-2}, \text{ or } 3x^2 - \frac{1}{x^2}$$

73. $y = (-4x)^3 = -64x^3$

$$\frac{dy}{dx} = -192x^2$$

75. $y = \sqrt[3]{8x} = \sqrt[3]{8}\sqrt[3]{x} = 2\sqrt[3]{x} = 2x^{1/3}$

$$\frac{dy}{dx} = \frac{2}{3}x^{-2/3}, \text{ or } \frac{2}{3x^{2/3}}, \text{ or } \frac{2}{3\sqrt[3]{x^2}}$$

77. $y = (x + 1)^3$

$$= x^3 + 3 \cdot x^2 \cdot 1 + 3 \cdot x \cdot 1^2 + 1^3$$

$$= x^3 + 3x^2 + 3x + 1$$

$$\frac{dy}{dx} = 3x^2 + 6x + 3$$

79. See answer section in the text.

81. $\boxed{tw}$

83. See the answer section in the text.

85. See the answer section in the text.

87. See the answer section in the text.

89. See the answer section in the text.

91. See the answer section in the text.

Exercise Set 2.6

1. $s(t) = t^3 + t$

a) $v(t) = s'(t) = 3t^2 + 1$

b) $a(t) = v'(t) = 6t$

c) $v(4) = 3 \cdot 4^2 + 1$

$= 3 \cdot 16 + 1$

$= 48 + 1$

$= 49$

The velocity when $t = 4$ sec is 49 ft/sec.

$a(4) = 6 \cdot 4 = 24$

The acceleration when $t = 4$ sec is 24 ft/sec^2.

3. $R(x) = 5x$

$C(x) = 0.001x^2 + 1.2x + 60$

a) $P(x) = R(x) - C(x)$

$= 5x - (0.001x^2 + 1.2x + 60)$

$= 5x - 0.001x^2 - 1.2x - 60$

$= -0.001x^2 + 3.8x - 60$

b) $R(x) = 5x$

$R(100) = 5 \cdot 100 = 500$

The total revenue from the sale of the first 100 units is \$500.

$C(x) = 0.001x^2 + 1.2x + 60$

$C(100) = 0.001(100)^2 + 1.2(100) + 60$

Substituting 100

$= 0.001(10,000) + 1.2(100) + 60$

$= 10 + 120 + 60$

$= 190$

The total cost of producing the first 100 units is \$190.

$P(x) = R(100) - C(100)$

$P(100) = 500 - 190$

$= 310$

The total profit from the production and sale of the first 100 units is \$310.

c) $R(x) = 5x$

so

$R'(x) = 5$

$C(x) = 0.001x^2 + 1.2x + 60$

so

$C'(x) = 0.002x + 1.2$

$P(x) = -0.001x^2 + 3.8x - 60$

so

$P'(x) = -0.002x + 3.8$

d) $R'(x) = 5$

$R'(100) = \$5$ per unit

$C'(x) = 0.002x + 1.2$

$C'(100) = 0.002(100) + 1.2$

$= 0.2 + 1.2$

$= \$1.4$, or \$1.40 per unit

$P'(100) = R'(100) - C'(100)$

$= \$5.00 - \$1.40$

$= \$3.60$ per unit

e) $\boxed{tw}$

5. $N(a) = -a^2 + 300a + 6$

a) The rate of change is given by $N'(a)$.

$N'(a) = -2a + 300$

b) Since a is measured in thousands of dollars, we find $N(10)$.

$N(10) = -(10)^2 + 300(10) + 6$

$= -100 + 3000 + 6$

$= 2906$

After spending \$10,000 on advertising, 2906 units will be sold.

c) $N'(10) = -2 \cdot 10 + 300$

$= -20 + 300$

$= 280$

The rate of change is 280 units/thousand dollars.

d) $\boxed{tw}$

7. $M(t) = -2t^2 + 100t + 180$

a) $M(5) = -2 \cdot 5^2 + 100 \cdot 5 + 180$

$= -2 \cdot 25 + 500 + 180$

$= -50 + 500 + 180$

$= 630$

The productivity is 630 units per month after 5 years of service.

$$M(10) = -2 \cdot 10^2 + 100 \cdot 10 + 180$$
$$= -2 \cdot 100 + 1000 + 180$$
$$= -200 + 1000 + 180$$
$$= 980$$

The productivity is 980 units per month after 10 years of service.

$$M(25) = -2 \cdot 25^2 + 100 \cdot 25 + 180$$
$$= -2 \cdot 625 + 2500 + 180$$
$$= -1250 + 2500 + 180$$
$$= 1430$$

The productivity is 1430 units per month after 25 years of service.

$$M(45) = -2 \cdot 45^2 + 100 \cdot 45 + 180$$
$$= -2 \cdot 2025 + 4500 + 180$$
$$= -4050 + 4500 + 180$$
$$= 630$$

The productivity is 630 units per month after 45 years of service.

b) The marginal productivity is given by $M'(t)$.
$$M'(t) = -4t + 100$$

c) $M'(5) = -4 \cdot 5 + 100 = -20 + 100 = 80$

At $t = 5$ the monthly marginal productivity is 80 units per year of service.

$M'(10) = -4 \cdot 10 + 100 = -40 + 100 = 60$

At $t = 10$ the monthly marginal productivity is 60 units per year of service.

$M'(25) = -4 \cdot 25 + 100 = -100 + 100 = 0$

At $t = 25$ the monthly marginal productivity is 0 units per year of service.

$M'(45) = -4 \cdot 45 + 100 = -180 + 100 = -80$

At $t = 45$ the monthly marginal productivity is -80 units per year of service.

d) $\boxed{tw}$

9. $D(p) = 100 - \sqrt{p} = 100 - p^{1/2}$

a) $\dfrac{dD}{dp} = -\dfrac{1}{2}p^{-1/2}$, or $-\dfrac{1}{2p^{1/2}}$, or $-\dfrac{1}{2\sqrt{p}}$

b) $D(25) = 100 - \sqrt{25} = 100 - 5 = 95$

The consumer will want to buy 95 units when the price is \$25 per unit.

c) At $p = 25$, $\dfrac{dD}{dp} = -\dfrac{1}{2\sqrt{25}} = -\dfrac{1}{2 \cdot 5} = -\dfrac{1}{10}$
dollars per unit

d) $\boxed{tw}$

11. a) $C(r) = 2\pi r$ Linear function, 2π is a constant

The rate of change of the circumference with respect to the radius is given by $C'(r)$.

$$C'(r) = 2\pi$$

b) $\boxed{tw}$

13. $T(t) = -0.1t^2 + 1.2t + 98.6$
$\qquad\qquad T$ is temperature (°F), and
$\qquad\qquad t$ is time (days)

a) $T'(t) = -0.1 \cdot 2t + 1.2 + 0$
$$= -0.2t + 1.2$$

b) $\qquad T(t) = -0.1t^2 + 1.2t + 98.6$
$$T(1.5) = -0.1(1.5)^2 + 1.2(1.5) + 98.6$$
$$\qquad\qquad\qquad\qquad \text{Substituting 1.5 for } t$$
$$= -0.1(2.25) + 1.2(1.5) + 98.6$$
$$= -0.225 + 1.8 + 98.6$$
$$= 100.175$$

At 1.5 days, the temperature is 100.175°F.

c) $\qquad T'(t) = -0.2t + 1.2$
$$T'(1.5) = -0.2(1.5) + 1.2$$
$$= -0.3 + 1.2$$
$$= 0.9$$

The rate of change of the temperature with respect to time at 1.5 days is 0.9°F per day.

d) $\boxed{tw}$

15. a) $\qquad T = W^{1.31}$
$$\dfrac{dT}{dW} = 1.31W^{1.31-1}$$
$$= 1.31W^{0.31}$$

b) $\boxed{tw}$

17. a) $R(Q) = Q^2\left(\dfrac{k}{2} - \dfrac{Q}{3}\right) = \dfrac{k}{2}Q^2 - \dfrac{Q^3}{3} = \dfrac{k}{2}Q^2 - \dfrac{1}{3}Q^3$

$\dfrac{dR}{dQ} = \dfrac{k}{2} \cdot 2Q - \dfrac{1}{3} \cdot 3Q^2 = kQ - Q^2$

b) $\boxed{tw}$

19. a) $A(t) = 0.08t + 19.7$

The rate of change of the median age A with respect to time t is given by $A'(t)$.

$$A'(t) = 0.08$$

b) $\boxed{tw}$

21. $\boxed{tw}$

23. See the answer section in the text.

25. See the answer section in the text.

Exercise Set 2.7

1. $y = x^3 \cdot x^8$, or x^{11}

Differentiate $y = x^3 \cdot x^8$ using the Product Rule (Theorem 5).

$$\dfrac{d}{dx}x^3 \cdot x^8 = x^3 \cdot 8x^7 + 3x^2 \cdot x^8$$
$$= 8x^{10} + 3x^{10}$$
$$= 11x^{10}$$

Differentiate $y = x^{11}$ using the Power Rule (Theorem 1).

$$\frac{d}{dx}x^{11} = 11x^{10}$$

3. $y = \dfrac{-1}{x}$ or $-1 \cdot x^{-1}$

Differentiate $y = \dfrac{1}{x}$ using the Quotient Rule (Theorem 6).

$$\frac{d}{dx}\frac{-1}{x} = \frac{x \cdot 0 - 1(-1)}{x^2}$$

$$= \frac{0+1}{x^2}$$

$$= \frac{1}{x^2}, \text{ or } x^{-2}$$

Differentiate $y = -1 \cdot x^{-1}$ using the Power Rule (Theorem 1) and Theorem 3.

$$\frac{d}{dx}-1 \cdot x^{-1} = -1 \cdot \frac{d}{dx}x^{-1} \qquad \text{Theorem 3}$$

$$= -1 \cdot (-1)x^{-1-1} \quad \text{Theorem 1}$$

$$= 1 \cdot x^{-2}$$

$$= x^{-2}, \text{ or } \frac{1}{x^2}$$

5. $y = \dfrac{x^8}{x^5}$ or x^3

Differentiate $y = \dfrac{x^8}{x^5}$ using the Quotient Rule (Theorem 6).

$$\frac{d}{dx}\frac{x^8}{x^5} = \frac{x^5 \cdot 8x^7 - 5x^4 \cdot x^8}{(x^5)^2}$$

$$= \frac{8x^{12} - 5x^{12}}{x^{10}}$$

$$= \frac{3x^{12}}{x^{10}}$$

$$= 3x^2$$

Differentiate $y = x^3$ using the Power Rule (Theorem 1).

$$\frac{d}{dx}x^3 = 3x^2$$

7. $y = (8x^5 - 3x^2 + 20)(8x^4 - 3\sqrt{x})$

$$= (8x^5 - 3x^2 + 20)(8x^4 - 3x^{1/2})$$

Using the Product Rule, we follow these steps:

(1) Write down the first factor.

(2) Multiply it by the derivative of the second factor.

(3) Write down the derivative of the first factor.

(4) Multiply it by the second factor.

(5) Add the result of (1) and (2) to the result of (3) and (4).

$$\frac{dy}{dx} = \overset{(1)}{(8x^5 - 3x^2 + 20)}\overset{(2)}{\left(32x^3 - \frac{3}{2}x^{-1/2}\right)} \overset{(5)}{+}$$

$$\underset{(3)}{(40x^4 - 6x)}\underset{(4)}{(8x^4 - 3x^{1/2})}$$

The derivative could also be expressed using radical symbols.

$$\frac{dy}{dx} = (8x^5 - 3x^2 + 20)\left(32x^3 - \frac{3}{2\sqrt{x}}\right) +$$

$$(40x^4 - 6x)(8x^4 - 3\sqrt{x})$$

9. $f(x) = x(300 - x)$

Using the Product Rule, we have

$$f'(x) = \overset{(1)}{x} \cdot \overset{(2)}{(-1)} \overset{(5)}{+} \overset{(3)}{1} \cdot \overset{(4)}{(300 - x)}$$

Follow steps listed below.

$$= -x + 300 - x$$

$$= 300 - 2x$$

Write down:

(1) First factor

(2) Derivative of second factor

(3) Derivative of first factor

(4) Second factor

(5) Plus sign

We could have multiplied the factors and then differentiated, avoiding the Product Rule.

$$f(x) = x(300 - x)$$

$$= 300x - x^2$$

$$f'(x) = 300 - 2x$$

11. $f(x) = (4\sqrt{x} - 6)(x^3 - 2x + 4)$, or

$$(4x^{1/2} - 6)(x^3 - 2x + 4)$$

$$f'(x) = (4x^{1/2} - 6)(3x^2 - 2) +$$

$$(2x^{-1/2})(x^3 - 2x + 4), \text{ or}$$

$$(4\sqrt{x} - 6)(3x^2 - 2) + \frac{2}{\sqrt{x}}(x^3 - 2x + 4)$$

13. $f(x) = (x + 3)^2 = (x + 3)(x + 3)$

$$f'(x) = (x + 3) \cdot 1 + 1 \cdot (x + 3)$$

$$= x + 3 + x + 3$$

$$= 2x + 6$$

15. $f(x) = (x^3 - 4x)^2 = (x^3 - 4x)(x^3 - 4x)$

$$f'(x) = (x^3 - 4x)(3x^2 - 4) + (3x^2 - 4)(x^3 - 4x)$$

$$= 3x^5 - 16x^3 + 16x + 3x^5 - 16x^3 + 16x$$

$$= 6x^5 - 32x^3 + 32x$$

17. $f(x) = 5x^{-3}(x^4 - 5x^3 + 10x - 2)$

$$f'(x) = 5x^{-3}(4x^3 - 15x^2 + 10) -$$

$$15x^{-4}(x^4 - 5x^3 + 10x - 2)$$

Simplifying, we get

$$f'(x) = 20 - 75x^{-1} + 50x^{-3} - 15 + 75x^{-1} -$$

$$150x^{-3} + 30x^{-4}$$

$$= 5 - 100x^{-3} + 30x^{-4}.$$

19. $f(x) = \left(x + \dfrac{2}{x}\right)(x^2 - 3)$, or $(x + 2x^{-1})(x^2 - 3)$

$f'(x) = (x + 2x^{-1})(2x) + (1 - 2x^{-2})(x^2 - 3)$, or

$$\left(x + \frac{2}{x}\right)(2x) + \left(1 - \frac{2}{x^2}\right)(x^2 - 3)$$

Simplifying, we get

$$f'(x) = 2x^2 + 4 + x^2 - 5 + \frac{6}{x^2}$$

$$= 3x^2 + \frac{6}{x^2} - 1.$$

21. $f(x) = \dfrac{x}{300 - x}$

Using the Quotient Rule, we have

$$f'(x) = \overset{(1)\quad (2)(3)\ (4)\ (5)}{\dfrac{(300 - x) \cdot 1 - (-1) \cdot x}{(300 - x)^2}}$$
$$\underset{(6)}{}$$

Follow steps listed below.

$$= \frac{300 - x + x}{(300 - x)^2}$$

$$= \frac{300}{(300 - x)^2}$$

Steps to follow:

(1) Write down the denominator.

(2) Multiply the denominator by the derivative of the numerator.

(3) Write a minus sign.

(4) Write down the derivative of the denominator.

(5) Multiply it by the numerator.

(6) Divide by the square of the denominator.

23. $f(x) = \dfrac{3x - 1}{2x + 5}$

Using the Quotient Rule, we have

$$f'(x) = \overset{(1)\quad (2)(3)(4)\quad (5)}{\dfrac{(2x + 5) \cdot 3 - 2 \cdot (3x - 1)}{(2x + 5)^2}}$$
$$\underset{(6)}{}$$

Follow steps listed below.

$$= \frac{6x + 15 - 6x + 2}{(2x + 5)^2}$$

$$= \frac{17}{(2x + 5)^2}$$

Steps to follow:

(1) Write down the denominator.

(2) Multiply the denominator by the derivative of the numerator.

(3) Write a minus sign.

(4) Write down the derivative of the denominator.

(5) Multiply it by the numerator.

(6) Divide by the square of the denominator.

25. $y = \dfrac{x^2 + 1}{x^3 - 1}$

Using the Quotient Rule, we have

$$\frac{dy}{dx} = \overset{(1)\quad (2)(3)(4)\qquad (5)}{\dfrac{(x^3 - 1) \cdot 2x - 3x^2 \cdot (x^2 + 1)}{(x^3 - 1)^2}}$$
$$\underset{(6)}{}$$

Follow numbered steps listed in Exercise 23.

$$= \frac{2x^4 - 2x - 3x^4 - 3x^2}{(x^3 - 1)^2}$$

$$= \frac{-x^4 - 3x^2 - 2x}{(x^3 - 1)^2}$$

27. $y = \dfrac{x}{1 - x}$

Using the Quotient Rule, we have

$$\frac{dy}{dx} = \frac{(1 - x) \cdot 1 - (-1) \cdot x}{(1 - x)^2}$$

$$= \frac{1 - x + x}{(1 - x)^2}$$

$$= \frac{1}{(1 - x)^2}$$

29. $y = \dfrac{x - 1}{x + 1}$

Using the Quotient Rule, we have

$$\frac{dy}{dx} = \frac{(x + 1) \cdot 1 - 1 \cdot (x - 1)}{(x + 1)^2}$$

$$= \frac{x + 1 - x + 1}{(x + 1)^2}$$

$$= \frac{2}{(x + 1)^2}$$

31. $f(x) = \dfrac{1}{x - 3}$

Using the Quotient Rule, we have

$$f'(x) = \frac{(x - 3) \cdot 0 - 1 \cdot 1}{(x - 3)^2}$$

$$= \frac{-1}{(x - 3)^2}$$

33. $f(x) = \dfrac{3x^2 + 2x}{x^2 + 1}$

$$f'(x) = \frac{(x^2 + 1) \cdot (6x + 2) - 2x \cdot (3x^2 + 2x)}{(x^2 + 1)^2}$$

Using the Quotient Rule

$$= \frac{6x^3 + 2x^2 + 6x + 2 - 6x^3 - 4x^2}{(x^2 + 1)^2}$$

$$= \frac{-2x^2 + 6x + 2}{(x^2 + 1)^2}$$

35. $f(x) = \dfrac{3x^2 - 5x}{x^8}$

Using the Quotient Rule:

$f'(x) = \dfrac{x^8(6x - 5) - 8x^7(3x^2 - 5x)}{(x^8)^2}$

$= \dfrac{6x^9 - 5x^8 - 24x^9 + 40x^8}{x^{16}} \quad [(x^8)^2 = x^{16}]$

$= \dfrac{-18x^9 + 35x^8}{x^{16}}$

$= \dfrac{x^8(-18x + 35)}{x^8 \cdot x^8}$ Factoring both numerator and denominator

$= \dfrac{-18x + 35}{x^8}$ Simplifying

Dividing first:

$f(x) = \dfrac{3x^2 - 5x}{x^8} = 3x^{-6} - 5x^{-7}$

$f'(x) = -18x^{-7} + 35x^{-8}$

This is equivalent to the result we found using the Quotient Rule.

37. $g(x) = \dfrac{4x + 3}{\sqrt{x}}$, or $\dfrac{4x + 3}{x^{1/2}}$

$g'(x) = \dfrac{\sqrt{x}(4) - \frac{1}{2}x^{-1/2}(4x + 3)}{(\sqrt{x})^2}$

$= \dfrac{4\sqrt{x} - \frac{1}{2}x^{-1/2}(4x + 3)}{x}$

Simplifying, we get

$g'(x) = \dfrac{4\sqrt{x} - \frac{4x + 3}{2\sqrt{x}}}{x}$

$= \dfrac{4\sqrt{x} \cdot \frac{2\sqrt{x}}{2\sqrt{x}} - \frac{4x + 3}{2\sqrt{x}}}{x}$ Subtracting in the numerator

$= \dfrac{\frac{8x - (4x + 3)}{2\sqrt{x}}}{x} = \dfrac{\frac{8x - 4x - 3}{2\sqrt{x}}}{x}$

$= \dfrac{\frac{4x - 3}{2\sqrt{x}}}{x} = \dfrac{4x - 3}{2\sqrt{x}} \cdot \dfrac{1}{x}$

$= \dfrac{4x - 3}{2x\sqrt{x}}$, or $\dfrac{4x - 3}{2x^{3/2}}$.

39. - 75.

77. $y = \dfrac{8}{x^2 + 4}$

$\dfrac{dy}{dx} = \dfrac{(x^2 + 4)(0) - 2x(8)}{(x^2 + 4)^2}$

$= \dfrac{-16x}{(x^2 + 4)^2}$

When $x = 0$, $\dfrac{dy}{dx} = \dfrac{-16 \cdot 0}{(0^2 + 4)^2} = 0$, so the slope of the tangent line at $(0, 2)$ is 0. We use the point-slope equation.

$y - y_1 = m(x - x_1)$
$y - 2 = 0(x - 0)$
$y - 2 = 0$
$y = 2$

When $x = -2$, $\dfrac{dy}{dx} = \dfrac{-16(-2)}{[(-2)^2 + 4]^2} = \dfrac{32}{64} = \dfrac{1}{2}$, so the slope of the tangent line at $(-2, 1)$ is $\dfrac{1}{2}$. We use the point-slope equation.

$y - y_1 = m(x - x_1)$
$y - 1 = \dfrac{1}{2}[x - (-2)]$
$y - 1 = \dfrac{1}{2}(x + 2)$
$y - 1 = \dfrac{1}{2}x + 1$
$y = \dfrac{1}{2}x + 2$

79. $D(p) = \dfrac{2p + 300}{10p + 11}$

(a) $D'(p) = \dfrac{(10p + 11)(2) - 10(2p + 300)}{(10p + 11)^2}$

$= \dfrac{20p + 22 - 20p - 3000}{(10p + 11)^2}$

$= \dfrac{-2978}{(10p + 11)^2}$

(b) $D'(4) = \dfrac{-2978}{(10 \cdot 4 + 11)^2} = \dfrac{-2978}{51^2} = -\dfrac{2978}{2601}$

81. $D(p) = 400 - p$

(a) $R(p) = p \cdot D(p) = p(400 - p) = 400p - p^2$

(b) $R'(p) = 400 - 2p$

83. $D(p) = \dfrac{4000}{p} + 3$

(a) $R(p) = p \cdot D(p) = p\left(\dfrac{4000}{p} + 3\right) = 4000 + 3p$

(b) $R'(p) = 3$

85. $A(x) = \dfrac{C(x)}{x}$ Average cost

$A'(x) = \dfrac{x \cdot C'(x) - 1 \cdot C(x)}{x^2}$ Marginal average cost

$= \dfrac{xC'(x) - C(x)}{x^2}$

87. $T(t) = \dfrac{4t}{t^2 + 1} + 98.6$

(a) $T'(t) = \dfrac{(t^2 + 1)(4) - 2t \cdot 4t}{(t^2 + 1)^2}$

$= \dfrac{4t^2 + 4 - 8t^2}{(t^2 + 1)^2}$

$= \dfrac{-4t^2 + 4}{(t^2 + 1)^2}$

(b) $T(2) = \dfrac{4 \cdot 2}{2^2 + 1} + 98.6$

$ = \dfrac{8}{5} + 98.6 = 1.6 + 98.6$

$ = 100.2° \text{ F}$

(c) $T'(2) = \dfrac{-4 \cdot 2^2 + 4}{(2^2 + 1)^2}$

$ = \dfrac{-16 + 4}{5^2}$

$ = -\dfrac{12}{25}, \text{ or } -0.48° \text{ F per hr}$

89. $f(x) = \dfrac{x^3}{\sqrt{x} - 5}, \text{ or } \dfrac{x^3}{x^{1/2} - 5}$

$f'(x) = \dfrac{(\sqrt{x} - 5) \cdot 3x^2 - \frac{1}{2}x^{-1/2} \cdot x^3}{(\sqrt{x} - 5)^2}$

 Using the Quotient Rule

$ = \dfrac{(\sqrt{x} - 5) \cdot 3x^2 - \dfrac{x^3}{2\sqrt{x}}}{(\sqrt{x} - 5)^2} \cdot \dfrac{2\sqrt{x}}{2\sqrt{x}}$

$ = \dfrac{2\sqrt{x}(\sqrt{x} - 5) \cdot 3x^2 - x^3}{2\sqrt{x}(\sqrt{x} - 5)^2}$

$ = \dfrac{(2x - 10\sqrt{x})3x^2 - x^3}{2\sqrt{x}(\sqrt{x} - 5)^2}$

$ = \dfrac{6x^3 - 30x^2\sqrt{x} - x^3}{2\sqrt{x}(\sqrt{x} - 5)^2}$

$ = \dfrac{5x^3 - 30x^2\sqrt{x}}{2\sqrt{x}(\sqrt{x} - 5)^2}$

91. $f(v) = \dfrac{3}{1 + v + v^2}$

$f'(v) = \dfrac{(1 + v + v^2) \cdot 0 - (1 + 2v) \cdot 3}{(1 + v + v^2)^2}$

 Using the Quotient Rule

$ = \dfrac{-3(1 + 2v)}{(1 + v + v^2)^2}$

93. $p(t) = \dfrac{t}{1 - t + t^2 - t^3}$

$p'(t) = \dfrac{(1 - t + t^2 - t^3) \cdot 1 - (-1 + 2t - 3t^2) \cdot t}{(1 - t + t^2 - t^3)^2}$

 Using the Quotient Rule

$ = \dfrac{1 - t + t^2 - t^3 + t - 2t^2 + 3t^3}{(1 - t + t^2 - t^3)^2}$

$ = \dfrac{1 - t^2 + 2t^3}{(1 - t + t^2 - t^3)^2}$

95. $h(x) = \dfrac{x^3 + 5x^2 - 2}{\sqrt{x}}, \text{ or } \dfrac{x^3 + 5x^2 - 2}{x^{1/2}}$

$h'(x) = \dfrac{\sqrt{x}(3x^2 + 10x) - \frac{1}{2}x^{-1/2}(x^3 + 5x^2 - 2)}{(\sqrt{x})^2}$

 Using the Quotient Rule

$ = \dfrac{\sqrt{x}(3x^2 + 10x) - \dfrac{x^3 + 5x^2 - 2}{2\sqrt{x}}}{x} \cdot \dfrac{2\sqrt{x}}{2\sqrt{x}}$

$ = \dfrac{2x(3x^2 + 10x) - (x^3 + 5x^2 - 2)}{2x\sqrt{x}}$

$ = \dfrac{6x^3 + 20x^2 - x^3 - 5x^2 + 2}{2x\sqrt{x}}$

$ = \dfrac{5x^3 + 15x^2 + 2}{2x\sqrt{x}}$

97. $f(x) = x(3x^3 + 6x - 2)(3x^4 + 7)$

$ = [x(3x^3 + 6x - 2)](3x^4 + 7)$ Grouping the fac-

 tors

$f'(x) = [x(3x^3 + 6x - 2)] \cdot \dfrac{d}{dx}(3x^4 + 7) +$

$ \left\{ \dfrac{d}{dx}[x(3x^3 + 6x - 2)] \right\} \cdot (3x^4 + 7)$

 Using the Product Rule on the

 entire expression

$ = [x(3x^3 + 6x - 2)](12x^3) +$

$ [x(9x^2 + 6) + 1 \cdot (3x^3 + 6x - 2)](3x^4 + 7)$

 Finding $\dfrac{d}{dx}(3x^4 + 7)$ and using

 the Product Rule on the

 expression in the brackets

$ = 12x^4(3x^3 + 6x - 2) + [x(9x^2 + 6) +$

$ (3x^3 + 6x - 2)](3x^4 + 7)$

99. $f(t) = (t^5 + 3) \cdot \dfrac{t^3 - 1}{t^3 + 1}$

$f'(t) = (t^5 + 3) \cdot \dfrac{(t^3 + 1) \cdot 3t^2 - 3t^2(t^3 - 1)}{(t^3 + 1)^2} +$

$ 5t^4 \cdot \dfrac{t^3 - 1}{t^3 + 1}$

$ = (t^5 + 3) \cdot \dfrac{3t^5 + 3t^2 - 3t^5 + 3t^2}{(t^3 + 1)^2} +$

$ 5t^4 \cdot \dfrac{t^3 - 1}{t^3 + 1}$

$ = \dfrac{6t^2(t^5 + 3)}{(t^3 + 1)^2} + \dfrac{5t^4(t^3 - 1)}{t^3 + 1}$

101. $f'(x) = \dfrac{(2x^2 + 3)(4x^3 - 7x + 2)}{x^7 - 2x^6 + 9}$

We will use the Quotient Rule. We must also use the Product Rule to find the derivative of the numerator.

$f'(x) = \{(x^7 - 2x^6 + 9)[(2x^2 + 3)(12x^2 - 7) +$

$4x(4x^3 - 7x + 2)] -$

$(7x^6 - 12x^5)(2x^2 + 3)(4x^3 - 7x + 2)\} / (x^7 - 2x^6 + 9)^2$

103. $\boxed{tw}$

105. See the answer section in the text.

107. See the answer section in the text.

109. See the answer section in the text.

Exercise Set 2.8

1. $y = (1 - x)^{55}$

$\dfrac{dy}{dx} = 55(1 - x)^{54} \cdot (-1)$ Using the Extended
Power Rule

Don't forget $\dfrac{d}{dx}(1 - x) = -1$.

$= -55(1 - x)^{54}$

3. $y = \sqrt{1 + 8x} = (1 + 8x)^{1/2}$

$\dfrac{dy}{dx} = \dfrac{1}{2}(1 + 8x)^{1/2 - 1} \cdot 8$ Using the Extended
Power Rule

Don't forget $\dfrac{d}{dx}(1 + 8x) = 8$.

$= 4(1 + 8x)^{-1/2}$, or $\dfrac{4}{\sqrt{1 + 8x}}$

5. $y = \sqrt{3x^2 - 4} = (3x^2 - 4)^{1/2}$

$\dfrac{dy}{dx} = \dfrac{1}{2}(3x^2 - 4)^{1/2 - 1} \cdot 6x$ Using the Extended
Power Rule

Don't forget $\dfrac{d}{dx}(3x^2 - 4) = 6x$.

$= 3x(3x^2 - 4)^{-1/2}$, or $\dfrac{3x}{\sqrt{3x^2 - 4}}$

7. $y = (3x^2 - 6)^{-40}$

$\dfrac{dy}{dx} = -40(3x^2 - 6)^{-40 - 1} \cdot (6x)$ Using the Extended
Power Rule

Don't forget $\dfrac{d}{dx}(3x^2 - 6) = 6x$.

$= -240x(3x^2 - 6)^{-41}$, or $-\dfrac{240x}{(3x^2 - 6)^{41}}$

9. $y = x\sqrt{2x + 3} = x(2x + 3)^{1/2}$

$\dfrac{dy}{dx} = x \cdot \dfrac{1}{2}(2x + 3)^{1/2 - 1} \cdot 2 + 1 \cdot (2x + 3)^{1/2}$

Using the Product Rule and the
Extended Power Rule

$= x(2x + 3)^{-1/2} + (2x + 3)^{1/2}$, or

$\dfrac{x}{\sqrt{2x + 3}} + \sqrt{2x + 3}$

The above answers are acceptable, but further simplification can be done.

$= \dfrac{x}{\sqrt{2x + 3}} + \sqrt{2x + 3} \cdot \dfrac{\sqrt{2x + 3}}{\sqrt{2x + 3}}$

Multiplying the second term by a form of 1

$= \dfrac{x}{\sqrt{2x + 3}} + \dfrac{2x + 3}{\sqrt{2x + 3}}$

$= \dfrac{3x + 3}{\sqrt{2x + 3}}$, or $\dfrac{3(x + 1)}{\sqrt{2x + 3}}$

11. $y = x^2\sqrt{x - 1} = x^2(x - 1)^{1/2}$

$\dfrac{dy}{dx} = x^2 \cdot \dfrac{1}{2}(x - 1)^{1/2 - 1} \cdot 1 + 2x(x - 1)^{1/2}$

Using the Product Rule and the
Extended Power Rule

$= \dfrac{1}{2}x^2(x - 1)^{-1/2} + 2x(x - 1)^{1/2}$

$\dfrac{x^2}{2\sqrt{x - 1}} + 2x\sqrt{x - 1}$

The above answers are acceptable, but further simplification can be done.

$= \dfrac{x^2}{2\sqrt{x - 1}} + 2x\sqrt{x - 1} \cdot \dfrac{2\sqrt{x - 1}}{2\sqrt{x - 1}}$

Multiplying the second
term by a form of 1

$= \dfrac{x^2}{2\sqrt{x - 1}} + \dfrac{4x(x - 1)}{2\sqrt{x - 1}}$

$= \dfrac{x^2 + 4x^2 - 4x}{2\sqrt{x - 1}}$

$= \dfrac{5x^2 - 4x}{2\sqrt{x - 1}}$

13. $y = \dfrac{1}{(3x + 8)^2} = (3x + 8)^{-2}$

$\dfrac{dy}{dx} = -2(3x + 8)^{-2 - 1} \cdot 3$

Using the Extended Power Rule

$= -6(3x + 8)^{-3}$, or $\dfrac{-6}{(3x + 8)^3}$

15. $f(x) = (1 + x^3)^3 - (1 + x^3)^4$

$f'(x) = 3(1 + x^3)^2 \cdot 3x^2 - 4(1 + x^3)^3 \cdot 3x^2$

Using the Difference Rule and the
Extended Power Rule

$= 3x^2(1 + x^3)^2[3 - 4(1 + x^3)]$

Factoring out $3x^2(1 + x^3)^2$

$= 3x^2(1 + x^3)^2(3 - 4 - 4x^3)$

$= 3x^2(1 + x^3)^2(-1 - 4x^3)$, or

$= -3x^2(1 + x^3)^2(1 + 4x^3)$

17. $f(x) = x^2 + (200 - x)^2$

$f'(x) = 2x + 2(200 - x)^1 \cdot (-1)$

Using the Sum Rule, the Power Rule,
and the Extended Power Rule

$= 2x - 400 + 2x$

$= 4x - 400$

19. $f(x) = (x+6)^{10}(x-5)^4$

$f'(x) = (x+6)^{10} \cdot 4(x-5)^3 \cdot 1 + 10(x+6)^9 \cdot 1 \cdot (x-5)^4$

Using the Product Rule and the Extended Power Rule

$= 4(x+6)^{10}(x-5)^3 + 10(x+6)^9(x-5)^4$

$= 2(x+6)^9(x-5)^3[2(x+6) + 5(x-5)]$

Factoring out $2(x+6)^9(x-5)^3$

$= 2(x+6)^9(x-5)^3(2x+12+5x-25)$

$= 2(x+6)^9(x-5)^3(7x-13)$

21. $f(x) = (x-4)^8(3-x)^4$

$f'(x) = (x-4)^8 \cdot 4(3-x)^3 \cdot (-1) +$
$\qquad 8(x-4)^7 \cdot 1 \cdot (3-x)^4$

Using the Product Rule and the Extended Power Rule

$= -4(x-4)^8(3-x)^3 + 8(x-4)^7(3-x)^4$

$= 4(x-4)^7(3-x)^3[-(x-4)+2(3-x)]$

Factoring out $4(x-4)^7(3-x)^3$

$= 4(x-4)^7(3-x)^3(-x+4+6-2x)$

$= 4(x-4)^7(3-x)^3(10-3x)$

23. $f(x) = -4x(2x-3)^3$

$f'(x) = -4x \cdot 3(2x-3)^2 \cdot 2 + (-4) \cdot (2x-3)^3$

Using the Product Rule and the Extended Power Rule

$= -24x(2x-3)^2 - 4(2x-3)^3$

$= -4(2x-3)^2[6x + (2x-3)]$

Factoring out $-4(2x-3)^2$

$= -4(2x-3)^2(8x-3)$, or

$\quad 4(2x-3)^2(3-8x)$

25. $f(x) = \sqrt{\dfrac{1-x}{1+x}} = \left(\dfrac{1-x}{1+x}\right)^{1/2}$

$f'(x) = \dfrac{1}{2}\left(\dfrac{1-x}{1+x}\right)^{1/2-1} \cdot \dfrac{(1+x)(-1) - (1)(1-x)}{(1+x)^2}$

Using the Extended Power Rule and the Quotient Rule

$= \dfrac{1}{2}\left(\dfrac{1-x}{1+x}\right)^{-1/2} \cdot \dfrac{-1-x-1+x}{(1+x)^2}$

$= \dfrac{1}{2}\left(\dfrac{1-x}{1+x}\right)^{-1/2} \cdot \dfrac{-2}{(1+x)^2}$

$= \left(\dfrac{1-x}{1+x}\right)^{-1/2} \cdot \dfrac{-1}{(1+x)^2}$

27. $f(x) = \left(\dfrac{3x-1}{5x+2}\right)^4$

$f'(x) = 4\left(\dfrac{3x-1}{5x+2}\right)^3 \cdot \dfrac{(5x+2)(3) - 5(3x-1)}{(5x+2)^2}$

Using the Extended Power Rule and the Quotient Rule

$= 4\left(\dfrac{3x-1}{5x+2}\right)^3 \cdot \dfrac{15x+6-15x+5}{(5x+2)^2}$

$= 4\left(\dfrac{3x-1}{5x+2}\right)^3 \cdot \dfrac{11}{(5x+2)^2}$

$= \left(\dfrac{3x-1}{5x+2}\right)^3 \cdot \dfrac{44}{(5x+2)^2}$

29. $f(x) = \sqrt[3]{x^4+3x^2} = (x^4+3x^2)^{1/3}$

$f'(x) = \dfrac{1}{3}(x^4+3x^2)^{1/3-1}(4x^3+6x)$

Using the Extended Power Rule

$= \dfrac{1}{3}(x^4+3x^2)^{-2/3}(4x^3+6x)$

31. $f(x) = (2x^3-3x^2+4x+1)^{100}$

$f'(x) = 100(2x^3-3x^2+4x+1)^{99}(6x^2-6x+4)$

Using the Extended Power Rule

33. $g(x) = \left(\dfrac{2x+3}{5x-1}\right)^{-4}$

$g'(x) = -4\left(\dfrac{2x+3}{5x-1}\right)^{-5} \cdot \dfrac{(5x-1)(2) - 5(2x+3)}{(5x-1)^2}$

Using the Extended Power Rule and the Quotient Rule

$= -4\left(\dfrac{2x+3}{5x-1}\right)^{-5} \cdot \dfrac{10x-2-10x-15}{(5x-1)^2}$

$= -4\left(\dfrac{2x+3}{5x-1}\right)^{-5} \cdot \dfrac{-17}{(5x-1)^2}$

$= \left(\dfrac{2x+3}{5x-1}\right)^{-5} \cdot \dfrac{68}{(5x-1)^2}$

35. $f(x) = \sqrt{\dfrac{x^2+1}{x^2-1}} = \left(\dfrac{x^2+1}{x^2-1}\right)^{1/2}$

$f'(x) = \dfrac{1}{2}\left(\dfrac{x^2+1}{x^2-1}\right)^{1/2-1} \cdot \dfrac{(x^2-1)(2x) - 2x(x^2+1)}{(x^2-1)^2}$

Using the Extended Power Rule and the Quotient Rule

$= \dfrac{1}{2}\left(\dfrac{x^2+1}{x^2-1}\right)^{-1/2} \cdot \dfrac{2x^3-2x-2x^3-2x}{(x^2-1)^2}$

$= \dfrac{1}{2}\left(\dfrac{x^2+1}{x^2-1}\right)^{-1/2} \cdot \dfrac{-4x}{(x^2-1)^2}$

$= \left(\dfrac{x^2+1}{x^2-1}\right)^{-1/2} \cdot \dfrac{-2x}{(x^2-1)^2}$

37. $f(x) = \dfrac{(2x+3)^4}{(3x-2)^5}$

$f'(x) = \dfrac{(3x-2)^5 \cdot 4(2x+3)^3(2) - 5(3x-2)^4(3)(2x+3)^4}{[(3x-2)^5]^2}$

Using the Quotient Rule and
the Extended Power Rule

$= \dfrac{(3x-2)^4(2x+3)^3[8(3x-2) - 15(2x+3)]}{(3x-2)^{10}}$

Factoring

$= \dfrac{(3x-2)^4(2x+3)^3(24x-16-30x-45)}{(3x-2)^{10}}$

$= \dfrac{(3x-2)^4}{(3x-2)^4} \cdot \dfrac{(2x+3)^3(-6x-61)}{(3x-2)^6}$

$= \dfrac{(2x+3)^3(-6x-61)}{(3x-2)^6}$

39. $f(x) = 12(2x+1)^{2/3}(3x-4)^{5/4}$

$f'(x) = 12(2x+1)^{2/3} \cdot \dfrac{5}{4}(3x-4)^{5/4-1}(3) +$

$12 \cdot \dfrac{2}{3}(2x+1)^{2/3-1}(2)(3x-4)^{5/4}$

Using the Product Rule and
the Extended Power Rule

$= 12 \cdot \dfrac{5}{4} \cdot 3(2x+1)^{2/3}(3x-4)^{1/4} +$

$12 \cdot \dfrac{2}{3} \cdot 2(2x+1)^{-1/3}(3x-4)^{5/4}$

$= 45(2x+1)^{2/3}(3x-4)^{1/4} +$

$16(2x+1)^{-1/3}(3x-4)^{5/4}$

We can also simplify by factoring:

$f'(x) = (2x+1)^{-1/3}(3x-4)^{1/4}[45(2x+1)+16(3x-4)]$

$= (2x+1)^{-1/3}(3x-4)^{1/4}(90x+45+48x-64)$

$= (2x+1)^{-1/3}(3x-4)^{1/4}(138x-19)$

41. $y = \sqrt{u} = u^{1/2}, \quad u = x^2 - 1$

$\dfrac{dy}{du} = \dfrac{1}{2}u^{1/2-1} = \dfrac{1}{2}u^{-1/2} = \dfrac{1}{2\sqrt{u}}$

$\dfrac{du}{dx} = 2x$

$\dfrac{dy}{dx} = \dfrac{dy}{du} \cdot \dfrac{du}{dx} = \dfrac{1}{2\sqrt{u}} \cdot 2x$

$= \dfrac{2x}{2\sqrt{x^2-1}}$ Substituting $x^2 - 1$ for u

$= \dfrac{x}{\sqrt{x^2-1}}$ Simplifying

43. $y = u^{50}, \ u = 4x^3 - 2x^2$

$\dfrac{dy}{du} = 50u^{49}$

$\dfrac{du}{dx} = 12x^2 - 4x$

$\dfrac{dy}{dx} = \dfrac{dy}{du} \cdot \dfrac{du}{dx}$

$= 50u^{49}(12x^2 - 4x)$

$= 50(4x^3 - 2x^2)^{49}(12x^2 - 4x)$ Substituting for u

45. $y = u(u+1), \ u = x^3 - 2x$

$\dfrac{dy}{du} = u \cdot 1 + 1 \cdot (u+1)$ Product Rule

$= u + u + 1$

$= 2u + 1$

$\dfrac{du}{dx} = 3x^2 - 2$

$\dfrac{dy}{dx} = \dfrac{dy}{du} \cdot \dfrac{du}{dx} = (2u+1)(3x^2-2)$

$= [2(x^3 - 2x) + 1](3x^2 - 2)$

Substituting for u

$= (2x^3 - 4x + 1)(3x^2 - 2)$

47. $y = \sqrt{x^2 + 3x} = (x^2 + 3x)^{1/2}$

$\dfrac{dy}{dx} = \dfrac{1}{2}(x^2 + 3x)^{-1/2}(2x+3)$

$= \dfrac{2x+3}{2\sqrt{x^2+3x}}$

When $x = 1$, $\dfrac{dy}{dx} = \dfrac{2 \cdot 1 + 3}{2\sqrt{1^2 + 3 \cdot 1}}$

$= \dfrac{2+3}{2\sqrt{4}}$

$= \dfrac{5}{2 \cdot 2}$

$= \dfrac{5}{4}$

Thus, the slope of the tangent line at $(1, 2)$ is $\dfrac{5}{4}$. We use point-slope equation.

$y - y_1 = m(x - x_1)$

$y - 2 = \dfrac{5}{4}(x - 1)$

$y - 2 = \dfrac{5}{4}x - \dfrac{5}{4}$

$y = \dfrac{5}{4}x + \dfrac{3}{4}$

49. $f(x) = \dfrac{x^2}{(1+x)^5}$, or $x^2(1+x)^{-5}$

a) $f'(x) = \dfrac{(1+x)^5 \cdot 2x - 5(1+x)^4 \cdot 1 \cdot x^2}{[(1+x)^5]^2}$

Using the Quotient Rule and the Extended Power Rule

$= \dfrac{2x(1+x)^5 - 5x^2(1+x)^4}{(1+x)^{10}}$

$= \dfrac{x(1+x)^4[2(1+x) - 5x]}{(1+x)^{10}}$

Factoring out $x(1+x)^4$ in the numerator

$= \dfrac{x(2-3x)}{(1+x)^6}$, or $\dfrac{2x - 3x^2}{(1+x)^6}$ Simplifying

b) $f'(x) = x^2 \cdot (-5)(1+x)^{-6} \cdot 1 + 2x(1+x)^{-5}$

Using the Product Rule and the Extended Power Rule

$= -5x^2(1+x)^{-6} + 2x(1+x)^{-5}$

$= x(1+x)^{-5}[-5x(1+x)^{-1} + 2]$

$= \dfrac{x}{(1+x)^5} \cdot \left[\dfrac{-5x}{1+x} + 2 \cdot \dfrac{1+x}{1+x}\right]$

$= \dfrac{x}{(1+x)^5} \cdot \dfrac{2-3x}{1+x}$

$= \dfrac{2x - 3x^2}{(1+x)^6}$

c) Same

51. $f(x) = 3x^2 + 2, \quad g(x) = 2x - 1$

$f \circ g(x) = f(g(x)) = f(2x - 1)$

$= 3(2x-1)^2 + 2$

$= 3(4x^2 - 4x + 1) + 2$

$= 12x^2 - 12x + 3 + 2$

$= 12x^2 - 12x + 5$

$g \circ f(x) = g(f(x)) = g(3x^2 + 2)$

$= 2(3x^2 + 2) - 1$

$= 6x^2 + 4 - 1$

$= 6x^2 + 3$

53. $f(x) = 4x^2 - 1, \quad g(x) = \dfrac{2}{x}$

$f \circ g(x) = f(g(x)) = f\left(\dfrac{2}{x}\right)$

$= 4\left(\dfrac{2}{x}\right)^2 - 1$

$= 4 \cdot \dfrac{4}{x^2} - 1$

$= \dfrac{16}{x^2} - 1$

$g \circ f(x) = g(f(x)) = g(4x^2 - 1)$

$= \dfrac{2}{4x^2 - 1}$

55. $f(x) = x^2 + 1, \quad g(x) = x^2 - 1$

$f \circ g(x) = f(g(x)) = f(x^2 - 1)$

$= (x^2 - 1)^2 + 1$

$= x^4 - 2x^2 + 1 + 1$

$= x^4 - 2x^2 + 2$

$g \circ f(x) = g(f(x)) = g(x^2 + 1)$

$= (x^2 + 1)^2 - 1$

$= x^4 + 2x^2 + 1 - 1$

$= x^4 + 2x^2$

57. $h(x) = (3x^2 - 7)^5$

Let $f(x) = x^5$ and $g(x) = 3x^2 - 7$.

$f \circ g(x) = f(g(x)) = f(3x^2 - 7) = (3x^2 - 7)^5$

Thus, $h(x) = f \circ g(x)$.

Answers may vary.

59. $h(x) = \dfrac{x^3 + 1}{x^3 - 1}$

Let $f(x) = \dfrac{x+1}{x-1}$ and $g(x) = x^3$.

$f \circ g(x) = f(g(x)) = f(x^3) = \dfrac{x^3 + 1}{x^3 - 1}$.

Thus, $h(x) = f \circ g(x)$.

Answers may vary.

61. Using the Chain Rule:

Let $y = f(u)$. Then

$\dfrac{dy}{dx} = \dfrac{dy}{du} \cdot \dfrac{du}{dx}$

$= 3u^2(8x^3)$

$= 3(2x^4 + 1)^2(8x^3)$ Substituting $2x^4 + 1$ for u

When $x = -1$, $\dfrac{dy}{dx} = 3[2(-1)^4 + 1]^2[8(-1)^3]$

$= 3(2 + 1)^2(-8)$

$= 3 \cdot 3^2(-8)$

$= -216$

Finding $f(g(x))$:

$f \circ g(x) = f(g(x)) = f(2x^4 + 1) = (2x^4 + 1)^3$

Then $(f \circ g)'(x) = 3(2x^4 + 1)^2(8x^3)$ and

$(f \circ g)'(-1) = -216$ as above.

63. Using the Chain Rule:

Let $y = f(u) = \sqrt[3]{u} = u^{1/3}$. Then

$\dfrac{dy}{dx} = \dfrac{dy}{du} \cdot \dfrac{du}{dx}$

$= \dfrac{1}{3}u^{-2/3} \cdot (-6x)$

$= -2x \cdot u^{-2/3}$

$= -2x(1 - 3x^2)^{-2/3}$ Substituting $1 - 3x^2$ for u

When $x = 2$, $\quad \dfrac{dy}{dx} = 2 \cdot 2(1 - 3 \cdot 2^2)^{-2/3}$

$$= -4(-11)^{-2/3} \approx -0.8087$$

Finding $f(g(x))$:

$f \circ g(x) = f(g(x)) = f(1 - 3x^2) = \sqrt[3]{1 - 3x^2}$, or $(1 - 3x^2)^{1/3}$

Then $(f \circ g)'(x) = \dfrac{1}{3}(1 - 3x^2)^{-2/3}(-6x) =$

$-2x(1 - 3x^2)^{-2/3}$ and

$(f \circ g)'(2) = -4(-11)^{-2/3} \approx -0.8087$ as above.

65. a) $\quad C(x) = 1000\sqrt{x^3 + 2}$

$$= 1000(x^3 + 2)^{1/2}$$

$C'(x) = 1000 \cdot \dfrac{1}{2}(x^3 + 2)^{-1/2} \cdot 3x^2$

   Using Theorem 3 and the Extended Power Rule

$$= 1500x^2(x^3 + 2)^{-1/2}, \text{ or } \dfrac{1500x^2}{\sqrt{x^3 + 2}}$$

$C'(10) = \dfrac{1500 \cdot 10^2}{\sqrt{10^3 + 2}} = \dfrac{1500 \cdot 100}{\sqrt{1000 + 2}} =$

$\dfrac{150,000}{\sqrt{1002}} \approx 4738.68$

b) $\boxed{tw}$

67. a) The marginal profit $P'(x)$ is given by $R'(x) - C'(x)$:

$P'(x) = R'(x) - C'(x)$

$$= \dfrac{2000x}{\sqrt{x^2 + 3}} - \dfrac{1500x^2}{\sqrt{x^3 + 2}}$$

b) $\boxed{tw}$

69. a) $A = \$1000(1 + i)^3$

$\dfrac{dA}{di} = \$1000 \cdot 3(1 + i)^2 \cdot 1$

   Using Theorem 3 and the Extended Power Rule

$$= \$3000(1 + i)^2$$

b) $\boxed{tw}$

71. $D(p) = \dfrac{80,000}{p}$, $p = 1.6t + 9$

(a) We substitute $1.6t + 9$ for p in $D(p)$:

$$D(t) = \dfrac{80,000}{1.6t + 9}$$

(b) $D'(t) = \dfrac{(1.6t + 9)(0) - 1.6(80,000)}{(1.6t + 9)^2}$ $\quad$ Quotient Rule

$$= \dfrac{-128,000}{(1.6t + 9)^2}$$

(c) $D'(100) = \dfrac{-128,000}{[1.6(100) + 9]^2}$

$$= \dfrac{-128,000}{169^2}$$

$$= \dfrac{-128,000}{28,561}$$

$$\approx -4.48 \text{ units/day}$$

73. $y = \sqrt[3]{x^3 - 6x + 1} = (x^3 - 6x + 1)^{1/3}$

$\dfrac{dy}{dx} = \dfrac{1}{3}(x^3 - 6x + 1)^{-2/3}(3x^2 - 6)$

   Using the Extended Power Rule

$$= \dfrac{x^2 - 2}{\sqrt[3]{(x^3 - 6x + 1)^2}} \quad \text{Simplifying}$$

75. $y = \dfrac{x}{\sqrt{x - 1}} = \dfrac{x}{(x - 1)^{1/2}}$

$\dfrac{dy}{dx} = \dfrac{(x - 1)^{1/2} \cdot 1 - \dfrac{1}{2}(x - 1)^{-1/2} \cdot 1 \cdot x}{[(x - 1)^{1/2}]^2}$

   Using the Quotient Rule and the Extended Power Rule

$$= \dfrac{\sqrt{x - 1} - \dfrac{x}{2\sqrt{x - 1}}}{x - 1} \cdot \dfrac{2\sqrt{x - 1}}{2\sqrt{x - 1}}$$

   Multiplying by a form of 1

$$= \dfrac{2(x - 1) - x}{2(x - 1)\sqrt{x - 1}}$$

$$= \dfrac{x - 2}{2(x - 1)^{3/2}}$$

$[(x - 1)\sqrt{x - 1} = (x - 1)^{2/2}(x - 1)^{1/2}$
$= (x - 1)^{3/2}]$

77. $u = \dfrac{(1 + 2v)^4}{v^4}$

$\dfrac{du}{dv} = \dfrac{v^4 \cdot 4(1 + 2v)^3 \cdot 2 - 4v^3(1 + 2v)^4}{(v^4)^2}$

   Using the Quotient Rule and the Extended Power Rule

$$= \dfrac{8v^4(1 + 2v)^3 - 4v^3(1 + 2v)^4}{v^8}$$

$$= \dfrac{4v^3(1 + 2v)^3[2v - (1 + 2v)]}{v^8}$$

$$= \dfrac{4v^3(1 + 2v)^3(-1)}{v^8}$$

$$= \dfrac{-4(1 + 2v)^3}{v^5} \quad \text{Simplifying}$$

79. $y = \dfrac{\sqrt{1-x^2}}{1-x} = \dfrac{(1-x^2)^{1/2}}{1-x}$

$$\frac{dy}{dx} = \frac{(1-x)\frac{1}{2}(1-x^2)^{-1/2}(-2x)-(-1)(1-x^2)^{1/2}}{(1-x)^2}$$

Using the Quotient Rule and the
Extended Power Rule

$$= \frac{\dfrac{-x(1-x)}{\sqrt{1-x^2}}+\sqrt{1-x^2}}{(1-x)^2} \cdot \frac{\sqrt{1-x^2}}{\sqrt{1-x^2}}$$

Multiplying by a form of 1

$$= \frac{-x(1-x)+(1-x^2)}{(1-x)^2\sqrt{1-x^2}}$$

$$= \frac{-x+x^2+1-x^2}{(1-x)^2\sqrt{1-x^2}}$$

$$= \frac{1-x}{(1-x)^2\sqrt{1-x^2}}$$

$$= \frac{1}{(1-x)\sqrt{1-x^2}}$$

Rationalizing the denominator, we get

$$\frac{dy}{dx} = \frac{1}{(1-x)\sqrt{1-x^2}} \cdot \frac{\sqrt{1-x^2}}{\sqrt{1-x^2}}$$

$$= \frac{\sqrt{1-x^2}}{(1-x)(1-x^2)}.$$

81. $y = \left(\dfrac{x^2-x-1}{x^2+1}\right)^3$

$$\frac{dy}{dx} = 3\left(\frac{x^2-x-1}{x^2+1}\right)^2 \cdot \frac{(x^2+1)(2x-1)-2x(x^2-x-1)}{(x^2+1)^2}$$

Using the Extended Power Rule and the
Quotient Rule

$$= \frac{3(x^2-x-1)^2}{(x^2+1)^2} \cdot \frac{2x^3-x^2+2x-1-2x^3+2x^2+2x}{(x^2+1)^2}$$

$$= \frac{3(x^2-x-1)^2(x^2+4x-1)}{(x^2+1)^4}$$

83. $s = \dfrac{\sqrt{t}-1}{\sqrt{t}+1} = \dfrac{t^{1/2}-1}{t^{1/2}+1}$

$$\frac{ds}{dt} = \frac{(t^{1/2}+1)\left(\frac{1}{2}t^{-1/2}\right)-\left(\frac{1}{2}t^{-1/2}\right)(t^{1/2}-1)}{(t^{1/2}+1)^2}$$

Using the Quotient Rule

$$= \frac{\frac{1}{2}t^0+\frac{1}{2}t^{-1/2}-\frac{1}{2}t^0+\frac{1}{2}t^{-1/2}}{(t^{1/2}+1)^2}$$

$$= \frac{t^{-1/2}}{(t^{1/2}+1)^2}$$

$$= \frac{1}{\sqrt{t}(\sqrt{t}+1)^2}$$

85. $\boxed{tw}$

87. See the answer section in the text.

89. $f(x) = x\sqrt{4-x^2} = x(4-x^2)^{1/2}$

$$f'(x) = x \cdot \frac{1}{2}(4-x^2)^{-1/2}(-2x) + (1)(4-x^2)^{1/2}$$

$$= \frac{-x^2}{\sqrt{4-x^2}} + \sqrt{4-x^2}$$

$$= \frac{-x^2+(4-x^2)}{\sqrt{4-x^2}}$$

$$= \frac{4-2x^2}{\sqrt{4-x^2}}$$

$y_1 = f(x) = x\sqrt{4-x^2}$ $y_2 = f'(x) = \dfrac{4-2x^2}{\sqrt{4-x^2}}$

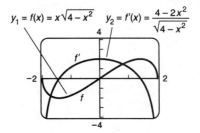

Exercise Set 2.9

1. $y = 3x + 5$

$$\frac{dy}{dx} = 3 \qquad \text{First derivative}$$

$$\frac{d^2y}{dx^2} = 0 \qquad \text{Second derivative}$$

3. $y = -\dfrac{1}{x} = -1 \cdot x^{-1}$

$$\frac{dy}{dx} = -1 \cdot (-1)x^{-1-1}$$

$$= x^{-2} \qquad \text{First derivative}$$

$$\frac{d^2y}{dx^2} = -2x^{-2-1}$$

$$= -2x^{-3}, \text{ or } -\frac{2}{x^3} \quad \text{Second derivative}$$

5. $y = x^{1/4}$

$$\frac{dy}{dx} = \frac{1}{4}x^{1/4-1}$$

$$= \frac{1}{4}x^{-3/4} \qquad \text{First derivative}$$

$$\frac{d^2y}{dx^2} = \frac{1}{4} \cdot \left(-\frac{3}{4}\right)x^{-3/4-1}$$

$$= -\frac{3}{16}x^{-7/4}, \text{ or } -\frac{3}{16x^{7/4}} \quad \text{Second derivative}$$

7. $y = x^4 + \dfrac{4}{x} = x^4 + 4x^{-1}$

$$\frac{dy}{dx} = 4x^{4-1} + 4 \cdot (-1) \cdot x^{-1-1}$$

$$= 4x^3 - 4x^{-2} \qquad \text{First derivative}$$

$$\frac{d^2y}{dx^2} = 4 \cdot 3x^{3-1} - 4 \cdot (-2) \cdot x^{-2-1}$$

$$= 12x^2 + 8x^{-3}, \text{ or } 12x^2 + \frac{8}{x^3}$$

Second derivative

9. $y = x^{-3}$

$\dfrac{dy}{dx} = -3x^{-3-1}$

$= -3x^{-4}$ First derivative

$\dfrac{d^2y}{dx^2} = -3 \cdot (-4) \cdot x^{-4-1}$

$= 12x^{-5}$, or $\dfrac{12}{x^5}$ Second derivative

11. $y = x^n$

$\dfrac{dy}{dx} = nx^{n-1}$ First derivative

$\dfrac{d^2y}{dx^2} = n \cdot (n-1) \cdot x^{(n-1)-1}$

$= n(n-1)x^{n-2}$ Second derivative

13. $y = x^4 - x^2$

$\dfrac{dy}{dx} = 4x^{4-1} - 2x^{2-1}$

$= 4x^3 - 2x$

$\dfrac{d^2y}{dx^2} = 4 \cdot 3x^{3-1} - 2$

$= 12x^2 - 2$

15. $y = \sqrt{x-1} = (x-1)^{1/2}$

$\dfrac{dy}{dx} = \dfrac{1}{2}(x-1)^{1/2-1} \cdot 1$

$= \dfrac{1}{2}(x-1)^{-1/2}$ First derivative

$\dfrac{d^2y}{dx^2} = \dfrac{1}{2} \cdot \left(-\dfrac{1}{2}\right)(x-1)^{-1/2-1} \cdot 1$

$= -\dfrac{1}{4}(x-1)^{-3/2}$, or $-\dfrac{1}{4(x-1)^{3/2}}$, or

$-\dfrac{1}{4\sqrt{(x-1)^3}}$ Second derivative

17. $y = ax^2 + bx + c$ x and y are variables;
                                           a, b, and c are constants

$\dfrac{dy}{dx} = a \cdot 2x + b + 0$

$= 2ax + b$ First derivative
                        $2a$ and b are constants

$\dfrac{d^2y}{dx^2} = 2a + 0$

$= 2a$ Second derivative

19. $y = (x^2 - 8x)^{43}$

$\dfrac{dy}{dx} = 43(x^2 - 8x)^{42}(2x - 8)$ First derivative

$\dfrac{d^2y}{dx^2} = 43(x^2 - 8x)^{42}(2) +$

$43 \cdot 42(x^2 - 8x)^{41}(2x-8)(2x-8)$

$= 2 \cdot 43(x^2 - 8x)^{41}[x^2 - 8x + 21(2x-8)^2]$

Factoring

$= 86(x^2 - 8x)^{41}[x^2 - 8x + 21(4x^2 - 32x + 64)]$

$= 86(x^2 - 8x)^{41}(x^2 - 8x + 84x^2 - 672x + 1344)$

$= 86(x^2 - 8x)^{41}(85x^2 - 680x + 1344)$

Second derivative

21. $y = (x^4 - 4x^2)^{50}$

$\dfrac{dy}{dx} = 50(x^4 - 4x^2)^{49}(4x^3 - 8x)$ First derivative

$\dfrac{d^2y}{dx^2} = 50(x^4 - 4x^2)^{49}(12x^2 - 8) +$

$50 \cdot 49(x^4 - 4x^2)^{48}(4x^3 - 8x)(4x^3 - 8x)$

$= 50 \cdot 4(x^4 - 4x^2)^{49}(3x^2 - 2) +$

$50 \cdot 49 \cdot 4 \cdot 4(x^4 - 4x^2)^{48}(x^3 - 2x)(x^3 - 2x)$

Factoring the terms

$= 200(x^4 - 4x^2)^{48}[(x^4 - 4x^2)(3x^2 - 2) +$

$49 \cdot 4(x^3 - 2x)^2]$

Factoring the expression

$= 200x^2(x^4 - 4x^2)^{48}[(x^2 - 4)(3x^2 - 2) +$

$196(x^2 - 2)^2]$ Factoring out x^2

$= 200x^2(x^4 - 4x^2)^{48}[3x^4 - 14x^2 + 8 +$

$196(x^4 - 4x^2 + 4)]$

$= 200x^2(x^4 - 4x^2)^{48}(3x^4 - 14x^2 + 8 + 196x^4 -$

$784x^2 + 784)$

$= 200x^2(x^4 - 4x^2)^{48}(199x^4 - 798x^2 + 792)$

Second derivative

23. $y = x^{2/3} + 4x$

$\dfrac{dy}{dx} = \dfrac{2}{3}x^{2/3-1} + 4$

$= \dfrac{2}{3}x^{-1/3} + 4$ First derivative

$\dfrac{d^2y}{dx^2} = \dfrac{2}{3}\left(-\dfrac{1}{3}\right)x^{-1/3-1}$

$= -\dfrac{2}{9}x^{-4/3}$ Second derivative

25. $y = (x - 8)^{3/4}$

$\dfrac{dy}{dx} = \dfrac{3}{4}(x - 8)^{3/4 - 1} \cdot 1$

$\qquad = \dfrac{3}{4}(x - 8)^{-1/4}$ First derivative

$\dfrac{d^2y}{dx^2} = \dfrac{3}{4}\left(-\dfrac{1}{4}\right)(x - 8)^{-1/4 - 1} \cdot 1$

$\qquad = -\dfrac{3}{16}(x - 8)^{-5/4}$ Second derivative

27. $y = \dfrac{1}{x^2} + \dfrac{2}{x^3} = x^{-2} + 2x^{-3}$

$\dfrac{dy}{dx} = -2x^{-3} - 6x^{-4}$ First derivative

$\dfrac{d^2y}{dx^2} = 6x^{-4} + 24x^{-5}$ Second derivative

29. $y = x^4$

$\dfrac{dy}{dx} = 4x^3$ First derivative

$\dfrac{d^2y}{dx^2} = 4 \cdot 3x^2$

$\qquad = 12x^2$ Second derivative

$\dfrac{d^3y}{dx^3} = 12 \cdot 2x$

$\qquad = 24x$ Third derivative

$\dfrac{d^4y}{dy^4} = 24$ Fourth derivative

31. $y = x^6 - x^3 + 2x$

$\dfrac{dy}{dx} = 6x^5 - 3x^2 + 2$ First derivative

$\dfrac{d^2y}{dx^2} = 30x^4 - 6x$ Second derivative

$\dfrac{d^3y}{dx^3} = 120x^3 - 6$ Third derivative

$\dfrac{d^4y}{dx^4} = 360x^2$ Fourth derivative

$\dfrac{d^5y}{dx^5} = 720x$ Fifth derivative

33. $y = (x^2 - 5)^{10}$

$\dfrac{dy}{dx} = 10(x^2 - 5)^9 \cdot 2x$

$\qquad = 20x(x^2 - 5)^9$

$\dfrac{d^2y}{dx^2} = 20x \cdot 9(x^2 - 5)^8 \cdot 2x + 20(x^2 - 5)^9$

$\qquad = 360x^2(x^2 - 5)^8 + 20(x^2 - 5)^9$

$\qquad = 20(x^2 - 5)^8[18x^2 + (x^2 - 5)]$

$\qquad = 20(x^2 - 5)^8(19x^2 - 5)$

35. $s(t) = t^3 + t^2 + 2t$

$v(t) = s'(t) = 3t^2 + 2t + 2$

$a(t) = s''(t) = 6t + 2$

37. $P(t) = 100,000(1 + 0.6t + t^2)$

$\qquad = 100,000 + 60,000t + 100,000t^2$

This function gives the number of
people in a population at time t

Whenever a quantity is a function of time, the first derivative gives the rate of change with respect to time and the second derivative gives the acceleration.

$P'(t) = 60,000 + 200,000t$

This function gives the rate of change
of the size of a population.

$P''(t) = 200,000$ Acceleration

This function gives the acceleration in
the size of the population.

39. $y = x^{-1} + x^{-2}$

$y' = -1 \cdot x^{-1-1} + (-2)x^{-2-1}$

$\qquad = -x^{-2} - 2x^{-3}$

$y'' = -(-2) \cdot x^{-2-1} - 2 \cdot (-3) \cdot x^{-3-1}$

$\qquad = 2x^{-3} + 6x^{-4}$

$y''' = 2 \cdot (-3) \cdot x^{-3-1} + 6 \cdot (-4) \cdot x^{-4-1}$

$\qquad = -6x^{-4} - 24x^{-5}$

41. $y = x\sqrt{1 + x^2} = x(1 + x^2)^{1/2}$

$y' = x \cdot \dfrac{1}{2}(1 + x^2)^{-1/2} \cdot 2x + 1 \cdot (1 + x^2)^{1/2}$

$\qquad = \dfrac{x^2}{(1 + x^2)^{1/2}} + (1 + x^2)^{1/2} \cdot \dfrac{(1 + x^2)^{1/2}}{(1 + x^2)^{1/2}}$

Multiplying the second term
by a form of 1

$\qquad = \dfrac{x^2 + 1 + x^2}{(1 + x^2)^{1/2}}$

$\qquad = \dfrac{2x^2 + 1}{(1 + x^2)^{1/2}}$

$y'' = \dfrac{(1+x^2)^{1/2} \cdot 4x - \dfrac{1}{2}(1+x^2)^{-1/2} \cdot 2x \cdot (2x^2+1)}{[(1+x^2)^{1/2}]^2}$

$\qquad = \dfrac{4x(1+x^2)^{1/2} - x(2x^2+1)(1+x^2)^{-1/2}}{1+x^2} \cdot \dfrac{(1+x^2)^{1/2}}{(1+x^2)^{1/2}}$

Multiplying by a form of 1

$\qquad = \dfrac{4x(1 + x^2) - x(2x^2 + 1)}{(1 + x^2)^{3/2}}$

$\qquad = \dfrac{4x + 4x^3 - 2x^3 - x}{(1 + x^2)^{3/2}}$

$\qquad = \dfrac{2x^3 + 3x}{(1 + x^2)^{3/2}}$

$$y''' = \frac{(1+x^2)^{3/2}(6x^2+3)-\frac{3}{2}(1+x^2)^{1/2}\cdot 2x\cdot(2x^3+3x)}{[(1+x^2)^{3/2}]^2}$$

$$= \frac{(1+x^2)^{3/2}(6x^2+3)-3x(1+x^2)^{1/2}(2x^3+3x)}{(1+x^2)^{6/2}}$$

$$= \frac{(1+x^2)^{1/2}[(1+x^2)(6x^2+3)-3x(2x^3+3x)]}{(1+x^2)^{1/2}(1+x^2)^{5/2}}$$

$$= \frac{6x^2+3+6x^4+3x^2-6x^4-9x^2}{(1+x^2)^{5/2}}$$

$$= \frac{3}{(1+x^2)^{5/2}}$$

43. $y = \dfrac{3x-1}{2x+3}$

$$y' = \frac{(2x+3)\cdot 3-2\cdot(3x-1)}{(2x+3)^2}$$

$$= \frac{6x+9-6x+2}{(2x+3)^2}$$

$$= \frac{11}{(2x+3)^2}, \text{ or } 11(2x+3)^{-2}$$

$$y'' = 11\cdot(-2)(2x+3)^{-3}\cdot 2$$

$$= -44(2x+3)^{-3}, \text{ or } \frac{-44}{(2x+3)^3}$$

$$y''' = -44\cdot(-3)\cdot(2x+3)^{-4}\cdot 2$$

$$= 264(2x+3)^{-4}, \text{ or } \frac{264}{(2x+3)^4}$$

45. $y = \dfrac{x}{\sqrt{x-1}}$, or $x(x-1)^{-1/2}$

$$y' = x\cdot\left(-\frac{1}{2}\right)(x-1)^{-3/2}\cdot 1+1\cdot(x-1)^{-1/2}$$

$$= \frac{-x}{2(x-1)^{3/2}}+\frac{1}{(x-1)^{1/2}}\cdot\frac{2(x-1)}{2(x-1)}$$

Multiplying the second term by a
form of 1

$$= \frac{-x+2x-2}{2(x-1)^{3/2}}$$

$$= \frac{x-2}{2(x-1)^{3/2}}$$

$$y'' = \frac{2(x-1)^{3/2}\cdot 1-2\cdot\frac{3}{2}(x-1)^{1/2}\cdot 1\cdot(x-2)}{[2(x-1)^{3/2}]^2}$$

$$= \frac{2(x-1)^{3/2}-3(x-2)(x-1)^{1/2}}{4(x-1)^3}\cdot\frac{(x-1)^{-1/2}}{(x-1)^{-1/2}}$$

Multiplying by a form of 1

$$= \frac{2(x-1)-3(x-2)}{4(x-1)^{5/2}}$$

$$= \frac{2x-2-3x+6}{4(x-1)^{5/2}}$$

$$= \frac{4-x}{4(x-1)^{5/2}}$$

$$y''' = \frac{4(x-1)^{5/2}(-1)-4\cdot\frac{5}{2}(x-1)^{3/2}\cdot 1\cdot(4-x)}{[4(x-1)^{5/2}]^2}$$

$$= \frac{-4(x-1)^{5/2}-10(4-x)(x-1)^{3/2}}{16(x-1)^5}\cdot\frac{(x-1)^{-1/2}}{(x-1)^{-1/2}}$$

Multiplying by a form of 1

$$= \frac{-4(x-1)^2-10(4-x)(x-1)}{16(x-1)^{9/2}}$$

$$= \frac{-4(x^2-2x+1)-10(-x^2+5x-4)}{16(x-1)^{9/2}}$$

$$= \frac{-4x^2+8x-4+10x^2-50x+40}{16(x-1)^{9/2}}$$

$$= \frac{6x^2-42x+36}{16(x-1)^{9/2}}$$

$$= \frac{6(x^2-7x+6)}{16(x-1)^{9/2}}$$

$$= \frac{6(x-6)(x-1)}{16(x-1)^{9/2}}$$

$$= \frac{3(x-6)}{8(x-1)^{7/2}}$$

47. $f(x) = \dfrac{x}{x-1}$

$$f'(x) = \frac{(x-1)\cdot 1-1\cdot x}{(x-1)^2}$$

$$= \frac{x-1-x}{(x-1)^2}$$

$$= \frac{-1}{(x-1)^2}$$

$$f''(x) = \frac{(x-1)^2\cdot 0-2(x-1)\cdot 1\cdot(-1)}{[(x-1)^2]^2}$$

$$= \frac{2(x-1)}{(x-1)^4}$$

$$= \frac{2}{(x-1)^3}$$

49. $f(x) = \dfrac{x-1}{x+2}$

$$f'(x) = \frac{(x+2)\cdot 1-1\cdot(x-1)}{(x+2)^2}$$

$$= \frac{x+2-x+1}{(x+2)^2}$$

$$= \frac{3}{(x+2)^2} \quad \text{First derivative}$$

Express $f'(x)$ as $3(x+2)^{-2}$. Then we have:

$$f''(x) = -6(x+2)^{-3}\cdot 1$$

$$= \frac{-6}{(x+2)^3} \quad \text{Second derivative}$$

Express $f''(x)$ as $-6(x+2)^{-3}$. Then we have:

$$f'''(x) = 18(x+2)^{-4}\cdot 1$$

$$= \frac{18}{(x+2)^4} \quad \text{Third derivative}$$

Express $f'''(x)$ as $18(x+2)^{-4}$. Then we have:

$$f^{(4)}(x) = -72(x+2)^{-5} \cdot 1$$

$$= \frac{-72}{(x+2)^5} \qquad \text{Fourth derivative}$$

Express $f^{(4)}(x)$ as $-72(x+2)^{-5}$. Then we have:

$$f^{(5)}(x) = 360(x+2)^{-6} \cdot 1$$

$$= \frac{360}{(x+2)^6} \qquad \text{Fifth derivative}$$

51. See the answer section in the text.

53. See the answer section in the text.

Chapter 3

Applications of Differentiation

Exercise Set 3.1

1. $f(x) = x^2 - 4x + 5$

First, find the critical points.

$f'(x) = 2x - 4$

$f'(x)$ exists for all real numbers. We solve $f'(x) = 0$:

$$2x - 4 = 0$$
$$2x = 4$$
$$x = 2$$

The only critical point is 2. We use 2 to divide the real number line into two intervals, A: $(-\infty, 2)$ and B: $(2, \infty)$:

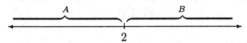

We use a test value in each interval to determine the sign of the derivative in each interval.

A: Test 0, $f'(0) = 2 \cdot 0 - 4 = -4 < 0$

B: Test 3, $f'(3) = 2 \cdot 3 - 4 = 2 > 0$

We see that $f(x)$ is decreasing on $(-\infty, 2)$ and increasing on $(2, \infty)$, and the change from decreasing to increasing indicates that a relative minimum occurs at $x = 2$. We substitute into the original equation to find $f(2)$:

$$f(2) = 2^2 - 4 \cdot 2 + 5 = 1$$

Thus, there is a relative minimum at $(2, 1)$. We use the information obtained to sketch the graph. Other function values are listed below.

x	$f(x)$
-2	17
-1	10
0	5
1	2
2	1
3	2
4	5
5	10

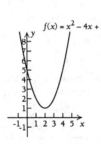

3. $f(x) = 5 + x - x^2$

First, find the critical points,

$f'(x) = 1 - 2x$

$f'(x)$ exists for all real numbers. We solve $f'(x) = 0$:

$$1 - 2x = 0$$
$$1 = 2x$$
$$\frac{1}{2} = x$$

The only critical point is $\frac{1}{2}$. We use $\frac{1}{2}$ to divide the real number line into two intervals, A: $\left(-\infty, \frac{1}{2}\right)$ and B: $\left(\frac{1}{2}, \infty\right)$:

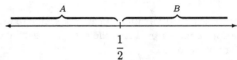

We use a test value in each interval to determine the sign of the derivative in each interval.

A: Test 0, $f'(0) = 1 - 2 \cdot 0 = 1 > 0$

B: Test 1, $f'(1) = 1 - 2 \cdot 1 = -1 < 0$

We see that $f(x)$ is increasing on $\left(-\infty, \frac{1}{2}\right)$ and decreasing on $\left(\frac{1}{2}, \infty\right)$, so there is a relative maximum at $x = \frac{1}{2}$. We find $f\left(\frac{1}{2}\right)$:

$$f\left(\frac{1}{2}\right) = 5 + \frac{1}{2} - \left(\frac{1}{2}\right)^2 = \frac{21}{4}$$

Thus, there is a relative maximum at $\left(\frac{1}{2}, \frac{21}{4}\right)$. We use the information obtained to sketch the graph. Other function values are listed below.

x	$f(x)$
-3	-7
-2	-1
-1	3
0	5
$\frac{1}{2}$	$\frac{21}{4}$
1	5
2	3
3	-1
4	-7

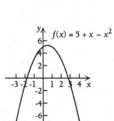

5. $f(x) = 1 + 6x + 3x^2$

First, find the critical points.

$f'(x) = 6 + 6x$

$f'(x)$ exists for all real numbers. We solve $f'(x) = 0$:

$$6 + 6x = 0$$
$$6x = -6$$
$$x = -1$$

The only critical point is -1. We use -1 to divide the real number line into two intervals, A: $(-\infty, -1)$ and B: $(-1, \infty)$:

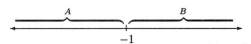

We use a test value in each interval to determine the sign of the derivative in each interval.

A: Test -2, $f'(-2) = 6 + 6(-2) = -6 < 0$

B: Test 0, $f'(0) = 6 + 6 \cdot 0 = 6 > 0$

We see that $f(x)$ is decreasing on $(-\infty, -1)$ and increasing on $(-1, \infty)$, so there is a relative minimum at $x = -1$. We find $f(-1)$:

$$f(-1) = 1 + 6(-1) + 3(-1)^2 = -2$$

Thus, there is a relative minimum at $(-1, -2)$. We use the information obtained to sketch the graph. Other function values are listed below.

x	$f(x)$
-3	10
-2	1
-1	-2
0	1
1	10
2	25

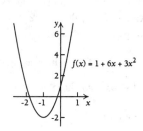

$f(x) = 1 + 6x + 3x^2$

7. $f(x) = x^3 - x^2 - x + 2$

First, find the critical points.

$$f'(x) = 3x^2 - 2x - 1$$

$f'(x)$ exists for all real numbers. We solve $f'(x) = 0$:

$$3x^2 - 2x - 1 = 0$$
$$(3x + 1)(x - 1) = 0$$
$$3x + 1 = 0 \quad \text{or} \quad x - 1 = 0$$
$$3x = -1 \quad \text{or} \qquad x = 1$$
$$x = -\frac{1}{3} \quad \text{or} \qquad x = 1$$

The critical points are $-\frac{1}{3}$ and 1. We use them to divide the real number line into three intervals, A: $\left(-\infty, -\frac{1}{3}\right)$, B: $\left(-\frac{1}{3}, 1\right)$, and C: $(1, \infty)$.

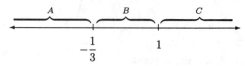

We use a test value in each interval to determine the sign of the derivative in each interval.

A: Test -1, $f'(-1) = 3(-1)^2 - 2(-1) - 1 =$
$3 + 2 - 1 = 4 > 0$

B: Test 0, $f'(0) = 3(0)^2 - 2(0) - 1 = -1 < 0$

C: Test 2, $f'(2) = 3(2)^2 - 2(2) - 1 = 12 - 4 - 1 =$
$7 > 0$

We see that $f(x)$ is increasing on $\left(-\infty, -\frac{1}{3}\right)$, decreasing on $\left(-\frac{1}{3}, 1\right)$, and increasing again on $(1, \infty)$, so there is a relative maximum at $x = -\frac{1}{3}$ and a relative minimum at $x = 1$. We find $f\left(-\frac{1}{3}\right)$:

$$f\left(-\frac{1}{3}\right) = \left(-\frac{1}{3}\right)^3 - \left(-\frac{1}{3}\right)^2 - \left(-\frac{1}{3}\right) + 2$$
$$= -\frac{1}{27} - \frac{1}{9} + \frac{1}{3} + 2$$
$$= \frac{59}{27}$$

Then we find $f(1)$:

$$f(1) = 1^3 - 1^2 - 1 + 2$$
$$= 1 - 1 - 1 + 2$$
$$= 1$$

There is a relative maximum at $\left(-\frac{1}{3}, \frac{59}{27}\right)$, and there is a relative minimum at $(1, 1)$. We use the information obtained to sketch the graph. Other function values are listed below.

x	$f(x)$
-2	-8
-1	1
0	2
2	4
3	17

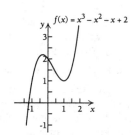

$f(x) = x^3 - x^2 - x + 2$

9. $f(x) = x^3 - 3x + 6$

First, find the critical points.

$$f'(x) = 3x^2 - 3$$

$f'(x)$ exists for all real numbers. We solve $f'(x) = 0$:

$$3x^2 - 3 = 0$$
$$x^2 - 1 = 0 \qquad \text{Dividing by 3}$$
$$(x + 1)(x - 1) = 0$$
$$x + 1 = 0 \quad \text{or} \quad x - 1 = 0$$
$$x = -1 \quad \text{or} \qquad x = 1$$

The critical points are -1 and 1. We use them to divide the real number line into three intervals, A: $(-\infty, -1)$, B: $(-1, 1)$, and C: $(1, \infty)$.

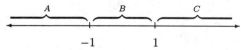

We use a test value in each interval to determine the sign of the derivative in each interval.

A: Test -2, $f'(-2) = 3(-2)^2 - 3 = 12 - 3 = 9 > 0$

B: Test 0, $f'(0) = 3 \cdot 0^2 - 3 = 0 - 3 = -3 < 0$

C: Test 2, $f'(2) = 3 \cdot 2^2 - 3 = 12 - 3 = 9 > 0$

We see that $f(x)$ is increasing on $(-\infty, -1)$, decreasing on $(-1, 1)$, and increasing again on $(1, \infty)$, so there is a relative maximum at $x = -1$ and a relative minimum at $x = 1$. We find $f(-1)$:

$$f(-1) = (-1)^3 - 3(-1) + 6 = -1 + 3 + 6 = 8$$

Then we find $f(1)$:

$$f(1) = 1^3 - 3 \cdot 1 + 6 = 1 - 3 + 6 = 4$$

There is a relative maximum at $(-1, 8)$, and there is a relative minimum at $(1, 4)$. We use the information obtained to sketch the graph. Other function values are listed below.

x	$f(x)$
-3	-12
-2	4
0	6
2	8
3	24

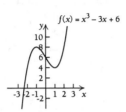

11. $f(x) = 3x^2 - 2x^3$

First, find the critical points.

$$f'(x) = 6x - 6x^2$$

$f'(x)$ exists for all real numbers. We solve $f'(x) = 0$:

$$6x - 6x^2 = 0$$
$$x - x^2 = 0 \qquad \text{Dividing by 6}$$
$$x(1 - x) = 0$$
$$x = 0 \ \text{ or } \ 1 - x = 0$$
$$x = 0 \ \text{ or } \qquad 1 = x$$

The critical points are 0 and 1. We use them to divide the real number line into three intervals, A: $(-\infty, 0)$, B: $(0, 1)$, and C: $(1, \infty)$.

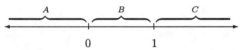

We use a test value in each interval to determine the sign of the derivative in each interval.

A: Test -1, $f'(-1) = 6(-1) - 6(-1)^2 = -6 - 6 = -12 < 0$

B: Test $\frac{1}{2}$, $f'\left(\frac{1}{2}\right) = 6\left(\frac{1}{2}\right) - 6\left(\frac{1}{2}\right)^2 = 3 - \frac{3}{2} = \frac{3}{2} > 0$

C: Test 2, $f'(2) = 6 \cdot 2 - 6 \cdot 2^2 = 12 - 24 = -12 < 0$

We see that $f(x)$ is decreasing on $(-\infty, 0)$, increasing on $(0, 1)$, and decreasing again on $(1, \infty)$, so there is a relative minimum at $x = 0$ and a relative maximum at $x = 1$. We find $f(0)$:

$$f(0) = 3 \cdot 0^2 - 2 \cdot 0^3 = 0 - 0 = 0$$

Then we find $f(1)$:

$$f(1) = 3 \cdot 1^2 - 2 \cdot 1^3 = 3 - 2 = 1$$

There is a relative minimum at $(0, 0)$, and there is a relative maximum at $(1, 1)$. We use the information obtained to sketch the graph. Other function values are listed below.

x	$f(x)$
-2	28
-1	5
2	-4
3	-27

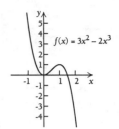

13. $f(x) = 2x^3$

First, find the critical points.

$$f'(x) = 6x^2$$

$f'(x)$ exists for all real numbers. We solve $f'(x) = 0$:

$$6x^2 = 0$$
$$x^2 = 0 \qquad \text{Dividing by 6}$$
$$x = 0$$

The only critical point is 0. We use 0 to divide the real number line into two intervals, A: $(-\infty, 0)$, and B: $(0, \infty)$.

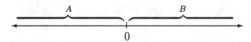

We use a test value in each interval to determine the sign of the derivative in each interval.

A: Test -1, $f'(-1) = 6(-1)^2 = 6 \cdot 1 = 6 > 0$

B: Test 1, $f'(1) = 6(1)^2 = 6 \cdot 1 = 6 > 0$

We see that $f(x)$ is increasing on both intervals, so the function has no relative extrema. We use the information obtained to sketch the graph. Some function values are listed below.

x	$f(x)$
-2	-16
-1	-2
0	0
1	2
2	16

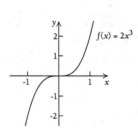

15. $f(x) = x^3 - 6x^2 + 10$

First, find the critical points.

$$f'(x) = 3x^2 - 12x$$

$f'(x)$ exists for all real numbers. We solve $f'(x) = 0$:

$$3x^2 - 12x = 0$$
$$x^2 - 4x = 0 \qquad \text{Dividing by 3}$$
$$x(x - 4) = 0$$
$$x = 0 \ \text{ or } \ x - 4 = 0$$
$$x = 0 \ \text{ or } \qquad x = 4$$

The critical points are 0 and 4. We use them to divide the real number line into three intervals, A: $(-\infty, 0)$, B: $(0, 4)$, and C: $(4, \infty)$.

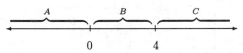

We use a test value in each interval to determine the sign of the derivative in each interval.

A: Test -1, $f'(-1) = 3(-1)^2 - 12(-1) = 3 + 12 = 15 > 0$

B: Test 1, $f'(1) = 3 \cdot 1^2 - 12 \cdot 1 = 3 - 12 = -9 < 0$

C: Test 5, $f'(5) = 3 \cdot 5^2 - 12 \cdot 5 = 75 - 60 = 15 > 0$

We see that $f(x)$ is increasing on $(-\infty, 0)$, decreasing on $(0, 4)$, and increasing again on $(4, \infty)$, so there is a relative maximum at $x = 0$ and a relative minimum at $x = 4$. We find $f(0)$:

$$f(0) = 0^3 - 6 \cdot 0^2 + 10 = 0 - 0 + 10 = 10$$

Then we find $f(4)$:

$$f(4) = 4^3 - 6 \cdot 4^2 + 10 = 64 - 96 + 10 = -22$$

There is a relative maximum at $(0, 10)$, and there is a relative minimum at $(4, -22)$. We use the information obtained to sketch the graph. Other function values are listed below.

x	$f(x)$
-2	-22
-1	3
1	5
2	-6
3	-17

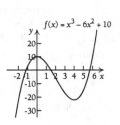

17. $f(x) = x^3 - x^4$

First, find the critical points.

$f'(x) = 3x^2 - 4x^3$

$f'(x)$ exists for all real numbers. We solve $f'(x) = 0$:

$$3x^2 - 4x^3 = 0$$
$$x^2(3 - 4x) = 0$$

$x^2 = 0$ or $3 - 4x = 0$

$x = 0$ or $3 = 4x$

$x = 0$ or $\dfrac{3}{4} = x$

The critical points are 0 and $\dfrac{3}{4}$. We use them to divide the real number line into three intervals, A: $(-\infty, 0)$, B: $\left(0, \dfrac{3}{4}\right)$, and C: $\left(\dfrac{3}{4}, \infty\right)$.

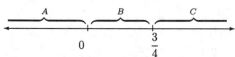

We use a test value in each interval to determine the sign of the derivative in each interval.

A: Test -1, $f'(-1) = 3(-1)^2 - 4(-1)^3 = 3 \cdot 1 - 4(-1) = 7 > 0$

B: Test $\dfrac{1}{2}$, $f'\left(\dfrac{1}{2}\right) = 3\left(\dfrac{1}{2}\right)^2 - 4\left(\dfrac{1}{2}\right)^3 = 3 \cdot \dfrac{1}{4} - 4 \cdot \dfrac{1}{8} = \dfrac{1}{4} > 0$

C: Test 1, $f'(1) = 3 \cdot 1^2 - 4 \cdot 1^3 = 3 \cdot 1 - 4 \cdot 1 = -1 < 0$

We see that $f(x)$ is increasing on $(-\infty, 0)$ and $\left(0, \dfrac{3}{4}\right)$ and is decreasing on $\left(\dfrac{3}{4}, \infty\right)$, so there is no relative extremum at $x = 0$ but there is a relative maximum at $x = \dfrac{3}{4}$. We find $f\left(\dfrac{3}{4}\right)$:

$$f\left(\dfrac{3}{4}\right) = \left(\dfrac{3}{4}\right)^3 - \left(\dfrac{3}{4}\right)^4 = \dfrac{27}{64} - \dfrac{81}{256} = \dfrac{27}{256}$$

There is a relative maximum at $\left(\dfrac{3}{4}, \dfrac{27}{256}\right)$. We use the information obtained to sketch the graph. Other function values are listed below.

x	$f(x)$
-2	-24
-1	-2
0	0
$\dfrac{1}{2}$	$\dfrac{1}{16}$
1	0
2	-8

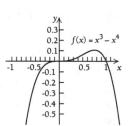

19. $f(x) = x^4 - 8x^2 + 3$

First, find the critical points.

$f'(x) = 4x^3 - 16x$

$f'(x)$ exists for all real numbers. We solve $f'(x) = 0$:

$$4x^3 - 16x = 0$$
$$x^3 - 4x = 0 \qquad \text{Dividing by 4}$$
$$x(x^2 - 4) = 0$$
$$x(x + 2)(x - 2) = 0$$

$x = 0$ or $x + 2 = 0$ or $x - 2 = 0$

$x = 0$ or $x = -2$ or $x = 2$

The critical points are -2, 0, and 2. We use them to divide the real number line into four intervals, A: $(-\infty, -2)$, B: $(-2, 0)$, C: $(0, 2)$, and D: $(2, \infty)$.

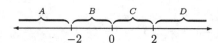

We use a test value in each interval to determine the sign of the derivative in each interval.

A: Test -3, $f'(-3) = 4(-3)^3 - 16(-3) = -108 + 48 = -60 < 0$

B: Test -1, $f'(-1) = 4(-1)^3 - 16(-1) = -4 + 16 = 12 > 0$

C: Test 1, $f'(1) = 4 \cdot 1^3 - 16 \cdot 1 = 4 - 16 = -12 < 0$

D: Test 3, $f'(3) = 4 \cdot 3^3 - 16 \cdot 3 = 108 - 48 = 60 > 0$

We see that $f(x)$ is decreasing on $(-\infty, -2)$, increasing on $(-2, 0)$, decreasing again on $(0, 2)$, and increasing again on $(2, \infty)$. Thus, there is a relative minimum at $x = -2$, a relative maximum at $x = 0$, and another relative minimum at $x = 2$.

We find $f(-2)$:

$$f(-2) = (-2)^4 - 8(-2)^2 + 3 = 16 - 32 + 3 = -13$$

Then we find $f(0)$:

$$f(0) = 0^4 - 8 \cdot 0^2 + 3 = 0 - 0 + 3 = 3$$

Finally, we find $f(2)$:

$$f(2) = 2^4 - 8 \cdot 2^2 + 3 = 16 - 32 + 3 = -13$$

There are relative minima at $(-2, -13)$ and $(2, -13)$, and there is a relative maximum at $(0, 3)$. We use the information obtained to sketch the graph. Other function values are listed below.

x	$f(x)$
-3	12
-1	-4
1	-4
3	12

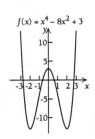

21. $f(x) = 1 - x^{2/3}$

First, find the critical points.

$$f'(x) = -\frac{2}{3}x^{-1/3} = -\frac{2}{3\sqrt[3]{x}}$$

$f'(x)$ does not exist for $x = 0$. The equation $f'(x) = 0$ has no solution, so the only critical point is 0. We use it to divide the real number line into two intervals: A: $(-\infty, 0)$ and B: $(0, \infty)$.

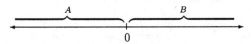

We use a test value in each interval to determine the sign of the derivative in each interval.

A: Test -1, $f'(-1) = -\dfrac{2}{3\sqrt[3]{-1}} = -\dfrac{2}{3(-1)} = \dfrac{2}{3} > 0$

B: Test 1, $f'(1) = -\dfrac{2}{3\sqrt[3]{1}} = -\dfrac{2}{3 \cdot 1} = -\dfrac{2}{3} < 0$

We see that $f(x)$ is increasing on $(-\infty, 0)$ and decreasing on $(0, \infty)$, so there is a relative maximum at $x = 0$.

We find $f(0)$:

$$f(0) = 1 - 0^{2/3} = 1 - 0 = 1$$

There is a relative maximum at $(0, 1)$. We use the information obtained to sketch the graph. Other function values are listed below.

x	$f(x)$
-27	-8
-8	-3
-1	0
1	0
8	-3
27	8

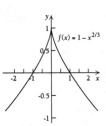

23. $f(x) = \dfrac{-8}{x^2 + 1} = -8(x^2 + 1)^{-1}$

First, find the critical points.

$$f'(x) = -8(-1)(x^2 + 1)^{-2}(2x)$$
$$= 16x(x^2 + 1)^{-2}$$
$$= \frac{16x}{(x^2 + 1)^2}$$

$f'(x)$ exists for all real numbers. We solve $f'(x) = 0$:

$$\frac{16x}{(x^2 + 1)^2} = 0$$
$$16x = 0 \quad \text{Multiplying by } (x^2 + 1)^2$$
$$x = 0$$

The only critical point is 0. We use it to divide the real number line into two intervals, A: $(-\infty, 0)$ and B: $(0, \infty)$.

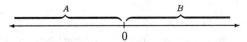

We use a test value in each interval to determine the sign of the derivative in each interval.

A: Test -1, $f'(-1) = \dfrac{16(-1)}{[(-1)^2 + 1]^2} = \dfrac{-16}{4} = -4 < 0$

B: Test 1, $f'(1) = \dfrac{16 \cdot 1}{(1^2 + 1)^2} = \dfrac{16}{4} = 4 > 0$

We see that $f(x)$ is decreasing on $(-\infty, 0)$ and increasing on $(0, \infty)$, so there is a relative minimum at $x = 0$.

We find $f(0)$:

$$f(0) = \frac{-8}{0^2 + 1} = \frac{-8}{1} = -8$$

There is a relative minimum at $(0, -8)$. We use the information obtained to sketch the graph. Other function values are listed below.

x	$f(x)$
-4	$-\dfrac{8}{17}$
-3	$-\dfrac{4}{5}$
-2	$-\dfrac{8}{5}$
-1	-4
1	-4
2	$-\dfrac{8}{5}$
3	$-\dfrac{4}{5}$
4	$-\dfrac{8}{17}$

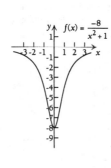

25. $f(x) = \dfrac{4x}{x^2 + 1}$

First, find the critical points.

$$f'(x) = \frac{(x^2+1)(4) - 2x(4x)}{(x^2+1)^2} \quad \text{Quotient Rule}$$

$$= \frac{4x^2 + 4 - 8x^2}{(x^2+1)^2}$$

$$= \frac{4 - 4x^2}{(x^2+1)^2}$$

$f'(x)$ exists for all real numbers. We solve $f'(x) = 0$:

$$\frac{4 - 4x^2}{(x^2+1)^2} = 0$$

$$4 - 4x^2 = 0 \quad \text{Multiplying by } (x^2+1)^2$$

$$1 - x^2 = 0 \quad \text{Dividing by 4}$$

$$(1 - x)(1 + x) = 0$$

$$1 - x = 0 \quad \text{or} \quad 1 + x = 0$$

$$1 = x \quad \text{or} \qquad x = -1$$

The critical points are -1 and 1. We use them to divide the real number line into three intervals, A: $(-\infty, -1)$, B: $(-1, 1)$, and C: $(1, \infty)$.

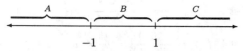

We use a test value in each interval to determine the sign of the derivative in each interval.

A: Test -2, $f'(-2) = \dfrac{4 - 4(-2)^2}{[(-2)^2 + 1]^2} = \dfrac{-12}{25} < 0$

B: Test 0, $f'(0) = \dfrac{4 - 4 \cdot 0^2}{(0^2 + 1)^2} = \dfrac{4}{1} = 4 > 0$

C: Test 2, $f'(2) = \dfrac{4 - 4 \cdot 2^2}{(2^2 + 1)^2} = \dfrac{-12}{25} < 0$

We see that $f(x)$ is decreasing on $(-\infty, -1)$, increasing on $(-1, 1)$, and decreasing again on $(1, \infty)$, so there is a relative minimum at $x = -1$ and a relative maximum at $x = 1$.

We find $f(-1)$:

$$f(-1) = \frac{4(-1)}{(-1)^2 + 1} = \frac{-4}{2} = -2$$

Then we find $f(1)$:

$$f(1) = \frac{4 \cdot 1}{1^2 + 1} = \frac{4}{2} = 2$$

There is a relative minimum at $(-1, -2)$, and there is a relative maximum at $(1, 2)$. We use the information obtained to sketch the graph. Other function values are listed below.

x	$f(x)$
-4	$-\dfrac{16}{17}$
-3	$-\dfrac{6}{5}$
-2	$-\dfrac{8}{5}$
0	0
2	$\dfrac{8}{5}$
3	$\dfrac{6}{5}$
4	$\dfrac{16}{17}$

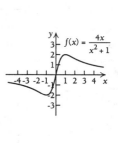

27. $f(x) = \sqrt[3]{x} = x^{1/3}$

First, find the critical points.

$$f'(x) = \frac{1}{3}x^{-2/3} = \frac{1}{3\sqrt[3]{x^2}}$$

$f'(x)$ does not exist for $x = 0$. The equation $f'(x) = 0$ has no solution, so the only critical point is 0. We use it to divide the real number line into two intervals, A: $(-\infty, 0)$, and B: $(0, \infty)$.

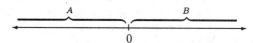

We use a test value in each interval to determine the sign of the derivative in each interval.

A: Test -1, $f'(-1) = \dfrac{1}{3\sqrt[3]{(-1)^2}} = \dfrac{1}{3 \cdot 1} = \dfrac{1}{3} > 0$

B: Test 1, $f'(1) = \dfrac{1}{3\sqrt[3]{1^2}} = \dfrac{1}{3 \cdot 1} = \dfrac{1}{3} > 0$

We see that $f(x)$ is increasing on both intervals, so the function has no relative extrema. We use the information obtained to sketch the graph. Some function values are listed below.

x	$f(x)$
-27	-3
-8	-2
-1	-1
0	0
1	1
8	2
27	3

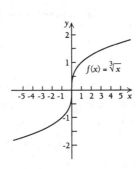

29. - 55.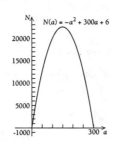

57. $N(a) = -a^2 + 300a + 6$, $0 \le a \le 300$

First, find the critical points.

$N'(a) = -2a + 300$

$N'(a)$ exists for all real numbers. We solve $N'(a) = 0$:

$$-2a + 300 = 0$$
$$-2a = -300$$
$$a = 150$$

The only critical point is 150. We use it to divide the interval $[0, 300]$ (the domain of $N(a)$) into two intervals, A: $[0, 150)$ and B: $(150, 300]$.

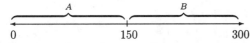

We use a test value in each interval to determine the sign of the derivative in each interval.

A: Test 0, $N'(0) = -2 \cdot 0 + 300 = 300 > 0$

B: Test 151, $N'(151) = -2 \cdot 151 + 300 = -2 < 0$

We see that $N(a)$ is increasing on $[0, 150)$ and decreasing on $(150, 300]$, so there is a relative maximum at $x = 150$. We find $N(150)$:

$$N(150) = -(150)^2 + 300(150) + 6$$
$$= -22,500 + 45,000 + 6$$
$$= 22,506$$

There is a relative maximum at $(150, 22, 506)$. We use the information obtained to sketch the graph. Other function values are listed below.

x	$f(x)$
0	6
50	$12,506$
100	$20,006$
200	$20,006$
250	$12,506$
300	6

59. $h = -0.002d^2 + 0.8d + 6.6$

Note that we consider only nonnegative values of d since the horizontal distance cannot be negative. In addition, we consider only those nonnegative values of d for which h is nonnegative since the height of the arrow cannot be negative. Thus, the domain of this function is $[0, 408.1]$. We find the critical points.

$h' = -0.004d + 0.8$

h' exists for all real numbers. We solve $h' = 0$:

$$-0.004d + 0.8 = 0$$
$$-0.004d = -0.8$$
$$d = 200$$

The only critical point is 200. We use it to divide the interval $[0, 408.1]$ into two intervals, A: $[0, 200)$ and B: $(200, 408.1]$.

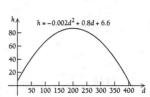

We use a test value in each interval to determine the sign of the derivative in each interval.

A: Test 100, $h' = -0.004(100) + 0.8 = 0.4 > 0$

B: Test 300, $h' = -0.004(300) + 0.8 = -0.4 < 0$

We see that h is increasing on $[0, 200)$ and decreasing on $(200, 408.1]$, so there is a relative maximum at $x = 200$. We find the value of h when $x = 200$.

$$h = -0.002(200)^2 + 0.8(200) + 6.6$$
$$= -80 + 160 + 6.6$$
$$= 86.6$$

There is a relative maximum at $(200, 86.6)$. We use this information to sketch the graph. Other function values are listed below.

d	h
0	6.6
100	66.6
300	66.6
400	6.6

61. $\boxed{tw}$

63. See the answer section in the text.

65. See the answer section in the text.

67. $\boxed{tw}$

Exercise Set 3.2

1. $f(x) = 2 - x^2$

a) Find $f'(x)$ and $f''(x)$.

$$f'(x) = -2x$$
$$f''(x) = -2$$

b) Find the critical points of f.

Since $f'(x)$ exists for all values of x, the only critical points are where $-2x = 0$.

$$-2x = 0$$
$$x = 0 \qquad \text{Critical point}$$

Find the function value at $x = 0$.

$$f(0) = 2 - 0^2 = 2$$

This gives the point $(0, 2)$ on the graph.

c) Use the Second-Derivative Test:

$$f''(x) = -2$$
$$f''(0) = -2 < 0$$

This tells us that $(0, 2)$ is a relative maximum. Then we can deduce that $f(x)$ is increasing on $(-\infty, 0)$ and decreasing on $(0, \infty)$.

d) Find possible inflection points.

The second derivative, $f''(x)$, exists and is -2 for all real numbers. Note that $f''(x)$ is never 0. Thus, there are no possible inflection points.

e) Since $f''(x)$ is always negative $(f''(x) = -2)$, f is concave down on the interval $(-\infty, \infty)$.

f) Sketch the graph using the preceding information. By solving $2 - x^2 = 0$ we can easily find the x-intercepts. They are $(-\sqrt{2}, 0)$ and $(\sqrt{2}, 0)$. Other function values can also be calculated.

x	$f(x)$
-2	-2
-1	1
1	1
2	-2

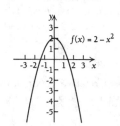

3. $f(x) = x^2 + x - 1$

a) Find $f'(x)$ and $f''(x)$.

$$f'(x) = 2x + 1$$
$$f''(x) = 2$$

b) Find the critical points of f.

Since $f'(x) = 2x + 1$ exists for all values of x, the only critical points are where $2x + 1 = 0$.

$$2x + 1 = 0$$
$$2x = -1$$
$$x = -\frac{1}{2} \qquad \text{Critical point}$$

Find the function value at $x = -\frac{1}{2}$.

$$f\left(-\frac{1}{2}\right) = \left(-\frac{1}{2}\right)^2 + \left(-\frac{1}{2}\right) - 1$$
$$= \frac{1}{4} - \frac{2}{4} - \frac{4}{4}$$
$$= -\frac{5}{4}$$

This gives the point $\left(-\frac{1}{2}, -\frac{5}{4}\right)$ on the graph.

c) Use the Second-Derivative Test:

$$f''(x) = 2$$
$$f''\left(-\frac{1}{2}\right) = 2 > 0$$

This tells us that $\left(-\frac{1}{2}, -\frac{5}{4}\right)$ is a relative minimum. Then we can deduce that $f(x)$ is decreasing on $\left(-\infty, \frac{1}{2}\right)$ and increasing on $\left(-\frac{1}{2}, \infty\right)$.

d) Find the possible inflection points. The second derivative, $f''(x)$, exists and is 2 for all real numbers. Note that $f''(x)$ is never 0. Thus, there are no possible inflection points.

e) Note that $f''(x)$ is always positive, $f''(x) = 2$. Thus, f is concave up on the interval $(-\infty, \infty)$.

f) Sketch the graph using the preceding information. Other function values can be calculated.

x	$f(x)$
-3	5
-2	1
-1	-1
0	-1
1	1
2	5

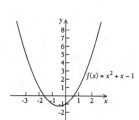

5. $f(x) = -4x^2 + 3x - 1$

a) Find $f'(x)$ and $f''(x)$.

$$f'(x) = -8x + 3$$
$$f''(x) = -8$$

b) Find the critical points of f.

Since $f'(x)$ exists for all values of x, the only critical points are where $-8x + 3 = 0$.

$$-8x + 3 = 0$$
$$-8x = -3$$
$$x = \frac{3}{8} \qquad \text{Critical point}$$

Then $f\left(\frac{3}{8}\right) = -4\left(\frac{3}{8}\right)^2 + 3\left(\frac{3}{8}\right) - 1$

$$= -\frac{9}{16} + \frac{9}{8} - 1$$
$$= -\frac{7}{16}.$$

This gives the point $\left(\frac{3}{8}, -\frac{7}{16}\right)$ on the graph.

c) Use the Second-Derivative Test:

$$f''(x) = -8$$
$$f''\left(\frac{3}{8}\right) = -8 < 0$$

This tells us that $\left(\frac{3}{8}, -\frac{7}{16}\right)$ is a relative maximum. Then we can deduce that $f(x)$ increasing on $\left(-\infty, \frac{3}{8}\right)$ and decreasing on $\left(\frac{3}{8}, \infty\right)$.

d) Find the possible inflection points. $f''(x)$ exists and is -8 for all real numbers. Note that $f''(x)$ is never 0. Thus, there are no possible inflection points.

e) Since $f''(x)$ is always negative, ($f''(x) = -8$), f is concave down on $(-\infty, \infty)$.

f) Sketch the graph using the preceding information. Other function values can also be calculated.

x	$f(x)$
-1	-8
0	-1
1	-2
2	-11

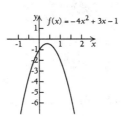

7. $f(x) = 2x^3 - 3x^2 - 36x + 28$

a) Find $f'(x)$ and $f''(x)$.

$f'(x) = 6x^2 - 6x - 36$

$f''(x) = 12x - 6$

b) Find the critical points of f.

Since $f'(x)$ exists for all values of x, the only critical points are where $6x^2 - 6x - 36 = 0$.

$6x^2 - 6x - 36 = 0$

$6(x + 2)(x - 3) = 0$

$x + 2 = 0$ or $x - 3 = 0$

$x = -2$ or $x = 3$ Critical points

Then $f(-2) = 2(-2)^3 - 3(-2)^2 - 36(-2) + 28$

$= -16 - 12 + 72 + 28$

$= 72,$

and $f(3) = 2 \cdot 3^3 - 3 \cdot 3^2 - 36 \cdot 3 + 28$

$= 54 - 27 - 108 + 28$

$= -53.$

These give the points $(-2, 72)$ and $(3, -53)$ on the graph.

c) Use the Second-Derivative Test:

$f''(-2) = 12(-2) - 6 = -30 < 0$, so $(-2, 72)$ is a relative maximum.

$f''(3) = 12 \cdot 3 - 6 = 30 > 0$, so $(3, -53)$ is a relative minimum.

Then if we use the points -2 and 3 to divide the real number line into three intervals, $(-\infty, -2)$, $(-2, 3)$, and $(3, \infty)$, we know that f is increasing on $(-\infty, -2)$, decreasing on $(-2, 3)$, and increasing again on $(3, \infty)$.

d) Find the possible inflection points.

$f''(x)$ exists for all valus of x, so we solve $f''(x) = 0$.

$12x - 6 = 0$

$12x = 6$

$x = \dfrac{1}{2}$ Possible inflection point

Then $f\left(\dfrac{1}{2}\right) = 2\left(\dfrac{1}{2}\right)^3 - 3\left(\dfrac{1}{2}\right)^2 - 36\left(\dfrac{1}{2}\right) + 28$

$= \dfrac{1}{4} - \dfrac{3}{4} - 18 + 28$

$= \dfrac{19}{2}.$

This gives the point $\left(\dfrac{1}{2}, \dfrac{19}{2}\right)$ on the graph.

e) To determine the concavity we use the possible inflection point, $\dfrac{1}{2}$, to divide the real number line into two invervals, A: $\left(-\infty, \dfrac{1}{2}\right)$ and B: $\left(\dfrac{1}{2}, \infty\right)$. Test a point in each interval.

A: Test 0, $f''(0) = 12 \cdot 0 - 6 = -6 < 0$

B: Test 1, $f''(1) = 12 \cdot 1 - 6 = 6 > 0$

Then f is concave down on $\left(-\infty, \dfrac{1}{2}\right)$ and concave up on $\left(\dfrac{1}{2}, \infty\right)$, so $\left(\dfrac{1}{2}, \dfrac{19}{2}\right)$ is an inflection point.

f) Sketch the graph using the preceding information. Other function values can also be calculated.

x	$f(x)$
-3	55
-1	59
0	28
1	-9
2	-40
4	-36

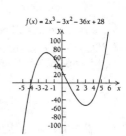

9. $f(x) = \dfrac{8}{3}x^3 - 2x + \dfrac{1}{3}$

a) Find $f'(x)$ and $f''(x)$.

$f'(x) = 8x^2 - 2$

$f''(x) = 16x$

b) Find the critical points of f.

Now $f'(x) = 8x^2 - 2$ exists for all values of x, so the only critical points of f are where $8x^2 - 2 = 0$.

$8x^2 - 2 = 0$

$8x^2 = 2$

$x^2 = \dfrac{2}{8}$

$x^2 = \dfrac{1}{4}$

$x = \pm\dfrac{1}{2}$ Critical points

Then $f\left(-\dfrac{1}{2}\right) = \dfrac{8}{3}\left(-\dfrac{1}{2}\right)^3 - 2\left(-\dfrac{1}{2}\right) + \dfrac{1}{3}$

$= -\dfrac{1}{3} + 1 + \dfrac{1}{3}$

$= 1,$

and $\quad f\left(\frac{1}{2}\right) = \frac{8}{3}\left(\frac{1}{2}\right)^3 - 2\left(\frac{1}{2}\right) + \frac{1}{3}$

$$= \frac{1}{3} - 1 + \frac{1}{3}$$

$$= -\frac{1}{3}$$

These give the points $\left(-\frac{1}{2}, 1\right)$ and $\left(\frac{1}{2}, -\frac{1}{3}\right)$ on the graph.

c) Use the Second-Derivative Test:

$f''\left(-\frac{1}{2}\right) = 16\left(-\frac{1}{2}\right) - 8 < 0$, so $\left(-\frac{1}{2}, 1\right)$ is a relative maximum.

$f''\left(\frac{1}{2}\right) = 16 \cdot \frac{1}{2} = 8 > 0$, so $\left(\frac{1}{2}, -\frac{1}{3}\right)$ is a relative minimum.

Then if we use the points $-\frac{1}{2}$ and $\frac{1}{2}$ to divide the real number line into three intervals, A: $\left(-\infty, -\frac{1}{2}\right)$, B: $\left(-\frac{1}{2}, \frac{1}{2}\right)$, and C: $\left(\frac{1}{2}, \infty\right)$, we know that f is increasing on $\left(-\infty, -\frac{1}{2}\right)$, decreasing on $\left(-\frac{1}{2}, \frac{1}{2}\right)$, and increasing again on $\left(\frac{1}{2}, \infty\right)$.

d) Find the possible inflection points.

Now $f''(x) = 16x$ exists for all values of x, so the only critical points of f' are where $16x = 0$.

$$16x = 0$$

$$x = 0 \quad \text{Possible inflection point}$$

Then $f(0) = \frac{8}{3} \cdot 0^3 - 2 \cdot 0 + \frac{1}{3} = \frac{1}{3}$, so $\left(0, \frac{1}{3}\right)$ is another point on the graph.

e) To determine the concavity we use the possible inflection point, 0, to divide the real number line into two invervals, A: $(-\infty, 0)$ and B: $(0, \infty)$. Test a point in each interval.

A: Test -1, $f''(-1) = 16(-1) = -16 < 0$

B: Test 1, $f''(1) = 16 \cdot 1 = 16 > 0$

Then $f'(x)$ is concave down on $(-\infty, 0)$ and concave up on $(0, \infty)$, so $\left(0, \frac{1}{3}\right)$ is an inflection point.

f) Sketch the graph using the preceding information. Other function values can also be calculated.

x	$f(x)$
-2	-17
-1	$-\frac{1}{3}$
0	$\frac{1}{3}$
1	1
2	$\frac{53}{3}$

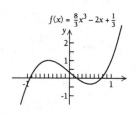

11. $f(x) = -x^3 + 3x^2 - 4$

a) Find $f'(x)$ and $f''(x)$.

$$f'(x) = -3x^2 + 6x$$

$$f''(x) = -6x + 6$$

b) Find the critical points of f.

Since $f'(x)$ exists for all values of x, the only critical points are where $-3x^2 + 6x = 0$.

$$-3x^2 + 6x = 0$$

$$-3x(x - 2) = 0$$

$$-3x = 0 \quad \text{or} \quad x - 2 = 0$$

$$x = 0 \quad \text{or} \quad x = 2 \quad \text{Critical points}$$

Then $f(0) = -0^3 + 3 \cdot 0^2 - 4 = -4$ and $f(2) = -2^3 + 3 \cdot 2^2 - 4 = -8 + 12 - 4 = 0$.

These give the points $(0, -4)$ and $(2, 0)$ on the graph.

c) Use the Second-Derivative Test:

$f''(0) = -6 \cdot 0 + 6 = 6 > 0$, so $(0, -4)$ is a relative minimum.

$f''(2) = -6 \cdot 2 + 6 = -12 + 6 = -6 < 0$, so $(2, 0)$ is a relative maximum.

Then if we use the points 0 and 2 to divide the real number line into three intervals, $(-\infty, 0)$, $(0, 2)$, and $(2, \infty)$, we know that f is decreasing on $(-\infty, 0)$, increasing on $(0, 2)$, and decreasing again on $(2, \infty)$.

d) Find the possible inflection points.

$f''(x)$ exists for all values of x, so we solve $f''(x) = 0$.

$$-6x + 6 = 0$$

$$-6x = -6$$

$$x = 1 \quad \text{Possible inflection point}$$

Then $f(1) = -1^3 + 3 \cdot 1^2 - 4 = -1 + 3 - 4 = -2$, so $(1, -2)$ is another point on the graph.

e) To determine the concavity we use the possible inflection point, 1, to divide the real number line into two invervals, A: $(-\infty, 1)$ and B: $(1, \infty)$. Test a point in each interval.

A: Test 0, $f''(0) = -6 \cdot 0 + 6 = 6 > 0$

B: Test 2, $f''(2) = -6 \cdot 2 + 6 = -6 < 0$

Then f is concave up on $(-\infty, 1)$ and concave down on $(1, \infty)$, so $(1, -2)$ is an inflection point.

f) Sketch the graph. Other function values can also be calculated.

x	$f(x)$
-2	16
-1	0
3	-4
4	-20

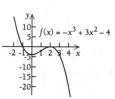

13. $f(x) = 3x^4 - 16x^3 + 18x^2$

a) $f'(x) = 12x^3 - 48x^2 + 36x$

$f''(x) = 36x^2 - 96x + 36$

b) Since $f'(x)$ exists for all values of x, the only critical points are where $f'(x) = 0$.

$$12x^3 - 48x^2 + 36x = 0$$
$$12x(x^2 - 4x + 3) = 0$$
$$12x(x - 1)(x - 3) = 0$$
$$12x = 0 \text{ or } x - 1 = 0 \text{ or } x - 3 = 0$$
$$x = 0 \text{ or } \quad x = 1 \text{ or } \quad x = 3$$

Then $f(0) = 3 \cdot 0^4 - 16 \cdot 0^3 + 18 \cdot 0^2 = 0$,
$$f(1) = 3 \cdot 1^4 - 16 \cdot 1^3 + 18 \cdot 1^2 = 5,$$
and $f(3) = 3 \cdot 3^4 - 16 \cdot 3^3 + 18 \cdot 3^2 = -27$.

These give the points $(0,0)$, $(1,5)$, and $(3,-27)$ on the graph.

c) Use the Second-Derivative Test:

$f''(0) = 36 \cdot 0^2 - 96 \cdot 0 + 36 = 36 > 0$, so $(0,0)$ is a relative minimum.

$f''(1) = 36 \cdot 1^2 - 96 \cdot 1 + 36 = -24 < 0$, so $(1,5)$ is a relative maximum.

$f''(3) = 36 \cdot 3^2 - 96 \cdot 3 + 36 = 72 > 0$, so $(3,-27)$ is a relative minimum.

Then if we use the points 0, 1, and 3 to divide the real number line into four intervals, $(-\infty, 0)$, $(0,1)$, $(1,3)$, and $(3,\infty)$, we know that f is decreasing on $(-\infty, 0)$ and on $(1,3)$ and is increasing on $(0,1)$ and $(3,\infty)$.

d) $f''(x)$ exists for all values of x, so the only possible inflection points are where $f''(x) = 0$.

$$36x^2 - 96x + 36 = 0$$
$$12(3x^2 - 8x + 3) = 0$$
$$3x^2 - 8x + 3 = 0$$

Using the quadratic formula, we find $x = \dfrac{4 \pm \sqrt{7}}{3}$, so $x \approx 0.45$ or $x \approx 2.22$ are possible inflection points.

Then $f(0.45) \approx 2.31$ and $f(2.22) \approx -13.48$, so $(0.45, 2.31)$ and $(2.22, -13.48)$ are two more points on the graph.

e) To determine the concavity we use the points 0.45 and 2.22 to divide the real number line into three invervals, A: $(-\infty, 0.45)$, B: $(0.45, 2.22)$, and C: $(2.22, \infty)$. Test a point in each interval.

A: Test 0, $f''(0) = 36 \cdot 0^2 - 96 \cdot 0 + 36 = 36 > 0$

B: Test 1, $f''(1) = 36 \cdot 1^2 - 96 \cdot 1 + 36 = $
$-24 < 0$

C: Test 3, $f''(3) = 36 \cdot 3^2 - 96 \cdot 3 + 36 = 72 > 0$

Then f is concave up on $(-\infty, 0.45)$, concave down on $(0.45, 2.22)$, and concave up on $(2.22, \infty)$, so $(0.45, 2.31)$ and $(2.22, -13.48)$ are inflection points.

f) Sketch the graph. Other function values can also be calculated.

x	$f(x)$
-1	37
2	-8
4	32

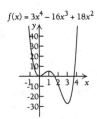

$f(x) = 3x^4 - 16x^3 + 18x^2$

15. $f(x) = (x + 1)^{2/3}$

a) $f'(x) = \dfrac{2}{3}(x+1)^{-1/3} = \dfrac{2}{3\sqrt[3]{x+1}}$

$f''(x) = -\dfrac{2}{9}(x+1)^{-4/3} = -\dfrac{2}{9\sqrt[3]{(x+1)^4}}$

b) Since $f'(-1)$ does not exist, -1 is a critical point. The equation $f'(x) = 0$ has no solution, so the only critical point is -1.

Now $f(-1) = (-1+1)^{2/3} = 0^{2/3} = 0$.

This gives the point $(-1, 0)$ on the graph.

c) We cannot use the Second-Derivative Test, because $f''(-1)$ is not defined. We will use the First-Derivative Test. Use -1 to divide the real number line into two intervals, A: $(-\infty, -1)$ and B: $(-1, \infty)$. Test a point in each interval.

A: Test -2, $f'(-2) = \dfrac{2}{3\sqrt[3]{-2+1}} = -\dfrac{2}{3} < 0$

B: Test 0, $f'(0) = \dfrac{2}{3\sqrt[3]{0+1}} = \dfrac{2}{3} > 0$

Since $f(x)$ is decreasing on $(-\infty, -1)$ and increasing on $(-1, \infty)$, there is a relative minimum at $(-1, 0)$.

d) Since $f''(-1)$ does not exist, -1 is a possible inflection point. The equation $f''(x) = 0$ has no solution, so the only possible inflection point is -1. We have already found $f(-1)$ in step (b).

e) To determine the concavity we use -1 to divide the real number line into two invervals as in step (c). Test a point in each interval.

A: Test -2, $f''(-2) = -\dfrac{2}{9\sqrt[3]{(-2+1)^4}} = -\dfrac{2}{9} < 0$

B: Test 0, $f''(0) = -\dfrac{2}{9\sqrt[3]{(0+1)^4}} = -\dfrac{2}{9} < 0$

Then f is concave down on both intervals, so there is no inflection point.

f) Sketch the graph using the preceding information. Other function values can also be calculated.

x	$f(x)$
-9	4
-2	1
0	1
7	4

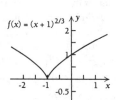

$f(x) = (x + 1)^{2/3}$

17. $f(x) = x^4 - 6x^2$

a) $f'(x) = 4x^3 - 12x$

$f''(x) = 12x^2 - 12$

b) Since $f'(x)$ exists for all values of x, the only critical points are where $4x^3 - 12x = 0$.

$$4x^3 - 12x = 0$$
$$4x(x^2 - 3) = 0$$
$$4x = 0 \text{ or } x^2 - 3 = 0$$
$$x = 0 \text{ or } \qquad x^2 = 3$$
$$x = 0 \text{ or } \qquad x = \pm\sqrt{3}$$

The critical points are $-\sqrt{3}$, 0, and $\sqrt{3}$.

$$f(-\sqrt{3}) = (-\sqrt{3})^4 - 6(-\sqrt{3})^2$$
$$= 9 - 6 \cdot 3 = 9 - 18 = -9$$
$$f(0) = 0^4 - 6 \cdot 0^2 = 0 - 0 = 0$$
$$f(\sqrt{3}) = (\sqrt{3})^4 - 6(\sqrt{3})^2$$
$$= 9 - 6 \cdot 3 = 9 - 18 = -9$$

These give the points $(-\sqrt{3}, -9)$, $(0, 0)$, and $(\sqrt{3}, -9)$ on the graph.

c) Use the Second-Derivative Test:

$$f''(-\sqrt{3}) = 12(-\sqrt{3})^2 - 12 = 12 \cdot 3 - 12 =$$

$24 > 0$, so $(-\sqrt{3}, -9)$ is a relative minimum.

$f''(0) = 12 \cdot 0^2 - 12 = -12 < 0$, so $(0,0)$ is a relative maximum.

$f''(\sqrt{3}) = 12(\sqrt{3})^2 - 12 = 12 \cdot 3 - 12 = 24 > 0$,

so $(\sqrt{3}, -9)$ is a relative minimum.

Then if we use the points $-\sqrt{3}$, 0, and $\sqrt{3}$ to divide the real number line into four intervals, $(-\infty, -\sqrt{3})$, $(-\sqrt{3}, 0)$, $(0, \sqrt{3})$, and $(\sqrt{3}, \infty)$, we know that f is decreasing on $(-\infty, -\sqrt{3})$ and on $(0, \sqrt{3})$ and is increasing on $(-\sqrt{3}, 0)$ and on $(\sqrt{3}, \infty)$.

d) Since $f''(x)$ exists for all values of x, the only possible inflection points are where $12x^2 - 12 = 0$.

$$12x^2 - 12 = 0$$
$$x^2 - 1 = 0$$
$$(x+1)(x-1) = 0$$
$$x + 1 = 0 \text{ or } x - 1 = 0$$
$$x = -1 \text{ or } \qquad x = 1$$

The possible inflection points are -1 and 1.

$$f(-1) = (-1)^4 - 6(-1)^2 = 1 - 6 \cdot 1 = -5$$
$$f(1) = 1^4 - 6 \cdot 1^2 = 1 - 6 \cdot 1 = -5$$

These give the points $(-1, -5)$ and $(1, -5)$ on the graph.

e) To determine the concavity we use the points -1 and 1 to divide the real number line into three invervals, A: $(-\infty, -1)$, B: $(-1, 1)$, and $(1, \infty)$. Test a point in each interval.

A: Test -2, $f''(-2) = 12(-2)^2 - 12 = 36 > 0$

B: Test 0, $f''(0) = 12 \cdot 0^2 - 12 = -12 < 0$

C: Test 2, $f''(2) = 12 \cdot 2^2 - 12 = 36 > 0$

We see that f is concave up on the intervals $(-\infty, -1)$ and $(1, \infty)$ and concave down on the interval $(-1, 1)$, so $(-1, -5)$ and $(1, -5)$ are inflection points.

f) Sketch the graph using the preceding information. By solving $x^4 - 6x^2 = 0$ we can find the x-intercepts. They are helpful in graphing.

$$x^4 - 6x^2 = 0$$
$$x^2(x^2 - 6) = 0$$
$$x^2 = 0 \text{ or } x^2 - 6 = 0$$
$$x = 0 \text{ or } \qquad x^2 = 6$$
$$x = 0 \text{ or } \qquad x = \pm\sqrt{6}$$

The x-intercepts are $(0, 0)$, $(-\sqrt{6}, 0)$, and $(\sqrt{6}, 0)$.

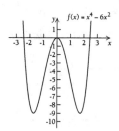

19. $f(x) = x^3 - 2x^2 - 4x + 3$

a) $f'(x) = 3x^2 - 4x - 4$

$f''(x) = 6x - 4$

b) Now $f'(x)$ exists for all values of x, so the only critical points of f are where $3x^2 - 4x - 4 = 0$.

$$3x^2 - 4x - 4 = 0$$
$$(3x + 2)(x - 2) = 0$$
$$3x + 2 = 0 \text{ or } x - 2 = 0$$
$$x = -\frac{2}{3} \text{ or } \qquad x = 2$$

The critical points are $-\frac{2}{3}$ and 2.

$$f\left(-\frac{2}{3}\right) = \left(-\frac{2}{3}\right)^3 - 2\left(-\frac{2}{3}\right)^2 - 4\left(-\frac{2}{3}\right) + 3$$
$$= -\frac{8}{27} - \frac{8}{9} + \frac{8}{3} + 3$$
$$= -\frac{8}{27} - \frac{24}{27} + \frac{72}{27} + \frac{81}{27}$$
$$= \frac{121}{27}$$
$$f(2) = 2^3 - 2 \cdot 2^2 - 4 \cdot 2 + 3$$
$$= 8 - 8 - 8 + 3$$
$$= -5$$

These give the points $\left(-\frac{2}{3}, \frac{121}{27}\right)$ and $(2, -5)$ on the graph.

c) Use the Second-Derivative Test:

$f''\left(-\frac{2}{3}\right) = 6\left(-\frac{2}{3}\right) - 4 = -4 - 4 = -8 < 0$, so

$\left(-\frac{2}{3}, \frac{121}{27}\right)$ is a relative maximum.

$f''(2) = 6 \cdot 2 - 4 = 12 - 4 = 8 > 0$, so $(2, -5)$ is a relative minimum.

Then if we use the points $-\dfrac{2}{3}$ and 2 to divide the real number line into three intervals, $\left(-\infty, -\dfrac{2}{3}\right)$, $\left(-\dfrac{2}{3}, 2\right)$, and $(2, \infty)$, we know that f is increasing on $\left(-\infty, -\dfrac{2}{3}\right)$ and on $(2, \infty)$ and is decreasing on $\left(-\dfrac{2}{3}, 2\right)$.

d) Now $f''(x)$ exists for all values of x, so the only possible inflection points are where $6x - 4 = 0$.

$$6x - 4 = 0$$
$$6x = 4$$
$$x = \frac{4}{6}$$
$$x = \frac{2}{3} \quad \text{Possible inflection point}$$

$$f\left(\frac{2}{3}\right) = \left(\frac{2}{3}\right)^3 - 2\left(\frac{2}{3}\right)^2 - 4\left(\frac{2}{3}\right) + 3$$
$$= \frac{8}{27} - \frac{8}{9} - \frac{8}{3} + 3$$
$$= \frac{8}{27} - \frac{24}{27} - \frac{72}{27} + \frac{81}{27}$$
$$= -\frac{7}{27}$$

This gives another point $\left(\dfrac{2}{3}, -\dfrac{7}{27}\right)$ on the graph.

e) To determine the concavity we use $\dfrac{2}{3}$ to divide the real number line into two invervals, A: $\left(-\infty, \dfrac{2}{3}\right)$ and B: $\left(\dfrac{2}{3}, \infty\right)$. Test a point in each interval.

A: Test 0, $f''(0) = 6 \cdot 0 - 4 = -4 < 0$

B: Test 1, $f''(1) = 6 \cdot 1 - 4 = 2 > 0$

We see that f is concave down on $\left(-\infty, \dfrac{2}{3}\right)$ and concave up on $\left(\dfrac{2}{3}, \infty\right)$, so $\left(\dfrac{2}{3}, -\dfrac{7}{27}\right)$ is an inflection point.

f) Sketch the graph using the preceding information. Other function values can also be calculated.

x	$f(x)$
-2	-5
-1	4
0	3
1	-2
3	0
4	19

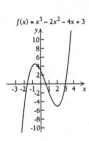

$f(x) = x^3 - 2x^2 - 4x + 3$

21. $f(x) = 3x^4 + 4x^3$

a) $f'(x) = 12x^3 + 12x^2$

$f''(x) = 36x^2 + 24x$

b) Since $f'(x)$ exists for all values of x, the only critical points of f are where $12x^3 + 12x^2 = 0$.

$$12x^3 + 12x^2 = 0$$
$$12x^2(x + 1) = 0$$
$$12x^2 = 0 \quad \text{or} \quad x + 1 = 0$$
$$x = 0 \quad \text{or} \quad x = -1$$

The critical points are 0 and -1.

$$f(0) = 3 \cdot 0^4 + 4 \cdot 0^3 = 0 + 0 = 0$$
$$f(-1) = 3(-1)^4 + 4(-1)^3 = 3 \cdot 1 + 4(-1)$$
$$= 3 - 4$$
$$= -1$$

These give the points $(0,0)$ and $(-1, -1)$ on the graph.

c) Use the Second-Derivative Test:

$f''(-1) = 36(-1)^2 + 24(-1) = 36 - 24 = 12 > 0$, so $(-1, -1)$ is a relative minimum.

$f''(0) = 36 \cdot 0^2 + 24 \cdot 0 = 0$, so this test fails. We will use the First-Derivative Test. Use 0 to divide the interval $(-1, \infty)$ into two intervals: A: $(-1, 0)$ and B: $(0, \infty)$. Test a point in each interval.

A: Test $-\dfrac{1}{2}$, $f'\left(-\dfrac{1}{2}\right) = 12\left(-\dfrac{1}{12}\right)^3 + 12\left(\dfrac{1}{2}\right)^2 =$

$\dfrac{3}{2} > 0$

B: Test 2, $f'(2) = 12 \cdot 2^3 + 12 \cdot 2^2 = 144 > 0$

Since f is increasing on both intervals, $(0,0)$ is not a relative extremum. Since $(-1, -1)$ is a relative minimum, we know that f is decreasing on $(-\infty, -1)$.

d) Now $f''(x)$ exists for all values of x, so the only possible inflection points are where $36x^2 + 24x = 0$.

$$36x^2 + 24x = 0$$
$$12x(3x + 2) = 0$$
$$12x = 0 \quad \text{or} \quad 3x + 2 = 0$$
$$x = 0 \quad \text{or} \quad x = -\frac{2}{3}$$

The possible inflection points are 0 and $-\dfrac{2}{3}$.

$$f(0) = 3 \cdot 0^4 + 4 \cdot 0^3 = 0 \quad \text{Already found in step (b)}$$

$$f\left(-\frac{2}{3}\right) = 3\left(-\frac{2}{3}\right)^4 + 4\left(-\frac{2}{3}\right)^3$$
$$= 3 \cdot \frac{16}{81} + 4 \cdot \left(-\frac{8}{27}\right)$$
$$= \frac{16}{27} - \frac{32}{27}$$
$$= -\frac{16}{27}$$

This gives one additional point $\left(-\dfrac{2}{3}, -\dfrac{16}{27}\right)$ on the graph.

e) To determine the concavity we use $-\dfrac{2}{3}$ and 0 to divide the real number line into three invervals, A: $\left(-\infty, -\dfrac{2}{3}\right)$, B: $\left(-\dfrac{2}{3}, 0\right)$, and C: $(0, \infty)$. Test a point in each interval.

A: Test -1, $f''(-1) = 36(-1)^2 + 24(-1) =$
$12 > 0$

D: Test $-\dfrac{1}{2}$, $f''\left(-\dfrac{1}{2}\right) = 36\left(-\dfrac{1}{2}\right)^2 + 24\left(-\dfrac{1}{2}\right) -$
$-3 < 0$

C: Test 1, $f''(1) = 36 \cdot 1^2 + 24 \cdot 1 = 60 > 0$

We see that f is concave up on the intervals $\left(-\infty, -\dfrac{2}{3}\right)$ and $(0, \infty)$ and concave down on the interval $\left(-\dfrac{2}{3}, 0\right)$, so $\left(-\dfrac{2}{3}, -\dfrac{16}{27}\right)$ and $(0, 0)$ are both inflection points.

f) Sketch the graph using the preceding information. By solving $3x^4 + 4x^3 = 0$ we can find x-intercepts. They are helpful in graphing.

$$3x^4 + 4x^3 = 0$$
$$x^3(3x + 4) = 0$$
$$x^3 = 0 \quad \text{or} \quad 3x + 4 = 0$$
$$x = 0 \quad \text{or} \qquad x = -\dfrac{4}{3}$$

The intercepts are $(0, 0)$ and $\left(-\dfrac{4}{3}, 0\right)$.

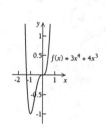

23. $f(x) = x^3 - 6x^2 - 135x$

a) $f'(x) = 3x^2 - 12x - 135$
$f''(x) = 6x - 12$

b) Since $f'(x)$ exists for all values of x, the only critical points of f are where $3x^2 - 12x - 135 = 0$.

$$3x^2 - 12x - 135 = 0$$
$$x^2 - 4x - 45 = 0$$
$$(x - 9)(x + 5) = 0$$
$$x - 9 = 0 \quad \text{or} \quad x + 5 = 0$$
$$x = 9 \quad \text{or} \qquad x = -5$$

The critical points are 9 and -5.

$$f(9) = 9^3 - 6 \cdot 9^2 - 135 \cdot 9$$
$$= 729 - 486 - 1215$$
$$= -972$$
$$f(-5) = (-5)^3 - 6(-5)^2 - 135(-5)$$
$$= -125 - 150 + 675$$
$$= 400$$

These give the points $(9, -972)$ and $(-5, 400)$ on the graph.

c) Use the Second-Derivative Test:

$f''(-5) = 6(-5) - 12 = -30 - 12 = -42 < 0$, so $(-5, 400)$ is a relative maximum.

$f''(9) = 6 \cdot 9 - 12 = 54 - 12 = 42 > 0$ so $(9, -972)$ is a relative minimum.

Then if we use the points -5 and 9 to divide the real number line into three intervals, $(-\infty, -5)$, $(-5, 9)$, and $(9, \infty)$, we know that f is increasing on $(-\infty, -5)$ and on $(9, \infty)$ and is decreasing on $(-5, 9)$.

d) Now $f''(x)$ exists for all values of x, so the only possible inflection points are where $6x - 12 = 0$.

$$6x - 12 = 0$$
$$6x = 12$$
$$x = 2 \quad \text{Possible inflection point}$$
$$f(2) = 2^3 - 6 \cdot 2^2 - 135 \cdot 2$$
$$= 8 - 24 - 270$$
$$= -286$$

This gives another point $(2, -286)$ on the graph.

e) To determine the concavity we use 2 to divide the real number line into two invervals, A: $(-\infty, 2)$ and B: $(2, \infty)$. Test a point in each interval.

A: Test 0, $f''(0) = 6 \cdot 0 - 12 = -12 < 0$
B: Test 3, $f''(3) = 6 \cdot 3 - 12 = 6 > 0$

We see that f is concave down on $(-\infty, 2)$ and concave up on $(2, \infty)$, so $(2, -286)$ is an inflection point.

f) Sketch the graph using the preceding information. Other function values can also be calculated.

x	$f(x)$
-11	-572
-10	-250
-9	0
-3	324
0	0
2	-286
5	-700
15	0
16	400

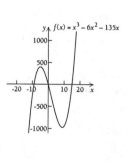

25. $f(x) = \dfrac{x}{x^2 + 1}$

a) $f'(x) = \dfrac{(x^2 + 1)(1) - 2x \cdot x}{(x^2 + 1)^2}$ \quad Quotient Rule

$$= \dfrac{x^2 + 1 - 2x^2}{(x^2 + 1)^2}$$
$$= \dfrac{1 - x^2}{(x^2 + 1)^2}$$

$$f''(x) = \frac{(x^2+1)^2(-2x) - 2(x^2+1)(2x)(1-x^2)}{[(x^2+1)^2]^2}$$

Quotient Rule

$$= \frac{(x^2+1)[(x^2+1)(-2x) - 4x(1-x^2)]}{(x^2+1)^4}$$

$$= \frac{-2x^3 - 2x - 4x + 4x^3}{(x^2+1)^3}$$

$$= \frac{2x^3 - 6x}{(x^2+1)^3}$$

b) Since $f'(x)$ exists for all real numbers, the only critical points are where $f'(x) = 0$.

$$\frac{1-x^2}{(x^2+1)^2} = 0$$

$$1 - x^2 = 0 \quad \text{Multiplying by } (x^2+1)^2$$

$$(1+x)(1-x) = 0$$

$$1 + x = 0 \quad \text{or} \quad 1 - x = 0$$

$$x = -1 \quad \text{or} \quad 1 = x \quad \text{Critical points}$$

Then $f(-1) = \dfrac{-1}{(-1)^2+1} = -\dfrac{1}{2}$ and $f(1) =$

$\dfrac{1}{1^2+1} = \dfrac{1}{2}$, so $\left(-1, -\dfrac{1}{2}\right)$ and $\left(1, \dfrac{1}{2}\right)$ are on the graph.

c) Use the Second-Derivative Test:

$$f''(-1) = \frac{2(-1)^3 - 6(-1)}{[(-1)^2+1]^3}$$

$$= \frac{-2+6}{2^3}$$

$$= \frac{4}{8} = \frac{1}{2} > 0, \text{ so } \left(-1, -\frac{1}{2}\right) \text{ is a}$$

relative minimum.

$$f''(1) = \frac{2 \cdot 1^3 - 6 \cdot 1}{[(1)^2+1]^3}$$

$$= \frac{2-6}{2^3}$$

$$= \frac{-4}{8} = -\frac{1}{2} < 0, \text{ so } \left(1, \frac{1}{2}\right) \text{ is a}$$

relative maximum.

Then if we use -1 and 1 to divide the real number line into three intervals, $(-\infty, -1)$, $(-1, 1)$, and $(1, \infty)$, we know that f is decreasing on $(-\infty, -1)$ and on $(1, \infty)$ and is increasing on $(-1, 1)$.

d) $f''(x)$ exists for all real numbers, so the only possible inflection points are where $f''(x) = 0$.

$$\frac{2x^3 - 6x}{(x^2+1)^3} = 0$$

$$2x(x^2 - 3) = 0$$

$$2x = 0 \quad \text{or} \quad x^2 - 3 = 0$$

$$x = 0 \quad \text{or} \quad x^2 = 3$$

$$x = 0 \quad \text{or} \quad x = \pm\sqrt{3}$$

Possible inflection points

$$f(-\sqrt{3}) = \frac{-\sqrt{3}}{(-\sqrt{3})^2+1} = -\frac{\sqrt{3}}{4}$$

$$f(0) = \frac{0}{0^2+1} = 0$$

$$f(\sqrt{3}) = \frac{\sqrt{3}}{(\sqrt{3})^2+1} = \frac{\sqrt{3}}{4}$$

These give the points $\left(-\sqrt{3}, -\dfrac{\sqrt{3}}{4}\right)$, $(0,0)$, and $\left(\sqrt{3}, \dfrac{\sqrt{3}}{4}\right)$ on the graph.

e) To determine the concavity we use $-\sqrt{3}$, 0, and $\sqrt{3}$ to divide the real number line into four intervals, A: $(-\infty, -\sqrt{3})$, B: $(-\sqrt{3}, 0)$, C: $(0, \sqrt{3})$, and D: $(\sqrt{3}, \infty)$. Test a point in each interval.

A: Test -2, $f''(-2) = \dfrac{-4}{125} < 0$

B: Test -1, $f''(-1) = \dfrac{1}{2} > 0$

C: Test 1, $f''(1) = \dfrac{-1}{2} < 0$

D: Test 2, $f''(2) = \dfrac{4}{125} > 0$

Then f is concave down on $(-\infty, -\sqrt{3})$ and on $(0, \sqrt{3})$ and is concave up on $(-\sqrt{3}, 0)$ and on $(\sqrt{3}, \infty)$, so $\left(-\sqrt{3}, -\dfrac{\sqrt{3}}{4}\right)$, $(0,0)$, and $\left(\sqrt{3}, \dfrac{\sqrt{3}}{4}\right)$ are all inflection points.

f) Sketch the graph using the preceding information. Other function values can also be calculated.

x	$f(x)$
-3	$-\dfrac{3}{10}$
-2	$-\dfrac{2}{5}$
2	$\dfrac{2}{5}$
3	$\dfrac{3}{10}$

$f(x) = \dfrac{x}{x^2+1}$

27. $f(x) = \dfrac{3}{x^2+1} = 3(x^2+1)^{-1}$

a) $f'(x) = 3(-1)(x^2+1)^{-2}(2x)$

$$= -6x(x^2+1)^{-2}, \text{ or } \frac{-6x}{(x^2+1)^2}$$

$$f''(x) = \frac{(x^2+1)^2(-6) - 2(x^2+1)(2x)(-6x)}{[(x^2+1)^2]^2}$$

$$= \frac{(x^2+1)[(x^2+1)(-6) - 2(2x)(-6x)]}{(x^2+1)^4}$$

$$= \frac{-6x^2 - 6 + 24x^2}{(x^2+1)^3}$$

$$= \frac{18x^2 - 6}{(x^2+1)^3}$$

b) Since $f'(x)$ exists for all real numbers, the only critical points are where $f'(x) = 0$.

$$\frac{-6x}{(x^2+1)^2} = 0$$

$$-6x = 0 \quad \text{Multiplying by } (x^2+1)^2$$

$$x = 0 \quad \text{Critical point}$$

Then $f(0) = \dfrac{3}{0^2+1} = 3$, so $(0,3)$ is on the graph.

c) Use the Second-Derivative Test:

$$f''(0) = \frac{18 \cdot 0^2 - 6}{(0^2+1)^3} = -6 < 0, \text{ so } (0,3) \text{ is a}$$

relative maximum.

Then if we use 0 to divide the real number line into two intervals, $(-\infty, 0)$ and $(0, \infty)$, we know that f is increasing on $(-\infty, 0)$ and decreasing on $(0, \infty)$.

d) $f''(x)$ exists for all real numbers, so the only possible inflection points are where $f''(x) = 0$.

$$\frac{18x^2 - 6}{(x^2+1)^3} = 0$$

$$18x^2 - 6 = 0 \qquad \text{Multiplying by } (x^2+1)^3$$

$$18x^2 = 6$$

$$x^2 = \frac{1}{3}$$

$$x = \pm\frac{1}{\sqrt{3}} \quad \begin{array}{l}\text{Possible inflection}\\ \text{points}\end{array}$$

$$f\left(-\frac{1}{\sqrt{3}}\right) = \frac{3}{\left(-\frac{1}{\sqrt{3}}\right)^2 + 1} = \frac{3}{\frac{1}{3}+1} = \frac{3}{\frac{4}{3}} =$$

$$\frac{3}{1} \cdot \frac{3}{4} = \frac{9}{4}$$

$$f\left(\frac{1}{\sqrt{3}}\right) = \frac{3}{\left(\frac{1}{\sqrt{3}}\right)^2 + 1} = \frac{3}{\frac{1}{3}+1} = \frac{3}{\frac{4}{3}} =$$

$$\frac{3}{1} \cdot \frac{3}{4} = \frac{9}{4}$$

These give the points $\left(-\dfrac{1}{\sqrt{3}}, \dfrac{9}{4}\right)$ and $\left(\dfrac{1}{\sqrt{3}}, \dfrac{9}{4}\right)$ on the graph.

e) To determine the concavity we use $-\dfrac{1}{\sqrt{3}}$ and

$\dfrac{1}{\sqrt{3}}$ to divide the real number line into three intervals, A: $\left(-\infty, -\dfrac{1}{\sqrt{3}}\right)$, B: $\left(-\dfrac{1}{\sqrt{3}}, \dfrac{1}{\sqrt{3}}\right)$, and

C: $\left(\dfrac{1}{\sqrt{3}}, \infty\right)$. Test a point in each interval.

A: Test -1, $f''(-1) = \dfrac{18(-1)^2 - 6}{[(-1)^2+1]^3} = \dfrac{12}{8} = \dfrac{3}{2} > 0$

C: Test 0, $f''(0) = \dfrac{18 \cdot 0^2 - 6}{(0^2+1)^3} = -6 < 0$

D: Test 1, $f''(1) = \dfrac{18 \cdot 1^2 - 6}{(1^2+1)^3} = \dfrac{12}{8} = \dfrac{3}{2} > 0$

Then f is concave up on $\left(-\infty, -\dfrac{1}{\sqrt{3}}\right)$, and on

$\left(\dfrac{1}{\sqrt{3}}, \infty\right)$ and is concave down on $\left(-\dfrac{1}{\sqrt{3}}, \dfrac{1}{\sqrt{3}}\right)$, so

$\left(-\dfrac{1}{\sqrt{3}}, \dfrac{9}{4}\right)$ and $\left(\dfrac{1}{\sqrt{3}}, \dfrac{9}{4}\right)$ are both inflection points.

f) Sketch the graph using the preceding information. Other function values can also be calculated.

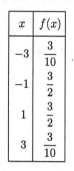

x	$f(x)$
-3	$\dfrac{3}{10}$
-1	$\dfrac{3}{2}$
1	$\dfrac{3}{2}$
3	$\dfrac{3}{10}$

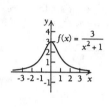

29. $f(x) = (x-1)^3$

a) $f'(x) = 3(x-1)^2(1) = 3(x-1)^2$

$f''(x) = 3 \cdot 2(x-1)(1) = 6(x-1)$

b) Since $f'(x)$ exists for all real numbers, the only critical points are where $f'(x) = 0$.

$$3(x-1)^2 = 0$$

$$(x-1)^2 = 0$$

$$x - 1 = 0$$

$$x = 1 \quad \text{Critical point}$$

Then $f(1) = (1-1)^3 = 0^3 = 0$, so $(1,0)$ is on the graph.

c) The Second-Derivative Test fails since $f''(1) = 0$, so we use the First-Derivative Test. Use 1 to divide the real number line into two intervals, A: $(-\infty, 1)$ and B: $(1, \infty)$. Test a point in each interval.

A: Test 0, $f'(0) = 3(0-1)^2 = 3 > 0$

B: Test 2, $f'(2) = 3(2-1)^2 = 3 > 0$

Since f is increasing on both intervals, $(1,0)$ is not a relative extremum.

d) $f''(x)$ exists for all real numbers, so the only possible inflection points are where $f''(x) = 0$.

$$6(x-1) = 0$$

$$x - 1 = 0$$

$$x = 1 \quad \text{Possible inflection point}$$

From step (b), we know that $(1,0)$ is on the graph.

e) To determine the concavity we use 1 to divide the real number line as in step (c).

A: Test 0, $f''(0) = 6(0-1) = -6 < 0$

B: Test 2, $f''(2) = 6(2-1) = 6 > 0$

Then f is concave down on $(-\infty, 1)$ and concave up on $(1, \infty)$, so $(1,0)$ is an inflection point.

f) Sketch the graph using the preceding information. Other function values can also be calculated.

x	$f(x)$
-3	-64
-2	-27
-1	-8
0	-1
2	1
3	8
4	27

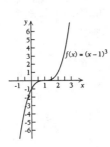

$f(x) = (x-1)^3$

31. $f(x) = x^2(1-x)^2$
$$= x^2(1 - 2x + x^2)$$
$$= x^2 - 2x^3 + x^4$$

a) $f'(x) = 2x - 6x^2 + 4x^3$

$f''(x) = 2 - 12x + 12x^2$

b) Since $f'(x)$ exists for all real numbers, the only critical points are where $f'(x) = 0$.
$$2x - 6x^2 + 4x^3 = 0$$
$$2x(1 - 3x + 2x^2) = 0$$
$$2x(1-x)(1-2x) = 0$$
$$2x = 0 \text{ or } 1 - x = 0 \text{ or } 1 - 2x = 0$$
$$x = 0 \text{ or } 1 = x \text{ or } 1 = 2x$$
$$x = 0 \text{ or } 1 = x \text{ or } \frac{1}{2} = x$$
$$\text{Critical points}$$

$$f(0) = 0^2(1-0)^2 = 0$$
$$f(1) = 1^2(1-1)^2 = 0$$
$$f\left(\frac{1}{2}\right) = \left(\frac{1}{2}\right)^2\left(1 - \frac{1}{2}\right)^2 = \frac{1}{4} \cdot \frac{1}{4} = \frac{1}{16}$$

Thus, $(0,0)$, $(1,0)$, and $\left(\frac{1}{2}, \frac{1}{16}\right)$ are on the graph.

c) Use the Second-Derivative Test:

$f''(0) = 2 - 12 \cdot 0 + 12 \cdot 0^2 = 2 > 0$, so $(0,0)$ is a relative minimum.

$f''\left(\frac{1}{2}\right) = 2 - 12 \cdot \frac{1}{2} + 12\left(\frac{1}{2}\right)^2 = 2 - 6 + 3 =$

$-1 < 0$, so $\left(\frac{1}{2}, \frac{1}{16}\right)$ is a relative maximum.

$f''(1) = 2 - 12 \cdot 1 + 12 \cdot 1^2 = 2 > 0$, so $(1,0)$ is a relative minimum.

Then if we use the points 0, $\frac{1}{2}$, and 1 to divide the real number line into four intervals, $(-\infty, 0)$, $\left(0, \frac{1}{2}\right)$, $\left(\frac{1}{2}, 1\right)$ and $(1, \infty)$, we know that f is decreasing on $(-\infty, 0)$ and on $\left(\frac{1}{2}, 1\right)$ and is increasing on $\left(0, \frac{1}{2}\right)$ and on $(1, \infty)$.

d) $f''(x)$ exists for all real numbers, so the only possible inflection points are where $f''(x) = 0$.
$$2 - 12x + 12x^2 = 0$$
$$2(1 - 6x + 6x^2) = 0$$

Using the quadratic formula we find
$$x = \frac{3 \pm \sqrt{3}}{6}$$

$x \approx 0.21$ or $x \approx 0.79$ Possible inflection points

$f(0.21) \approx 0.03$ and $f(0.79) \approx 0.03$, so $(0.21, 0.03)$ and $(0.79, 0.03)$ are on the graph.

e) To determine the concavity we use 0.21 and 0.79 to divide the real number line into three intervals, A: $(-\infty, 0.21)$, B: $(0.21, 0.79)$, and C: $(0.79, \infty)$.

A: Test 0, $f''(0) = 2 - 12 \cdot 0 + 12 \cdot 0^2 = 2 > 0$

B: Test 0.5, $f''(0.5) = 2 - 12(0.5) + 12(0.5)^2 = -1 < 0$

C: Test 1, $f''(1) = 2 - 12 \cdot 1 + 12 \cdot 1^2 = 2 > 0$

Then f is concave up on $(-\infty, 0.21)$ and on $(0.79, \infty)$ and is concave down on $(0.21, 0.79)$, so $(0.21, 0.03)$ and $(0.79, 0.03)$ are both inflection points.

f) Sketch the graph using the preceding information. Other function values can also be calculated.

x	$f(x)$
-2	36
-1	4
2	4
3	36

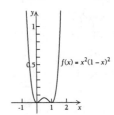

$f(x) = x^2(1-x)^2$

33. $f(x) = 20x^3 - 3x^5$

a) $f'(x) = 60x^2 - 15x^4$

$f''(x) = 120x - 60x^3$

b) Since $f'(x)$ exists for all real numbers, the only critical points are where $f'(x) = 0$.
$$60x^2 - 15x^4 = 0$$
$$15x^2(4 - x^2) = 0$$
$$15x^2(2 + x)(2 - x) = 0$$
$$15x^2 = 0 \text{ or } 2 + x = 0 \text{ or } 2 - x = 0$$
$$x = 0 \text{ or } x = -2 \text{ or } 2 = x$$
$$\text{Critical points}$$

$$f(0) = 20 \cdot 0^3 - 3 \cdot 0^5 = 0$$
$$f(-2) = 20(-2)^3 - 3(-2)^5 = -160 + 96 = -64$$
$$f(2) = 20 \cdot 2^3 - 3 \cdot 2^5 = 160 - 96 = 64$$

Thus, $(0,0)$, $(-2, -64)$, and $(2, 64)$ are on the graph.

c) Use the Second-Derivative Test:

$f''(-2) = 120(-2) - 60(-2)^3 = -240 + 480 = 240 > 0$, so $(-2, -64)$ is a relative minimum.

$f''(2) = 120 \cdot 2 - 60 \cdot 2^3 = 240 - 480 = -240 < 0$, so $(2, 64)$ is a relative maximum.

$f''(0) = 120 \cdot 0 - 60 \cdot 0^3 = 0$, so we will use the First-Derivative Test on $x = 0$. Use 0 to divide the interval $(-2, 2)$ into two intervals, A: $(-2, 0)$, and B: $(0, 2)$. Test a point in each interval.

A: Test -1, $f'(-1) = 60(-1)^2 - 15(-1)^4 =$

$60 - 15 = 45 > 0$

B: Test 1, $f'(1) = 60 \cdot 1^2 - 15 \cdot 1^4 = 60 - 15 = 45 > 0$

Then f is increasing on both intervals, so $(0, 0)$ is not a relative extremum.

If we use the points -2 and 2 to divide the real number line into three intervals, $(-\infty, -2)$, $(-2, 2)$, and $(2, \infty)$, we know that f is decreasing on $(-\infty, -2)$ and on $(2, \infty)$ and is increasing on $(-2, 2)$.

d) $f''(x)$ exists for all real numbers, so the only possible inflection points are where $f''(x) = 0$.

$$120x - 60x^3 = 0$$
$$60x(2 - x^2) = 0$$
$$60x = 0 \quad \text{or} \quad 2 - x^2 = 0$$
$$x = 0 \quad \text{or} \quad 2 = x^2$$
$$x = 0 \quad \text{or} \quad \pm\sqrt{2} = x \quad \text{Possible inflection}$$
$$\text{points}$$

$f(-\sqrt{2}) = 20(-\sqrt{2})^3 - 3(-\sqrt{2})^5 =$
$\qquad -40\sqrt{2} + 12\sqrt{2} = -28\sqrt{2}$

$f(0) = 0$ from step (b)

$f(\sqrt{2}) = 20(\sqrt{2})^3 - 3(\sqrt{2})^5 =$
$\qquad 40\sqrt{2} - 12\sqrt{2} = 28\sqrt{2}$

Thus, $(-\sqrt{2}, -28\sqrt{2})$ and $(\sqrt{2}, 28\sqrt{2})$ are also on the graph.

e) To determine the concavity we use $-\sqrt{2}$, 0, and $\sqrt{2}$ to divide the real number line into four intervals, A: $(-\infty, -\sqrt{2})$, B: $(-\sqrt{2}, 0)$, C: $(0, \sqrt{2})$, and D: $(\sqrt{2}, \infty)$.

A: Test -2, $f''(-2) = 120(-2) - 60(-2)^3 =$

$240 > 0$

B: Test -1, $f''(-1) = 120(-1) - 60(-1)^3 =$

$-60 < 0$

C: Test 1, $f''(1) = 120 \cdot 1 - 60 \cdot 1^3 = 60 > 0$

D: Test 2, $f''(2) = 120 \cdot 2 - 60 \cdot 2^3 = -240 < 0$

Then f is concave up on $(-\infty, -\sqrt{2})$ and on $(0, \sqrt{2})$ and is concave down on $(-\sqrt{2}, 0)$ and on $(\sqrt{2}, \infty)$, so $(-\sqrt{2}, -28\sqrt{2})$, $(0, 0)$, and $(\sqrt{2}, 28\sqrt{2})$ are all inflection points.

f) Sketch the graph using the preceding information. Other function values can also be calculated.

x	$f(x)$
-3	189
-2	-64
-1	-17
1	17
2	64
3	-189

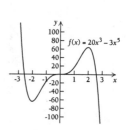

35. $f(x) = x\sqrt{4 - x^2} = x(4 - x^2)^{1/2}$

a) $f'(x) = x \cdot \dfrac{1}{2}(4 - x^2)^{-1/2}(-2x) + 1 \cdot (4 - x^2)^{1/2}$

$\qquad = \dfrac{-x^2}{\sqrt{4 - x^2}} + \sqrt{4 - x^2}$

$\qquad = \dfrac{-x^2 + 4 - x^2}{\sqrt{4 - x^2}}$

$\qquad = \dfrac{4 - 2x^2}{\sqrt{4 - x^2}}$, or $(4 - 2x^2)(4 - x^2)^{-1/2}$

$f''(x) = (4 - 2x^2)\left(-\dfrac{1}{2}\right)(4 - x^2)^{-3/2}(-2x) +$

$\qquad (-4x)(4 - x^2)^{-1/2}$

$\qquad = \dfrac{x(4 - 2x^2)}{(4 - x^2)^{3/2}} - \dfrac{4x}{(4 - x^2)^{1/2}}$

$\qquad = \dfrac{x(4 - 2x^2) - 4x(4 - x^2)}{(4 - x^2)^{3/2}}$

$\qquad = \dfrac{4x - 2x^3 - 16x + 4x^3}{(4 - x^2)^{3/2}}$

$\qquad = \dfrac{2x^3 - 12x}{(4 - x^2)^{3/2}}$

b) $f'(x)$ does not exist where $4 - x^2 = 0$. Solve:

$$4 - x^2 = 0$$
$$(2 + x)(2 - x) = 0$$
$$2 + x = 0 \quad \text{or} \quad 2 - x = 0$$
$$x = -2 \quad \text{or} \quad 2 = x$$

Note that $f(x)$ is not defined for $x < -2$ or $x > 2$. (For these values $4 - x^2 < 0$.) Therefore, relative extrema canot occur at $x = -2$ or $x = 2$, because there is no open interval containing -2 or 2 on which the function is defined. For this reason, we do not consider -2 and 2 further in our discussion of relative extrema.

Critical points occur where $f'(x) = 0$. Solve:

$$\dfrac{4 - 2x^2}{\sqrt{4 - x^2}} = 0$$
$$4 - 2x^2 = 0$$
$$4 = 2x^2$$
$$2 = x^2$$
$$\pm\sqrt{2} = x \quad \text{Critical points}$$

$$f(-\sqrt{2}) = -\sqrt{2}\sqrt{4 - (-\sqrt{2})^2} = -\sqrt{2} \cdot \sqrt{2} = -2$$

$$f(\sqrt{2}) = \sqrt{2}\sqrt{4 - (\sqrt{2})^2} = \sqrt{2} \cdot \sqrt{2} = 2$$

Then $(-\sqrt{2}, -2)$ and $(\sqrt{2}, 2)$ are on the graph.

c) Use the Second-Derivative Test:

$$f''(-\sqrt{2}) = \frac{2(-\sqrt{2})^3 - 12(-\sqrt{2})}{[4 - (-\sqrt{2})^2]^{3/2}} =$$

$$\frac{-4\sqrt{2} + 12\sqrt{2}}{2^{3/2}} = \frac{8\sqrt{2}}{2\sqrt{2}} = 4 > 0, \quad \text{so } (-\sqrt{2}, -2) \text{ is a}$$

relative minimum.

$$f''(\sqrt{2}) = \frac{2(\sqrt{2})^3 - 12(\sqrt{2})}{[4 - (\sqrt{2})^2]^{3/2}} =$$

$$\frac{4\sqrt{2} - 12\sqrt{2}}{2^{3/2}} = \frac{-8\sqrt{2}}{2\sqrt{2}} = -4 < 0, \quad \text{so } (\sqrt{2}, 2) \text{ is a rel-}$$

ative maximum.

If we use the points $-\sqrt{2}$ and $\sqrt{2}$ to divide the interval $(-2, 2)$ into three intervals, $(-2, -\sqrt{2})$, $(-\sqrt{2}, \sqrt{2})$, and $(\sqrt{2}, 2)$, we know that f is decreasing on $(-2, -\sqrt{2})$ and on $(\sqrt{2}, 2)$ and is increasing on $(-\sqrt{2}, \sqrt{2})$.

d) $f''(x)$ does not exist where $4 - x^2 = 0$. From step (b) we know that this occurs at $x = -2$ and at $x = 2$. However, just as relative extrema cannot occur at $(-2, 0)$ and $(2, 0)$, they cannot be inflection points either. Inflection points could occur where $f''(x) = 0$.

$$\frac{2x^3 - 12x}{(4 - x^2)^{3/2}} = 0$$

$$2x^3 - 12x = 0$$

$$2x(x^2 - 6) = 0$$

$$2x = 0 \quad \text{or} \quad x^2 - 6 = 0$$

$$x = 0 \quad \text{or} \quad x^2 = 6$$

$$x = 0 \quad \text{or} \quad x = \pm\sqrt{6}$$

Note that $f(x)$ is not defined for $x = \pm\sqrt{6}$. Therefore, the only possible inflection point is $x = 0$.

$$f(0) = 0\sqrt{4 - 0^2} = 0 \cdot 2 = 0$$

Then $(0, 0)$ is on the graph.

e) To determine the concavity we use 0 to divide the interval $(-2, 2)$ into two intervals, A: $(-2, 0)$ and B: $(0, 2)$.

A: Test -1, $f''(-1) = \dfrac{2(-1)^3 - 12(-1)}{[4 - (-1)^2]^{3/2}} =$

$$\frac{10}{3^{3/2}} > 0$$

B: Test 1, $f''(1) = \dfrac{2 \cdot 1^3 - 12 \cdot 1}{(4 - 1^2)^{3/2}} = \dfrac{-10}{3^{3/2}} < 0$

Then f is concave up on $(-2, 0)$ and concave down on $(0, 2)$, so $(0, 0)$ is an inflection point.

f) Sketch the graph using the preceding information. Other function values can also be calculated.

x	$f(x)$
$-\sqrt{3}$	$-\sqrt{3}$
-1	$-\sqrt{3}$
1	$\sqrt{3}$
$\sqrt{3}$	$\sqrt{3}$

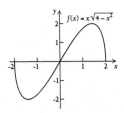

37. $f(x) = (x-1)^{1/3} - 1$

a) $f'(x) = \dfrac{1}{3}(x-1)^{-2/3}$, or $\dfrac{1}{3(x-1)^{2/3}}$

$$f''(x) = \frac{1}{3}\left(-\frac{2}{3}\right)(x-1)^{-5/3}$$

$$= -\frac{2}{9}(x-1)^{-5/3}, \text{ or } -\frac{2}{9(x-1)^{5/3}}$$

b) $f'(x)$ does not exist for $x = 1$. The equation $f'(x) = 0$ has no solution, so $x = 1$ is the only critical point. $f(1) = (1-1)^{1/3} - 1 = 0 - 1 = -1$, so $(1, -1)$ is on the graph.

c) Use the First-Derivative Test: Use 1 to divide the real number line into two intervals, A: $(-\infty, 1)$ and B: $(1, \infty)$. Test a point in each interval.

A: Test 0, $f'(0) = \dfrac{1}{3(0-1)^{2/3}} = \dfrac{1}{3 \cdot 1} = \dfrac{1}{3} > 0$

B: Test 2, $f'(2) = \dfrac{1}{3(2-1)^{2/3}} = \dfrac{1}{3 \cdot 1} = \dfrac{1}{3} > 0$

Then f is increasing on both intervals, so $(1, -1)$ is not a relative extremum.

d) $f''(x)$ does not exist for $x = 1$. The equation $f''(x) = 0$ has no solution, so $x = 1$ is the only possible inflection point. From step (b) we know $(1, -1)$ is on the graph.

e) To determine the concavity we use 1 to divide the real number line as in step (c). Test a point in each interval.

A: Test 0, $f''(0) = -\dfrac{2}{9(0-1)^{5/3}} = -\dfrac{2}{9(-1)} =$

$$\frac{2}{9} > 0$$

B: Test 2, $f''(2) = -\dfrac{2}{9(2-1)^{5/3}} = -\dfrac{2}{9 \cdot 1} =$

$$-\frac{2}{9} < 0$$

Then f is concave up on $(-\infty, 1)$ and concave down on $(1, \infty)$, so $(1, -1)$ is an inflection point.

f) Sketch the graph using the preceding information. Other function values can also be calculated.

x	$f(x)$
-7	-3
0	-2
2	0
9	1

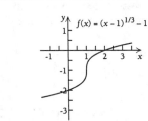

39. $f(x) = x^3 + 3x + 1$

$\quad f'(x) = 3x^2 + 3$

$\quad f''(x) = 6x$

$f''(x)$ exists for all values of x, so we solve $f''(x) = 0$.

$\quad 6x = 0$

$\quad\quad x = 0 \quad$ Possible inflection point

We use the possible inflection point, 0, to divide the real number line into two intervals, A: $(-\infty, 0)$ and B: $(0, \infty)$. Test a point in each interval.

A: Test -1, $f''(-1) = 6(-1) = -6 < 0$

B: Test 1, $f''(1) = 6 \cdot 1 = 6 > 0$

Then f is concave down on $(-\infty, 0)$ and concave up on $(0, \infty)$. We find that $f(0) = 0^3 + 3 \cdot 0 + 1 = 1$, so $(0, 1)$ is an inflection point.

41. $f(x) = \dfrac{4}{3}x^3 - 2x^2 + x$

$\quad f'(x) = 4x^2 - 4x + 1$

$\quad f''(x) = 8x - 4$

$f''(x)$ exists for all values of x, so we solve $f''(x) = 0$.

$\quad 8x - 4 = 0$

$\quad\quad 8x = 4$

$\quad\quad\quad x = \dfrac{1}{2} \quad$ Possible inflection point

We use the possible inflection point, $\dfrac{1}{2}$, to divide the real number line into two intervals, A: $\left(-\infty, \dfrac{1}{2}\right)$ and B: $\left(\dfrac{1}{2}, \infty\right)$. Test a point in each interval.

A: Test 0, $f''(0) = 8 \cdot 0 - 4 = -4 < 0$

B: Test 1, $f''(1) = 8 \cdot 1 - 4 = 4 > 0$

Then f is concave down on $\left(-\infty, \dfrac{1}{2}\right)$ and concave up on $\left(\dfrac{1}{2}, \infty\right)$. We find that $f\left(\dfrac{1}{2}\right) = \dfrac{4}{3}\left(\dfrac{1}{2}\right)^3 - 2\left(\dfrac{1}{2}\right)^2 + \dfrac{1}{2} = \dfrac{1}{6}$, so $\left(\dfrac{1}{2}, \dfrac{1}{6}\right)$ is an inflection point.

43. - 83.

85. $R(x) = 50x - 0.5x^2$

$\quad C(x) = 4x + 10$

$\quad P(x) = R(x) - C(x)$

$\quad\quad = (50x - 0.5x^2) - (4x + 10)$

$\quad\quad = -0.5x^2 + 46x - 10$

We will restrict the domains of all three functions to $x \geq 0$ since a negative number of units cannot be produced and sold.

First graph $R(x) = 50x - 0.5x^2$

a) $\quad R'(x) = 50 - x$

$\quad\quad R''(x) = -1$

b) Since $R'(x)$ exists for all $x \geq 0$, the only critical points are where $50 - x = 0$.

$50 - x = 0$

$\quad 50 = x \quad$ Critical point

Find the function value at $x = 50$.

$\quad R(50) = 50 \cdot 50 - 0.5(50)^2$

$\quad\quad = 2500 - 1250$

$\quad\quad = 1250$

This gives the point $(50, 1250)$ on the graph.

c) Use the Second-Derivative Test:

$R''(50) = -1 < 0$, so $(50, 1250)$ is a relative maximum.

Then if we use 50 to divide the interval $(0, \infty)$ into two intervals, $(0, 50)$ and $(50, \infty)$, we know that R is increasing on $(0, 50)$ and decreasing on $(50, \infty)$.

d) Since $R''(x)$ exists for all $x \geq 0$, and is always negative $(R''(x) = -1)$, the equation $R''(x) = 0$ has no solution. Thus there are no posssible inflection points.

e) Since $R''(x) < 0$ for all $x \geq 0$, R is concave down on the interval $(0, \infty)$.

f) Sketch the graph using the preceding information. The x-intercepts of R are easily found by solving $50x - 0.5x^2 = 0$.

$\quad 50x - 0.5x^2 = 0$

$\quad 500x - 5x^2 = 0$

$\quad 5x(100 - x) = 0$

$\quad 5x = 0 \quad$ or $\quad 100 - x = 0$

$\quad\quad x = 0 \quad$ or $\quad\quad 100 = x$

The x-intercepts are $(0, 0)$ and $(100, 0)$.

Next we graph $C(x) = 4x + 10$. This is a linear function with slope 4 and y-intercept $(0, 10)$.

Finally we graph $P(x) = -0.5x^2 + 46x - 10$.

a) $\quad P'(x) = -x + 46$

$\quad\quad P''(x) = -1$

b) Since $P'(x)$ exists for all $x \geq 0$, the only critical points are where $-x + 46 = 0$.

$\quad -x + 46 = 0$

$\quad\quad 46 = x \quad$ Critical point

Find the function value at $x = 46$.

$\quad P(46) = -0.5(46)^2 + 46 \cdot 46 - 10$

$\quad\quad = -1058 + 2116 - 10$

$\quad\quad = 1048$

This gives the point $(46, 1048)$ on the graph.

c) Use the Second-Derivative Test:

$P''(46) = -1 < 0$, so $(46, 1048)$ is a relative maximum.

Then if we use 46 to divide the interval $(0, \infty)$ into two intervals, $(0, 46)$ and $(46, \infty)$, we know that P is increasing on $(0, 46)$ and decreasing on $(46, \infty)$.

d) Since $P''(x)$ exists for all $x \geq 0$, and is always negative $(P''(x) = -1)$, the equation $P''(x) = 0$ has no solution. Thus there are no possible inflection points.

e) Since $P''(x) < 0$ for all $x \geq 0$, P is concave down on the interval $(0, \infty)$.

f) Sketch the graph using the preceding information.

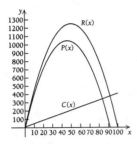

87. $V(r) = k(20r^2 - r^3),\ 0 \leq r \leq 20$

$V'(r) = k(40r - 3r^2)$

$V''(r) = k(40 - 6r)$

$V'(r)$ exists for all r in $[0, 20]$, so the only critical points occur where $V'(r) = 0$.

$k(40r - 3r^2) = 0$

$40r - 3r^2 = 0$

$r(40 - 3r) = 0$

$r = 0\ $ or $\ 40 - 3r = 0$

$r = 0\ $ or $\ \dfrac{40}{3} = r\quad$ Critical points

Use the Second-Derivative Test:

$V''(0) = k(40 - 6 \cdot 0) = 40k > 0$

$V''\left(\dfrac{40}{3}\right) = k\left(40 - 6 \cdot \dfrac{40}{3}\right) = k(40 - 80) =$

$-40k < 0$

Since $V''\left(\dfrac{40}{3}\right) < 0$, we know that there is a relative maximum at $x = \dfrac{40}{3}$. Thus the maximum velocity is required to remove an object whose radius is $\dfrac{40}{3}$, or $13\dfrac{1}{3}$ mm.

89. ☐ *tw*

91. ☐ *tw*

93. See the answer section in the text.

95. See the answer section in the text.

97. See the answer section in the text.

99. ☐ *tw*

Exercise Set 3.3

1. Find $\displaystyle\lim_{x \to \infty} \dfrac{2x - 4}{5x}$.

   We will use some algebra and the fact that as $x \to \infty$, $\dfrac{b}{ax^n} \to 0$, for any positive integer n.

$\displaystyle\lim_{x \to \infty} \dfrac{2x - 4}{5x}$

$= \displaystyle\lim_{x \to \infty} \dfrac{2x - 4}{5x} \cdot \dfrac{(1/x)}{(1/x)}\quad$ Multiplying by a form of 1

$= \displaystyle\lim_{x \to \infty} \dfrac{2x \cdot \dfrac{1}{x} - 4 \cdot \dfrac{1}{x}}{5x \cdot \dfrac{1}{x}}$

$= \displaystyle\lim_{x \to \infty} \dfrac{2 - \dfrac{4}{x}}{5}$

$= \dfrac{2 - 0}{5}\qquad$ As $x \to \infty$, $\dfrac{4}{x} \to 0$.

$= \dfrac{2}{5}$

3. Find $\displaystyle\lim_{x \to \infty} \left(5 - \dfrac{2}{x}\right)$.

   We will use the fact that as $x \to \infty$, $\dfrac{b}{ax^n} \to 0$, for any positive integer n.

$\displaystyle\lim_{x \to \infty} \left(5 - \dfrac{2}{x}\right)$

$= 5 - 0\qquad$ As $x \to \infty$, $\dfrac{2}{x} \to 0$.

$= 5$

5. Find $\displaystyle\lim_{x \to \infty} \dfrac{2x - 5}{4x + 3}$.

   We will use some algebra and the fact that as $x \to \infty$, $\dfrac{b}{ax^n} \to 0$, for any positive integer n.

$\displaystyle\lim_{x \to \infty} \dfrac{2x - 5}{4x + 3}$

$= \displaystyle\lim_{x \to \infty} \dfrac{2x - 5}{4x + 3} \cdot \dfrac{(1/x)}{(1/x)}\quad$ Multiplying by a form of 1

$= \displaystyle\lim_{x \to \infty} \dfrac{2x \cdot \dfrac{1}{x} - 5 \cdot \dfrac{1}{x}}{4x \cdot \dfrac{1}{x} + 3 \cdot \dfrac{1}{x}}$

$= \displaystyle\lim_{x \to \infty} \dfrac{2 - \dfrac{5}{x}}{4 + \dfrac{3}{x}}$

$= \dfrac{2 - 0}{4 + 0}\qquad$ As $x \to \infty$, $\dfrac{5}{x} \to 0$ and $\dfrac{3}{x} \to 0$.

$= \dfrac{2}{4} = \dfrac{1}{2}$

7. Find $\displaystyle\lim_{x \to \infty} \dfrac{2x^2 - 5}{3x^2 - x + 7}$.

   We will use some algebra and the fact that as $x \to \infty$, $\dfrac{b}{ax^n} \to 0$, for any positive integer n.

$$\lim_{x \to \infty} \frac{2x^2 - 5}{3x^2 - x + 7}$$

$$= \lim_{x \to \infty} \frac{2x^2 - 5}{3x^2 - x + 7} \cdot \frac{(1/x^2)}{(1/x^2)} \quad \text{Multiplying by a form of 1}$$

$$= \lim_{x \to \infty} \frac{2x^2 \cdot \dfrac{1}{x^2} - 5 \cdot \dfrac{1}{x^2}}{3x^2 \cdot \dfrac{1}{x^2} - x \cdot \dfrac{1}{x^2} + 7 \cdot \dfrac{1}{x^2}}$$

$$= \lim_{x \to \infty} \frac{2 - \dfrac{5}{x^2}}{3 - \dfrac{1}{x} + \dfrac{7}{x^2}}$$

$$= \frac{2 - 0}{3 - 0 + 0} \quad \text{As } x \to \infty, \; \frac{5}{x^2} \to 0, \; \frac{1}{x} \to 0,$$

$$\text{and } \frac{7}{x^2} \to 0$$

$$= \frac{2}{3}$$

9. Find $\lim\limits_{x \to \infty} \dfrac{4 - 3x}{5 - 2x^2}$.

We divide the numerator and the denominator by x^2, the highest power of x in the denominator.

$$\lim_{x \to \infty} \frac{4 - 3x}{5 - 2x^2} = \lim_{x \to \infty} \frac{\dfrac{4}{x^2} - \dfrac{3}{x}}{\dfrac{5}{x^2} - 2}$$

$$= \frac{0 - 0}{0 - 2}$$

$$= 0$$

11. Find $\lim\limits_{x \to \infty} \dfrac{8x^4 - 3x^2}{5x^2 + 6x}$.

We divide the numerator and the denominator by x^2, the highest power of x in the denominator.

$$\lim_{x \to \infty} \frac{8x^4 - 3x^2}{5x^2 + 6x} = \lim_{x \to \infty} \frac{8x^2 - 3}{5 + \dfrac{6}{x}}$$

$$= \frac{\lim\limits_{x \to \infty} 8x^2 - 3}{5 + 0} = \infty$$

13. Find $\lim\limits_{x \to \infty} \dfrac{6x^4 - 5x^2 + 7}{8x^6 + 4x^3 - 8x}$.

We divide the numerator and the denominator by x^6, the highest power of x in the denominator.

$$\lim_{x \to \infty} \frac{6x^4 - 5x^4 + 7}{8x^6 + 4x^3 - 8x} = \lim_{x \to \infty} \frac{\dfrac{6}{x^2} - \dfrac{5}{x^4} + \dfrac{7}{x^6}}{8 + \dfrac{4}{x^3} - \dfrac{8}{x^5}}$$

$$= \frac{0 - 0 + 0}{8 + 0 - 0}$$

$$= 0$$

15. Find $\lim\limits_{x \to \infty} \dfrac{11x^5 + 4x^3 - 6x^2 + 2}{6x^3 + 5x^2 + 3x - 1}$.

We divide the numerator and the denominator by x^3, the highest power of x in the denominator.

$$\lim_{x \to \infty} \frac{11x^5 + 4x^3 - 6x + 2}{6x^3 + 5x^2 + 3x - 1} = \lim_{x \to \infty} \frac{11x^2 + 4 - \dfrac{6}{x^2} + \dfrac{2}{x^3}}{6 + \dfrac{5}{x} + \dfrac{3}{x^2} - \dfrac{1}{x^3}}$$

$$= \frac{\lim\limits_{x \to \infty} 11x^2 + 4 - 0 + 0}{6 + 0 + 0 - 0}$$

$$= \infty$$

17. $f(x) = \dfrac{4}{x}$, or $4x^{-1}$

a) *Intercepts.* Since the numerator is the constant 4, there are no x-intercepts. The number 0 is not in the domain of the function, so there are no y-intercepts.

b) *Asymptotes.*

Vertical. The denominator is 0 for $x = 0$, so the line $x = 0$ is a vertical asymptote.

Horizontal. The degree of the numerator is less than the degree of the denominator, so $y = 0$ is a horizontal asymptote.

Oblique. There is no oblique asymptote since the degree of the numerator is not one more than the degree of the denominator.

c) *Derivatives.*

$$f'(x) = -4x^{-2} = -\frac{4}{x^2}$$

$$f''(x) = 8x^{-3} = \frac{8}{x^3}$$

d) *Critical points.* The number 0 is not in the domain of f. Now $f'(x)$ exists for all values of x except 0. The equation $f'(x) = 0$ has no solution, so there are no critical points.

e) *Increasing, decreasing, relative extrema.* Use 0 to divide the real number line into two intervals, A: $(-\infty, 0)$ and B: $(0, \infty)$. Test a point in each interval.

A: Test -1, $f'(-1) = -\dfrac{4}{(-1)^2} = -4 < 0$

B: Test 1, $f'(1) = -\dfrac{4}{1^2} = -4 < 0$

Then f is decreasing on both intervals. Since there are no critical points, there are no relative extrema.

f) *Inflection points.* $f''(0)$ does not exist, but because $f(0)$ does not exist there cannot be an inflection point at 0. The equation $f''(x) = 0$ has no solution, so there are no inflection points.

g) *Concavity.* Use 0 to divide the real number line as in step (e). Note that for any $x < 0$, $x^3 < 0$, so

$$f''(x) = \frac{8}{x^3} < 0$$

and for any $x > 0$, $x^3 > 0$, so

$$f''(x) = \frac{8}{x^3} > 0.$$

Then f is concave down on $(-\infty, 0)$ and concave up on $(0, \infty)$.

h) *Sketch.* Use the preceding information to sketch the graph. Compute function values as needed.

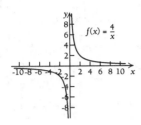

19. $f(x) = \dfrac{-2}{x - 5}$

a) *Intercepts.* Since the numerator is the constant -2, there are no x-intercepts. To find the y-intercepts we compute $f(0)$:

$$f(0) = \frac{-2}{0 - 5} = \frac{-2}{-5} = \frac{2}{5}$$

Then $\left(0, \dfrac{2}{5}\right)$ is the y-intercept.

b) *Asymptotes.*

Vertical. The denominator is 0 for $x = 5$, so the line $x = 5$ is a vertical asymptote.

Horizontal. The degree of the numerator is less than the degree of the denominator, so $y = 0$ is a horizontal asymptote.

Oblique. There is no oblique asymptote since the degree of the numerator is not one more than the degree of the denominator.

c) *Derivatives.*

$$f'(x) = 2(x - 5)^{-2} = \frac{2}{(x - 5)^2}$$

$$f''(x) = -4(x - 5)^{-3} = -\frac{4}{(x - 5)^3}$$

d) *Critical points.* $f'(5)$ does not exist, but because $f(5)$ does not exist, $x = 5$ is not a critical point. The equation $f'(x) = 0$ has no solution, so there are no critical points.

e) *Increasing, decreasing, relative extrema.* Use 5 to divide the real number line into two intervals, A: $(-\infty, 5)$ and B: $(5, \infty)$. Test a point in each interval.

A: Test 0, $f'(0) = \dfrac{2}{(0 - 5)^2} = \dfrac{2}{25} > 0$

B: Test 6, $f'(6) = \dfrac{2}{(6 - 5)^2} = 2 > 0$

Then f is increasing on both intervals. Since there are no critical points, there are no relative extrema.

f) *Inflection points.* $f''(5)$ does not exist, but because $f(5)$ does not exist there cannot be an inflection point at 5. The equation $f''(x) = 0$ has no solution, so there are no inflection points.

g) *Concavity.* Use 5 to divide the real number line as in step (e). Note that for any $x < 5$, $(x - 5)^3 < 0$, so

$$f''(x) = -\frac{4}{(x - 5)^3} > 0$$

and for any $x > 5$, $(x - 5)^3 > 0$, so

$$f''(x) = -\frac{4}{(x - 5)^3} < 0.$$

Then f is concave up on $(-\infty, 5)$ and concave down on $(5, \infty)$.

h) *Sketch.* Use the preceding information to sketch the graph. Compute function values as needed.

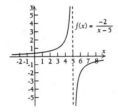

21. $f(x) = \dfrac{1}{x - 3}$

a) *Intercepts.* Since the numerator is the constant 1, there are no x-intercepts.

$f(0) = \dfrac{1}{0 - 3} = -\dfrac{1}{3}$, so $\left(0, -\dfrac{1}{3}\right)$ is the y-intercept.

b) *Asymptotes.*

Vertical. The denominator is 0 for $x = 3$, so the line $x = 3$ is a vertical asymptote.

Horizontal. The degree of the numerator is less than the degree of the denominator, so $y = 0$ is a horizontal asymptote.

Oblique. There is no oblique asymptote since the degree of the numerator is not one more than the degree of the denominator.

c) *Derivatives.*

$$f'(x) = -(x - 3)^{-2} = -\frac{1}{(x - 3)^2}$$

$$f''(x) = 2(x - 3)^{-3} = \frac{2}{(x - 3)^3}$$

d) *Critical points.* $f'(3)$ does not exist, but because $f(3)$ does not exist, $x = 3$ is not a critical point. The equation $f'(x) = 0$ has no solution, so there are no critical points.

e) *Increasing, decreasing, relative extrema.* Use 3 to divide the real number line into two intervals, A: $(-\infty, 3)$ and B: $(3, \infty)$. Test a point in each interval.

A: Test 0, $f'(0) = -\dfrac{1}{(0 - 3)^2} = -\dfrac{1}{9} < 0$

B: Test 4, $f'(4) = -\dfrac{1}{(4 - 3)^2} = -1 < 0$

Then f is decreasing on both intervals. Since there are no critical points, there are no relative extrema.

f) *Inflection points.* $f''(3)$ does not exist, but because $f(3)$ does not exist there cannot be an inflection point at 3. The equation $f''(x) = 0$ has no solution, so there are no inflection points.

g) *Concavity.* Use 3 to divide the real number line as in step (e). Note that for any $x < 3$, $(x-3)^3 < 0$, so

$$f''(x) = \frac{2}{(x-3)^3} < 0$$

and for any $x > 3$, $(x-3)^3 > 0$, so

$$f''(x) = \frac{2}{(x-3)^3} > 0.$$

Then f is concave down on $(-\infty, 3)$ and concave up on $(3, \infty)$.

h) *Sketch.* Use the preceding information to sketch the graph. Compute function values as needed.

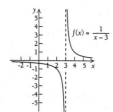

23. $f(x) = \dfrac{-2}{x+5}$

a) *Intercepts.* Since the numerator is the constant -2, there are no x-intercepts.

$f(0) = \dfrac{-2}{0+5} = -\dfrac{2}{5}$, so $\left(0, -\dfrac{2}{5}\right)$ is the y-intercept.

b) *Asymptotes.*

Vertical. The denominator is 0 for $x = -5$, so the line $x = -5$ is a vertical asymptote.

Horizontal. The degree of the numerator is less than the degree of the denominator, so $y = 0$ is a horizontal asymptote.

Oblique. There is no oblique asymptote since the degree of the numerator is not one more than the degree of the denominator.

c) *Derivatives.*

$$f'(x) = 2(x+5)^{-2} = \frac{2}{(x+5)^2}$$

$$f''(x) = -4(x+5)^{-3} = \frac{-4}{(x+5)^3}$$

d) *Critical points.* $f'(-5)$ does not exist, but because $f(-5)$ does not exist, $x = -5$ is not a critical point. The equation $f'(x) = 0$ has no solution, so there are no critical points.

e) *Increasing, decreasing, relative extrema.* Use -5 to divide the real number line into two intervals, A: $(-\infty, -5)$ and B: $(-5, \infty)$. Test a point in each interval.

A: Test -6, $f'(0) = \dfrac{2}{(-6+5)^2} = 2 > 0$

B: Test 0, $f'(0) = \dfrac{2}{(0+5)^2} = \dfrac{2}{25} > 0$

Then f is increasing on both intervals. Since there are no critical points, there are no relative extrema.

f) *Inflection points.* $f''(-5)$ does not exist, but because $f(-5)$ does not exist there cannot be an inflection point at -5. The equation $f''(x) = 0$ has no solution, so there are no inflection points.

g) *Concavity.* Use -5 to divide the real number line as in step (e). Test a point in each interval.

A: Test -6, $f''(-6) = \dfrac{-4}{(-6+5)^3} = 4 > 0$

B: Test 0, $f''(0) = \dfrac{-4}{(0+5)^3} = -\dfrac{4}{125} < 0$

Then f is concave up on $(-\infty, -5)$ and concave down on $(-5, \infty)$.

h) *Sketch.* Use the preceding information to sketch the graph. Compute function values as needed.

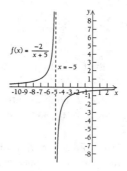

25. $f(x) = \dfrac{2x+1}{x}$

a) *Intercepts.* To find the x-intercepts, solve $f(x) = 0$.

$$\frac{2x+1}{x} = 0$$
$$2x + 1 = 0$$
$$2x = -1$$
$$x = -\frac{1}{2}$$

Since $x = -\dfrac{1}{2}$ does not make the denominator 0, the x-intercept is $\left(-\dfrac{1}{2}, 0\right)$. The number 0 is not in the domain of f, so there are no y-intercepts.

b) *Asymptotes.*

Vertical. The denominator is 0 for $x = 0$, so the line $x = 0$ is a vertical asymptote.

Horizontal. The numerator and denominator have the same degree, so $y = \dfrac{2}{1}$, or $y = 2$, is a horizontal asymptote.

Oblique. There is no oblique asymptote since the degree of the numerator is not one more than the degree of the denominator.

c) *Derivatives.*

$$f'(x) = -\frac{1}{x^2}$$

$$f''(x) = 2x^{-3}, \text{ or } \frac{2}{x^3}$$

d) *Critical points.* $f'(0)$ does not exist, but because $f(0)$ does not exist $x = 0$ is not a critical point. The equation $f'(x) = 0$ has no solution, so there are no critical points.

e) *Increasing, decreasing, relative extrema.* Use 0 to divide the real number line into two intervals, A: $(-\infty, 0)$ and B: $(0, \infty)$. Test a point in each interval.

A: Test -1, $f'(-1) = -\dfrac{1}{(-1)^2} = -1 < 0$

B: Test 1, $f'(1) = -\dfrac{1}{1^2} = -1 < 0$

Then f is decreasing on both intervals. Since there are no critical points, there are no relative extrema.

f) *Inflection points.* $f''(0)$ does not exist, but because $f(0)$ does not exist there cannot be an inflection point at 0. The equation $f''(x) = 0$ has no solution, so there are no inflection points.

g) *Concavity.* Use 0 to divide the real number line as in step (e). Test a point in each interval.

A: Test -1, $f''(-1) = \dfrac{2}{(-1)^3} = -2 < 0$

B: Test 1, $f''(1) = \dfrac{2}{1^3} = 2 > 0$

Then f is concave down on $(-\infty, 0)$ and concave up on $(0, \infty)$.

h) *Sketch.* Use the preceding information to sketch the graph. Compute function values as needed.

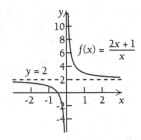

$$f(x) = \frac{2x+1}{x}$$

27. $f(x) = x + \dfrac{9}{x} = \dfrac{x^2 + 9}{x}$

a) *Intercepts.* The equation $f(x) = 0$ has no real number solution, so there are no x-intercepts. The number 0 is not in the domain of the function, so there are no y-intercepts.

b) *Asymptotes.*

Vertical. The denominator is 0 for $x = 0$, so the line $x = 0$ is a vertical asymptote.

Horizontal. The degree of the numerator is greater than the degree of the denominator, so there are no horizontal asymptotes.

Oblique. As $|x|$ gets very large, $f(x) = x + \dfrac{9}{x}$ approaches x, so $y = x$ is an oblique asymptote.

c) *Derivatives.*
$$f'(x) = 1 - 9x^{-2} = 1 - \frac{9}{x^2}$$
$$f''(x) = 18x^{-3} = \frac{18}{x^3}$$

d) *Critical points.* $f'(0)$ does not exist, but because $f(0)$ does not exist $x = 0$ is not a critical point. Solve $f'(x) = 0$.
$$1 - \frac{9}{x^2} = 1$$
$$1 = \frac{9}{x^2}$$
$$x^2 = 9$$
$$x = \pm 3$$

Thus, -3 and 3 are critical points. $f(-3) = -6$ and $f(3) = 6$, so $(-3, -6)$ and $(3, 6)$ are on the graph.

e) *Increasing, decreasing, relative extrema.* Use -3, 0, and 3 to divide the real number line into four intervals, A: $(-\infty, -3)$, B: $(-3, 0)$, C: $(0, 3)$, and D: $(3, \infty)$. Test a point in each interval.

A: Test -4, $f'(-4) = 1 - \dfrac{9}{(-4)^2} = \dfrac{7}{16} > 0$

B: Test -1, $f'(-1) = 1 - \dfrac{9}{(-1)^2} = -8 < 0$

C: Test 1, $f'(1) = 1 - \dfrac{9}{1^2} = -8 < 0$

D: Test 4, $f'(4) = 1 - \dfrac{9}{4^2} = \dfrac{7}{16} > 0$

Then f is increasing on $(-\infty, -3)$ and on $(3, \infty)$ and is decreasing on $(-3, 0)$ and on $(0, 3)$. Thus, there is a relative maximum at $(-3, -6)$ and a relative minimum at $(3, 6)$.

f) *Inflection points.* $f''(0)$ does not exist, but because $f(0)$ does not exist there cannot be an inflection point at 0. The equation $f''(x) = 0$ has no solution, so there are no inflection points.

g) *Concavity.* Use 0 to divide the real number line into two intervals, A: $(-\infty, 0)$ and B: $(0, \infty)$. Test a point in each interval.

A: Test -1, $f''(-1) = \dfrac{18}{(-1)^3} = -18 < 0$

B: Test 1, $f''(1) = \dfrac{18}{1^3} = 18 > 0$

Then f is concave down on $(-\infty, 0)$ and concave up on $(0, \infty)$.

h) *Sketch.* Use the preceding information to sketch the graph. Compute other function values as needed.

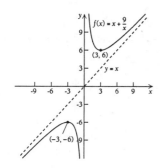

29. $f(x) = \dfrac{2}{x^2}$

a) *Intercepts.* Since the numerator is the constant 2, there are no x-intercepts. The number 0 is not in the domain of the function, so there are no y-intercepts.

b) *Asymptotes.*

   Vertical. The denominator is 0 for $x = 0$, so the line $x = 0$ is a vertical asymptote.

   Horizontal. The degree of the numerator is less than the degree of the denominator, so $y = 0$ is a horizontal asymptote.

   Oblique. There is no oblique asymptote since the degree of the numerator is not one more than the degree of the denominator.

c) *Derivatives.*
$$f'(x) = -4x^{-3} = -\frac{4}{x^3}$$
$$f''(x) = 12x^{-4} = \frac{12}{x^4}$$

d) *Critical points.* $f'(0)$ does not exist, but because $f(0)$ does not exist $x = 0$ is not a critical point. The equation $f'(x) = 0$ has no solution, so there are no critical points.

e) *Increasing, decreasing, relative extrema.* Use 0 to divide the real number line into two intervals, A: $(-\infty, 0)$ and B: $(0, \infty)$.

   A: Test -1, $f'(-1) = -\dfrac{4}{(-1)^3} = 4 > 0$

   B: Test 1, $f'(1) = -\dfrac{4}{1^3} = -4 < 0$

   Then f is increasing on $(-\infty, 0)$ and is decreasing on $(0, \infty)$. Since there are no critical points, there are no relative extrema.

f) *Inflection points.* $f''(0)$ does not exist, but because $f(0)$ does not exist there cannot be an inflection point at 0. The equation $f''(x) = 0$ has no solution, so there are no inflection points.

g) *Concavity.* Use 0 to divide the real number line as in step (e). Test a point in each interval.

   A: Test -1, $f''(-1) = \dfrac{12}{(-1)^4} = 12 > 0$

   B: Test 1, $f''(1) = \dfrac{12}{1^4} = 12 > 0$

   Then f is concave up on both intervals.

h) *Sketch.* Use the preceding information to sketch the graph. Compute function values as needed.

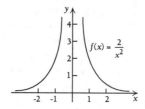

31. $f(x) = \dfrac{x}{x - 3}$

a) *Intercepts.* The numerator is 0 for $x = 0$ and this value of x does not make the denominator 0, so $(0, 0)$ is the x-intercept. $f(0) = \dfrac{0}{0 - 3} = 0$, so the y-intercept is the x-intercept $(0, 0)$.

b) *Asymptotes.*

   Vertical. The denominator is 0 for $x = 3$, so the line $x = 3$ is a vertical asymptote.

   Horizontal. The numerator and the denominator have the same degree, so $y = \dfrac{1}{1}$, or $y = 1$, is a horizontal asymptote.

   Oblique. There is no oblique asymptote since the degree of the numerator is not one more than the degree of the denominator.

c) *Derivatives.*
$$f'(x) = -\frac{3}{(x - 3)^2}$$
$$f''(x) = 6(x - 3)^{-3} = \frac{6}{(x - 3)^3}$$

d) *Critical points.* $f'(3)$ does not exist, but because $f(3)$ does not exist $x = 3$ is not a critical point. The equation $f'(x) = 0$ has no solution, so there are no critical points.

e) *Increasing, decreasing, relative extrema.* Use 3 to divide the real number line into two intervals, A: $(-\infty, 3)$ and B: $(3, \infty)$. Test a point in each interval.

   A: Test 0, $f'(0) = -\dfrac{1}{3} < 0$

   B: Test 4, $f'(4) = -3 < 0$

   Then f is decreasing on both intervals. Since there are no critical points, there are no relative extrema.

f) *Inflection points.* $f''(3)$ does not exist, but because $f(3)$ does not exist there cannot be an inflection point at 3. The equation $f''(x) = 0$ has no solution, so there are no inflection points.

g) *Concavity.* Use 3 to divide the real number line as in step (e).

   A: Test 0, $f''(0) = -\dfrac{2}{9} < 0$

   B: Test 4, $f''(4) = 6 > 0$

   Then f is concave down on $(-\infty, 3)$ and concave up on $(3, \infty)$.

h) *Sketch.* Use the preceding information to sketch the graph. Compute function values as needed.

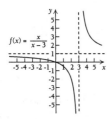

$f(x) = \dfrac{x}{x-3}$

33. $f(x) = \dfrac{1}{x^2 + 3}$

a) *Intercepts.* Since the numerator is the constant 1, there are no x-intercepts.

$f(0) = \dfrac{1}{0^2 + 3} = \dfrac{1}{3}$, so $\left(0, \dfrac{1}{3}\right)$ is the y-intercept.

b) *Asymptotes.*

Vertical. $x^2 + 3 = 0$ has no real number solutions, so there are no vertical asymptotes.

Horizontal. The degree of the numerator is less than the degree of the denominator, so $y = 0$ is a horizontal asymptote.

Oblique. There is no oblique asymptote since the degree of the numerator is not one more than the degree of the denominator.

c) *Derivatives.*

$$f'(x) = -\dfrac{2x}{(x^2 + 3)^2}$$

$$f''(x) = \dfrac{6x^2 - 6}{(x^2 + 3)^3}$$

d) *Critical points.* $f'(x)$ exists for all real numbers. Solve $f'(x) = 0$.

$$-\dfrac{2x}{(x^2 + 3)^2} = 0$$
$$-2x = 0$$
$$x = 0 \quad \text{Critical point}$$

From step (a) we already know $\left(0, \dfrac{1}{3}\right)$ is on the graph.

e) *Increasing, decreasing, relative extrema.* Use 0 to divide the real number line into two intervals, A: $(-\infty, 0)$ and B: $(0, \infty)$. Test a point in each interval.

A: Test -1, $f'(-1) = \dfrac{1}{8} > 0$

B: Test 1, $f'(1) = -\dfrac{1}{8} < 0$

Then f is increasing on $(-\infty, 0)$ and decreasing on $(0, \infty)$. Thus, $\left(0, \dfrac{1}{3}\right)$ is a relative maximum.

f) *Inflection points.* $f''(x)$ exists for all real numbers. Solve $f''(x) = 0$.

$$\dfrac{6x^2 - 6}{(x^2 + 3)^3} = 0$$
$$6x^2 - 6 = 0$$
$$6(x + 1)(x - 1) = 0$$
$$x = -1 \quad \text{or} \quad x = 1 \quad \text{Possible inflection points}$$

$f(-1) = \dfrac{1}{4}$ and $f(1) = \dfrac{1}{4}$, so $\left(-1, \dfrac{1}{4}\right)$ and $\left(1, \dfrac{1}{4}\right)$ are on the graph.

g) *Concavity.* Use -1 and 1 to divide the real number line into three intervals, A: $(-\infty, -1)$, B: $(-1, 1)$, and C: $(1, \infty)$. Test a point in each interval.

A: Test -2, $f''(-2) = \dfrac{18}{343} > 0$

B: Test 0, $f''(0) = -\dfrac{2}{9} < 0$

C: Test 2, $f''(2) = \dfrac{18}{343} > 0$

Then f is concave up on $(-\infty, -1)$ and on $(1, \infty)$ and is concave down on $(-1, 1)$. Thus, $\left(-1, \dfrac{1}{4}\right)$ and $\left(1, \dfrac{1}{4}\right)$ are both inflection points.

h) *Sketch.* Use the preceding information to sketch the graph. Compute other function values as needed.

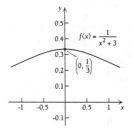

35. $f(x) = \dfrac{x - 1}{x + 2}$

a) *Intercepts.* The numerator is 0 for $x = 1$ and this value of x does not make the denominator 0, so $(1, 0)$ is the x-intercept.

$f(0) = \dfrac{0 - 1}{0 + 2} = -\dfrac{1}{2}$, so $\left(0, -\dfrac{1}{2}\right)$ is the y-intercept.

b) *Asymptotes.*

Vertical. The denominator is 0 for $x = -2$, so the line $x = -2$ is a vertical asymptote.

Horizontal. The numerator and the denominator have the same degree, so $y = \dfrac{1}{1}$, or $y = 1$, is a horizontal asymptote.

Oblique. There is no oblique asymptote since the degree of the numerator is not one more than the degree of the denominator.

c) *Derivatives.*

$$f'(x) = \dfrac{3}{(x + 2)^2}$$

$$f''(x) = \dfrac{-6}{(x + 2)^3}$$

d) *Critical points.* $f'(-2)$ does not exist, but because $f(-2)$ does not exist $x = -2$ is not a critical point. The equation $f'(x) = 0$ has no solution, so there are no critical points.

e) *Increasing, decreasing, relative extrema.* Use -2 to divide the real number line into two intervals, A: $(-\infty, -2)$ and B: $(-2, \infty)$. Test a point in each interval.

A: Test -3, $f'(-3) = 3 > 0$

B: Test -1, $f'(-1) = 3 > 0$

Then f is increasing on both intervals. Since there are no critical points, there are relative extrema.

f) *Inflection points.* $f''(-2)$ does not exist, but because $f(-2)$ does not exist there cannot be an inflection point at -2. The equation $f''(x) = 0$ has no solution, so there are no inflection points.

g) *Concavity.* Use -2 to divide the real number line as in step (e). Test a point in each interval.

A: Test -3, $f''(-3) = 6 > 0$

B: Test -1, $f''(-1) = -6 < 0$

Then f is concave up on $(-\infty, -2)$ and concave down on $(-2, \infty)$.

h) *Sketch.* Use the preceding information to sketch the graph. Compute function values as needed.

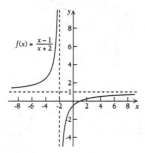

37. $f(x) = \dfrac{x^2 - 4}{x + 3}$

a) *Intercepts.* The numerator $x^2 - 4 = (x+2)(x-2)$ is 0 for $x = -2$ and or $x = 2$, and neither of these values makes the denominator 0. Thus, the x-intercepts are $(-2, 0)$ and $(2, 0)$.

$$f(0) = \frac{0^2 - 4}{0 + 3} = -\frac{4}{3}, \text{ so } \left(0, -\frac{4}{3}\right) \text{ is the}$$

y-intercept.

b) *Asymptotes.*

Vertical. The denominator is 0 for $x = -3$, so the line $x = -3$ is a vertical asymptote.

Horizontal. The degree of the numerator is greater than the degree of the denominator, so there are no horizontal asymptotes.

Oblique.

$$f(x) = x - 3 + \frac{5}{x+3} \qquad
\begin{array}{r}
x - 3 \\
x+3 \overline{\smash{\big)}\ x^2 - 4} \\
\underline{x^2 + 3x} \\
-3x - 4 \\
\underline{-3x - 9} \\
5
\end{array}$$

As $|x|$ gets very large, $f(x)$ approaches $x - 3$, so $y = x - 3$ is an oblique asymptote.

c) *Derivatives.*

$$f'(x) = \frac{x^2 + 6x + 4}{(x + 3)^2}$$

$$f''(x) = \frac{10}{(x + 3)^3}$$

d) *Critical points.* $f'(-3)$ does not exist, but because $f(-3)$ does not exist $x = -3$ is not a critical point. Solve $f'(x) = 0$.

$$\frac{x^2 + 6x + 4}{(x + 3)^2} = 0$$

$$x^2 + 6x + 4 = 0$$

$$x = -3 \pm \sqrt{5} \quad \text{Using the quadratic formula}$$

$x \approx -5.24$ or $x \approx -0.76$ Critical points

$f(-5.24) \approx -10.47$ and $f(-0.76) \approx -1.53$, so $(-5.24, -10.47)$ and $(-0.76, -1.53)$ are on the graph.

e) *Increasing, decreasing, relative extrema.* Use -5.24, -3, and -0.76 to divide the real number line into four intervals, A: $(-\infty, -5.24)$, B: $(-5.24, -3)$, C: $(-3, -0.76)$, and D: $(-0.76, \infty)$. Test a point in each interval.

A: Test -6, $f'(-6) = \dfrac{4}{9} > 0$

B: Test -4, $f'(-4) = -4 < 0$

C: Test -2, $f'(-2) = -4 < 0$

D: Test 0, $f'(0) = \dfrac{4}{9} > 0$

Then f is increasing on $(-\infty, -5.24)$ and on $(-0.76, \infty)$ and is decreasing on $(-5.24, -3)$ and on $(-3, -0.76)$. Thus, $(-5.24, -10.47)$ is a relative maximum and $(-0.76, -1.53)$ is a relative minimum.

f) *Inflection points.* $f''(-3)$ does not exist, but because $f(-3)$ does not exist there cannot be an inflection point at -3. The equation $f''(x) = 0$ has no solution, so there are no inflection points.

g) *Concavity.* Use -3 to divide the real number line into two intervals, A: $(-\infty, -3)$ and B: $(-3, \infty)$. Test a point in each interval.

A: Test -4, $f''(-4) = -10 < 0$

B: Test -2, $f''(-2) = 10 > 0$

Then f is concave down on $(-\infty, -3)$ and concave up on $(-3, \infty)$.

h) *Sketch.* Use the preceding information to sketch the graph. Compute other function values as needed.

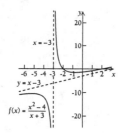

39. $f(x) = \dfrac{x-1}{x^2 - 2x - 3}$

a) *Intercepts.* The numerator is 0 for $x = 1$, and this value of x does not make the denominator 0. Then $(1, 0)$ is the x-intercept.

$f(0) = \dfrac{0-1}{0^2 - 2\cdot 0 - 3} = \dfrac{1}{3}$, so $\left(0, \dfrac{1}{3}\right)$ is the y-intercept.

b) *Asymptotes.*

Vertical. The denominator $x^2 - 2x - 3 = (x+1)(x-3)$ is 0 for $x = -1$ or $x = 3$. Then the lines $x = -1$ and $x = 3$ are vertical asymptotes.

Horizontal. The degree of the numerator is less than the degree of the denominator, so $y = 0$ is a horizontal asymptote.

Oblique. There is no oblique asymptote since the degree of the numerator is not one more than the degree of the denominator.

c) *Derivatives.*
$$f'(x) = \frac{-x^2 + 2x - 5}{(x^2 - 2x - 3)^2}$$
$$f''(x) = \frac{2x^3 - 6x^2 + 30x - 26}{(x^2 - 2x - 3)^3}$$

d) *Critical points.* $f'(-1)$ and $f'(3)$ do not exist, but because $f(-1)$ and $f(3)$ do not exist $x = -1$ and $x = 3$ are not critical points. The equation $f'(x) = 0$ has no real number solution, so there are no critical points.

e) *Increasing, decreasing, relative extrema.* Use -1 and 3 to divide the real number line into three intervals, A: $(-\infty, -1)$, B: $(-1, 3)$, and C: $(3, \infty)$. Test a point in each interval.

A: Test -2, $f'(-2) = -\dfrac{13}{25} < 0$

B: Test 0, $f'(0) = -\dfrac{5}{9} < 0$

C: Test 4, $f'(4) = -\dfrac{13}{25} < 0$

Then f is decreasing on all three intervals. Since there are no critical points, there are no relative extrema.

f) *Inflection points.* $f''(-1)$ and $f''(3)$ do not exist, but because $f(-1)$ and $f(3)$ do not exist there cannot be an inflection point at -1 or at 3. Solve $f''(x) = 0$.

$$\frac{2x^3 - 6x^2 + 30x - 26}{(x^2 - 2x - 3)^3} = 0$$
$$2x^3 - 6x^2 + 30x - 26 = 0$$
$$(x-1)(2x^2 - 4x + 26) = 0$$
$$x - 1 = 0 \quad \text{or} \quad 2x^2 - 4x + 26 = 0$$
$$\downarrow$$
$$x = 1 \qquad \text{No real number solution}$$

$f(1) = 0$, so $(1, 0)$ is on the graph and is a possible inflection point.

g) *Concavity.* Use -1, 1, and 3 to divide the real number line into four intervals, A: $(-\infty, -1)$, B: $(-1, 1)$, C: $(1, 3)$, and D $(3, \infty)$. Test a point in each interval.

A: Test -2, $f''(-2) = -\dfrac{126}{125} < 0$

B: Test 0, $f''(0) = \dfrac{26}{27} > 0$

C: Test 2, $f''(2) = -\dfrac{26}{27} < 0$

D: Test 4, $f''(4) = \dfrac{126}{125} > 0$

Then f is concave down on $(-\infty, -1)$ and on $(1, 3)$ and is concave up on $(-1, 1)$ and on $(3, \infty)$. Thus, $(1, 0)$ is an inflection point.

h) *Sketch.* Use the preceding information to sketch the graph. Compute other function values as needed.

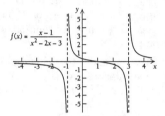

41. $f(x) = \dfrac{2x^2}{x^2 - 16}$

a) *Intercepts.* The numerator is 0 for $x = 0$, and this value of x does not make the denominator 0, so $(0, 0)$ is the x-intercept.

$f(0) = 0$, so the y-intercept is the x-intercept $(0, 0)$.

b) *Asymptotes.*

Vertical. The denominator $x^2 - 16 = (x+4)(x-4)$ is 0 for $x = -4$ or $x = 4$, so the lines $x = -4$ and $x = 4$ are vertical asymptotes.

Horizontal. The numerator and denominator have the same degree, so $y = \dfrac{2}{1}$, or $y = 2$, is a horizontal asymptote.

Oblique. There is no oblique asymptote since the degree of the numerator is not one more than the degree of the denominator.

c) *Derivatives.*
$$f'(x) = \frac{-64x}{(x^2 - 16)^2}$$
$$f''(x) = \frac{192x^2 + 1024}{(x^2 - 16)^3}$$

d) *Critical points.* $f'(-4)$ and $f'(4)$ do not exist, but because $f(-4)$ and $f(4)$ do not exist $x = -4$ and $x = 4$ are not critical points. Solve $f'(x) = 0$.

$$\frac{-64x}{(x^2 - 16)^2} = 0$$
$$-64x = 0$$
$$x = 0 \quad \text{Critical point}$$

From step (a) we already know that $(0, 0)$ is on the graph.

e) *Increasing, decreasing, relative extrema.*   Use -4, 0, and 4 to divide the real number line into four intervals, A: $(-\infty, -4)$, B: $(-4, 0)$, C: $(0, 4)$, and D: $(4, \infty)$.

A: Test -5, $f'(-5) = \dfrac{320}{81} > 0$

B: Test -1, $f'(-1) = \dfrac{64}{225} > 0$

C: Test 1, $f'(1) = -\dfrac{64}{225} < 0$

D: Test 5, $f'(5) = -\dfrac{320}{81} < 0$

Then f is increasing on $(-\infty, -4)$ and on $(-4, 0)$ and is decreasing on $(0, 4)$ and on $(4, \infty)$. Thus, there is a relative maximum at $(0, 0)$.

f) *Inflection points.* $f''(-4)$ and $f''(4)$ do not exist, but because $f(-4)$ and $f(4)$ do not exist there cannot be an inflection point at -4 or 4. The equation $f''(x) = 0$ has no real-number solution, so there are no inflection points.

g) *Concavity.*   Use -4 and 4 to divide the real number line into three intervals, A: $(-\infty, -4)$, B: $(-4, 4)$, and C: $(4, \infty)$. Test a point in each interval.

A: Test -5, $f''(-5) = \dfrac{5824}{729} > 0$

B: Test 0, $f''(0) = -\dfrac{1}{4} < 0$

C: Test 5, $f''(5) = -\dfrac{5824}{729} > 0$

Then f is concave up on $(-\infty, -4)$ and on $(4, \infty)$ and is concave down on $(-4, 4)$.

h) *Sketch.*   Use the preceding information to sketch the graph. Compute other function values as needed.

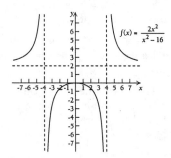

43. $f(x) = \dfrac{1}{x^2 - 1}$

a) *Intercepts.*   Since the numerator is the constant 1, there are no x-intercepts.

$f(0) = \dfrac{1}{0^2 - 1} = -1$, so the y-intercept is $(0, -1)$.

b) *Asymptotes.*

Vertical.   The denominator $x^2 - 1 = (x + 1)(x - 1)$ is 0 for $x = -1$ or $x = 1$, so the lines $x = -1$ and $x = 1$ are vertical asymptotes.

Horizontal.   The degree of the numerator is less than the degree of the denominator, so $y = 0$ is a horizontal asymptote.

Oblique.   There is no oblique asymptote since the degree of the numerator is not one more than the degree of the denominator.

c) *Derivatives.*
$$f'(x) = \frac{-2x}{(x^2 - 1)^2}$$
$$f''(x) = \frac{2(3x^2 + 1)}{(x^2 - 1)^3}$$

d) *Critical points.* $f'(-1)$ and $f'(1)$ do not exist, but because $f(-1)$ and $f(1)$ do not exist $x = -1$ and $x = 1$ are not critical points. Solve $f'(x) = 0$.
$$\frac{-2x}{(x^2 - 1)^2} = 0$$
$$-2x = 0$$
$$x = 0 \quad \text{Critical point}$$

From step (a) we already know that $(0, -1)$ is on the graph.

e) *Increasing, decreasing, relative extrema.*   Use -1, 0, and 1 to divide the real number line into four intervals, A: $(-\infty, -1)$, B: $(-1, 0)$, C: $(0, 1)$, and D: $(1, \infty)$. Test a point in each interval.

A: Test -2, $f'(-2) = \dfrac{4}{9} > 0$

B: Test $-\dfrac{1}{2}$, $f'\left(-\dfrac{1}{2}\right) = \dfrac{16}{9} > 0$

C: Test $\dfrac{1}{2}$, $f'\left(\dfrac{1}{2}\right) = -\dfrac{16}{9} < 0$

D: Test 2, $f'(2) = -\dfrac{4}{9} < 0$

Then f is increasing on $(-\infty, -1)$ and on $(-1, 0)$ and is decreasing on $(0, 1)$ and on $(1, \infty)$. Thus, there is a relative maximum at $(0, -1)$.

f) *Inflection points.* $f''(-1)$ and $f''(1)$ do not exist, but because $f(-1)$ and $f(1)$ do not exist there cannot be an inflection point at -1 or at 1. The equation $f''(x) = 0$ has no real-number solution, so there are no inflection points.

g) *Concavity.*   Use -1 and 1 to divide the real number line into three intervals, A: $(-\infty, -1)$, B: $(-1, 1)$, and C: $(1, \infty)$. Test a point in each interval.

A: Test -2, $f''(-2) = \dfrac{26}{9} > 0$

B: Test 0, $f''(0) = -2 < 0$

C: Test 2, $f''(2) = \dfrac{26}{9} > 0$

Then f is concave up on $(-\infty, -1)$ and on $(1, \infty)$ and is concave down on $(-1, 1)$.

h) *Sketch.*   Use the preceding information to sketch the graph. Compute other function values as needed.

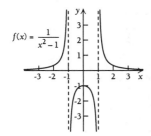

$f(x) = \dfrac{1}{x^2 - 1}$

45. $f(x) = \dfrac{x^2 + 1}{x}$

a) *Intercepts.* Since the numerator has no real-number solutions, there are no x-intercepts.

$f(0)$ does not exist, so there is no y-intercept.

b) *Asymptotes.*

Vertical. The denominator x is 0 for $x = 0$, so the line $x = 0$ is a vertical asymptote.

Horizontal. The degree of the numerator is not less than or equal to the degree of the denominator, so there is no horizontal asymptote.

Oblique. The degree of the numerator is one more than the degree of the denominator, so there is an oblique asymptote. When we divide $x^2 + 1$ by x we have $f(x) = \dfrac{x^2 + 1}{x} = x + \dfrac{1}{x}$. As $|x|$ gets very large, $\dfrac{1}{x}$ approaches 0. Thus, $y = x$ is an oblique asymptote.

c) *Derivatives.*

$$f'(x) = \frac{x^2 - 1}{x^2}$$

$$f''(x) = \frac{2}{x^3}$$

d) *Critical points.* $f'(0)$ does not exist, but because $f(0)$ does not exist 0 is not a critical point. Solve $f'(x) = 0$.

$$\frac{x^2 - 1}{x^2} = 0$$

$$x^2 - 1 = 0$$

$$(x + 1)(x - 1) = 0$$

$x = -1$ or $x = 1$ Critical points

$f(-1) = -2$ and $f(1) = 2$, so $(-1, -2)$ and $(1, 2)$ are on the graph.

e) *Increasing, decreasing, relative extrema.* Use -1, 0, and 1 to divide the real number line into four intervals, A: $(-\infty, -1)$, B: $(-1, 0)$, C: $(0, 1)$, and D: $(1, \infty)$. Test a point in each interval.

A: Test -2, $f'(-2) = \dfrac{3}{4} > 0$

B: Test $-\dfrac{1}{2}$, $f'\left(-\dfrac{1}{2}\right) = -3 < 0$

C: Test $\dfrac{1}{2}$, $f'\left(\dfrac{1}{2}\right) = -3 < 0$

D: Test 2, $f'(2) = \dfrac{3}{4} > 0$

Then f is increasing on $(-\infty, -1)$ and on $(1, \infty)$ and is decreasing on $(-1, 0)$ and $(0, 1)$. Thus, there is a relative maximum at $(-1, -2)$ and a relative minimum at $(1, 2)$.

f) *Inflection points.* $f''(0)$ does not exist, but because $f(0)$ does not exist there cannot be an inflection point at 0. The equation $f''(x) = 0$ has no solution, so there are no inflection points.

g) *Concavity.* Use 0 to divide the real number line into two intervals, A: $(-\infty, 0)$ and B: $(0, \infty)$. Test a point in each interval.

A: Test -1, $f''(-1) = -2 < 0$

B: Test 1, $f''(1) = 2 > 0$

Then f is concave down on $(-\infty, 0)$ and is concave up on $(0, \infty)$.

h) *Sketch.* Use the preceding information to sketch the graph. Compute other function values as needed.

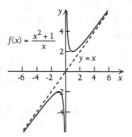

$f(x) = \dfrac{x^2 + 1}{x}$

47. $V(t) = 50 - \dfrac{25t^2}{(t + 2)^2}$

a) $V(0) = 50 - \dfrac{25 \cdot 0^2}{(0 + 2)^2} = 50 - 0 = \$50$

$V(5) = 50 - \dfrac{25 \cdot 5^2}{(5 + 2)^2} = 50 - \dfrac{625}{49} \approx \$37.24$

$V(10) = 50 - \dfrac{25 \cdot 10^2}{(10 + 2)^2} = 50 - \dfrac{2500}{144} \approx \$32.64$

$V(70) = 50 - \dfrac{25 \cdot 70^2}{(70 + 2)^2} = 50 - \dfrac{122,500}{5184} \approx \$26.37$

b) $V'(t) = -\dfrac{(t + 2)^2(50t) - 2(t + 2)(25t^2)}{(t + 2)^4}$

$= -\dfrac{(t + 2)[(t + 2)(50t) - 2(25t^2)]}{(t + 2)^4}$

$= -\dfrac{50t^2 + 100t - 50t^2}{(t + 2)^3}$

$= -\dfrac{100t}{(t + 2)^3}$

$V''(t) = -\dfrac{(t + 2)^3(100) - 3(t + 2)^2(100t)}{(t + 2)^6}$

$= -\dfrac{(t + 2)^2[(t + 2)(100) - 3(100t)]}{(t + 2)^4}$

$= -\dfrac{100t + 200 - 300t}{(t + 2)^6}$

$= -\dfrac{200 - 200t}{(t + 2)^4}$, or $\dfrac{200t - 200}{(t + 2)^4}$

$V'(t)$ exists for all values of t in $[0, \infty)$. Solve $V'(t) = 0$.

$$-\frac{100t}{(t+2)^3} = 0$$

$$-100t = 0$$

$$t = 0 \quad \text{Critical point}$$

Use the Second-Derivative Test:

$$V''(0) = \frac{200 \cdot 0 - 200}{(0+2)^4} = -\frac{200}{16} < 0$$

Thus, there is a relative maximum at $t = 0$.

$V(0) = \$50$, so the maximum value of the product is 50.

c) Using the techniques of this section we find the following additional information:

Intercepts. No t-intercepts in $[0, \infty)$; V-intercept is $(0, 50)$.

Asymptotes. There are no vertical asymptotes in $[0, \infty)$. The line $V = 25$ is a horizontal asymptote. There is no oblique asymptote.

Increasing, decreasing, relative extrema. $V(t)$ is decreasing on $[0, \infty)$. The only relative extremum is at $(0, 50)$.

Inflection points, concavity. $V(t)$ is concave down on $[0, 1)$ and concave up on $(1, \infty)$, and $(1, 47.22)$ is an inflection point.

We use this information and compute other function values as needed to sketch the graph.

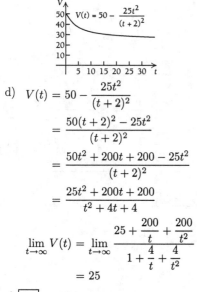

d) $V(t) = 50 - \dfrac{25t^2}{(t+2)^2}$

$$= \frac{50(t+2)^2 - 25t^2}{(t+2)^2}$$

$$= \frac{50t^2 + 200t + 200 - 25t^2}{(t+2)^2}$$

$$= \frac{25t^2 + 200t + 200}{t^2 + 4t + 4}$$

$$\lim_{t \to \infty} V(t) = \lim_{t \to \infty} \frac{25 + \dfrac{200}{t} + \dfrac{200}{t^2}}{1 + \dfrac{4}{t} + \dfrac{4}{t^2}}$$

$$= 25$$

e) $\boxed{tw}$

49. $C(p) = \dfrac{\$48,000}{100 - p}$

We will only consider the interval $[0, 100)$ since it is not possible to remove less than 0% or more than 100% of the pollutants and $C(p)$ is not defined for $p = 100$.

a) $C(0) = \dfrac{\$48,000}{100 - 0} = \$480$

$C(20) = \dfrac{\$48,000}{100 - 20} = \$600$

$C(80) = \dfrac{\$48,000}{100 - 80} = \$2400$

$C(90) = \dfrac{\$48,000}{100 - 90} = \$4800$

b) $\displaystyle\lim_{p \to 100^-} C(p) = \lim_{p \to 100^-} \frac{\$48,000}{100 - p} = \infty$

c) $\boxed{tw}$

d) Using the techniques of this section we find the following additional information.

Intercepts. No p-intercept; $(0, 480)$ is the C-intercept.

Asymptotes.   Vertical. $p = 100$

            Horizontal. $C = 0$

            Oblique.   None

Increasing, decreasing, relative extrema. $C(p)$ is increasing on $[0, 100)$. There are no relative extrema.

Inflection points, concavity. $C(p)$ is concave up on $[0, 100)$. There is no inflection point.

We use this information and compute other function values as needed to sketch the graph.

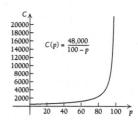

e) $\boxed{tw}$

51. a) $E(9) = 9 \cdot \dfrac{4}{9} = 4.00$

$E(8) = 9 \cdot \dfrac{4}{8} = 4.50$

$E(7) = 9 \cdot \dfrac{4}{7} \approx 5.14$

$E(6) = 9 \cdot \dfrac{4}{6} = 6.00$

$E(5) = 9 \cdot \dfrac{4}{5} = 7.20$

$E(4) = 9 \cdot \dfrac{4}{4} = 9.00$

$E(3) = 9 \cdot \dfrac{4}{3} = 12.00$

$E(2) = 9 \cdot \dfrac{4}{2} = 18.00$

$E(1) = 9 \cdot \dfrac{4}{1} = 36.00$

$E\left(\dfrac{2}{3}\right) = 9 \cdot \dfrac{4}{\frac{2}{3}} = 9 \cdot \left(4 \cdot \dfrac{3}{2}\right) = 54.00$

$$E\left(\frac{1}{3}\right) = 9 \cdot \frac{4}{\frac{1}{3}} = 9 \cdot \left(4 \cdot \frac{3}{1}\right) = 108.00$$

We complete the table.

Innings pitched (i)	Earned-run average (E)
9	4.00
8	4.50
7	5.14
6	6.00
5	7.20
4	9.00
3	12.00
2	18.00
1	36.00
$\frac{2}{3}$	54.00
$\frac{1}{3}$	108.00

b) As i approaches 0 from the right, the values of $E(i)$ increase without bound, so
$$\lim_{i \to 0^+} E(i) = \infty.$$

c) $\boxed{tw}$

53. $\boxed{tw}$

55.

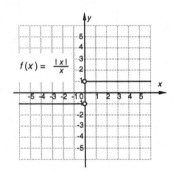

$f(x) = \frac{|x|}{x}$

$\lim_{x \to 0^-} \frac{|x|}{x} = -1$, $\lim_{x \to 0^+} \frac{|x|}{x} = 1$, so $\lim_{x \to 0} \frac{|x|}{x}$ does not exist.

57. We divide the numerator and the denominator by x^2, the highest power of x in the denominator.

$$\lim_{x \to \infty} \frac{-6x^3 + 7x}{2x^2 - 3x - 10} = \lim_{x \to \infty} \frac{-6x + \frac{7}{x}}{2 - \frac{3}{x} - \frac{10}{x^2}}$$
$$= \frac{\lim_{x \to \infty} -6x + 0}{2 - 0 - 0}$$
$$= -\infty$$

(The numerator increases without bound negatively while the denominator approaches 2.)

59. $\lim_{x \to 1} \dfrac{x^3 - 1}{x^2 - 1} = \lim_{x \to 1} \dfrac{(x-1)(x^2 + x + 1)}{(x-1)(x+1)} =$

$\lim_{x \to 1} \dfrac{x^2 + x + 1}{x + 1} = \dfrac{1 + 1 + 1}{1 + 1} = \dfrac{3}{2}$

61. We divide the numerator and denominator by x, the highest power of x in the denominator.

$$\lim_{x \to -\infty} \frac{2x^4 + x}{x + 1} = \lim_{x \to -\infty} \frac{2x^3 + 1}{1 + \frac{1}{x}} =$$
$$\frac{\lim_{x \to -\infty} 2x^3 + 1}{1 + 0} = -\infty$$

63. See the answer section in the text.

65. See the answer section in the text.

67. See the answer section in the text.

69. $\boxed{tw}$

Exercise Set 3.4

1. a) Over the interval $[20, 80]$ the function has a maximum value when the speed is 41 mph.

b) Over the interval $[20, 80]$ the function has a minimum value when the speed is 80 mph.

c) When the speed is 70 mph, the function value, or miles per gallon, is 13.5 mpg.

d) When the speed is 55 mph, the function value, or miles per gallon, is 16.5 mpg.

e) The mileage at 70 mph is 13.5 mpg. The mileage at 55 mph is 16.5 mpg.

The increase in mileage is $16.5 - 13.5$, or 3 mpg.

The percent of increase is

$\dfrac{3 \quad \text{(Increase)}}{13.5 \quad \text{(Original mpg)}}$, or $\approx 0.22 = 0.22\%$.

3. $f(x) = 5 + x - x^2$; $[0, 2]$

a) Find $f'(x)$.

$f'(x) = 1 - 2x$

b) Find the critical points. The derivative exists for all real numbers. Thus we merely solve $f'(x) = 0$.

$$1 - 2x = 0$$
$$1 = 2x$$
$$\frac{1}{2} = x$$

c) Determine the critical point and the endpoints. These points are

$0, \dfrac{1}{2}$, and 2.

d) Find the function values at the points in part (c):

$$f(0) = 5 + 0 - 0^2 = 5$$

$$f\left(\frac{1}{2}\right) = 5 + \frac{1}{2} - \left(\frac{1}{2}\right)^2 = 5 + \frac{1}{2} - \frac{1}{4} = 5\frac{1}{4}$$
                                   Maximum

$$f(2) = 5 + 2 - 2^2 = 5 + 2 - 4 = 3$$
                                   Minimum

The largest of these values, $5\frac{1}{4}$, is the maximum. It occurs at $x = \frac{1}{2}$. The smallest of these values, 3, is the minimum. It occurs at $x = 2$

Thus, on the interval $[0, 2]$, the absolute maximum $= 5\frac{1}{4}$ at $x = \frac{1}{2}$ and the absolute minimum $= 3$ at $x = 2$.

5. $f(x) = x^3 - x^2 - x + 2$; $[0, 2]$

a) Find $f'(x)$.

$$f'(x) = 3x^2 - 2x - 1$$

b) Find the critical points. The derivative exists for all real numbers. Thus we merely solve $f'(x) = 0$.

$$3x^2 - 2x - 1 = 0$$

$$(3x + 1)(x - 1) = 0 \quad \text{Factoring}$$

$$3x + 1 = 0 \quad \text{or} \quad x - 1 = 0 \quad \text{Principle of}$$
                                   zero products

$$x = -\frac{1}{3} \quad \text{or} \quad x = 1$$

Note that $-\frac{1}{3}$ is not in the interval $[0, 2]$, so 1 is the only critical point.

c) Deterine the critical point and the endpoints. These points are 0, 1, and 2.

d) Find the function values at the points in part (c):

$$f(0) = 0^3 - 0^2 - 0 + 2 = 2$$

$$f(1) = 1^3 - 1^2 - 1 + 2 = 1 - 1 - 1 + 2 = 1$$
                                   Minimum

$$f(2) = 2^3 - 2^2 - 2 + 2 = 8 - 4 - 2 + 2 = 4$$
                                   Maximum

The largest of these values, 4, is the maximum. It occurs at $x = 2$. The smallest of these values, 1, is the minimum. It occurs at $x = 1$.

Thus, on the interval $[0, 2]$, the absolute maximum $= 4$ at $x = 2$ and the absolue minimum $= 1$ at $x = 1$.

7. $f(x) = x^3 - x^2 - x + 2$; $[-1, 0]$

As in Exercise 5, the derivative is 0 at $-\frac{1}{3}$ and 1. But only $-\frac{1}{3}$ is in the interval $[-1, 0]$, so there is only one critical point, $-\frac{1}{3}$. The critical point and the endpoints are -1, $-\frac{1}{3}$, and 0.

$$f(-1) = (-1)^3 - (-1)^2 - (-1) + 2$$

$$= -1 - 1 + 1 + 2 = 1 \quad \text{Minimum}$$

$$f\left(-\frac{1}{3}\right) = \left(-\frac{1}{3}\right)^3 - \left(-\frac{1}{3}\right)^2 - \left(-\frac{1}{3}\right) + 2$$

$$= -\frac{1}{27} - \frac{1}{9} + \frac{1}{3} + 2$$

$$= -\frac{1}{27} - \frac{3}{27} + \frac{9}{27} + 2$$

$$= 2\frac{5}{27} = \frac{59}{27} \qquad \text{Maximum}$$

$$f(0) = 0^3 - 0^2 - 0 + 2 = 2$$

The largest of these values, $\frac{59}{27}$, is the maximum. It occurs at $x = -\frac{1}{3}$. The smallest of these values, 1, is the minimum. It occurs at $x = -1$.

Thus, on the interval $[-1, 0]$, the absolute maximum $= \frac{59}{27}$ at $x = -\frac{1}{3}$ and the absolute minimum $= 1$ at $x = -1$.

9. $f(x) = 3x - 2$; $[-1, 1]$

a) Find $f'(x)$.

$$f'(x) = 3$$

b) and c)

The derivative exists and is 3 for all real numbers. Note that $f'(x)$ is never 0. Thus, there are no critical points for $f(x) = 3x - 2$, and the maximum and minimum values occur at the endpoints.

d) Find the function values at the endpoints.

$$f(-1) = 3(-1) - 2 = -3 - 2 = -5 \quad \text{Minimum}$$

$$f(1) = 3 \cdot 1 - 2 = 3 - 2 = 1 \quad \text{Maximum}$$

Thus, on the interval $[-1, 1]$, the absolute maximum $= 1$ at $x = 1$ and the absolute minimum $= -5$ at $x = -1$.

11. $f(x) = 7 - 4x$; $[-2, 5]$

a) Find $f'(x)$

$$f'(x) = -4$$

b) and c)

The derivative exists and is -4 for all real numbers. Note that $f'(x)$ is never 0. Thus, there are no critical points for $f(x) = 7 - 4x$, and the maximum and minimum values occur at the endpoints.

d) Find the function values at the endpoints.

$$f(-2) = 7 - 4(-2) = 7 + 8 = 15 \quad \text{Maximum}$$

$$f(5) = 7 - 4 \cdot 5 = 7 - 20 = -13 \quad \text{Minimum}$$

Thus, on the interval $[-2, 5]$, the absolute maximum $= 15$ at $x = -2$ and the absolute minimum $= -13$ at $x = 5$.

13. $f(x) = -5$; $[-1, 1]$

Note that $f'(x) = 0$ for all real numbers. Thus, all points in $[-1, 1]$ are critical points. For all x, $f(x) = -5$.

Thus, absolute maximum $=$ absolute minimum $= -5$ for all x in $[-1, 1]$.

15. $f(x) = x^2 - 6x - 3$; $[-1, 5]$

a) $f'(x) = 2x - 6$

b) The derivative exists for all real numbers. We solve $f'(x) = 0$.

$$2x - 6 = 0$$
$$2x = 6$$
$$x = 3$$

c) The critical point and the endpoints are -1, 3, and 5.

d) $f(-1) = (-1)^2 - 6(-1) - 3 = 1 + 6 - 3 = 4$

   $\qquad\qquad\qquad\qquad\qquad\qquad$ Maximum

   $f(3) = 3^2 - 6 \cdot 3 - 3 = 9 - 18 - 3 = -12$

   $\qquad\qquad\qquad\qquad\qquad\qquad$ Minimum

   $f(5) = 5^2 - 6 \cdot 5 - 3 = 25 - 30 - 3 = -8$

   On the interval $[-1, 5]$, the absolute maximum $= 4$ at $x = -1$ and the absolute minimum $= -12$ at $x = 3$.

17. $f(x) = 3 - 2x - 5x^2$; $[-3, 3]$

a) $f'(x) = -2 - 10x$

b) The derivative exists for all real numbers. We solve $f'(x) = 0$.

$$-2 - 10x = 0$$
$$-10x = 2$$
$$x = -\frac{1}{5}$$

c) The critical point and the endpoints are -3, $-\frac{1}{5}$, and 3.

d) $f(-3) = 3 - 2(-3) - 5(-3)^2 = 3 + 6 - 45 =$
   $\qquad -36$

   $f\left(-\frac{1}{5}\right) = 3 - 2\left(-\frac{1}{5}\right) - 5\left(-\frac{1}{5}\right)^2 =$

   $\qquad 3 + \frac{2}{5} - \frac{1}{5} = 3\frac{1}{5}$ $\qquad$ Maximum

   $f(3) = 3 - 2 \cdot 3 - 5 \cdot 3^2 = 3 - 6 - 45 = -48$

   $\qquad\qquad\qquad\qquad\qquad\qquad$ Minimum

   On the interval $[-3, 3]$, the absolute maximum $= 3\frac{1}{5}$ at $x = -\frac{1}{5}$ and the absolute minimum $= -48$ at $x = 3$.

19. $f(x) = x^3 - 3x^2$; $[0, 5]$

a) $f'(x) = 3x^2 - 6x$

b) The derivative exists for all real numbers. We solve $f'(x) = 0$.

$$3x^2 - 6x = 0$$
$$3x(x - 2) = 0$$
$$3x = 0 \text{ or } x - 2 = 0$$
$$x = 0 \text{ or } \qquad x = 2$$

c) The critical point and endpoints are 0, 2, and 5.

d) $f(0) = 0^3 - 3 \cdot 0^2 = 0 - 0 = 0$

   $f(2) = 2^3 - 3 \cdot 2^2 = 8 - 12 = -4$ $\qquad$ Minimum

   $f(5) = 5^3 - 3 \cdot 5^2 = 125 - 75 = 50$ $\qquad$ Maximum

On the interval $[0, 5]$, the absolute maximum $= 50$ at $x = 5$ and the absolute minimum $= -4$ at $x = 2$.

21. $f(x) = x^3 - 3x$; $[-5, 1]$

a) $f'(x) = 3x^2 - 3$

b) The derivative exists for all real numbers. We solve $f'(x) = 0$.

$$3x^2 - 3 = 0$$
$$3x(x^2 - 1) = 0$$
$$3(x + 1)(x - 1) = 0$$
$$x + 1 = 0 \quad \text{or} \quad x - 1 = 0$$
$$x = -1 \text{ or} \qquad x = 1$$

c) The critical point and endpoints are -5, -1, and 1.

d) $f(-5) = (-5)^3 - 3(-5) = -125 + 15 = -110$

   $\qquad\qquad\qquad\qquad\qquad\qquad$ Minimum

   $f(-1) = (-1)^3 - 3(-1) = -1 + 3 = 2$ $\quad$ Maximum

   $f(1) = 1^3 - 3 \cdot 1 = 1 - 3 = -2$

   On the interval $[-5, 1]$, the absolute maximum $= 2$ at $x = -1$ and the absolute minimum $= -110$ at $x = -5$.

23. $f(x) = 1 - x^3$; $[-8, 8]$

a) $f'(x) = -3x^2$

b) The derivative exists for all real numbers. We solve $f'(x) = 0$.

$$-3x^2 = 0$$
$$x^2 = 0$$
$$x = 0$$

c) The critical point and endpoints are -8, 0, and 8.

d) $f(-8) = 1 - (-8)^3 = 1 + 512 = 513$ $\quad$ Maximum

   $f(0) = 1 - 0^3 = 1 - 0 = 1$

   $f(8) = 1 - 8^3 = 1 - 512 = -511$ $\qquad$ Minimum

   On the interval $[-8, 8]$, the absolute maximum $= 513$ at $x = -8$ and the absolute minimum $= -511$ at $x = 8$.

25. $f(x) = 12 + 9x - 3x^2 - x^3$; $[-3, 1]$

a) $f'(x) = 9 - 6x - 3x^2$

b) The derivative exists for all real numbers. We solve $f'(x) = 0$.

$$9 - 6x - 3x^2 = 0$$
$$3(3 - 2x - x^2) = 0$$
$$3(3 + x)(1 - x) = 0$$
$$3 + x = 0 \quad \text{or} \quad 1 - x = 0$$
$$x = -3 \text{ or} \qquad 1 = x$$

c) The critical points are the endpoints, -3 and 1.

d) $f(-3) = 12 + 9(-3) - 3(-3)^2 - (-3)^3$

   $\qquad = 12 - 27 - 27 + 27$

   $\qquad = -15$ $\qquad\qquad\qquad$ Minimum

$$f(1) = 12 + 9 \cdot 1 - 3 \cdot 1^2 - 1^3$$
$$= 12 + 9 - 3 - 1$$
$$= 17 \qquad\qquad \text{Maximum}$$

On the interval $[-3, 1]$, the absolute maximum $= 17$ at $x = 1$ and the absolute minimum $= -15$ at $x = -3$.

27. $f(x) = x^4 - 2x^3$; $[-2, 2]$

a) $f'(x) = 4x^3 - 6x^2$

b) The derivative exists for all real numbers. We solve $f'(x) = 0$.

$$4x^3 - 6x^2 = 0$$
$$2x^2(2x - 3) = 0$$
$$2x^2 = 0 \quad\text{or}\quad 2x - 3 = 0$$
$$x = 0 \quad\text{or}\qquad x = \frac{3}{2}$$

c) The critical points and endpoints are -2, 0, $\frac{3}{2}$, and 2.

d) $f(-2) = (-2)^4 - 2(-2)^3 = 16 + 16 = 32$
$\qquad\qquad\qquad\qquad\qquad\qquad$ Maximum

$\quad f(0) = 0^4 - 2 \cdot 0^3 = 0 - 0 = 0$

$\quad f\left(\frac{3}{2}\right) = \left(\frac{3}{2}\right)^4 - 2\left(\frac{3}{2}\right)^3 = \frac{81}{16} - \frac{27}{4} = -\frac{27}{16}$
$\qquad\qquad\qquad\qquad\qquad\qquad$ Minimum

$\quad f(2) = 2^4 - 2 \cdot 2^3 = 16 - 16 = 0$

On the interval $[-2, 2]$, the absolute maximum $= 32$ at $x = -2$ and the absolute minimum $= -\frac{27}{16}$ at $x = \frac{3}{2}$.

29. $f(x) = x^4 - 2x^2 + 5$; $[-2, 2]$

a) $f'(x) = 4x^3 - 4x$

b) The derivative exists for all real numbers. We solve $f'(x) = 0$.

$$4x^3 - 4x = 0$$
$$4x(x^2 - 1) = 0$$
$$4x(x + 1)(x - 1) = 0$$
$$4x = 0 \quad\text{or}\quad x + 1 = 0 \quad\text{or}\quad x - 1 = 0$$
$$x = 0 \quad\text{or}\quad x = -1 \quad\text{or}\quad x = 1$$

c) The critical points and endpoints are -2, -1, 0, 1, and 2.

d) $f(-2) = (-2)^4 - 2(-2)^2 + 5 = 16 - 8 + 5 = 13$
$\qquad\qquad\qquad\qquad\qquad\qquad$ Maximum

$\quad f(-1) = (-1)^4 - 2(-1)^2 + 5 = 1 - 2 + 5 = 4$
$\qquad\qquad\qquad\qquad\qquad\qquad$ Minimum

$\quad f(0) = 0^4 - 2 \cdot 0^2 + 5 = 0 - 0 + 5 = 5$

$\quad f(1) = 1^4 - 2 \cdot 1^2 + 5 = 1 - 2 + 5 = 4$
$\qquad\qquad\qquad\qquad\qquad\qquad$ Minimum

$\quad f(2) = 2^4 - 2 \cdot 2^2 + 5 = 16 - 8 + 5 = 13$
$\qquad\qquad\qquad\qquad\qquad\qquad$ Maximum

On the interval $[-2, 2]$, the absolute maximum $= 13$ at $x = -2$ and $x = 2$ and the absolute minimum $= 4$ at $x = -1$ and $x = 1$.

31. $f(x) = (x + 3)^{2/3} - 5$; $[-4, 5]$

a) $f'(x) = \frac{2}{3}(x + 3)^{-1/3} = \frac{2}{3(x + 3)^{1/3}}$

b) The derivative does not exist for $x = -3$. The equation $f'(x) = 0$ has no solution, so -3 is the only critical point.

c) The critical points and endpoints are -4, -3, and 5.

d) $f(-4) = (-4 + 3)^{2/3} - 5 = (-1)^{2/3} - 5 =$
$\qquad\qquad 1 - 5 = -4$

$\quad f(-3) = (-3 + 3)^{2/3} - 5 = 0^{2/3} - 5 =$
$\qquad\qquad 0 - 5 = -5 \qquad\qquad \text{Minimum}$

$\quad f(5) = (5 + 3)^{2/3} - 5 = 8^{2/3} - 5 =$
$\qquad\qquad 4 - 5 = -1 \qquad\qquad \text{Maximum}$

On the interval $[-4, 5]$, the absolute maximum $= -1$ at $x = 5$ and the absolute minimum $= -5$ at $x = -3$.

33. $f(x) = x + \frac{1}{x}$; $[1, 20]$

Express the function as $f(x) = x + x^{-1}$.

a) $f'(x) = 1 - x^{-2} = 1 - \frac{1}{x^2}$

b) The derivative exists for all x in $[1, 20]$. We solve $f'(x) = 0$.

$$1 - \frac{1}{x^2} = 0$$
$$1 = \frac{1}{x^2}$$
$$x^2 = 1$$
$$x = \pm 1$$

Only 1 is in the interval $[1, 20]$.

c) The critical point and endpoints are 1 and 20.

d) $f(-1) = 1 + \frac{1}{1} = 1 + 1 = 2$ Minimum

$\quad f(20) = 20 + \frac{1}{20} = 20\frac{1}{20}$ Maximum

On the interval $[1, 20]$, the absolute maximum $= 20\frac{1}{20}$ at $x = 20$ and the absolute minimum $= 2$ at $x = 1$.

35. $f(x) = \frac{x^2}{x^2 + 1}$; $[-2, 2]$

a) $f'(x) = \frac{(x^2 + 1)(2x) - 2x(x^2)}{(x^2 + 1)^2}$ Quotient Rule

$\qquad = \frac{2x^3 + 2x - 2x^3}{(x^2 + 1)^2}$

$\qquad = \frac{2x}{(x^2 + 1)^2}$

b) The derivative exists for all real numbers. We solve $f'(x) = 0$.

$$\frac{2x}{(x^2 + 1)^2} = 0$$
$$2x = 0$$
$$x = 0$$

c) The critical point and endpoints are -2, 0, and 2.

d) $f(-2) = \dfrac{(-2)^2}{(-2)^2 + 1} = \dfrac{4}{4+1} = \dfrac{4}{5}$ Maximum

$\quad f(0) = \dfrac{0^2}{0^2 + 1} = \dfrac{0}{1} = 0$ Minimum

$\quad f(2) = \dfrac{2^2}{2^2 + 1} = \dfrac{4}{4+1} = \dfrac{4}{5}$ Maximum

On the interval $[-2, 2]$, the absolute maximum $= \dfrac{4}{5}$ at $x = -2$ and $x = 2$ and the absolute minimum $= 0$ at $x = 0$.

37. $f(x) = (x+1)^{1/3}$; $[-2, 26]$

a) $f'(x) = \dfrac{1}{3}(x+1)^{-2/3} = \dfrac{1}{3(x+1)^{2/3}}$

b) The derivative does not exist for $x = -1$. The equation $f'(x) = 0$ has no solution, so -1 is the only critical point.

c) The critical point and endpoints are -2, -1, and 26.

d) $f(-2) = (-2+1)^{1/3} = (-1)^{1/3} = -1$ Minimum

$\quad f(-1) = (-1+1)^{1/3} = 0^{1/3} = 0$

$\quad f(26) = (26+1)^{1/3} = 27^{1/3} = 3$ Maximum

On the interval $[-2, 26]$, the absolute maximum $= 3$ at $x = 26$ and the absolute minimum $= -1$ at $x = -2$.

39. - 47.

49. $f(x) = x(70 - x)$
$\qquad = 70x - x^2$

When no interval is specified, we use the real line $(-\infty, \infty)$.

a) Find $f'(x)$.

$\quad f'(x) = 70 - 2x$

b) Find the critical points.

The derivative exists for all real numbers. Thus we solve $f'(x) = 0$.

$\quad 70 - 2x = 0$

$\qquad -2x = -70$

$\qquad\quad x = 35$

e) Since there is only one critical point, we can apply Max-Min Principle 2.

Find $f''(x)$.

$f''(x) = -2$

Now the second derivative is constant, so $f''(35) = -2$, and since this is negative, we have a maximum at $x = 35$.

Find the function value at $x = 35$.

$\quad f(35) = 35(70 - 35)$

$\qquad\quad = 35(35)$

$\qquad\quad = 1225$

Thus the absolute maximum $= 1225$ at $x = 35$. The function has no minimum value.

51. $f(x) = 2x^2 - 40x + 400$

When no interval is specified, we use the real line $(-\infty, \infty)$.

a) Find $f'(x)$.

$\quad f'(x) = 4x - 40$

b) Find the critical points.

The derivative exists for all real numbers. Thus we solve $f'(x) = 0$.

$\quad 4x - 40 = 0$

$\qquad 4x = 40$

$\qquad\; x = 10$

e) Since there is only one critical point, we can apply Max-Min Principle 2.

Find $f''(x)$.

$f''(x) = 4$

Now the second derivative is constant, so $f''(10) = 4$, and since this is positive, we have a minimum at $x = 10$.

Find the function value at $x = 10$.

$\quad f(10) = 2 \cdot 10^2 - 40 \cdot 10 + 400$

$\qquad\quad = 200 - 400 + 400$

$\qquad\quad = 200$

Thus the absolute minimum $= 200$ at $x = 10$. The function has no maximum value.

53. $f(x) = x - \dfrac{4}{3}x^3$; $(0, \infty)$

a) Find $f'(x)$.

$\quad f'(x) = 1 - 4x^2$

b) Find the critical points.

The derivative exists for all real numbers. Thus we solve $f'(x) = 0$.

$\quad 1 - 4x^2 = 0$

$\qquad -4x^2 = -1$

$\qquad\;\; x^2 = \dfrac{1}{4}$

$\qquad\;\;\; x = \pm\dfrac{1}{2}$

e) The interval is not closed. The only critical point in $(0, \infty)$ is $\dfrac{1}{2}$. Thus, we can apply the second derivative

$\qquad f''(x) = -8x$

to determine whether we have a maximum or a minimum. Now $f''(x)$ is negative for all values of x in $(0, \infty)$, so there is a maximum at $x = \dfrac{1}{2}$.

$f''\left(\dfrac{1}{2}\right) = -8 \cdot \left(\dfrac{1}{2}\right) = -4 < 0$

Find the function value at $x = \dfrac{1}{2}$.

$f\left(\dfrac{1}{2}\right) = \dfrac{1}{2} - \dfrac{4}{3}\left(\dfrac{1}{2}\right)^3 = \dfrac{1}{2} - \dfrac{4}{3}\cdot\dfrac{1}{8} = \dfrac{1}{2} - \dfrac{1}{6} = \dfrac{1}{3}$

Thus the absolute maximum $= \dfrac{1}{3}$ at $x = \dfrac{1}{2}$. The function has no minimum value.

55. $f(x) = 17x - x^2$

When no interval is specified, we use the real line $(-\infty, \infty)$.

a) Find $f'(x)$.

$f'(x) = 17 - 2x$

b) Find the critical points.

The derivative exists for all real numbers. Thus we solve $f'(x) = 0$.

$$17 - 2x = 0$$
$$-2x = -17$$
$$x = \frac{17}{2}$$

e) Since there is only one critical point, we can apply Max-Min Principle 2.

Find $f''(x)$.

$f''(x) = -2$

Now the second derivative is constant, so

$f''\left(\dfrac{17}{2}\right) = -2$, and since this is negative, we have a maximum at $x = \dfrac{17}{2}$.

Find the function value at $x = \dfrac{17}{2}$.

$$f\left(\frac{17}{2}\right) = 17 \cdot \frac{17}{2} - \left(\frac{17}{2}\right)^2$$
$$= \frac{289}{2} - \frac{289}{4}$$
$$= \frac{578}{4} - \frac{289}{4}$$
$$= \frac{289}{4}$$

Thus the absolute maximum $= \dfrac{289}{4}$ at $x = \dfrac{17}{2}$. The function has no minimum value.

57. $f(x) = \dfrac{1}{3}x^3 - 3x;\ [-2, 2]$

a) Find $f'(x)$.

$f'(x) = x^2 - 3$

b) Find the critical points.

The derivative exists for all real numbers. Thus we solve $f'(x) = 0$.

$$x^2 - 3 = 0$$
$$x^2 = 3$$
$$x = \pm\sqrt{3} \approx \pm 1.732$$

Both critical points are in the interval $[-2, 2]$.

c) If the interval is closed and there is more than one critical point, then use Max-Min Principle 1.

The critical points and the endpoints are -2, $-\sqrt{3}$, $\sqrt{3}$, and 2.

Next we find the function values at these points.

$$f(-2) = \frac{1}{3}(-2)^3 - 3(-2) = -\frac{8}{3} + 6 = \frac{10}{3} = 3.3\overline{3}$$

$$f(-\sqrt{3}) = \frac{1}{3}(-\sqrt{3})^3 - 3(-\sqrt{3})$$
$$= \frac{1}{3}(-3\sqrt{3}) + 3\sqrt{3}$$
$$[(-\sqrt{3})^3 = (-\sqrt{3})(-\sqrt{3})(-\sqrt{3}) = -3\sqrt{3}]$$
$$= -\sqrt{3} + 3\sqrt{3}$$
$$= 2\sqrt{3} \approx 2(1.732) = 3.464$$
$$\text{Maximum}$$

$$f(\sqrt{3}) = \frac{1}{3}(\sqrt{3})^3 - 3\sqrt{3}$$
$$= \frac{1}{3}(3\sqrt{3}) - 3\sqrt{3}$$
$$[(\sqrt{3})^3 = \sqrt{3}\cdot\sqrt{3}\cdot\sqrt{3} = 3\sqrt{3}]$$
$$= \sqrt{3} - 3\sqrt{3}$$
$$= -2\sqrt{3} \approx -2(1.732) = -3.464$$
$$\text{Minimum}$$

$$f(2) = \frac{1}{3}\cdot 2^3 - 3\cdot 2 = \frac{8}{3} - 6 = -\frac{10}{3} = -3.3\overline{3}$$

The largest of these values, $2\sqrt{3} \approx 3.464$, is the maximum. It occurs at $x = -\sqrt{3}$. The smallest of these values, $-2\sqrt{3} \approx -3.464$, is the minimum. It occurs at $x = \sqrt{3}$.

Thus the absolute maximum $= 2\sqrt{3}$ at $x = -\sqrt{3}$ and the absolute minimum $= -2\sqrt{3}$ at $x = \sqrt{3}$.

59. $f(x) = -0.001x^2 + 4.8x - 60$

When no interval is specified, we use the real line $(-\infty, \infty)$.

a) Find $f'(x)$.

$f'(x) = -0.002x + 4.8$

b) Find the critical points.

The derivative exists for all real numbers. Thus we solve $f'(x) = 0$.

$$-0.002x + 4.8 = 0$$
$$-0.002x = -4.8$$
$$x = 2400$$

e) Since there is only one critical point, we can apply Max-Min Principle 2.

Find $f''(x)$.

$f''(x) = -0.002$

Now the second derivative is constant, so $f''(2400) = -0.002$, and since this is negative, we have a maximum at $x = 2400$.

Find the function value at $x = 2400$.

$$f(2400) = -0.001(2400)^2 + 4.8(2400) - 60$$
$$= -5760 + 11{,}520 - 60$$
$$= 5700$$

Thus the absolute maximum $= 5700$ at $x = 2400$. The function has no minimum value.

61. $f(x) = -\frac{1}{3}x^3 + 6x^2 - 11x - 50$; $(0,3)$

a) Find $f'(x)$.
$$f'(x) = -x^2 + 12x - 11$$

b) Find the critical points.

The derivative exists for all real numbers. Thus we solve $f'(x) = 0$.
$$-x^2 + 12x - 11 = 0$$
$$x^2 - 12x + 11 = 0$$
$$(x-11)(x-1) = 0$$
$$x - 11 = 0 \quad \text{or} \quad x - 1 = 0$$
$$x = 11 \quad \text{or} \quad x = 1$$

e) The interval is not closed. The only critical point in $(0,3)$ is 1. Thus, we can apply the second derivative
$$f''(x) = -2x + 12$$

to determine whether we have a maximum or a minimum.
$$f''(1) = -2 \cdot 1 + 12 = -2 + 12 = 10 > 0$$

Since the second derivative is positive when $x = 1$, there is a minimum at $x = 1$.

Find the function value at $x = 1$.
$$f(1) = -\frac{1}{3} \cdot 1^3 + 6 \cdot 1^2 - 11 \cdot 1 - 50$$
$$= -\frac{1}{3} + 6 - 11 - 50$$
$$= -55\frac{1}{3}$$

Thus the absolute minimum $= -55\frac{1}{3}$ at $x = 1$. The function has no maximum value in $(0,3)$.

63. $f(x) = 15x^2 - \frac{1}{2}x^3$; $[0,30]$

a) Find $f'(x)$.
$$f'(x) = 30x - \frac{3}{2}x^2$$

b) Find the critical points.

The derivative exists for all real numbers. Thus we solve $f'(x) = 0$.
$$30x - \frac{3}{2}x^2 = 0$$
$$60x - 3x^2 = 0$$
$$3x(20 - x) = 0$$
$$3x = 0 \quad \text{or} \quad 20 - x = 0$$
$$x = 0 \quad \text{or} \quad x = 20$$

Both critical points are in the interval $[0,30]$.

e) If the interval is closed and there is more than one critical point, then use Max-Min Principle 1.

The critical points and the endpoints are 0, 20, and 30.

Next we find the function values at these points.

$$f(0) = 15 \cdot 0^2 - \frac{1}{2} \cdot 0^3$$
$$= 0 \qquad \text{Minimum}$$
$$f(20) = 15 \cdot 20^2 - \frac{1}{2} \cdot 20^3$$
$$= 15 \cdot 400 - \frac{1}{2} \cdot 8000$$
$$= 6000 - 4000$$
$$= 2000 \qquad \text{Maximum}$$
$$f(30) = 15 \cdot 30^2 - \frac{1}{2} \cdot 30^3$$
$$= 15 \cdot 900 - \frac{1}{2} \cdot 27,000$$
$$= 13,500 - 13,500$$
$$= 0 \qquad \text{Minimum}$$

The largest of these values, 2000, is the maximum. It occurs at $x = 20$. The smallest of these values is 0. It occurs twice, at $x = 0$ and $x = 30$. Thus, on the interval $[0,30]$, the absolute maximum $= 2000$ at $x = 20$ and the absolute minimum $= 0$ at $x = 0$ and $x = 30$.

65. $f(x) = 2x + \frac{72}{x}$; $(0,\infty)$

$$f(x) = 2x + 72x^{-1}$$

a) Find $f'(x)$.
$$f'(x) = 2 - 72x^{-2}$$
$$= 2 - \frac{72}{x^2}$$

b) Find the critical points.

Now $f'(x)$ exists for all values of x in $(0,\infty)$. Thus the only critical points are those for which $f'(x) = 0$.
$$2 - \frac{72}{x^2} = 0$$
$$2 = \frac{72}{x^2}$$
$$2x^2 = 72 \quad \text{Multiplying by } x^2, \text{ since } x \neq 0$$
$$x^2 = 36$$
$$x = \pm 6$$

e) The only critical point in $(0,\infty)$ is 6. Thus, we can apply the second derivative
$$f''(x) = 144x^{-3} = \frac{144}{x^3}$$

to determine whether we have a maximum or minimum.
$$f''(6) = \frac{144}{6^3} = \frac{144}{216} > 0$$

Since the second derivative is positive when $x = 6$, there is a minimum at $x = 6$.

Find the function value at $x = 6$.
$$f(6) = 2 \cdot 6 + \frac{72}{6}$$
$$= 12 + 12$$
$$= 24$$

Thus the absolute minimum $= 24$ at $x = 6$. The function has no maximum value in $(0,\infty)$.

67. $f(x) = x^2 + \dfrac{432}{x}; \ (0, \infty)$

$f(x) = x^2 + 432x^{-1}$

a) Find $f'(x)$.

$f'(x) = 2x - 432x^{-2}$

$ = 2x - \dfrac{432}{x^2}$

b) Find the critical points.

Now $f'(x)$ exists for all values of x in $(0, \infty)$. Thus the only critical points are those for which $f'(x) = 0$.

$2x - \dfrac{432}{x^2} = 0$

$2x = \dfrac{432}{x^2}$

$2x^3 = 432 \qquad$ Multiplying by x^2, since $x \neq 0$

$x^3 = 216$

$x = 6$

e) Since there is only one critical point, we can use Max-Min Principle 2 to determine whether we have a maximum or a minimum.

Find $f''(x)$.

$f''(x) = 2 + 864x^{-3}$

$ = 2 + \dfrac{864}{x^3}$

Since

$f''(6) = 2 + \dfrac{864}{6^3} = 2 + \dfrac{864}{216} = 2 + 4 = 6$

and $6 > 0$, there is a minimum at $x = 6$.

Find the function value at $x = 6$.

$f(6) = 6^2 + \dfrac{432}{6}$

$ = 36 + 72$

$ = 108$

Thus the absolute minimum $= 108$ at $x = 6$. The function has no maximum value.

69. $f(x) = 2x^4 - x; \ [-1, 1]$

a) Find $f'(x)$.

$f'(x) = 8x^3 - 1$

b) Find the critical points.

The derivative exists for all real numbers. Thus we solve $f'(x) = 0$.

$8x^3 - 1 = 0$

$8x^3 = 1$

$x^3 = \dfrac{1}{8}$

$x = \dfrac{1}{2}$

There is one critical point. It is in the interval $[-1, 1]$.

d) On a closed interval Max-Min Principle 1 can always be used.

The critical point and the endpoints are -1, $\dfrac{1}{2}$, and 1.

We find the function values at these points.

$f(-1) = 2(-1)^4 - (-1) = 2 + 1 = 3 \quad$ Maximum

$f\left(\dfrac{1}{2}\right) = 2\left(\dfrac{1}{2}\right)^4 - \dfrac{1}{2} = \dfrac{1}{8} - \dfrac{1}{2} = -\dfrac{3}{8} \quad$ Minimum

$f(1) = 2(1)^4 - 1 = 2 - 1 = 1$

The largest of these values, 3, is the maximum. It occurs at $x = -1$. The smallest of these values, $-\dfrac{3}{8}$, is the minimum. It occurs at $x = \dfrac{1}{2}$.

Thus the absolute maximum $= 3$ at $x = -1$ and the absolute minimum $= -\dfrac{3}{8}$ at $x = \dfrac{1}{2}$.

71. $f(x) = \sqrt[3]{x}; \ [0, 8]$

$f(x) = x^{1/3}$

a) Find $f'(x)$.

$f'(x) = \dfrac{1}{3}x^{-2/3}$

$ = \dfrac{1}{3\sqrt[3]{x^2}}$

b) Find the critical points.

The derivative does not exist for $x = 0$. Thus, 0 is a critical point. Since $f'(x) = 0$ has no solution, there are no other critical points.

d) On a closed interval Max-Min Principle 1 can always be used.

The critical point and the endpoints are 0 and 8.

We find the function values at these points.

$f(0) = \sqrt[3]{0} = 0 \quad$ Minimum

$f(8) = \sqrt[3]{8} = 2 \quad$ Maximum

Thus the absolute maximum $= 2$ at $x = 8$ and the absolute minimum $= 0$ at $x = 0$.

73. $f(x) = (x + 1)^3$

When no interval is specified, we use the real line $(-\infty, \infty)$.

a) Find $f'(x)$.

$f'(x) = 3(x + 1)^2 \cdot 1 = 3(x + 1)^2$

b) Find the critical points.

The derivative exists for all real numbers. Thus we solve $f'(x) = 0$.

$3(x + 1)^2 = 0$

$(x + 1)^2 = 0$

$x + 1 = 0$

$x = -1$

e) Since there is only one critical point and there are no endpoints, we can try to apply Max-Min Principle 2 using the second derivative:

$f''(x) = 6(x + 1).$

Now

$f''(-1) = 6(-1 + 1) = 0,$

so Max-Min Principle 2 fails. We cannot use Max-Min Principle 1 because there are no endpoints. We note that $f'(x) = 3(x+1)^2$ is never negative. Thus, $f(x)$ is increasing everywhere except at $x = -1$, so there is no maximum or minimum. At $x = -1$, the function has a point of inflection.

75. $f(x) = 2x - 3;\ [-1, 1]$

a) Find $f'(x)$.

$f'(x) = 2$

b) and c)

The derivative exists and is 2 for all real numbers. Note the $f'(x)$ is never 0. Thus, there are no critical points for $f(x) = 2x - 3$, and the maximum and minimum values occur at the endpoints, -1 and 1.

Find the function values at the endpoints.

$$f(x) = 2x - 3$$
$$f(-1) = 2(-1) - 3 = -2 - 3 = -5 \quad \text{Minimum}$$
$$f(1) = 2 \cdot 1 - 3 = 2 - 3 = -1 \quad \text{Maximum}$$

Thus, on the interval $[-1, 1]$, the absolute maximum $= -1$ at $x = 1$ and the absolute minimum $= -5$ at $x = -1$.

77. $f(x) = 2x - 3$

Find $f'(x)$.

$f'(x) = 2$

The derivative exists and is 2 for all real numbers. Note that $f'(x)$ is never 0. Thus, there are no critical points for $f(x) = 2x - 3$. The interval is $(-\infty, \infty)$. There are no endpoints. We also observe that $f'(x)$ is always positive. Thus $f(x)$ is increasing everywhere, so there is no maximum or minimum.

79. $f(x) = x^{2/3} = \sqrt[3]{x^2};\ [-1, 1]$

Find $f'(x)$.

$$f'(x) = \frac{2}{3}x^{-1/3} = \frac{2}{3\sqrt[3]{x}}$$

The derivative exists for all real numbers except 0. Thus, 0 is a critical point. Since there are no values of x for which $f'(x) = 0$, there are no other critical points.

On a closed interval Max-Min Principal 1 can always be used. The only critical point is 0. The endpoints are -1 and 1. We find the function value at each of these points.

$$f(-1) = \sqrt[3]{(-1)^2} = \sqrt[3]{1} = 1 \quad \text{Maximum}$$
$$f(0) = \sqrt[3]{0^2} = \sqrt[3]{0} = 0 \quad \text{Minimum}$$
$$f(1) = \sqrt[3]{1^2} = \sqrt[3]{1} = 1 \quad \text{Maximum}$$

Thus the absolute maximum $= 1$ at $x = -1$ and $x = 1$ and the absolute minimum $= 0$ at $x = 0$.

81. $f(x) = \frac{1}{3}x^3 - x + \frac{2}{3};\ (-\infty, \infty)$

Find $f'(x)$.

$f'(x) = x^2 - 1$

Find the critical points.

The derivative exists for all real numbers. We solve $f'(x) = 0$.

$$x^2 - 1 = 0$$
$$(x - 1)(x + 1) = 0$$
$$x - 1 = 0 \quad \text{or} \quad x + 1 = 0$$
$$x = 1 \quad \text{or} \quad x = -1$$

The case of finding maximum and minimum values when more than one critical point occurs in an interval which is not closed can only be solved with a detailed graph or by techniques beyond the scope of this book.

Let us study the graph.

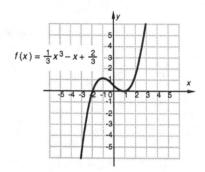

$$f(x) = \frac{1}{3}x^3 - x + \frac{2}{3}$$

From the graph we observe that on the interval $(-\infty, \infty)$, there is no absolute maximum or absolute minimum.

83. $f(x) = \frac{1}{3}x^3 - 2x^2 + x;\ [0, 4]$

Find $f'(x)$.

$f'(x) = x^2 - 4x + 1$

Find the critical points.

The derivative exists for all real numbers. We solve $f'(x) = 0$.

$$x^2 - 4x + 1 = 0$$
$$x = \frac{-(-4) \pm \sqrt{(-4)^2 - 4 \cdot 1 \cdot 1}}{2 \cdot 1} \quad \text{Using the}$$
$$\text{quadratic formula}$$
$$= \frac{4 \pm \sqrt{12}}{2}$$
$$= \frac{4 \pm 2\sqrt{3}}{2}$$
$$= 2 \pm \sqrt{3} \qquad (2 + \sqrt{3} \approx 3.7,\ 2 - \sqrt{3} \approx 0.3)$$

On a closed interval Max-Min Principle 1 can always be used. The endpoints are 0 and 4. The critical points are $2 - \sqrt{3}$ and $2 + \sqrt{3}$. We find the function value at each of these points.

$$f(x) = \frac{1}{3}x^3 - 2x^2 + x = x\left(\frac{1}{3}x^2 - 2x + 1\right)$$

$$f(0) = 0\left(\frac{1}{3} \cdot 0^2 - 2 \cdot 0 + 1\right) = 0$$

$$f(2 - \sqrt{3}) = (2 - \sqrt{3})\left[\frac{1}{3}(2 - \sqrt{3})^2 - 2(2 - \sqrt{3}) + 1\right]$$

$$= (2 - \sqrt{3})\left[\frac{1}{3}(7 - 4\sqrt{3}) - 2(2 - \sqrt{3}) + 1\right]$$

$$= (2 - \sqrt{3})\left(\frac{7}{3} - \frac{4}{3}\sqrt{3} - 4 + 2\sqrt{3} + 1\right)$$

$$= (2 - \sqrt{3})\left(-\frac{2}{3} + \frac{2}{3}\sqrt{3}\right)$$

$$= -\frac{2}{3}(2 - \sqrt{3})(1 - \sqrt{3})$$

$$= -\frac{2}{3}(2 - 3\sqrt{3} + 3)$$

$$= -\frac{2}{3}(5 - 3\sqrt{3})$$

$$= -\frac{10}{3} + 2\sqrt{3} \approx 0.131 \qquad \text{Maximum}$$

$$f(2 + \sqrt{3}) = (2 + \sqrt{3})\left[\frac{1}{3}(2 + \sqrt{3})^2 - 2(2 + \sqrt{3}) + 1\right]$$

$$= (2 + \sqrt{3})\left[\frac{1}{3}(7 + 4\sqrt{3}) - 2(2 + \sqrt{3}) + 1\right]$$

$$= (2 + \sqrt{3})\left(\frac{7}{3} + \frac{4}{3}\sqrt{3} - 4 - 2\sqrt{3} + 1\right)$$

$$= (2 + \sqrt{3})\left(-\frac{2}{3} - \frac{2}{3}\sqrt{3}\right)$$

$$= -\frac{2}{3}(2 + \sqrt{3})(1 + \sqrt{3})$$

$$= -\frac{2}{3}(2 + 3\sqrt{3} + 3)$$

$$= -\frac{2}{3}(5 + 3\sqrt{3})$$

$$= -\frac{10}{3} - 2\sqrt{3} \approx -6.797 \qquad \text{Minimum}$$

$$f(4) = 4\left(\frac{1}{3} \cdot 4^2 - 2 \cdot 4 + 1\right)$$

$$= 4\left(\frac{16}{3} - 8 + 1\right)$$

$$= 4\left(-\frac{5}{3}\right)$$

$$= -\frac{20}{3} \approx -6.667$$

Thus the absolute maximum $= -\frac{10}{3} + 2\sqrt{3}$ at $x = 2 - \sqrt{3}$
and the absolute minimum $= -\frac{10}{3} - 2\sqrt{3}$ at $x = 2 + \sqrt{3}$.

85. $f(x) = x^4 - 2x^2$

Find $f'(x)$.

$f'(x) = 4x^3 - 4x$

Find the critical points.

The derivative exists for all real numbers. We solve $f'(x) = 0$.

$$4x^3 - 4x = 0$$

$$4x(x^2 - 1) = 0$$

$$4x(x - 1)(x + 1) = 0$$

$$4x = 0 \quad \text{or} \quad x - 1 = 0 \quad \text{or} \quad x + 1 = 0$$

$$x = 0 \quad \text{or} \qquad x = 1 \quad \text{or} \qquad x = -1$$

The case of finding maximum and minimum values when more than one critical point occurs in an interval which is not closed can only be solved with a detailed graph or by techniques beyond the scope of this book.

Let us study the graph.

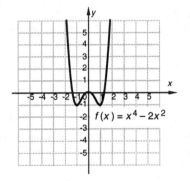

We observe from the graph that the function has no maximum value. The function has an absolute minimum value $= -1$ when $x = 1$ and $x = -1$.

87. - 95. ▮▮▮

97. $M(t) = -2t^2 + 100t + 180, \ 0 \le t \le 40$

a) $M'(t) = -4t + 100$

b) $M'(t)$ exists for all real numbers. We solve $M'(t) = 0$.

$$-4t + 100 = 0$$

$$t = 25$$

c) The critical point and endpoints are 0, 25, and 40.

d)
$$M(0) = -2 \cdot 0^2 + 100 \cdot 0 + 180 = 180$$
$$M(25) = -2 \cdot 25^2 + 100 \cdot 25 + 180 = 1430$$
$$\qquad\qquad\qquad\qquad\qquad \text{Maximum}$$
$$M(40) = -2 \cdot 40^2 + 100 \cdot 40 + 180 = 980$$

The maximum productivity for $0 \le t \le 40$ is 1430 units per month at $t = 25$ years of service.

99. $P(x) = \dfrac{1500}{x^2 - 6x + 10}$

a)
$$P'(x) = \frac{(x^2 - 6x + 10)(0) - (2x - 6)(1500)}{(x^2 - 6x + 10)^2}$$
$$\qquad\qquad\qquad\qquad\qquad \text{Quotient Rule}$$
$$= \frac{-3000x + 9000}{(x^2 - 6x + 10)^2}$$

b) Since $x^2 - 6x + 10 = 0$ has no real-number solutions, $P'(x)$ exists for all real numbers. We solve $P'(x) = 0$.

$$\frac{-3000x + 9000}{x^2 - 6x + 10} = 0$$

$$-3000x + 9000 = 0$$

$$-3000x = -9000$$

$$x = 3$$

e) We use Max-Min Principle 2.

$f''(x) =$

$\dfrac{(x^2-6x+10)^2(-3000)-2(x^2-6x+10)(2x-6)(-3000x+9000)}{[(x^2-6x+10)^2]^2}$

$=\dfrac{(x^2-6x+10)[(x^2-6x+10)(-3000)-2(2x-6)(-3000x+9000)]}{(x^2-6x+10)^4}$

$=\dfrac{-3000x^2+18,000x-30,000+12,000x^2-72,000x+108,000}{(x^2-6x+10)^3}$

$=\dfrac{9000x^2-54,000x+78,000}{(x^2-6x+10)^3}$

$f''(3) = \dfrac{9000\cdot 3^2 - 54,000\cdot 3 + 78,000}{(3^2-6\cdot 3+10)^3}$

$= -3000 < 0$, so there is an absolute maximum at $x = 3$.

Then total profit is a maximum when 3 units are produced and sold. $P(3) = 1500$, so the maximum total profit is \$1500.

101. $y = -6.1x^2 + 752x + 22,620$

a) $\dfrac{dy}{dx} = -12.2x + 752$

b) The derivative exists for all real numbers.

We solve $\dfrac{dy}{dx} = 0$.

$-12.2x + 752 = 0$

$-12.2x = -752$

$x = \dfrac{3760}{61}$

e) We use Max-Min Principle 2.

$\dfrac{d^2y}{dx^2} = -12.2$

Since $\dfrac{d^2y}{dx^2} < 0$ for all values of x, there is an absolute maximum at $x = \dfrac{3760}{61}$. Thus, the most accidents occur at a travel speed of $\dfrac{3760}{61} \approx 61.64$ mph.

103. $g(x) = x\sqrt{x+3};\ [-3, 3]$

$= x(x+3)^{1/2}$

Find $g'(x)$.

$g'(x) = x\cdot\dfrac{1}{2}(x+3)^{-1/2}\cdot 1 + 1\cdot(x+3)^{1/2}$

$= \dfrac{x}{2(x+3)^{1/2}} + (x+3)^{1/2}\cdot\dfrac{2(x+3)^{1/2}}{2(x+3)^{1/2}}$

Multiplying the second term by a form of 1

$= \dfrac{x}{2(x+3)^{1/2}} + \dfrac{2(x+3)}{2(x+3)^{1/2}}$

$= \dfrac{3x+6}{2(x+3)^{1/2}}$, or $\dfrac{3(x+2)}{2\sqrt{x+3}}$

Find the critical points.

The derivative exists for all real numbers except -3. Thus, -3 is a critical point. To check for other critical points we solve $g'(x) = 0$, when $x \neq -3$.

$\dfrac{3(x+2)}{2\sqrt{x+3}} = 0$ \quad Assuming $x \neq -3$

$3(x+2) = 0$ \quad Multiplying by $2\sqrt{x+3}$, since $x \neq -3$

$x + 2 = 0$

$x = -2$

On a closed interval Max-Min Principle 1 can always be used. The critical points are -3 and -2. The endpoints are -3 and 3. We find the function value at each of these points.

$g(x) = x\sqrt{x+3}$

$g(-3) = -3\sqrt{-3+3} = -3\sqrt{0} = -3\cdot 0 = 0$

$g(-2) = -2\sqrt{-2+3} = -2\sqrt{1} = -2\cdot 1 = -2$ \quad Minimum

$g(3) = 3\sqrt{3+3} = 3\sqrt{6}$ \quad Maximum

Thus the absolute maximum is $3\sqrt{6}$ at $x = 3$ and the absolute minimum is -2 at $x = -2$.

105. $C(x) = (2x+4) + \left(\dfrac{2}{x-6}\right),\ x > 6$

$= 2x + 4 + 2(x-6)^{-1}$

Find $C'(x)$.

$C'(x) = 2 + 0 - 2(x-6)^{-2}\cdot 1$

$= 2 - \dfrac{2}{(x-6)^2}$

Find the critical points.

The derivative exists for all real numbers in the interval $(6, \infty)$. Thus, we solve $C'(x) = 0$.

$2 - \dfrac{2}{(x-6)^2} = 0$

$2(x-6)^2 - 2 = 0$ \quad Multiplying by $(x-6)^2$ since $x \neq 6$

$(x-6)^2 - 1 = 0$

$x^2 - 12x + 36 - 1 = 0$

$x^2 - 12x + 35 = 0$

$(x-5)(x-7) = 0$

$x - 5 = 0$ \quad or \quad $x - 7 = 0$

$x = 5$ \quad or \quad $x = 7$

The only critical point in $(6, \infty)$ is 7.

We can apply the second derivative

$C''(x) = 4(x-6)^{-3}$, or $\dfrac{4}{(x-6)^3}$

to determine whether we have a maximum or a minimum. Now $C''(x)$ is positive for all values of x in $(6, \infty)$, thus there is a minimum at $x = 7$.

The firm should use 7 "quality units" to minimize its total cost of service.

107. $\boxed{tw}$

109.

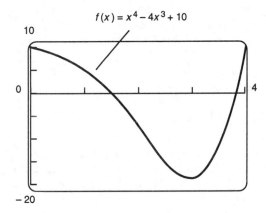

$f(x) = x^4 - 4x^3 + 10$

Minimum $= -17$ at $x = 3$; maximum $= 10$ at $x = 0$ and $x = 4$.

111.

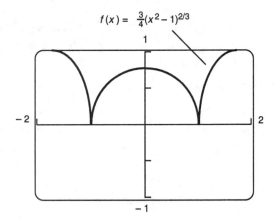

$f(x) = \frac{3}{4}(x^2 - 1)^{2/3}$

Minimum $= 0$ at $x = -1$ and $x = 1$

113. a) Using a grapher we fit the linear equation $y = x + 8.857$. We substitute 7 for x to find the pressure at 7 minutes.

$$y = 7 + 8.857 = 15.857$$

The pressure is 15.857 mm.

b) Rounding the coefficients to 3 decimal places, we have $y = 0.117x^4 - 1.520x^3 + 6.193x^2 - 7.018x + 10$.

Using the table feature we find that, when $x = 7$, $y = 24.857$ mm. (If we use the rounded coefficients above, we get 23.888 mm.) Using the trace feature, we estimate that the smallest contraction on the interval $[0, 10]$ was about 7.62 mm.

Exercise Set 3.5

---

1. Express $Q = xy$ as a function of one variable. First solve $x + y = 50$ for y.

$$x + y = 50$$
$$y = 50 - x$$

Then substitute $50 - x$ for y in $Q = xy$.

$Q = xy$

$Q = x(50 - x)$ Substituting

$\quad = 50x - x^2$

Find $Q'(x)$, where $Q(x) = 50x - x^2$.

$Q'(x) = 50 - 2x$

This derivative exists for all values of x; thus the only critical points are where

$Q'(x) = 50 - 2x = 0$

$\qquad\qquad 50 = 2x$

$\qquad\qquad 25 = x.$

Since there is only one critical point, we can use the second derivative to determine whether we have a maximum. Note that

$$Q''(x) = -2,$$

which is a constant. Thus $Q''(25)$ is negative, so $Q(25)$ is a maximum.

Now

$Q(x) = x(50 - x)$

$Q(25) = 25(50 - 25)$ Substituting

$\quad\quad = 25 \cdot 25$

$\quad\quad = 625$ Maximum

Thus the maximum product is 625 when $x = 25$. If $x = 25$ then $y = 50 - 25$, or 25. The two numbers are 25 and 25.

3. Since $Q = x(50-x)$ has no minimum, there is no minimum product. (See Exercise 1.)

5. Let x be one number and y be the other. Since the difference of the numbers must be 4,

$$x - y = 4.$$

The product, Q, of the two numbers is given b $Q = xy$.

We must minimize $Q = xy$, where $x - y = 4$.

Express $Q = xy$ as a function of one variable. First solve $x - y = 4$ for y.

$x - y = 4$

$x - 4 = y$

Then substitute $x - 4$ for y in $Q = xy$.

$Q = xy$

$Q = x(x - 4)$ Substituting

$\quad = x^2 - 4x$

Find $Q'(x)$, where $Q(x) = x^2 - 4x$.

$Q'(x) = 2x - 4$

This derivative exists for all values of x; thus the only critical points are where

$Q'(x) = 2x - 4 = 0$

$\qquad\qquad 2x = 4$

$\qquad\qquad x = 2.$

Since there is only one critical point, we can use the second derivative to determine whether we have a maximum. Note that

$$Q''(x) = 2,$$

which is a constant. Thus $Q''(2)$ is positive, so $Q(2)$ is a minimum.

Now

$$Q(x) = x(x - 4)$$
$$Q(2) = 2(2 - 4) \quad \text{Substituting}$$
$$= 2(-2)$$
$$= -4 \qquad \text{Minimum}$$

Thus the minimum product is -4 when $x = 2$. Substitute 2 for x in $x - 4 = y$ to find y.

$$x - 4 = y$$
$$2 - 4 = y \quad \text{Substituting 2 for } x$$
$$-2 = y$$

The two numbers which have the minimum product are 2 and -2.

7. Maximize $Q = xy^2$, where x and y are positive numbers, such that $x + y^2 = 1$.

Express Q as a function of one variable. First solve $x + y^2 = 1$ for y^2.

$$x + y^2 = 1$$
$$y^2 = 1 - x$$

Then substitute $1 - x$ for y^2 in $Q = xy^2$.

$$Q = xy^2$$
$$Q = x(1 - x) \quad \text{Substituting}$$
$$= x - x^2$$

Find $Q'(x)$, where $Q(x) = x - x^2$.

$$Q'(x) = 1 - 2x$$

This derivative exists for all values of x; thus the only critical points are where

$$Q'(x) = 1 - 2x = 0$$
$$-2x = -1$$
$$x = \frac{1}{2}.$$

Since there is only one critical point, we can use the second derivative to determine whether we have a maximum. Note that

$$Q''(x) = -2,$$

which is a constant. Thus $Q''\left(\frac{1}{2}\right)$ is negative, so $Q\left(\frac{1}{2}\right)$ is a maximum. Now

$$Q(x) = x(1 - x)$$
$$Q\left(\frac{1}{2}\right) = \frac{1}{2}\left(1 - \frac{1}{2}\right) \quad \text{Substituting}$$
$$= \frac{1}{2}\left(\frac{1}{2}\right)$$
$$= \frac{1}{4} \qquad \text{Maximum}$$

Substitute $\frac{1}{2}$ for x in $x + y^2 = 1$ and solve for y.

$$x + y^2 = 1$$
$$\frac{1}{2} + y^2 = 1$$
$$y^2 = \frac{1}{2}$$
$$y = \sqrt{\frac{1}{2}} \qquad x \text{ and } y \text{ must be positive.}$$

Thus the maximum value of Q is $\frac{1}{4}$ when $x = \frac{1}{2}$ and $y = \sqrt{\frac{1}{2}}$.

9. Minimize $Q = x^2 + y^2$, where $x + y = 20$.

Express Q as a function of one variable. First solve $x + y = 20$ for y.

$$x + y = 20$$
$$y = 20 - x$$

Then substitute $20 - x$ for y in $Q = x^2 + y^2$.

$$Q = x^2 + y^2$$
$$Q = x^2 + (20 - x)^2 \qquad \text{Substituting}$$
$$= x^2 + 400 - 40x + x^2$$
$$= 2x^2 - 40x + 400$$

Find $Q'(x)$, where $Q(x) = 2x^2 - 40x + 400$.

$$Q'(x) = 4x - 40$$

This derivative exists for all values of x; thus the only critical points are where

$$Q'(x) = 4x - 40 = 0$$
$$4x = 40$$
$$x = 10.$$

Since there is only one critical point, we can use the second derivative to determine whether we have a minimum. Note that

$$Q''(x) = 4,$$

which is a constant. Thus $Q''(10)$ is positive, so $Q(10)$ is a minimum. Now

$$Q(x) = x^2 + (20 - x)^2$$
$$Q(10) = 10^2 + (20 - 10)^2 \quad \text{Substituting}$$
$$= 100 + 100$$
$$= 200 \qquad \text{Minimum}$$

Substitute 10 for x in $y = 20 - x$ to find y.

$$y = 20 - x$$
$$y = 20 - 10 \quad \text{Substituting}$$
$$y = 10$$

Thus the minimum value of Q is 200 when $x = 10$ and $y = 10$.

11. Maximize $Q = xy$, where x and y are positive numbers such that $\frac{4}{3}x^2 + y = 16$.

Express Q as a function of one variable. First solve $\frac{4}{3}x^2 + y = 16$ for y.

$$\frac{4}{3}x^2 + y = 16$$
$$y = 16 - \frac{4}{3}x^2$$

Then substitute $16 - \frac{4}{3}x^2$ for y in $Q = xy$.

$$Q = xy$$
$$Q = x\left(16 - \frac{4}{3}x^2\right) \quad \text{Substituting}$$
$$= 16x - \frac{4}{3}x^3$$

Find $Q'(x)$, where $Q(x) = 16x - \frac{4}{3}x^3$.

$$Q'(x) = 16 - 4x^2$$

This derivative exists for all values of x; thus the only critical points are where

$$Q'(x) = 16 - 4x^2 = 0$$
$$-4x^2 = -16$$
$$x^2 = 4$$
$$x = \pm 2.$$

Since there is only one critical point in the domain of Q, 2, we can use the second derivative to determine whether we have a maximum. Note that,

$$Q''(x) = -8x,$$

and

$$Q''(2) = -8 \cdot 2 = -16.$$

Since $Q''(2)$ is negative, $Q(2)$ is a maximum. Now

$$Q(x) = x\left(16 - \frac{4}{3}x^2\right)$$
$$Q(2) = 2\left(16 - \frac{4}{3} \cdot 2^2\right) \quad \text{Substituting}$$
$$= 2\left(16 - \frac{16}{3}\right)$$
$$= 2\left(\frac{48}{3} - \frac{16}{3}\right)$$
$$= 2 \cdot \frac{32}{3}$$
$$= \frac{64}{3} \qquad \text{Maximum}$$

Substitute 2 for x in $y = 16 - \frac{4}{3}x^2$ to find y.

$$y = 16 - \frac{4}{3}x^2$$
$$y = 16 - \frac{4}{3} \cdot 2^2$$
$$y = 16 - \frac{16}{3}$$
$$y = \frac{48}{3} - \frac{16}{3}$$
$$y = \frac{32}{3}$$

Thus the maximum value of Q is $\frac{64}{3}$ when $x = 2$ and $y = \frac{32}{3}$.

13.

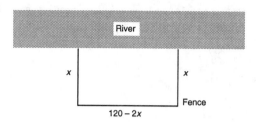

Let x be the width. Then $120 - 2x$ represents the length. The area is given by

$$A = l \cdot w$$
$$A = (120 - 2x)x \quad \text{Substituting}$$
$$= 120x - 2x^2$$

We must maximize $A(x) = 120x - 2x^2$ on the interval $(0, 60)$. We consider the interval $(0, 60)$ because x is the length of one side and cannot be negative. Since there is only 120 yd of fencing, x cannot be greater than 60. Also, x cannot be 60 because the length of the lot would be 0.

We first find $A'(x)$, where $A(x) = 120x - 2x^2$.

$$A'(x) = 120 - 4x$$

This derivative exists for all values of x in $(0, 60)$. Thus the only critical points are where

$$A'(x) = 120 - 4x = 0$$
$$-4x = -120$$
$$x = 30.$$

Since there is only one critical point in the interval, we can use the second derivative to determine whether we have a maximum. Note that

$$A''(x) = -4,$$

which is a constant. Thus $A''(30)$ is negative, so $A(30)$ is a maximum.

Now

$$A(x) = (120 - 2x)x$$
$$A(30) = (120 - 2 \cdot 30) \cdot 30 \quad \text{Substituting}$$
$$= 60 \cdot 30$$
$$= 1800 \qquad \text{Maximum}$$

Note that when $x = 30$, $120 - 2x = 120 - 2 \cdot 30$, or 60. The maximum area is 1800 yd^2 when the dimensions are 30 yd by 60 yd.

15. Let x represent the length and y the width. It is helpful to draw a picture.

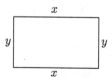

The perimeter is $x + x + y + y$, or $2x + 2y$. Thus

$$2x + 2y = 54$$

or

$$x + y = 27 \qquad \text{Multiplying by } \frac{1}{2}$$

The area is given by $A = xy$.

First solve $x + y = 27$ for y.

$$x + y = 27$$
$$y = 27 - x$$

Then substitute $27 - x$ for y in $A = xy$.

$$A = xy$$
$$A = x(27 - x) \quad \text{Substituting}$$
$$= 27x - x^2$$

We must maximize $A(x) = 27x - x^2$ on the interval $(0, 27)$. We consider the interval $(0, 27)$ because x is the length of one side and cannot be negative. Since the perimeter cannot exceed 54 ft, x cannot be greater than 27. Also, x cannot be 27 because the width of the room would be 0.

Find $A'(x)$, where $A(x) = 27x - x^2$.

$$A'(x) = 27 - 2x$$

This derivative exists for all values of x in $(0, 27)$. Thus the only critical points are where

$$A'(x) = 27 - 2x = 0$$
$$-2x = -27$$
$$x = \frac{27}{2}, \text{ or } 13.5$$

Since there is only one critical point in the interval, we can use the second derivative to determine whether we have a maximum. Note that

$$A''(x) = -2,$$

which is a constant. Thus $A''(13.5)$ is negative, so $A(13.5)$ is a maximum.

Now

$$A(x) = x(27 - x)$$
$$A(13.5) = 13.5(27 - 13.5) \quad \text{Substituting}$$
$$= 13.5(13.5)$$
$$= 182.25 \quad\quad\quad \text{Maximum}$$

Note that when $x = 13.5$, $y = 27 - 13.5$, or 13.5. The maximum area is 182.25 ft^2 when the dimensions are 13.5 ft by 13.5 ft.

17.

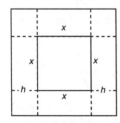 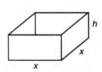 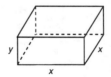

When squares of length h on a side are cut out of the corners, we are left with a square base of length x. The volume of the resulting box is

$$V = lwh = x \cdot x \cdot h.$$

We want to express V in terms of one variable. Note that the overall length of a side of the cardboard is 30 in. We see from the drawing that

$$h + x + h = 30,$$
or
$$x + 2h = 30.$$

Solving for h we get

$$2h = 30 - x$$
$$h = \frac{1}{2}(30 - x) = \frac{1}{2} \cdot 30 - \frac{1}{2}x = 15 - \frac{1}{2}x.$$

Thus

$$V = x \cdot x \cdot \left(15 - \frac{1}{2}x\right) = x^2\left(15 - \frac{1}{2}x\right) = 15x^2 - \frac{1}{2}x^3.$$ We must maximize $V(x)$ on the interval $(0, 30)$. We first find $V'(x)$.

$$V'(x) = 30x - \frac{3}{2}x^2.$$

Now $V'(x)$ exists for all x in the interval $(0, 30)$, so we set it equal to 0 to find the critical points:

$$V'(x) = 30x - \frac{3}{2}x^2 = 0$$
$$x\left(30 - \frac{3}{2}x\right) = 0$$
$$x = 0 \text{ or } 30 - \frac{3}{2}x = 0$$
$$x = 0 \text{ or } \quad -\frac{3}{2}x = -30$$
$$x = 0 \text{ or } \quad\quad x = 20$$

The only critical point in $(0, 30)$ is 20. Thus we can use the second derivative,

$$V''(x) = 30 - 3x,$$

to determine whether we have a maximum. Since

$$V''(20) = 30 - 3 \cdot 20 = 30 - 60 = -30,$$

$V''(20)$ is negative, so $V(20)$ is a maximum, and

$$V(20) = 15 \cdot 20^2 - \frac{1}{2} \cdot 20^3$$
$$= 15 \cdot 400 - \frac{1}{2} \cdot 8000$$
$$= 6000 - 4000$$
$$= 2000$$

The maximum volume is 2000 in^3. The dimensions are $x = 20$ in. by $x = 20$ in. by $h = 15 - \frac{1}{2} \cdot 20 = 5$ in.

19. We first make a drawing.

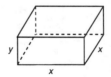

The surface area of the open-top, square-based, rectangular box is

$$S = x^2 + 4xy \quad\quad x^2 \text{ is area of base; } xy \text{ is area}$$
$$\text{of one side, } 4xy \text{ is total area}$$
$$\text{of 4 sides}$$

The volume must be 62.5 cubic inches, and is given by

$$V = l \cdot w \cdot h = x \cdot x \cdot y = x^2y = 62.5.$$

To express S in terms of one variable, we solve $x^2y = 62.5$ for y:

$$y = \frac{62.5}{x^2}$$

Then

$$S(x) = x^2 + 4x\left(\frac{62.5}{x^2}\right) \quad \text{Substituting } \frac{62.5}{x^2} \text{ for } y$$

$$= x^2 + \frac{250}{x}, \text{ or } x^2 + 250x^{-1}$$

Now S is defined only for positive numbers, and the problem dictates that the length x be positive, so we minimize S on the interval $(0, \infty)$.

We first find $S'(x)$.

$$S'(x) = 2x - 250x^{-2}$$

$$= 2x - \frac{250}{x^2}$$

Since $S'(x)$ exists for all x in $(0, \infty)$, the only critical points are where $S'(x) = 0$. We solve the following:

$$2x - \frac{250}{x^2} = 0$$

$$x^2\left(2x - \frac{250}{x^2}\right) = x^2 \cdot 0 \quad \text{Multiplying by } x^2 \text{ to clear fractions}$$

$$2x^3 - 250 = 0$$

$$2x^3 = 250$$

$$x^3 = 125$$

$$x = 5$$

Since there is only one critical point, we use the second derivative to determine whether we have a minimum. Note that

$$S''(x) = 2 + 500x^{-3}$$

$$= 2 + \frac{500}{x^3}$$

Since $S''(x)$ is positive for all positive values of x, we have a minimum at $x = 5$.

We find y when $x = 2.5$:

$$y = \frac{62.5}{x^2}$$

$$= \frac{62.5}{5^2} \quad \text{Substituting 5 for } x$$

$$= \frac{62.5}{25}$$

$$= 2.5$$

Thus the surface area is minimized when $x = 5$ in. and $y = 2.5$ in. We find the minimum surface area by substituting 5 for x and 2.5 for y in

$$S = x^2 + 4xy$$

$$S = 5^2 + 4 \cdot 5 \cdot 2.5$$

$$= 25 + 50$$

$$= 75.$$

The minimum surface area is 75 in^2 when the dimensions are 5 in. by 5 in. by 2.5 in.

21. $R(x) = 50x - 0.5x^2$

$C(x) = 4x + 10$

$$\begin{aligned} P(x) &= R(x) - C(x) \quad \text{Profit = Revenue − Cost} \\ &= (50x - 0.5x^2) - (4x + 10) \quad \text{Substituting} \\ &= -0.5x^2 + 46x - 10 \end{aligned}$$

To find the maximum value of $P(x)$, we first find $P'(x)$.

$$P'(x) = -x + 46$$

This derivative exists for all values of x, but we are actually interested in only the numbers x in $[0, \infty)$ since we cannot produce a negative number of units. Thus we solve $P'(x) = 0$.

$$P'(x) = -x + 46 = 0$$

$$-x = -46$$

$$x = 46$$

Since there is only one critical point, we can use the second derivative to determine whether we have a maximum. Note that

$$P''(x) = -1, \text{ a constant.}$$

Thus $P''(46)$ is negative, so $P(46)$ is a maximum.

The maximum profit is given by

$$\begin{aligned} P(46) &= -0.5(46)^2 + 46 \cdot 46 - 10 \\ &= -1058 + 2116 - 10 \\ &= 1048 \end{aligned}$$

Thus the maximum profit is \$1048 when 46 units are produced and sold.

23. $R(x) = 2x$

$C(x) = 0.01x^2 + 0.6x + 30$

$$\begin{aligned} P(x) &= R(x) - C(x) \\ &= 2x - (0.01x^2 + 0.6x + 30) \\ &= -0.01x^2 + 1.4x - 30 \end{aligned}$$

To find the maximum value of $P(x)$, we first find $P'(x)$.

$$P'(x) = -0.02x + 1.4$$

This derivative exists for all values of x, but we are actually interested in only the numbers x in $[0, \infty)$ since we cannot produce a negative number of units. Thus we solve $P'(x) = 0$.

$$P'(x) = -0.02x + 1.4 = 0$$

$$-0.02x = -1.4$$

$$x = 70$$

Since there is only one critical point, we can use the second derivative to determine whether we have a maximum. Note that

$$P''(x) = -0.02, \text{ a constant.}$$

Thus $P''(70)$ is negative, so $P(70)$ is a maximum.

The maximum profit is given by

$$\begin{aligned} P(70) &= -0.01(70)^2 + 1.4(70) - 30 \\ &= -49 + 98 - 30 \\ &= 19 \end{aligned}$$

Thus the maximum profit is \$19 when 70 units are produced and sold.

25. $R(x) = 9x - 2x^2$

$C(x) = x^3 - 3x^2 + 4x + 1$

$R(x)$ and $C(x)$ are in thousands of dollars, and x is in thousands of units.

$$P(x) = R(x) - C(x)$$
$$= (9x - 2x^2) - (x^3 - 3x^2 + 4x + 1)$$
$$= -x^3 + x^2 + 5x - 1$$

To find the maximum value of $P(x)$, we first find $P'(x)$.

$$P'(x) = -3x^2 + 2x + 5$$

This derivative exists for all values of x, but we are actually interested in only the numbers x in $[0, \infty)$ since we cannot produce a negative number of units. Thus we solve $P'(x) = 0$.

$$P'(x) = -3x^2 + 2x + 5 = 0$$
$$3x^2 - 2x - 5 = 0$$
$$(3x - 5)(x + 1) = 0$$
$$3x - 5 = 0 \quad \text{or} \quad x + 1 = 0$$
$$x = \frac{5}{3} \quad \text{or} \quad x = -1$$

There is only one critical point in the interval $[0, \infty)$. We use the second derivative to determine whether we have a maximum. Note that

$$P''(x) = -6x + 2.$$

Find $P''\left(\frac{5}{3}\right)$.

$$P''\left(\frac{5}{3}\right) = -6 \cdot \frac{5}{3} + 2 = -10 + 2 = -8$$

$P''\left(\frac{5}{3}\right) < 0$, so $P\left(\frac{5}{3}\right)$ is a maximum.

The maximum profit is given by

$$P(x) = -x^3 + x^2 + 5x - 1$$

$$P\left(\frac{5}{3}\right) = -\left(\frac{5}{3}\right)^3 + \left(\frac{5}{3}\right)^2 + 5 \cdot \frac{5}{3} - 1$$

$$= -\frac{125}{27} + \frac{25}{9} + \frac{25}{3} - 1$$

$$= -\frac{125}{27} + \frac{75}{27} + \frac{225}{27} - \frac{27}{27}$$

$$= \frac{148}{27}$$

Note that $\frac{5}{3}$ thousand ≈ 1.667 thousand $= 1667$ and that $\frac{148}{27}$ thousand ≈ 5.481 thousand $= 5481$.

Thus the maximum profit is approximately $5481 when approximately 1667 units are produced and sold.

27. $p = 150 - 0.5x$ Price per suit

$C(x) = 4000 + 0.25x^2$ Total cost

a) $\underbrace{\text{Total}}_{\text{revenue}} = \underbrace{\text{Number}}_{\text{of suits}} \cdot \underbrace{\text{Price}}_{\text{per suit}}$

$R(x) = x \cdot (150 - 0.5x)$

          Substituting x for number of suits and $150 - 0.5x$ for price per suit

$R(x) = 150x - 0.5x^2$

b) $\underbrace{\text{Total}}_{\text{profit}} = \underbrace{\text{Total}}_{\text{revenue}} \cdot \underbrace{\text{Total}}_{\text{cost}}$

$$P(x) = R(x) - C(x)$$
$$P(x) = (150x - 0.5x^2) - (4000 + 0.25x^2)$$
$$P(x) = 150x - 0.5x^2 - 4000 - 0.25x^2$$
$$P(x) = -0.75x^2 + 150x - 4000$$

c) To determine the number of suits required to maximize profit, we first find $P'(x)$.

$$P'(x) = -1.5x + 150$$

The derivative exists for all real numbers, but we are only interested in numbers x in $[0, \infty)$ because a negative number of suits cannot be produced. Thus we solve:

$$P'(x) = -1.5x + 150 = 0$$
$$-1.5x = -150$$
$$x = 100$$

Since there is only one critical point, we use the second derivative to determine whether we have a maximum. Note that

$$P''(x) = -1.5, \text{ a constant.}$$

Thus $P''(100)$ is negative, so $P(100)$ is a maximum. The company must produce and sell 100 suits to maximize profit.

d) The maximum profit is given by substituting 100 for x in the profit function.

$$P(x) = -0.75x^2 + 150x - 4000$$
$$P(100) = -0.75 \cdot 100^2 + 150 \cdot 100 - 4000$$
$$= -7500 + 15,000 - 4000$$
$$= 3500$$

Thus the maximum profit is $3500.

e) The price per suit is given by

$$p = 150 - 0.5x$$
$$p = 150 - 0.5(100) \quad \text{Substituting 100 for } x$$
$$= 150 - 50$$
$$= 100$$

The price per suit must be $100.

29. Let x = the amount by which the price of $6 should be increased (if x is negative, the price would be decreased). We first express total revenue R as a function of x.

$$R(x) = (\text{Revenue from tickets}) + (\text{Revenue from concessions})$$

$$= (\text{Number of people}) \cdot (\text{Ticket price}) + \$1.50(\text{Number of people})$$

$$= (70,000 - 10,000x)(6 + x) + 1.50(70,000 - 10,000x)$$

$$= 420,000 + 70,000x - 60,000x - 10,000x^2 + 105,000 - 15,000x$$

$$R(x) = -10,000x^2 - 5000x + 525,000$$

To find x such that $R(x)$ is a maximum, we first find $R'(x)$:

$R'(x) = -20,000x - 5000.$

This derivative exists for all real numbers x; thus the only critical points are where $R'(x) = 0$.

$$R'(x) = -20,000x - 5000 = 0$$
$$-20,000x = 5000$$
$$x = -0.25$$

Since this is the only critical point, we can use the second derivative,

$$R''(x) = -20,000,$$

to determine whether we have a maximum.

Since $R''(-0.25)$ is negative, $R(-0.25)$ is a maximum. Thus to maximize revenue the university should charge

$\$6 + (-\$0.25)$ or $\$5.75$ per ticket.

That is, this reduced ticket price will get more people into the theater,

$70,000 - 10,000(-0.25)$, or $72,500$,

and will result in maximum revenue.

31. Let $x = $ the number of additional trees per acre which should be planted. Then the number of trees planted per acre is represented by $(20 + x)$ and the yield per tree by $(30 - x)$.

Total yield per acre	=	Yield per tree	·	Number of trees per acre

$$Y(x) = (30 - x)(20 + x)$$
$$= 600 + 10x - x^2$$

To find x such that $Y(x)$ is a maximum we first find $Y'(x)$.

$Y'(x) = 10 - 2x$

This derivative exists for all real numbers x; thus the only critical points are where $Y'(x) = 0$, so we solve that equation.

$$10 - 2x = 0$$
$$-2x = -10$$
$$x = 5$$

Since this is the only critical point, we can use the second derivative,

$$Y''(x) = -2,$$

to determine whether we have a maximum. Since $Y''(5)$ is negative, $Y(5)$ is a maximum. Thus to maximize yield, the apple farm should plant

$20 + 5$, or 25 trees per acre.

33. The volume of the box is given by

$V = x \cdot x \cdot y = x^2 y = 320.$

The area of the base is x^2. The cost of the base is $15x^2$ cents.

The area of the top is x^2. The cost of the top is $10x^2$ cents.

The area of each side is xy; the total area for the four sides is $4xy$. The cost of the four sides is $2.5(4xy)$ cents.

The total cost in cents is given by

$$C = 15x^2 + 10x^2 + 2.5(4xy)$$
$$= 25x^2 + 10xy$$

To express C in terms of one variable, we solve $x^2 y = 320$ for y:

$$y = \frac{320}{x^2}$$

Then

$$C(x) = 25x^2 + 10x\left(\frac{320}{x^2}\right) \quad \text{Substituting } \frac{320}{x^2} \text{ for } y$$
$$= 25x^2 + \frac{3200}{x}, \text{ or } 25x^2 + 3200x^{-1}$$

Now C is defined only for positive numbers, and the problem dictates that the length x be positive, so we are minimizing C on the interval $(0, \infty)$.

We first find $C'(x)$.

$$C'(x) = 50x - 3200x^{-2}$$
$$= 50x - \frac{3200}{x^2}$$

Since $C'(x)$ exists for all x in $(0, \infty)$, the only critical points are where $C'(x) = 0$. Thus we solve the following equation:

$$50x - \frac{3200}{x^2} = 0$$
$$x^2\left(50x - \frac{3200}{x^2}\right) = x^2 \cdot 0 \quad \begin{array}{l} \text{Multiplying by } x^2 \text{ to} \\ \text{clear fractions} \end{array}$$
$$50x^3 - 3200 = 0$$
$$50x^3 = 3200$$
$$x^3 = 64$$
$$x = 4$$

This is the only critical point, so we can use the second derivative to determine whether we have a minimum.

$$C''(x) = 50 + 6400x^{-3}$$
$$= 50 + \frac{6400}{x^3}$$

Note that this is positive for all positive values of x. Thus we have a minimum at $x = 4$. We find y when $x = 4$:

$$y = \frac{320}{x^2}$$
$$y = \frac{320}{4^2} \quad \text{Substituting 4 for } x$$
$$y = \frac{320}{16}$$
$$y = 20$$

Thus the cost is minimized when the dimensions are 4 ft by 4 ft by 20 ft.

35. Let A represent the amount deposited in savings accounts and i represent the interest rate paid on the money deposited. If A is directly proportional to i, then there is some positive constant k such that $A = ki$. The interest earned by the bank is represented by $18\%A$, or $0.18A$. The interest paid by the bank is represented by iA. The profit received by the bank is given by

$P = 18\%A - iA.$

We first express $P = 18\%A - iA$ as a function of one variable by substituting ki for A.

$$P = 18\%(ki) - i(ki)$$
$$P = 0.18ki - ki^2 \qquad k \text{ is a constant}$$

We maximize P on the interval $(0, \infty)$. We first find $P'(i)$.

$$P'(i) = 0.18k - 2ki$$

Since $P'(i)$ exists for all i in $(0, \infty)$, the only critical points are where $P'(i) = 0$. We solve the following equation.

$$0.18k - 2ki = 0$$
$$-2ki = -0.18k$$
$$i = \frac{-0.18k}{-2k}$$
$$i = 0.09$$

Since there is only one critical point, we can use the second derivative to determine whether we have a maximum. Note that $P''(i) = -2k$, which is a negative constant $(k > 0)$. Thus $P''(0.09)$ is negative, so $P(0.09)$ is a maximum. To maximize profit, the bank should pay 9% on its savings accounts.

37. Let x = the lot size. Now the inventory costs are given by

$$C(x) = \begin{array}{c} \text{Yearly} \\ \text{carrying} \\ \text{costs} \end{array} + \begin{array}{c} \text{Yearly} \\ \text{reorder} \\ \text{costs} \end{array}$$

We consider each separately:

Yearly carrying costs: The average amount held in stock is $x/2$, and it costs \$20 per pool table for storage. Thus

$$\begin{array}{c} \text{Yearly} \\ \text{carrying} \\ \text{costs} \end{array} = \begin{array}{c} \text{Yearly cost} \\ \text{per item} \end{array} \cdot \begin{array}{c} \text{Average number} \\ \text{of items} \end{array}$$
$$= 20 \cdot \frac{x}{2}$$
$$= 10x$$

Yearly reorder costs: Now x = lot size, and suppose there are N reorders each year. Then $Nx = 100$, and $N = 100/x$. If there is a fixed cost of \$40 for each order plus \$16 for each pool table, the cost of each order is $40 + 16x$. Thus

$$\begin{array}{c} \text{Yearly} \\ \text{reorder} \\ \text{costs} \end{array} = \begin{array}{c} \text{Cost of each} \\ \text{order} \end{array} \cdot \begin{array}{c} \text{Number of} \\ \text{reorders} \end{array}$$
$$= (40 + 16x)\left(\frac{100}{x}\right)$$
$$= \frac{4000}{x} + 1600$$

Hence

$$C(x) = 10x + \frac{4000}{x} + 1600$$
$$= 10x + 4000x^{-1} + 1600$$

We want to find a minimum value of C on the interval $[1, 100]$. We first find $C'(x)$.

$$C'(x) = 10 - 4000x^{-2}$$
$$= 10 - \frac{4000}{x^2}$$

Now $C'(x)$ exists for all x in $[1, 100]$, so the only critical points are where $C'(x) = 0$. We solve $C'(x) = 0$.

$$10 - \frac{4000}{x^2} = 0$$
$$10 = \frac{4000}{x^2}$$
$$10x^2 = 4000$$
$$x^2 = 400$$
$$x = \pm 20$$

Now there is only one critical point in the interval $[1, 100]$, $x = 20$, so we can use the second derivative to see if we have a minimum:

$$C''(x) = 8000x^{-3} = \frac{8000}{x^3}$$

Now $C''(x)$ is positive for all x in $[1, 100]$, so we do have a minimum at $x = 20$. Thus to minimize inventory costs, the store should order pool tables $100/20$, or 5 times a year. The lot size is 20.

39. Let x = the lot size. Now the inventory costs are given by

$$C(x) = \begin{array}{c} \text{Yearly} \\ \text{carrying} \\ \text{costs} \end{array} + \begin{array}{c} \text{Yearly} \\ \text{reorder} \\ \text{costs} \end{array}$$

We consider each separately:

Yearly carrying costs: The average amount held in stock is $x/2$, and it costs \$8 per calculator for storage. Thus

$$\begin{array}{c} \text{Yearly} \\ \text{carrying} \\ \text{costs} \end{array} = \begin{array}{c} \text{Yearly cost} \\ \text{per item} \end{array} \cdot \begin{array}{c} \text{Average number} \\ \text{of items} \end{array}$$
$$= 8 \cdot \frac{x}{2}$$
$$= 4x$$

Yearly reorder costs: Now x = lot size, and suppose there are N reorders each year. Then $Nx = 360$, and $N = 360/x$. If there is a fixed cost of \$10 for each order plus \$8 for each calculator, the cost of each order is $10 + 8x$. Thus

$$\begin{array}{c} \text{Yearly} \\ \text{reorder} \\ \text{costs} \end{array} = \begin{array}{c} \text{Cost of each} \\ \text{order} \end{array} \cdot \begin{array}{c} \text{Number of} \\ \text{reorders} \end{array}$$
$$= (10 + 8x)\left(\frac{360}{x}\right)$$
$$= \frac{3600}{x} + 2880$$

Hence

$$C(x) = 4x + \frac{3600}{x} + 2880$$
$$= 4x + 3600x^{-1} + 2880$$

We want to find a minimum value of C on the interval $[1, 360]$. We first find $C'(x)$.

$$C'(x) = 4 - 3600x^{-2}$$
$$= 4 - \frac{3600}{x^2}$$

Now $C'(x)$ exists for all x in $[1, 360]$, so the only critical points are where $C'(x) = 0$. We solve $C'(x) = 0$.

$$4 - \frac{3600}{x^2} = 0$$

$$4 = \frac{3600}{x^2}$$

$$4x^2 = 3600$$

$$x^2 = 900$$

$$x = \pm 30$$

Now there is only one critical point in the interval $[1, 360]$, $x = 30$, so we can use the second derivative to see if we have a minimum:

$$C''(x) = 7200x^{-3} = \frac{7200}{x^3}$$

Now $C''(x)$ is positive for all x in $[1, 360]$, so we do have a minimum at $x = 30$. Thus to minimize inventory costs, the store should order calculators $360/30$, or 12, times per year. The lot size is 30.

41.   Yearly
    carrying $= 9 \cdot \dfrac{x}{2} = 4.5x$
    costs

   Yearly
   reorder $= (10 + 8x) \cdot \dfrac{360}{x} = \dfrac{3600}{x} + 2880$
   costs

Hence

$$C(x) = 4.5x + \frac{3600}{x} + 2800$$

and

$$C'(x) = 4.5 - \frac{3600}{x^2}$$

Now $C'(x)$ exists for all x in $[1, 360]$, so the only critical points are where $C'(x) = 0$. We solve $C'(x) = 0$.

$$4.5 - \frac{3600}{x^2} = 0$$

$$4.5 = \frac{3600}{x^2}$$

$$4.5x^2 = 3600$$

$$x^2 = 800$$

$$x = \pm\sqrt{800} \approx 28.3$$

Now there is only one critical point in the interval $[1, 360]$, $x = \sqrt{800} \approx 28.3$, so we can use the second derivative to see if we have a minimum:

$$C''(x) = 7200x^{-3} = \frac{7200}{x^3}$$

Now $C''(x)$ is positive for all x in $[1, 360]$, so we do have a minimum at $x \approx 28.3$. Since it does not make sense to reorder 28.3 calculators each time, we consider the two integers closest to 28.3, which are 28 and 29.

$C(28) \approx \$3134.57$ and $C(29) \approx \$3134.64$

Then the lot size that will minimize cost is 28. The store should reorder $360/28 \approx 13$ times per year.

43. Case I.

If y is the length, the girth is $x + x + x + x$, or $4x$.

Case II.

If x is the length, the girth is $x + y + x + y$, or $2x + 2y$.

Case I.

The combined length and girth is

$y + 4x = 84$.

The volume is

$V = x \cdot x \cdot y$.

We want to express V in terms of one variable. We solve $y + 4x = 84$ for y.

$$y + 4x = 84$$

$$y = 84 - 4x$$

Thus

$V = x \cdot x \cdot (84 - 4x)$, or $84x^2 - 4x^3$.

To maximize $V(x)$ we first find $V'(x)$.

$$V'(x) = 168x - 12x^2$$

Now $V'(x)$ exists for all x so we set it equal to 0 to find the critical points:

$$V'(x) = 168x - 12x^2 = 0$$

$$12x(14 - x) = 0$$

$$12x = 0 \ \text{ or } \ 14 - x = 0$$

$$x = 0 \ \text{ or } \qquad x = 14$$

Since $x \neq 0$, we only consider $x = 14$. We can use the second derivative to determine that $V(14)$ is a maximum. If $x = 14$, $y = 84 - 4 \cdot 14$, or 28. The dimensions are 14 in. by 14 in. by 28 in. The volume is $14 \times 14 \times 28$, or 5488 in^3.

Case II.

The combined length and girth is

$x + 2x + 2y = 84$

or

$3x + 2y = 84$

The volume is

$V = x \cdot x \cdot y$.

We want to express V in terms of one variable. We solve $3x + 2y = 84$ for y.

$$3x + 2y = 84$$

$$2y = 84 - 3x$$

$$y = 42 - 1.5x$$

Thus

$V = x \cdot x \cdot (42 - 1.5x)$, or $42x^2 - 1.5x^3$.

To maximize $V(x)$ we first find $V'(x)$.

$$V'(x) = 84x - 4.5x^2$$

Now $V'(x)$ exists for all x so we set it equal to 0 to find the critical points:

$$V'(x) = 84x - 4.5x^2 = 0$$

$$3x(28 - 1.5x) = 0$$

$$3x = 0 \ \text{ or } \ 28 - 1.5x = 0$$

$$x = 0 \ \text{ or } \qquad -1.5x = -28$$

$$x = 0 \ \text{ or } \qquad x \approx 18.7$$

Since $x \neq 0$, we only consider 18.7. We can use the second derivative to determine that $V(18.7)$ is a maximum. If $x = 18.7$, $y = 42 - 1.5(18.7)$, or 13.95. The dimensions are

18.7 in. by 18.7 in. by 13.95 in. The volume is
$18.7 \times 18.7 \times 13.95 \approx 4878$ in^3.

Comparing Case I and Case II we see that the maximum volume is 5488 in^3 when the dimensions are 14 in. by 14 in. by 28 in.

45.

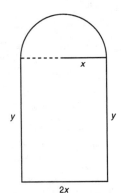

(2x represents the width of the window and the diameter of the circle.)

Consider a circle whose radius is x. The circumference of the circle is

$$C = \pi(2x). \qquad (C = \pi d)$$

The perimeter of the semicircle is

$$\frac{1}{2}C = \frac{1}{2}\pi(2x) = \pi x.$$

The perimeter of the three sides of the rectangle which form the remaining part of the total perimeter of the window is

$$2x + y + y \text{ or } 2x + 2y.$$

The total perimeter of the window is

$$\pi x + 2x + 2y = 24.$$

Maximizing the amount of light is the same as maximizing the area of the window. The area of the circle is

$$A = \pi x^2. \qquad (A = \pi r^2)$$

The area of the semicircle is

$$\frac{1}{2}A = \frac{1}{2}\pi x^2.$$

The area of the rectangle is

$$2x \cdot y.$$

The total area of the Norman window is

$$A = \frac{1}{2}\pi x^2 + 2xy.$$

To express A in terms of one variable, we solve $\pi x + 2x + 2y = 24$ for y:

$$\pi x + 2x + 2y = 24$$
$$2y = 24 - \pi x - 2x$$
$$y = 12 - \frac{\pi}{2}x - x$$

Then

$$A(x) = \frac{1}{2}\pi x^2 + 2x\left(12 - \frac{\pi}{2}x - x\right) \quad \text{Substituting}$$

$$= \frac{1}{2}\pi x^2 + 24x - \pi x^2 - 2x^2$$

$$= \left(-\frac{1}{2}\pi - 2\right)x^2 + 24x$$

We maximize A on the interval $(0, \infty)$. We first find $A'(x)$.

$$A'(x) = (-\pi - 4)x + 24$$

Since $A'(x)$ exists for all x in $(0, \infty)$, the only critical points are where $A'(x) = 0$. Thus we solve the following equation:

$$(-\pi - 4)x + 24 = 0$$
$$(-\pi - 4)x = -24$$
$$x = \frac{-24}{-\pi - 4}$$
$$x = \frac{24}{\pi + 4} \qquad \text{Multiplying by } \frac{-1}{-1}$$

This is the only critical point, so we can use the second derivative to determine whether we have a maximum.

$$A''(x) = -\pi - 4$$

Note that $-\pi - 4$ is a constant. It is negative for all values of x. Thus $A''\left(\dfrac{24}{\pi + 4}\right)$ is negative, and we have a maximum at $x = \dfrac{24}{\pi + 4}$.

We find y when $x = \dfrac{24}{\pi + 4}$.

$$y = 12 - \frac{\pi}{2}x - x$$

$$y = 12 - \frac{\pi}{2}\left(\frac{24}{\pi + 4}\right) - \left(\frac{24}{\pi + 4}\right)$$

$$\text{Substituting } \frac{24}{\pi + 4} \text{ for } x$$

$$= 12\left(\frac{\pi + 4}{\pi + 4}\right) - \frac{\pi}{2}\left(\frac{24}{\pi + 4}\right) - \left(\frac{24}{\pi + 4}\right)$$

$$\text{⌐————————— Multiplying by 1}$$

$$= \frac{12\pi + 48 - 12\pi - 24}{\pi + 4}$$

$$= \frac{24}{\pi + 4}$$

To maximize the amount of light through the window the dimensions must be

$$x = \frac{24}{\pi + 4} \text{ ft and } y = \frac{24}{\pi + 4} \text{ ft.}$$

47. Let x be the positive number. Then $\dfrac{1}{x}$ is the reciprocal of the number, and x^2 is the square of the number.

The sum, S, of the reciprocal and five times the square is given by

$$S = \frac{1}{x} + 5x^2, \text{ or } x^{-1} + 5x^2.$$

We minimize $S(x)$ on the interval $(0, \infty)$. We first find $S'(x)$.

$$S'(x) = -x^{-2} + 10x$$
$$= -\frac{1}{x^2} + 10x$$

Since $S'(x)$ exists for all x in $(0, \infty)$, the only critical points are where $S'(x) = 0$. We solve the following equation.

$$-\frac{1}{x^2} + 10x = 0$$

$$x^2\left(-\frac{1}{x^2} + 10x\right) = x^2 \cdot 0 \quad \begin{array}{l}\text{Multiplying by } x^2 \text{ to} \\ \text{clear fractions}\end{array}$$

$$-1 + 10x^3 = 0$$

$$x^3 = \frac{1}{10}$$

$$x = \sqrt[3]{\frac{1}{10}}$$

Since there is only one critical point, we can use the second derivative,

$$S''(x) = 2x^{-3} + 10$$

$$= \frac{2}{x^3} + 10$$

to determine whether we have a minimum. Since $S''(x)$ is positive for all positive values of x, the sum is a minimum when $x = \sqrt[3]{\frac{1}{10}}$.

49.

$$\begin{array}{l}\text{Yearly} \\ \text{carrying} \\ \text{costs}\end{array} = a \cdot \frac{x}{2} = \frac{ax}{2}$$

$$\begin{array}{l}\text{Yearly} \\ \text{reorder} \\ \text{costs}\end{array} = (b + cx)\frac{Q}{x} = \frac{bQ}{x} + cQ$$

Hence

$$C(x) = \frac{ax}{2} + \frac{bQ}{x} + cQ$$

$$= \frac{a}{2}x + bQx^{-1} + cQ$$

We want to find a minimum value of C on the interval $[1, Q]$. We first find $C'(x)$.

$$C'(x) = \frac{a}{2} - bQx^{-2} = \frac{a}{2} - \frac{bQ}{x^2}$$

Now $C'(x)$ exists for all x in $[1, Q]$, so the only critical points are where $C'(x) = 0$. We solve $C'(x) = 0$.

$$\frac{a}{2} - \frac{bQ}{x^2} = 0$$

$$\frac{a}{2} = \frac{bQ}{x^2}$$

$$ax^2 = 2bQ$$

$$x^2 = \frac{2bQ}{a}$$

$$x = \pm\sqrt{\frac{2bQ}{a}}$$

Now there is only one critical point in the interval $[1, Q]$, $x = \sqrt{\frac{2bQ}{a}}$, so we can use the second derivative to see if we have a minimum:

$$C''(x) = 2bQx^{-3} = \frac{2bQ}{x^3}$$

Now $C''(x)$ is positive for all x in $[1, Q]$, so we do have a minimum at $x = \sqrt{\frac{2bQ}{a}}$. Thus to minimize inventory

costs, the store should order $\sqrt{\frac{2bQ}{a}}$ units of a product $\frac{Q}{\sqrt{\frac{2bQ}{a}}}$ times per year.

51. The circumference of the circle is

$$x = \pi d, \text{ or } x = 2\pi r. \qquad (C = \pi d)$$

Solving the equation $x = 2\pi r$ for r, we get $r = \frac{x}{2\pi}$.

$$A_c = \pi\left(\frac{x}{2\pi}\right)^2 \qquad (A = \pi r^2)$$

$$= \pi \cdot \frac{x^2}{4\pi^2}$$

$$= \frac{x^2}{4\pi}$$

The perimeter of the square is $24 - x$.

The length of a side of the square is $\frac{24 - x}{4}$.

The area of the square is

$$A_s = \left(\frac{24 - x}{4}\right)^2 \qquad (A = s^2)$$

$$= \frac{576 - 48x + x^2}{16}$$

$$= \frac{x^2 - 48x + 576}{16}$$

The sum of the areas is

$$A = \frac{x^2}{4\pi} + \frac{x^2 - 48x + 576}{16}$$

$$= \frac{1}{4\pi}x^2 + \frac{1}{16}x^2 - 3x + 36$$

$$= \left(\frac{1}{4\pi} + \frac{1}{16}\right)x^2 - 3x + 36$$

$$= \left(\frac{4}{16\pi} + \frac{\pi}{16\pi}\right)x^2 - 3x + 36$$

$$= \frac{4 + \pi}{16\pi}x^2 - 3x + 36$$

We minimize A on the interval $(0, 24)$. We first find $A'(x)$.

$$A'(x) = 2\left(\frac{4 + \pi}{16\pi}\right)x - 3$$

$$= \frac{4 + \pi}{8\pi}x - 3$$

Since $A'(x)$ exists for all x in $(0, 24)$, the only critical point is where $A'(x) = 0$. We solve the following equation.

$$\frac{4 + \pi}{8\pi}x - 3 = 0$$

$$\frac{4 + \pi}{8\pi}x = 3$$

$$(4 + \pi)x = 24\pi$$

$$x = \frac{24\pi}{4 + \pi} \approx 10.56$$

Since there is only one critical point, we can use the second derivative to determine whether we have a minimum. Note that $A''(x) = \frac{4 + \pi}{8\pi}$, which is a positive constant. Thus $A''\left(\frac{24\pi}{4 + \pi}\right)$ is positive, so $A\left(\frac{24\pi}{4 + \pi}\right)$ is a minimum and $x \approx 10.56$ in. and $24 - x \approx 13.44$ in. There is no maximum if the string is to be cut. One would interpret the maximum to be at the endpoint, with the string uncut and used to form a circle.

53. First find the distance over water (from C to S).

$$(CS)^2 = 3^2 + x^2$$
$$CS = \sqrt{9 + x^2}$$

The distance from S to A is $8 - x$.

Letting r represent the energy rate over land and $1.28r$ represent the energy rate over water, we have the following equation for total energy.

Total energy	=	Energy rate over water	$\cdot$	Distance over water	+	Energy rate over land	$\cdot$	Distance over land

$$TE = 1.28r \cdot \sqrt{9 + x^2} + r \cdot (8 - x)$$
$$= 1.28r(9 + x^2)^{1/2} + 8r - rx$$
$$\qquad\qquad r \text{ is a positive constant}$$

We minimize the total energy over the interval $[0, 8]$. We find $TE'(x)$.

$$TE'(x) = 1.28r \cdot \frac{1}{2}(9 + x^2)^{-1/2} \cdot 2x - r$$

$$= \frac{1.28rx}{(9 + x^2)^{1/2}} - r$$

Since $TE'(x)$ exists for all x in $[0, 8]$, the only critical point is where $TE'(x) = 0$. We solve the following equation.

$$\frac{1.28rx}{(9 + x^2)^{1/2}} - r = 0$$

$$\frac{1.28rx}{(9 + x^2)^{1/2}} = r$$

$$1.28rx = r(9 + x^2)^{1/2}$$

$$1.28x = (9 + x^2)^{1/2}$$

$$(1.28x)^2 = [(9 + x^2)^{1/2}]^2$$

$$1.6384x^2 = 9 + x^2$$

$$0.6384x^2 = 9$$

$$x^2 = \frac{9}{0.6384}$$

$$x^2 \approx 14.0977$$

$$x \approx 3.75$$

Since there is only one critical point, we can use the second derivative to determine whether we have a maximum. Note that $TE''(x) = \frac{11.52r}{(9 + x^2)^{3/2}}$ and that $TE''(x) > 0$ for all x in $[0, 8]$. Thus $TE(3.75)$ is a minimum and S should be about $8 - 3.75$, or 4.25 miles downshore from A.

55. $C(x) = 8x + 20 + \dfrac{x^3}{100}$

a) $C'(x) = 8 + 0 + \dfrac{1}{100} \cdot 3x^2$

$\qquad C'(x) = 8 + \dfrac{3}{100}x^2$

b) $A(x) = \dfrac{C(x)}{x} = \dfrac{8x + 20 + \dfrac{x^3}{100}}{x}$

$\qquad = 8 + \dfrac{20}{x} + \dfrac{x^2}{100},$

$\qquad$ or $8 + 20x^{-1} + \dfrac{1}{100}x^2$

c) $A'(x) = -20x^{-2} + \dfrac{1}{100} \cdot 2x$

$\qquad A'(x) = -\dfrac{20}{x^2} + \dfrac{1}{50}x$

d) Minimize $A(x)$ over $(0, \infty)$. Since $A'(x)$ exists for all x in $(0, \infty)$, we merely solve

$$-\frac{20}{x^2} + \frac{1}{50}x = 0$$

$$\frac{x}{50} = \frac{20}{x^2}$$

$$x^3 = 1000 \qquad (x \neq 0)$$

$$x = 10$$

Since there is only one critical point, we can use the second derivative to determine whether we have a minimum. Note that $A''(x) = \dfrac{40}{x^3} + \dfrac{1}{50}$ and that $A''(10) = \dfrac{3}{50} > 0$. Thus $A(10)$ is a minimum.

$$A(10) = 8 + \frac{20}{10} + \frac{10^2}{100} = 8 + 2 + 1 = 11$$

The minimum average cost is 11 when $x = 10$.

$$C'(10) = 8 + \frac{3}{100} \cdot 10^2 = 8 + 3 = 11$$

The marginal cost at $x = 10$ is 11.

e) Thus $A(10) = C'(10) = 11$.

57. Minimize $Q = x^3 + 2y^3$, where x and y are positive numbers, such that $x + y = 1$.

Express Q as a function of one variable. First solve $x + y = 1$ for y.

$$x + y = 1$$
$$y = 1 - x$$

Then substitute $1 - x$ for y in $Q = x^3 + 2y^3$.

$$Q = x^3 + 2y^3$$
$$Q = x^3 + 2(1 - x)^3 \qquad \text{Substituting}$$
$$= x^3 + 2(1 - 3x + 3x^2 - x^3)$$
$$= x^3 + 2 - 6x + 6x^2 - 2x^3$$
$$= -x^3 + 6x^2 - 6x + 2$$

Find $Q'(x)$.

$$Q'(x) = -3x^2 + 12x - 6$$

This derivative exists for all x in the interval $(0, \infty)$; thus the only critical points are where $Q'(x) = 0$.

$$Q'(x) = -3x^2 + 12x - 6 = 0$$
$$x^2 - 4x + 2 = 0$$

We now use the quadratic formula.

$$x = \frac{-(-4) \pm \sqrt{(-4)^2 - 4 \cdot 1 \cdot 2}}{2 \cdot 1}$$

$$= \frac{4 \pm \sqrt{16 - 8}}{2} - \frac{4 \pm \sqrt{8}}{2}$$

$$= \frac{4 \pm 2\sqrt{2}}{2} = 2 \pm \sqrt{2}$$

When $x = 2 + \sqrt{2}$, $y = 1 - (2 + \sqrt{2}) = -1 - \sqrt{2}$.

When $x = 2 - \sqrt{2}$, $y = 1 - (2 - \sqrt{2}) = -1 + \sqrt{2}$.

Since x and y are restricted to positive numbers, we only consider $x = 2 - \sqrt{2}$ and $y = -1 + \sqrt{2}$. The minimum value of Q is found by substituting.

$$Q = x^3 + 2y^3$$
$$= (2 - \sqrt{2})^3 + 2(-1 + \sqrt{2})^3 \quad \text{Substituting}$$
$$= (20 - 14\sqrt{2}) + 2(5\sqrt{2} - 7)$$
$$= 20 - 14\sqrt{2} + 10\sqrt{2} - 14$$
$$= 6 - 4\sqrt{2}$$

59. $\boxed{tw}$

Exercise Set 3.6

1. For $y = f(x) = x^2$, $x = 2$, and $\Delta x = 0.01$,

$$\Delta y = f(x + \Delta x) - f(x)$$
$$\Delta y = f(2 + 0.01) - f(2) \quad \text{Substituting 2 for } x \text{ and } 0.01 \text{ for } \Delta x$$
$$= f(2.01) - f(2)$$
$$= (2.01)^2 - 2^2$$
$$= 4.0401 - 4$$
$$= 0.0401$$

$$f'(x)\Delta x = 2x \cdot \Delta x \quad [f(x) = x^2;\ f'(x) = 2x]$$
$$f'(2)\Delta x = 2 \cdot 2(0.01) \quad \text{Substituting 2 for } x \text{ and } 0.01 \text{ for } \Delta x$$
$$= 0.04$$

3. For $y = f(x) = x + x^2$, $x = 3$, and $\Delta x = 0.04$,

$$\Delta y = f(x + \Delta x) - f(x)$$
$$\Delta y = f(3 + 0.04) - f(3) \quad \text{Substituting 3 for } x \text{ and } 0.04 \text{ for } \Delta x$$
$$= f(3.04) - f(3)$$
$$= [3.04 + (3.04)^2] - (3 + 3^2)$$
$$= (3.04 + 9.2416) - (3 + 9)$$
$$= 12.2816 - 12$$
$$= 0.2816$$

$$f'(x)\Delta x = (1 + 2x)\Delta x \quad [f(x) = x + x^2, f'(x) = 1 + 2x]$$
$$f'(3)\Delta x = (1 + 2 \cdot 3)0.04 \quad \text{Substituting 3 for } x \text{ and } 0.04 \text{ for } \Delta x$$
$$= 7(0.04)$$
$$= 0.28$$

5. For $y = f(x) = \frac{1}{x^2}$, or x^{-2}, $x = 1$, and $\Delta x = 0.5$,

$$\Delta y = f(x + \Delta x) - f(x)$$
$$\Delta y = f(1 + 0.5) - f(1) \quad \text{Substituting 1 for } x \text{ and } 0.5 \text{ for } \Delta x$$
$$= f(1.5) - f(1)$$
$$= \frac{1}{(1.5)^2} - \frac{1}{1^2}$$
$$= \frac{1}{2.25} - 1$$
$$\approx -0.556$$

$$f'(x)\Delta x = -2x^{-3} \cdot \Delta x \quad [f(x) = x^{-2};\ f'(x) = -2x^{-3}]$$
$$= -\frac{2}{x^3} \cdot \Delta x$$
$$f'(1)\Delta x = -\frac{2}{1^3} \cdot 0.5 \quad \text{Substituting 1 for } x \text{ and } 0.5 \text{ for } \Delta x$$
$$= -2 \cdot 0.5$$
$$= -1$$

7. For $y = f(x) = 3x - 1$, $x = 4$, and $\Delta x = 2$,

$$\Delta y = f(x + \Delta x) - f(x)$$
$$\Delta y = f(4 + 2) - f(4) \quad \text{Substituting 4 for } x \text{ and } 2 \text{ for } \Delta x$$
$$= f(6) - f(4)$$
$$= (3 \cdot 6 - 1) - (3 \cdot 4 - 1)$$
$$= 17 - 11$$
$$= 6$$

$$f'(x)\Delta x = 3 \cdot \Delta x \quad [f(x) = 3x - 1;\ f'(x) = 3]$$
$$f'(4)\Delta x = 3 \cdot 2 \quad \text{Substituting 4 for } x \text{ and } 2 \text{ for } \Delta x$$
$$= 6$$

9. $C(x) = 0.01x^2 + 0.6x + 30$

$$\Delta C = C(x + \Delta x) - C(x)$$
$$\Delta C = C(70 + 1) - C(70) \quad \text{Substituting 70 for } x \text{ and } 1 \text{ for } \Delta x$$
$$= C(71) - C(70)$$
$$= [0.01(71^2) + 0.6(71) + 30] - [0.01(70^2) + 0.6(70) + 30]$$
$$= [50.41 + 42.6 + 30] - [49 + 42 + 30]$$
$$= 123.01 - 121$$
$$= \$2.01$$

$$C'(x) = 0.02x + 0.6$$
$$C'(70) = 0.02(70) + 0.6 \quad \text{Substituting 70 for } x$$
$$= 1.4 + 0.6$$
$$= \$2.00$$

11. $R(x) = 2x$

$\Delta R = R(x + \Delta x) - R(x)$

$\Delta R = R(70 + 1) - R(70)$ Substituting 70 for x
                                 and 1 for Δx

$\quad = R(71) - R(70)$

$\quad = 2 \cdot 71 - 2 \cdot 70$

$\quad = 142 - 140$

$\quad = \$2$

$R'(x) = 2$ $R'(x)$ is a constant function

$R'(70) = \$2$

13. $C(x) = 0.01x^2 + 0.6x + 30$

     $R(x) = 2x$

   a) $P(x) = R(x) - C(x)$

            $= 2x - (0.01x^2 + 0.6x + 30)$

            $= 2x - 0.01x^2 - 0.6x - 30$

            $= -0.01x^2 + 1.4x - 30$

   b) $\Delta P = P(x + \Delta x) - P(x)$

       $\Delta P = P(70 + 1) - P(70)$ Substituting 70 for
                                    x and 1 for Δx

          $= P(71) - P(70)$

          $= (-0.01 \cdot 71^2 + 1.4 \cdot 71 - 30) -$

                  $(-0.01 \cdot 70^2 + 1.4 \cdot 70 - 30)$

          $= (-50.41 + 99.4 - 30) - (-49 + 98 - 30)$

          $= 18.99 - 19$

          $= -\$0.01$

      $P'(x) = -0.02x + 1.4$

      $P'(70) = -0.02 \cdot 70 + 1.4$

                  Substituting 70 for x

         $= -1.4 + 1.4$

         $= \$0$

15. We first think of the number closest to 19 that is a perfect square. This is 16. We will approximate how y, or $\sqrt{x}$, changes when 16 changes by $\Delta x = 3$. Let

   $y = f(x) = \sqrt{x}$, or $x^{1/2}$

   Then

      $\Delta y = \sqrt{x + \Delta x} - \sqrt{x}$ $[\Delta y = f(x + \Delta x) - f(x)]$

          $= \sqrt{x + \Delta x} - y$ Substituting y for $\sqrt{x}$

   so

   $y + \Delta y = \sqrt{x + \Delta x}$ Adding y

   Now

   $\Delta y \approx f'(x)\Delta x$

   $\Delta y \approx \dfrac{1}{2}x^{-1/2}\Delta x$ Substituting $\dfrac{1}{2}x^{-1/2}$ for $f'(x)$

       $\approx \dfrac{1}{2\sqrt{x}}\Delta x$

   Let $x = 16$ and $\Delta x = 3$. Then

   $\Delta y \approx \dfrac{1}{2\sqrt{16}} \cdot 3$ Substituting 16 for x and
                             3 for Δx

      $\approx \dfrac{3}{8}$, or 0.375

So $\sqrt{19} = \sqrt{16 + 3}$

           $= \sqrt{x + \Delta x}$

           $= y + \Delta y$

           $= \sqrt{x} + \Delta y$ Substituting $\sqrt{x}$ for y

           $\approx \sqrt{16} + 0.375$

           $\approx 4 + 0.375$

           ≈ 4.375

To six decimal places, $\sqrt{19} = 4.358899$. Thus the approximation 4.375 is fairly close.

17. We first think of the number closest to 102 that is a perfect square. This is 100. We will approximate how y, or $\sqrt{x}$, changes when 100 changes by $\Delta x = 2$. Let

   $y = f(x) = \sqrt{x}$, or $x^{1/2}$

   Then

      $\Delta y = \sqrt{x + \Delta x} - \sqrt{x}$ $[\Delta y = f(x + \Delta x) - f(x)]$

          $= \sqrt{x + \Delta x} - y$ Substituting y for $\sqrt{x}$

   so

   $y + \Delta y = \sqrt{x + \Delta x}$ Adding y

   Now

   $\Delta y \approx f'(x)\Delta x$

   $\Delta y \approx \dfrac{1}{2}x^{-1/2}\Delta x$ Substituting $\dfrac{1}{2}x^{-1/2}$ for $f'(x)$

       $\approx \dfrac{1}{2\sqrt{x}}\Delta x$

   Let $x = 100$ and $\Delta x = 2$. Then

   $\Delta y \approx \dfrac{1}{2\sqrt{100}} \cdot 2$ Substituting 100 for x and
                             2 for Δx

      $\approx \dfrac{1}{\sqrt{100}}$

      $\approx \dfrac{1}{10}$, or 0.1

So $\sqrt{102} = \sqrt{100 + 2}$

            $= \sqrt{x + \Delta x}$

            $= y + \Delta y$

            $= \sqrt{x} + \Delta y$ Substituting $\sqrt{x}$ for y

            $\approx \sqrt{100} + 0.1$

            $\approx 10 + 0.1$

            ≈ 10.1

To six decimal places, $\sqrt{102} = 10.099505$. Thus the approximation 10.1 is fairly close.

19. We think of the number closest to 10 that is a perfect cube. This is 8. We will approximate how y, or $\sqrt[3]{x}$, changes when 8 changes by $\Delta x = 2$. Let

   $y = f(x) = \sqrt[3]{x}$, or $x^{1/3}$

   Then

      $\Delta y = \sqrt[3]{x + \Delta x} - \sqrt[3]{x}$ $[\Delta y = f(x + \Delta x) - f(x)]$

          $= \sqrt[3]{x + \Delta x} - y$ Substituting y for $\sqrt[3]{x}$

   so

   $y + \Delta y = \sqrt[3]{x + \Delta x}$ Adding y

Now

$$\Delta y \approx f'(x)\Delta x$$

$$\Delta y \approx \frac{1}{3}x^{-2/3}\Delta x \quad \text{Substituting } \frac{1}{3}x^{-2/3} \text{ for } f'(x)$$

$$\approx \frac{1}{3\sqrt[3]{x^2}}\Delta x$$

Let $x = 8$ and $\Delta x = 2$. Then

$$\Delta y \approx \frac{1}{3\sqrt[3]{8^2}}\cdot 2 \quad \begin{array}{l}\text{Substituting 8 for } x \text{ and}\\ 2 \text{ for } \Delta x\end{array}$$

$$\approx \frac{1}{3\sqrt[3]{64}}\cdot 2$$

$$\approx \frac{1}{3\cdot 4}\cdot 2$$

$$\approx \frac{1}{6}, \text{ or } 0.167$$

So $\sqrt[3]{10} = \sqrt[3]{8+2}$

$$= \sqrt[3]{x+\Delta x}$$

$$= y + \Delta y$$

$$= \sqrt[3]{x} + \Delta y \quad \text{Substituting } \sqrt[3]{x} \text{ for } y$$

$$\approx \sqrt[3]{8} + 0.167$$

$$\approx 2 + 0.167$$

$$\approx 2.167$$

To six decimal places, $\sqrt[3]{10} = 2.154435$. Thus the approximation 2.167 is fairly close.

21. $y = (2x^3 + 1)^{3/2}$

$$\frac{dy}{dx} = \frac{3}{2}(2x^3+1)^{1/2}\cdot 6x^2 = 9x^2(2x^3+1)^{1/2}$$

$$dy = 9x^2(2x^3+1)^{1/2}dx$$

23. $y = \sqrt[5]{x+27} = (x+27)^{1/5}$

$$\frac{dy}{dx} = \frac{1}{5}(x+27)^{-4/5}$$

$$dy = \frac{1}{5}(x+27)^{-4/5}dx$$

25. $y = x^4 - 2x^3 + 5x^2 + 3x - 4$

$$\frac{dy}{dx} = 4x^3 - 6x^2 + 10x + 3$$

$$dy = (4x^3 - 6x^2 + 10x + 3)dx$$

27. $dy = (4x^3 - 6x^2 + 10x + 3)dx$

$$dy = (4\cdot 2^3 - 6\cdot 2^2 + 10\cdot 2 + 3)0.1$$
$$\qquad\qquad \text{Substituting 2 for } x \text{ and } 0.1 \text{ for } \Delta x$$

$$= (32 - 24 + 20 + 3)0.1$$

$$= 31(0.1)$$

$$= 3.1$$

29. $A(x) = \dfrac{13x + 100}{x}$, $x = 100$, $\Delta x = 101 - 100 = 1$

$$A'(x) = \frac{x\cdot 13 - 1\cdot(13x+100)}{x^2} = \frac{13x - 13x - 100}{x^2}$$

$$= \frac{-100}{x^2}$$

$$\Delta A \approx A'(x)\Delta x$$

$$\approx -\frac{100}{x^2}\Delta x$$

$$\approx -\frac{100}{100^2}\cdot 1$$

$$\approx -0.01$$

The average cost changes by about $-\$0.01$. (This is a decrease in cost.)

31. $N(a) = -a^2 + 300a + 6$, $a = 100$, $\Delta a = 101 - 100 = 1$

$$N'(a) = -2a + 300$$

$$\Delta N \approx N'(a)\Delta a$$

$$\approx (-2a + 300)\Delta a$$

$$\approx (-2\cdot 100 + 300)\cdot 1$$

$$\approx 100$$

About 100 more units will be sold.

33. $V = \dfrac{4}{3}\pi r^3$, $r = 1$, $\Delta r = 1.2 - 1 = 0.2$

$$V' = \frac{4}{3}\pi\cdot 3r^2 = 4\pi r^2$$

$$\Delta V \approx V'(r)\Delta r$$

$$\Delta V \approx 4\pi r^2\Delta r$$

$$\approx 4(3.14)(1)^2(0.2)$$

$$\approx 2.512$$

The change in volume is approximately 2.512 cm^3.

35.

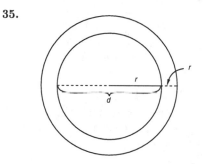

The circumference of the earth, which is the original length of the rope, is given by $C = \pi d$, or $C(r) = 2\pi r$. We need to find the change in the length of the radius, Δr, when the length of the rope is increased 10 ft.

$$\Delta C \approx C'(r)\Delta r \quad \begin{array}{l}\Delta C \text{ represents the change in}\\ \text{the length of the rope.}\end{array}$$

$$10 \approx 2\pi\cdot\Delta r \quad \begin{array}{l}\text{Substituting 10 for } \Delta C \text{ and}\\ 2\pi \text{ for } C'(r)\end{array}$$

$$\frac{10}{2\pi} = \Delta r$$

$$\frac{5}{\pi} = \Delta r$$

$$1.6 \approx \Delta r$$

Thus the rope is raised approximately 1.6 ft above the earth.

37. $\boxed{tw}$

Exercise Set 3.7

1. Differential implicitly to find $\dfrac{dy}{dx}$.

$$xy - x + 2y = 3$$

$$\frac{d}{dx}(xy - x + 2y) = \frac{d}{dx}3$$

Differentiating both sides with respect to x

$$\frac{d}{dx}xy - \frac{d}{dx}x + \frac{d}{dx}2y = 0$$

$$x \cdot \frac{dy}{dx} + 1 \cdot y - 1 + 2 \cdot \frac{dy}{dx} = 0$$

$$x \cdot \frac{dy}{dx} + 2 \cdot \frac{dy}{dx} = 1 - y$$

Getting all terms involving dy/dx alone on one side

$$(x + 2)\frac{dy}{dx} = 1 - y$$

$$\frac{dy}{dx} = \frac{1 - y}{x + 2}$$

Find the slope of the tangent line to the curve at $\left(-5, \dfrac{2}{3}\right)$.

$$\frac{dy}{dx} = \frac{1 - y}{x + 2}$$

$$\frac{dy}{dx} = \frac{1 - \dfrac{2}{3}}{-5 + 2} \qquad \text{Replacing } x \text{ by } -5 \text{ and } y \text{ by } \frac{2}{3}$$

$$= \frac{\dfrac{1}{3}}{-3} = \frac{1}{3} \cdot \left(-\frac{1}{3}\right) = -\frac{1}{9}$$

3. Differentiate implicitly to find $\dfrac{dy}{dx}$.

$$x^2 + y^2 = 1$$

$$\frac{d}{dx}(x^2 + y^2) = \frac{d}{dx}1 \quad \begin{array}{l}\text{Differentiating both}\\ \text{sides with respect to } x\end{array}$$

$$\frac{d}{dx}x^2 + \frac{d}{dx}y^2 = 0$$

$$2x + 2y\frac{dy}{dx} = 0$$

$$2y\frac{dy}{dx} = -2x$$

$$\frac{dy}{dx} = \frac{-2x}{2y}$$

$$\frac{dy}{dx} = -\frac{x}{y}$$

Find the slope of the tangent line to the curve at $\left(\dfrac{1}{2}, \dfrac{\sqrt{3}}{2}\right)$.

$$\frac{dy}{dx} = -\frac{x}{y}$$

$$= -\frac{\dfrac{1}{2}}{\dfrac{\sqrt{3}}{2}} \qquad \text{Replacing } x \text{ by } \frac{1}{2} \text{ and } y \text{ by } \frac{\sqrt{3}}{2}$$

$$= -\frac{1}{2} \cdot \frac{2}{\sqrt{3}}$$

$$= -\frac{1}{\sqrt{3}}$$

5. Differentiate implicitly to find $\dfrac{dy}{dx}$.

$$x^2y - 2x^3 - y^3 + 1 = 0$$

$$\frac{d}{dx}(x^2y - 2x^3 - y^3 + 1) = \frac{d}{dx}0$$

Differentiating both sides with respect to x

$$\frac{d}{dx}x^2y - \frac{d}{dx}2x^3 - \frac{d}{dx}y^3 + \frac{d}{dx}1 = 0$$

$$x^2\frac{dy}{dx} + 2x \cdot y - 6x^2 - 3y^2\frac{dy}{dx} + 0 = 0$$

$$x^2 \cdot \frac{dy}{dx} - 3y^2 \cdot \frac{dy}{dx} = 6x^2 - 2xy$$

Getting all terms involving dy/dx alone on one side

$$(x^2 - 3y^2)\frac{dy}{dx} = 6x^2 - 2xy$$

$$\frac{dy}{dx} = \frac{6x^2 - 2xy}{x^2 - 3y^2}$$

Find the slope of the tangent line to the curve at $(2, -3)$.

$$\frac{dy}{dx} = \frac{6x^2 - 2xy}{x^2 - 3y^2}$$

$$= \frac{6 \cdot 2^2 - 2 \cdot 2 \cdot (-3)}{2^2 - 3 \cdot (-3)^2} \qquad \begin{array}{l}\text{Replacing } x \text{ by } 2 \text{ and } y \text{ by}\\ -3\end{array}$$

$$= \frac{24 + 12}{4 - 27}$$

$$= \frac{36}{-23}$$

$$= -\frac{36}{23}$$

7. Differentiate implicitly to find $\dfrac{dy}{dx}$.

$$2xy + 3 = 0$$

$$\frac{d}{dx}(2xy + 3) = \frac{d}{dx}0$$

Differentiating both sides with respect to x

$$\frac{d}{dx}2xy + \frac{d}{dx}3 = 0$$

$$2\left(x \cdot \frac{dy}{dx} + 1 \cdot y\right) + 0 = 0$$

$$2x \cdot \frac{dy}{dx} + 2y = 0$$

$$2x \cdot \frac{dy}{dx} = -2y$$

$$\frac{dy}{dx} = \frac{-2y}{2x}$$

$$\frac{dy}{dx} = -\frac{y}{x}$$

9. Differentiate implicitly to find $\dfrac{dy}{dx}$.

$$x^2 - y^2 = 16$$

$$\frac{d}{dx}(x^2 - y^2) = \frac{d}{dx}16 \quad \text{Differentiating both sides with respect to } x$$

$$\frac{d}{dx}x^2 - \frac{d}{dx}y^2 = 0$$

$$2x - 2y \cdot \frac{dy}{dx} = 0$$

$$-2y \cdot \frac{dy}{dx} = -2x$$

$$\frac{dy}{dx} = \frac{-2x}{-2y}$$

$$\frac{dy}{dx} = \frac{x}{y}$$

11. Differentiate implicitly to find $\dfrac{dy}{dx}$.

$$y^5 = x^3$$

$$\frac{d}{dx}y^5 = \frac{d}{dx}x^3 \quad \text{Differentiating both sides with respect to } x$$

$$5y^4 \cdot \frac{dy}{dx} = 3x^2$$

$$\frac{dy}{dx} = \frac{3x^2}{5y^4}$$

13. Differentiate implicitly to find $\dfrac{dy}{dx}$.

$$x^2y^3 + x^3y^4 = 11$$

$$\frac{d}{dx}(x^2y^3 + x^3y^4) = \frac{d}{dx}11$$

Differentiating both sides with respect to x

$$\frac{d}{dx}x^2y^3 + \frac{d}{dx}x^3y^4 = 0$$

$$x^2 \cdot 3y^2 \cdot \frac{dy}{dx} + 2x \cdot y^3 + x^3 \cdot 4y^3 \cdot \frac{dy}{dx} + 3x^2 \cdot y^4 = 0$$

$$3x^2y^2 \cdot \frac{dy}{dx} + 4x^3y^3 \cdot \frac{dy}{dx} = -2xy^3 - 3x^2y^4$$

$$(3x^2y^2 + 4x^3y^3)\frac{dy}{dx} = -2xy^3 - 3x^2y^4$$

$$\frac{dy}{dx} = \frac{-2xy^3 - 3x^2y^4}{3x^2y^2 + 4x^3y^3}$$

$$= \frac{xy^2(-2y - 3xy^2)}{xy^2(3x + 4x^2y)}$$

$$= \frac{-2y - 3xy^2}{3x + 4x^2y}$$

15. Differentiate implicitly to find $\dfrac{dp}{dx}$.

$$p^2 + p + 2x = 40$$

$$\frac{d}{dx}(p^2 + p + 2x) = \frac{d}{dx}40 \quad \text{Differentiating both sides with respect to } x$$

$$2p \cdot \frac{dp}{dx} + \frac{dp}{dx} + 2 \cdot 1 = 0$$

$$(2p + 1)\frac{dp}{dx} = -2$$

$$\frac{dp}{dx} = \frac{-2}{2p + 1}$$

17. Differentiate implicitly to find $\dfrac{dp}{dx}$.

$$(p + 4)(x + 3) = 48$$

$$px + 3p + 4x + 12 = 48$$

$$px + 3p + 4x = 36$$

$$\frac{d}{dx}(px + 3p + 4x) = \frac{d}{dx}36$$

Differentiating both sides with respect to x

$$p \cdot 1 + \frac{dp}{dx} \cdot x + 3 \cdot \frac{dp}{dx} + 4 \cdot 1 = 0$$

$$(x + 3)\frac{dp}{dx} = -p - 4$$

$$\frac{dp}{dx} = \frac{-p - 4}{x + 3}, \text{ or } -\frac{p + 4}{x + 3}$$

19.
$$A^3 + B^3 = 9$$

$$\frac{d}{dt}(A^3 + B^3) = \frac{d}{dt}9 \quad \text{Differentiating with respect to } t$$

$$3A^2 \cdot \frac{dA}{dt} + 3B^2 \cdot \frac{dB}{dt} = 0$$

$$3A^2 \cdot \frac{dA}{dt} = -3B^2 \cdot \frac{dB}{dt}$$

$$\frac{dA}{dt} = \frac{-3B^2}{3A^2} \cdot \frac{dB}{dt}$$

$$\frac{dA}{dt} = -\frac{B^2}{A^2} \cdot \frac{dB}{dt}$$

Find B when $A = 2$:
$$A^3 + B^3 = 9$$
$$2^3 + B^3 = 9$$
$$8 + B^3 = 9$$
$$B^3 = 1$$
$$B = 1$$

$$\frac{dA}{dt} = -\frac{B^2}{A^2} \cdot \frac{dB}{dt}$$

$$= -\frac{1^2}{2^2} \cdot 3$$

Substituting 2 for A, 1 for B, and 3 for dB/dt

$$= -\frac{3}{4}$$

21. $R(x) = 50x - 0.5x^2$

$$\frac{dR}{dt} = \frac{d}{dt}(50x - 0.5x^2) \quad \text{Differentiating with respect to time}$$

$$= 50 \cdot \frac{dx}{dt} - x \cdot \frac{dx}{dt}$$

$$= 50 \cdot 20 - 30 \cdot 20 \quad \text{Substituting 20 for } dx/dt \text{ and 30 for } x$$

$$= 1000 - 600$$

$$= \$400 \text{ per day}$$

$C(x) = 4x + 10$

$$\frac{dC}{dt} = \frac{d}{dt}(4x + 10) \quad \text{Differentiating with respect to time}$$

$$= 4 \cdot \frac{dx}{dt} + 0$$

$$= 4 \cdot \frac{dx}{dt}$$

$$= 4 \cdot 20 \quad \text{Substituting 20 for } dx/dt$$

$$= \$80 \text{ per day}$$

Since $P(x) = R(x) - C(x)$,

$$\frac{dP}{dt} = \frac{dR}{dt} - \frac{dC}{dt}$$

$$= \$400 \text{ per day} - \$80 \text{ per day}$$

$$= \$320 \text{ per day}$$

23. $R(x) = 2x$

$$\frac{dR}{dt} = \frac{d}{dt}(2x) \quad \text{Differentiating with respect to time}$$

$$= 2 \cdot \frac{dx}{dt}$$

$$= 2 \cdot 8 \quad \text{Substituting 8 for } dx/dt$$

$$= \$16 \text{ per day}$$

$C(x) = 0.01x^2 + 0.6x + 30$

$$\frac{dC}{dt} = \frac{d}{dt}(0.01x^2 + 0.6x + 30) \quad \text{Differentiating with respect to time}$$

$$= \frac{d}{dt}0.01x^2 + \frac{d}{dt}0.6x + \frac{d}{dt}30$$

$$= 0.02x \cdot \frac{dx}{dt} + 0.6 \cdot \frac{dx}{dt} + 0$$

$$= 0.02 \cdot 20 \cdot 8 + 0.6 \cdot 8 \quad \text{Substituting 20 for } x \text{ and 8 for } dx/dt$$

$$= 3.2 + 4.8$$

$$= \$8 \text{ per day}$$

Since $P(x) = R(x) - C(x)$,

$$\frac{dP}{dt} = \frac{dR}{dt} - \frac{dC}{dt}$$

$$= \$16 \text{ per day} - \$8 \text{ per day}$$

$$= \$8 \text{ per day}$$

25. $V = \frac{4}{3}\pi r^3$, $\frac{dr}{dt} = 0.03 \frac{\text{cm}}{\text{day}}$

We take the derivative of both sides with respect to t.

$$\frac{dV}{dt} = \frac{4}{3}\pi \cdot 3r^2 \cdot \frac{dr}{dt}$$

$$= 4\pi r^2 \cdot \frac{dr}{dt}$$

$$= 4\pi \cdot (1.2 \text{ cm})^2 \cdot \left(0.03 \frac{\text{cm}}{\text{day}}\right)$$

$$= 4\pi \cdot (1.44 \text{ cm}^2) \cdot \left(0.03 \frac{\text{cm}}{\text{day}}\right)$$

$$= 0.1728\pi \frac{\text{cm}^3}{\text{day}}$$

$$\approx 0.54 \frac{\text{cm}^3}{\text{day}}$$

27. $V = \frac{p}{4Lv}(R^2 - r^2)$ We will assume r, p, L, and v are constants.

a) $$\frac{dV}{dt} = \frac{d}{dt}\left[\frac{p}{4Lv}(R^2 - r^2)\right]$$

$$= \frac{p}{4Lv} \cdot \frac{d}{dt}(R^2 - r^2)$$

$$= \frac{p}{4Lv}\left[\frac{d}{dt}R^2 - \frac{d}{dt}r^2\right]$$

$$= \frac{p}{4Lv}\left(2R \cdot \frac{dR}{dt} - 0\right)$$

$$= \frac{pR}{2Lv} \cdot \frac{dR}{dt}$$

$$= \frac{100 \cdot R}{2 \cdot 1 \cdot 0.05} \cdot \frac{dR}{dt} \quad \text{Substituting 100 for } p, \text{ 1 for } L, \text{ and 0.05 for } v$$

$$= 1000R \cdot \frac{dR}{dt}$$

b) $\dfrac{dV}{dt} = 1000R \cdot \dfrac{dR}{dt}$

$= 1000 \cdot (0.0075) \cdot (-0.0015)$

Substituting 0.0075 for R and
-0.0015 for dR/dt

$= -0.01125 \,\dfrac{\text{mm}^3}{\text{min}}$

29. $D^2 = x^2 + y^2$

After 1 hour,

$D^2 = 25^2 + 60^2$ Substituting 25 for x and
60 for y

$D^2 = 625 + 3600$

$D^2 = 4225$

$D = 65$

We also know that $\dfrac{dx}{dt} = 25\,\dfrac{\text{mi}}{\text{hr}}$ and that $\dfrac{dy}{dt} = 60\,\dfrac{\text{mi}}{\text{hr}}$.

We differentiate implicity to find $\dfrac{dD}{dt}$.

$D^2 = x^2 + y^2$

$\dfrac{d}{dt}D^2 = \dfrac{d}{dt}(x^2 + y^2)$

$\dfrac{d}{dt}D^2 = \dfrac{d}{dt}x^2 + \dfrac{d}{dt}y^2$

$2D \cdot \dfrac{dD}{dt} = 2x \cdot \dfrac{dx}{dt} + 2y \cdot \dfrac{dy}{dt}$

$2 \cdot 65 \cdot \dfrac{dD}{dt} = 2 \cdot 25 \cdot 25\,\dfrac{\text{mi}}{\text{hr}} + 2 \cdot 60 \cdot 60\,\dfrac{\text{mi}}{\text{hr}}$ Substituting

$\dfrac{dD}{dt} = \dfrac{1250 + 7200}{130}\,\dfrac{\text{mi}}{\text{hr}}$

$\dfrac{dD}{dt} = 65\,\dfrac{\text{mi}}{\text{hr}}$

31. Differentiate implicity to find $\dfrac{dy}{dx}$.

$\sqrt{x} + \sqrt{y} = 1$

$\dfrac{d}{dx}(x^{1/2} + y^{1/2}) = \dfrac{d}{dx}1$ Differentiating both
sides with respect to x

$\dfrac{1}{2}x^{-1/2} + \dfrac{1}{2}y^{-1/2}\dfrac{dy}{dx} = 0$

$\dfrac{1}{2}y^{-1/2}\dfrac{dy}{dx} = -\dfrac{1}{2}x^{-1/2}$

$y^{-1/2}\dfrac{dy}{dx} = -x^{-1/2}$

$\dfrac{dy}{dx} = -\dfrac{x^{-1/2}}{y^{-1/2}}$

$\dfrac{dy}{dx} = -\dfrac{y^{1/2}}{x^{1/2}}$, or $-\dfrac{\sqrt{y}}{\sqrt{x}}$

33. Differentiate implicity to find $\dfrac{dy}{dx}$.

$y^3 = \dfrac{x-1}{x+1}$

$\dfrac{d}{dx}y^3 = \dfrac{d}{dx}\dfrac{x-1}{x+1}$ Differentiating both sides
with respect to x

$3y^2 \cdot \dfrac{dy}{dx} = \dfrac{(x+1) \cdot 1 - 1 \cdot (x-1)}{(x+1)^2}$

$3y^2 \cdot \dfrac{dy}{dx} = \dfrac{x+1-x+1}{(x+1)^2}$

$3y^2 \cdot \dfrac{dy}{dx} = \dfrac{2}{(x+1)^2}$

$\dfrac{dy}{dx} = \dfrac{2}{3y^2(x+1)^2}$

35. Differentiate implicity to find $\dfrac{dy}{dx}$.

$x^{3/2} + y^{2/3} = 1$

$\dfrac{d}{dx}(x^{3/2} + y^{2/3}) = \dfrac{d}{dx}1$ Differentiating both
sides with respect
to x

$\dfrac{3}{2}x^{1/2} + \dfrac{2}{3}y^{-1/3} \cdot \dfrac{dy}{dx} = 0$

$\dfrac{2}{3}y^{-1/3} \cdot \dfrac{dy}{dx} = -\dfrac{3}{2}x^{1/2}$

$\dfrac{dy}{dx} = \dfrac{-\dfrac{3}{2}x^{1/2}}{\dfrac{2}{3}y^{-1/3}}$

$\dfrac{dy}{dx} = -\dfrac{9}{4}x^{1/2}y^{1/3}$, or

$-\dfrac{9}{4}\sqrt{x}\sqrt[3]{y}$

37. Differentiate implicity to find $\dfrac{dy}{dx}$.

$xy + x - 2y = 4$

$\dfrac{d}{dx}(xy + x - 2y) = \dfrac{d}{dx}4$

$x \cdot \dfrac{dy}{dx} + 1 \cdot y + 1 - 2 \cdot \dfrac{dy}{dx} = 0$

$(x-2)\dfrac{dy}{dx} = -1 - y$

$\dfrac{dy}{dx} = \dfrac{-1-y}{x-2}$, or $\dfrac{1+y}{2-x}$

Differentiate implicity to find $\dfrac{d^2y}{dx^2}$.

$$\frac{dy}{dx} = \frac{1+y}{2-x}$$

$$\frac{d}{dx}\frac{dy}{dx} = \frac{d}{dx}\frac{1+y}{2-x}$$

$$\frac{d^2y}{dx^2} = \frac{(2-x)\cdot\dfrac{dy}{dx} - (-1)\cdot(1+y)}{(2-x)^2}$$

$$= \frac{(2-x)\cdot\dfrac{1+y}{2-x} + 1 + y}{(2-x)^2}$$

$$\text{Substituting } \frac{1+y}{2-x} \text{ for } \frac{dy}{dx}$$

$$= \frac{1+y+1+y}{(2-x)^2}$$

$$= \frac{2+2y}{(2-x)^2}$$

39. Differentiate implicity to find $\dfrac{dy}{dx}$.

$$x^2 - y^2 = 5$$

$$\frac{d}{dx}(x^2 - y^2) = \frac{d}{dx}5$$

$$2x - 2y\cdot\frac{dy}{dx} = 0$$

$$-2y\cdot\frac{dy}{dx} = -2x$$

$$\frac{dy}{dx} = \frac{-2x}{-2y}$$

$$\frac{dy}{dx} = \frac{x}{y}$$

Differentiate implicity to find $\dfrac{d^2y}{dx^2}$.

$$\frac{dx}{dy} = \frac{x}{y}$$

$$\frac{d}{dx}\frac{dy}{dx} = \frac{d}{dx}\frac{x}{y}$$

$$\frac{d^2y}{dx^2} = \frac{y\cdot 1 - \dfrac{dy}{dx}\cdot x}{y^2}$$

$$= \frac{y - \dfrac{x}{y}\cdot x}{y^2} \qquad \text{Substituting } \frac{x}{y} \text{ for } \frac{dy}{dx}$$

$$= \frac{\dfrac{y^2}{y} - \dfrac{x^2}{y}}{y^2}$$

$$= \frac{y^2 - x^2}{y}\cdot\frac{1}{y^2}$$

$$= \frac{y^2 - x^2}{y^3}$$

41. $\boxed{tw}$

43. See the answer section in the text.

45. See the answer section in the text.

47. See the answer section in the text.

Chapter 4

Exponential and Logarithmic Functions

Exercise Set 4.1

1. Graph: $y = 4^x$

First we find some function values.

Note: For

$$x = -2, y = 4^{-2} = \frac{1}{4^2} = \frac{1}{16} = 0.0625$$

$$x = -1, y = 4^{-1} = \frac{1}{4} = 0.25$$

$x = 0, \quad y = 4^0 = 1$

$x = 1, \quad y = 4^1 = 4$

$x = 2, \quad y = 4^2 = 16$

x	y
-2	0.0625
-1	0.25
0	1
1	4
2	16

Plot these points and connect them with a smooth curve.

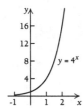

3. Graph: $y = (0.4)^x$

First we find some function values.

Note: For

$$x = -2, y = (0.4)^{-2} = \frac{1}{(0.4)^2} = 6.25$$

$$x = -1, y = (0.4)^{-1} = \frac{1}{0.4} = 2.5$$

$x = 0, \quad y = (0.4)^0 = 1$

$x = 1, \quad y = (0.4)^1 = 0.4$

$x = 2, \quad y = (0.4)^2 = 0.16$

x	y
-2	6.25
-1	2.5
0	1
1	0.4
2	0.16

Plot these points and connect them with a smooth curve.

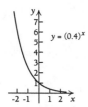

5. Graph: $x = 4^y$

First we find some function values.

Note: For

$$x = -2, y = 4^{-2} = \frac{1}{4^2} = \frac{1}{16}$$

$$x = -1, y = 4^{-1} = \frac{1}{4}$$

$x = 0, \quad y = 4^0 = 1$

$x = 1, \quad y = 4^1 = 4$

$x = 2, \quad y = 4^2 = 16$

x	y
$\frac{1}{16}$	-2
$\frac{1}{4}$	-1
1	0
4	1
16	2

Plot these points and connect them with a smooth curve.

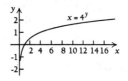

7. $f(x) = e^{3x}$

$f'(x) = 3e^{3x} \qquad \left[\dfrac{d}{dx}e^{f(x)} = f'(x)e^{f(x)}\right]$

9. $f(x) = 5e^{-2x}$

$f'(x) = 5 \cdot \underbrace{(-2)e^{-2x}}_{} \qquad \left[\dfrac{d}{dx}[c \cdot f(x)] = c \cdot f'(x)\right]$

$\qquad\qquad\qquad \left[\dfrac{d}{dx}e^{f(x)} = f'(x)e^{f(x)}\right]$

$\qquad = -10e^{-2x}$

11. $f(x) = 3 - e^{-x}$

$f'(x) = 0 - \underbrace{(-1)e^{-x}}_{}$

$\qquad\qquad\qquad \left[\dfrac{d}{dx}e^{f(x)} = f'(x)e^{f(x)}\right]$

$\qquad = e^{-x}$

13. $f(x) = -7e^x$

$f'(x) = -7e^x \qquad \left[\dfrac{d}{dx} - 7e^x = -7 \cdot \dfrac{d}{dx}e^x\right]$

$\qquad\qquad\qquad \left[\dfrac{d}{dx}e^x = e^x\right]$

15. $f(x) = \dfrac{1}{2}e^{2x}$

$f'(x) = \dfrac{1}{2} \cdot \underbrace{2e^{2x}}_{} \qquad \left[\dfrac{d}{dx}[c \cdot f(x)] = c \cdot f'(x)\right]$

$\qquad\qquad\qquad \left[\dfrac{d}{dx}e^{f(x)} = f'(x)e^{f(x)}\right]$

$\qquad = e^{2x}$

17. $f(x) = x^4 e^x$

$f'(x) = x^4 \cdot e^x + 4x^3 \cdot e^x$ Using the Product Rule

$\qquad = x^3 e^x(x+4)$

19. $f(x) = \dfrac{e^x}{x^4}$

$f'(x) = \dfrac{x^4 \cdot e^x - 4x^3 \cdot e^x}{x^8}$ Using the Quotient Rule

$ = \dfrac{x^3 e^x (x-4)}{x^3 \cdot x^5}$ Factoring both numerator and denominator

$ = \dfrac{x^3}{x^3} \cdot \dfrac{e^x(x-4)}{x^5}$

$ = \dfrac{e^x(x-4)}{x^5}$ Simplifying

21. $f(x) = e^{-x^2 + 7x}$

$f'(x) = (-2x + 7)e^{-x^2 + 7x}$ $\left[\dfrac{d}{dx}e^{f(x)} = f'(x)e^{f(x)}\right]$

 $[f(x) = -x^2 + 7x,$

 $f'(x) = -2x + 7]$

23. $f(x) = e^{-x^2/2}$

$ = e^{(-1/2)x^2}$

$f'(x) = \left(-\dfrac{1}{2} \cdot 2x\right)e^{(-1/2)x^2}$

$ = -xe^{-x^2/2}$

25. $y = e^{\sqrt{x-7}}$

$ = e^{(x-7)^{1/2}}$

$\dfrac{dy}{dx} = \dfrac{1}{2}(x-7)^{-1/2} \cdot e^{(x-7)^{1/2}}$

$\phantom{\dfrac{dy}{dx}} = \dfrac{e^{(x-7)^{1/2}}}{2(x-7)^{1/2}}$

$\phantom{\dfrac{dy}{dx}} = \dfrac{e^{\sqrt{x-7}}}{2\sqrt{x-7}}$

27. $y = \sqrt{e^x - 1}$

$ = (e^x - 1)^{1/2}$

$\dfrac{dy}{dx} = \dfrac{1}{2}(e^x - 1)^{-1/2} \cdot e^x$ [Extended Power Rule;

$\phantom{\dfrac{dy}{dx} = xx}$ $\dfrac{d}{dx}(e^x - 1) = e^x - 0 = e^x]$

$\phantom{\dfrac{dy}{dx}} = \dfrac{e^x}{2\sqrt{e^x - 1}}$

29. $y = xe^{-2x} + e^{-x} + x^3$

$\dfrac{dy}{dx} = x \cdot (-2) \cdot e^{-2x} + 1 \cdot e^{-2x} + (-1) \cdot e^{-x} + 3x^2$

$\phantom{\dfrac{dy}{dx}} = -2xe^{-2x} + e^{-2x} - e^{-x} + 3x^2$

$\phantom{\dfrac{dy}{dx}} = (1 - 2x)e^{-2x} - e^{-x} + 3x^2$

31. $y = 1 - e^{-x}$

$\dfrac{dy}{dx} = 0 - (-1)e^{-x}$

$\phantom{\dfrac{dy}{dx}} = e^{-x}$

33. $y = 1 - e^{-kx}$

$\dfrac{dy}{dx} = 0 - (-k)e^{-kx}$

$\phantom{\dfrac{dy}{dx}} = ke^{-kx}$

35. Graph: $f(x) = e^{2x}$

Using a calculator we first find some function values.

Note: For

$x = -2,\ f(-2) = e^{2(-2)} = e^{-4} = 0.0183$

$x = -1,\ f(-1) = e^{2(-1)} = e^{-2} = 0.1353$

$x = 0,\ f(0) = e^{2 \cdot 0} = e^0 = 1$

$x = 1,\ f(1) = e^{2 \cdot 1} = e^2 = 7.3891$

$x = 2,\ f(2) = e^{2 \cdot 2} = e^4 = 54.598$

x	$f(x)$
-2	0.0183
-1	0.1353
0	1
1	7.3891
2	54.598

Plot these points and connect them with a smooth curve.

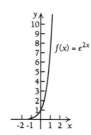

Derivatives. $f'(x) = 2e^{2x}$ and $f''(x) = 4e^{2x}$.

Critical points of f. Since $f'(x) > 0$ for all real numbers x, we know that the derivative exists for all real numbers and there is no solution of the equation $f'(x) = 0$. There are no critical points and therefore no maximum or minimum values.

Increasing. Since $f'(x) > 0$ for all real numbers x, the function f is increasing over the entire real line, $(-\infty, \infty)$.

Inflection points. Since $f''(x) > 0$ for all real numbers x, the equation $f''(x) = 0$ has no solution and there are no points of inflection.

Concavity. Since $f''(x) > 0$ for all real numbers x, the function f' is increasing and the graph is concave up over the entire real line.

37. Graph: $f(x) = e^{-2x}$

Using a calculator we first find some function values.

Note: For

$x = -2,\ f(-2) = e^{-2(-2)} = e^4 = 54.598$

$x = -1,\ f(-1) = e^{-2(-1)} = e^2 = 7.3891$

$x = 0,\ f(0) = e^{-2 \cdot 0} = e^0 = 1$

$x = 1,\ f(1) = e^{-2 \cdot 1} = e^{-2} = 0.1353$

$x = 2,\ f(2) = e^{-2 \cdot 2} = e^{-4} = 0.0183$

x	$f(x)$
-2	54.598
-1	7.3891
0	1
1	0.1353
2	0.0183

Plot these points and connect them with a smooth curve.

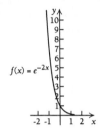

Derivatives. $f'(x) = -2e^{-2x}$ and $f''(x) = 4e^{-2x}$.

Critical points of f. Since $f'(x) < 0$ for all real numbers x, we know that the derivative exists for all real numbers and there is no solution of the equation $f'(x) = 0$. There are no critical points and therefore no maximum or minimum values.

Decreasing. Since $f'(x) < 0$ for all real numbers x, the function f is decreasing over the entire real line, $(-\infty, \infty)$.

Inflection points. Since $f''(x) > 0$ for all real numbers x, the equation $f''(x) = 0$ has no solution and there are no points of inflection.

Concavity. Since $f''(x) > 0$ for all real numbers x, the function f' is increasing and the graph is concave up over the entire real line.

39. Graph: $f(x) = 3 - e^{-x}$, for nonnegative values of x.

Using a calculator we first find some function values.

Note: For
$x = 0$, $f(0) = 3 - e^{-0} = 3 - 1 = 2$
$x = 1$, $f(1) = 3 - e^{-1} = 3 - 0.3679 = 2.6321$
$x = 2$, $f(2) = 3 - e^{-2} = 3 - 0.1353 = 2.8647$
$x = 3$, $f(3) = 3 - e^{-3} = 3 - 0.0498 = 2.9502$
$x = 4$, $f(4) = 3 - e^{-4} = 3 - 0.0183 = 2.9817$
$x = 6$, $f(6) = 3 - e^{-6} = 3 - 0.0025 = 2.9975$

x	$f(x)$
0	2
1	2.6321
2	2.8647
3	2.9502
4	2.9817
6	2.9975

Plot these points and connect them with a smooth curve.

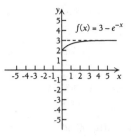

Derivatives. $f'(x) = e^{-x}$ and $f''(x) = -e^{-x}$.

Critical points of f. Since $f'(x) > 0$ for all real numbers x, we know that the derivative exists for all real numbers and there is no solution of the equation $f'(x) = 0$. There are no critical points and therefore no maximum or minimum values.

Increasing. Since $f'(x) > 0$ for all real numbers x, the function f is increasing over the entire real line, $(-\infty, \infty)$.

Inflection points. Since $f''(x) < 0$ for all real numbers x, the equation $f''(x) = 0$ has no solution and there are no points of inflection.

Concavity. Since $f''(x) < 0$ for all real numbers x, the function f' is decreasing and the graph is concave down over the entire real line.

41. - 45. ▮▮

47. We first find the slope of the tangent line at $(0, 1)$, $f'(0)$:
$$f(x) = e^x$$
$$f'(x) = e^x$$
$$f'(0) = e^0 = 1$$

Then we find the equation of the line with slope 1 and containing the point $(0, 1)$:
$$y - y_1 = m(x - x_1) \quad \text{Point-slope equation}$$
$$y - 1 = 1(x - 0)$$
$$y - 1 = x$$
$$y = x + 1$$

49. ▮▮

51. $C(t) = 100 - 50e^{-t}$

a) $C'(t) = 0 - 50 \cdot (-1) \cdot e^{-t}$
$\qquad = 50e^{-t}$

b) $C'(0) = 50e^{-0} = 50e^0 = 50 \cdot 1 = \$50$ million

c) $C'(4) = 50e^{-4} = 50(0.018316) \approx \$0.916$ million

d) $\boxed{tw}$

53. $x = D(p) = 408e^{-0.003p}$

a) $x = D(120) = 480e^{-0.003(120)}$
$\qquad = 480e^{-0.36}$
$\qquad \approx 480(0.697676)$
$\qquad \approx 335$

$$x = D(180) = 480e^{-0.003(180)}$$
$$= 480e^{-0.54}$$
$$\approx 480(0.582748)$$
$$\approx 280$$
$$x = D(340) = 480e^{-0.003(340)}$$
$$= 480e^{-1.02}$$
$$\approx 480(0.360595)$$
$$\approx 173$$

b) We plot the points $(120, 335)$, $(180, 280)$, and $(340, 173)$ and other points as needed. Then we connect the points with a smooth curve.

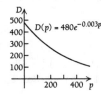

c) $D'(p) = 480(-0.003)e^{-0.003p}$
$$= -1.44e^{-0.003p}$$

d) $\boxed{tw}$

55. $C(t) = 10t^2e^{-t}$

a) $C(0) = 10 \cdot 0^2 \cdot e^{-0} = 0$ ppm

$C(1) = 10 \cdot 1^2 \cdot e^{-1}$
$$\approx 10(0.367879)$$
$$\approx 3.7 \text{ ppm}$$

$C(2) = 10 \cdot 2^2 \cdot e^{-2}$
$$\approx 40(0.135335)$$
$$\approx 5.4 \text{ ppm}$$

$C(3) = 10 \cdot 3^2 \cdot e^{-3}$
$$\approx 90(0.049787)$$
$$\approx 4.48 \text{ ppm}$$

$C(10) = 10 \cdot 10^2 \cdot e^{-10}$
$$\approx 1000(0.000045)$$
$$\approx 0.05 \text{ ppm}$$

b) We plot the points $(0, 0)$, $(1, 3.7)$, $(2, 5.4)$, $(3, 4.48)$, and $(10, 0.05)$ and other points as needed. Then we connect the points with a smooth curve.

c) $C'(t) = 10t^2(-1)e^{-t} + 20te^{-t}$
$$= -10te^{-t}(t - 2), \text{ or } 10te^{-t}(2 - t)$$

d) We find the maximum value of $C(t)$ on $[0, \infty)$.

$C'(t)$ exists for all t in $[0, \infty)$. We solve $C'(t) = 0$.

$$10te^{-t}(2 - t) = 0$$
$$t(2 - t) = 0 \qquad (10e^t \neq 0)$$
$$t = 0 \text{ or } 2 - t = 0$$
$$t = 0 \text{ or } \qquad 2 = t$$

Since the function has two critical points and the interval is not closed, we must examine the graph to find the maximum value. We see that the maximum value of the concentration is about 5.4 ppm. It occurs at $t = 2$ hr.

e) $\boxed{tw}$

57. $y = (e^{3x} + 1)^5$
$$\frac{dy}{dx} = 5(e^{3x} + 1)^4 \cdot 3e^{3x}$$
$$= 15e^{3x}(e^{3x} + 1)^4$$

59. $y = \dfrac{e^{3t} - e^{7t}}{e^{4t}}$
$$= \frac{e^{3t}(1 - e^{4t})}{e^{3t} \cdot e^t} \qquad \text{Factoring}$$
$$= \frac{1 - e^{4t}}{e^t} \qquad \text{Simplifying}$$
$$\frac{dy}{dt} = \frac{e^t(-4e^{4t}) - e^t(1 - e^{4t})}{(e^t)^2} \qquad \begin{array}{l}\text{Using the}\\\text{Quotient Rule}\end{array}$$
$$= \frac{e^t(-4e^{4t} - 1 + e^{4t})}{e^t \cdot e^t}$$
$$= \frac{-3e^{4t} - 1}{e^t}$$
$$= -3e^{3t} - e^{-t}$$

61. $y = \dfrac{e^x}{x^2 + 1}$
$$\frac{dy}{dx} = \frac{(x^2 + 1)e^x - 2x \cdot e^x}{(x^2 + 1)^2}$$
$$= \frac{e^x(x^2 - 2x + 1)}{(x^2 + 1)^2}$$
$$= \frac{e^x(x - 1)^2}{(x^2 + 1)^2}$$

63. $f(x) = e^{\sqrt{x}} + \sqrt{e^x}$
$$= e^{x^{1/2}} + e^{x/2}$$
$$f'(x) = \frac{1}{2}x^{-1/2}e^{x^{1/2}} + \frac{1}{2}e^{x/2}$$
$$= \frac{e^{\sqrt{x}}}{2\sqrt{x}} + \frac{\sqrt{e^x}}{2}$$

65. $f(x) = e^{x/2} \cdot \sqrt{x - 1}$
$$= e^{x/2} \cdot (x - 1)^{1/2}$$

$$f'(x) = e^{x/2} \cdot \frac{1}{2}(x-1)^{-1/2} + \frac{1}{2}e^{x/2} \cdot (x-1)^{1/2}$$

$$= \frac{1}{2}e^{x/2}\left((x-1)^{-1/2} + (x-1)^{1/2}\right)$$

$$= \frac{1}{2}e^{x/2}\left(\frac{1}{\sqrt{x-1}} + \sqrt{x-1}\right)$$

$$= \frac{1}{2}e^{x/2}\left(\frac{1}{\sqrt{x-1}} + \sqrt{x-1} \cdot \frac{\sqrt{x-1}}{\sqrt{x-1}}\right)$$

$$= \frac{1}{2}e^{x/2}\left(\frac{1}{\sqrt{x-1}} + \frac{x-1}{\sqrt{x-1}}\right)$$

$$= \frac{1}{2}e^{x/2}\left(\frac{x}{\sqrt{x-1}}\right)$$

67. $f(x) = \dfrac{e^x - e^{-x}}{e^x + e^{-x}}$

$$f'(x) = \frac{(e^x + e^{-x})(e^x + e^{-x}) - (e^x - e^{-x})(e^x - e^{-x})}{(e^x + e^{-x})^2}$$

$$= \frac{(e^{2x} + e^0 + e^0 + e^{-2x}) - (e^{2x} - e^0 - e^0 + e^{-2x})}{(e^x + e^{-x})^2}$$

$$= \frac{e^{2x} + 1 + 1 + e^{-2x} - e^{2x} + 1 + 1 - e^{-2x}}{(e^x + e^{-x})^2}$$

$$= \frac{4}{(e^x + e^{-x})^2}$$

69. $e = \lim_{t \to 0} f(t); \ f(t) = (1+t)^{1/t}$

$$f(1) = (1+1)^{1/1} = 2^1 = 2$$

$$f(0.5) = (1+0.5)^{1/0.5} = 1.5^2 = 2.25$$

$$f(0.2) = (1+0.2)^{1/0.2} = 1.2^5 = 2.48832$$

$$f(0.1) = (1+0.1)^{1/0.1} = 1.1^{10} \approx 2.59374$$

$$f(0.001) = (1+0.001)^{1/0.001} = (1.001)^{1000}$$

$$\approx 2.71692$$

71. $f(x) = x^2 e^{-x}; \ [0,4]$

$$f'(x) = x^2 \cdot (-1)e^{-x} + 2x \cdot e^{-x}$$

$$= -x^2 e^{-x} + 2x e^{-x}$$

Since $f'(x)$ exists for all x in $[0,4]$, the only critical points are where $f'(x) = 0$.

$$f'(x) = -x^2 e^{-x} + 2x e^{-x} = 0$$

$$-e^{-x}(x^2 - 2x) = 0$$

$$x^2 - 2x = 0 \quad (e^{-x} \neq 0)$$

$$x(x-2) = 0$$

$$x = 0 \quad \text{or} \quad x - 2 = 0$$

$$x = 0 \quad \text{or} \qquad x = 2$$

Use Max-Min Principle 1. Find the function values at $x = 0$, $x = 2$, and $x = 4$.

$$f(x) = x^2 e^{-x}$$

$$f(0) = 0^2 \cdot e^{-0} = 0 \cdot 1 = 0$$

$$f(2) = 2^2 \cdot e^{-2} \approx 4(0.1353) \approx 0.54$$

$$f(4) = 4^2 \cdot e^{-4} \approx 16(0.0183) \approx 0.29$$

The maximum value of $f(x)$ is $4e^{-2}$, or approximately 0.54, at $x = 2$.

73. $\boxed{tw}$

75. Graph: $y = x^2 e^{-x}$

From Exercise 71 we see that there is a relative maximum at $(2, 4e^{-2})$, or $(2, 0.54)$. Using our work in Exercise 71 and the First-Derivative Test we find that there is a relative minimum at $(0, 0)$. Also observe that $y \geq 0$ for all x.

For

$$x = -3, \quad y = (-3)^2 e^{-(-3)} = 9e^3 \approx 180.77$$

$$x = -2, \quad y = (-2)^2 e^{-(-2)} = 4e^2 \approx 29.56$$

$$x = -1, \quad y = (-1)^2 e^{-(-1)} = e \approx 2.72$$

$$x = -0.5, \ y = (-0.5)^2 e^{-(-0.5)} = 0.25e^{0.5} \approx 0.41$$

$$x = 0, \quad y = 0^2 e^{-0} = 0 \cdot 1 = 0$$

$$x = 1, \quad y = 1^2 e^{-1} = e^{-1} \approx 0.37$$

$$x = 2, \quad y = 2^2 e^{-2} = 4e^{-2} \approx 0.54$$

$$x = 3, \quad y = 3^2 e^{-3} = 9e^{-3} \approx 0.45$$

$$x = 4, \quad y = 4^2 e^{-4} = 16e^{-4} \approx 0.29$$

$$x = 5, \quad y = 5^2 e^{-5} = 25e^{-5} \approx 0.17$$

$$x = 10, \quad y = 10^2 e^{-10} = 100e^{-10} \approx 0.005$$

Plot these points and connect them with a smooth curve.

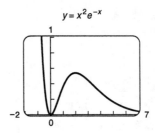

$y = x^2 e^{-x}$

77. See page 295 in the text. The graphs of $f(x)$, $f'(x)$, and $f''(x)$ are all the graph of $y = e^x$.

79.

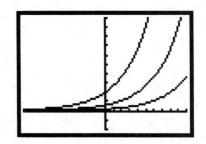

81. See the answer section in the text.

Exercise Set 4.2

1. $\log_2 8 = 3$ Logarithmic equation

     $2^3 = 8$ Exponential equation; 2 is the base, 3 is the exponent

3. $\log_8 2 = \dfrac{1}{3}$ Logarithmic equation

     $8^{1/3} = 2$ Exponential equation; 8 is the base, 1/3 is the exponent

5. $\log_a K = J$ Logarithmic equation

   $a^J = K$ Exponential equation;
   a is the base, J is the exponent

7. $\log_b T = v$ Logarithmic equation

   $b^v = T$ Exponential equation;
   b is the base, v is the exponent

9. $e^M = b$ Exponential equation;
   e is the base, M is the exponent

   $\log_e b = M$ Logarithmic equation

   or $\ln b = M$ $\ln b$ is the abbreviation for $\log_e b$

11. $10^2 = 100$ Exponential equation; 10 is
   the base, 2 is the exponent

   $\log_{10} 100 = 2$ Logarithmic equation

13. $10^{-1} = 0.1$ Exponential equation; 10 is
   the base, -1 is the exponent

   $\log_{10} 0.1 = -1$ Logarithmic equation

15. $M^p = V$ Exponential equation; M is
   the base, p is the exponent

   $\log_M V = p$ Logarithmic equation

17. $\log_b 15 = \log_b 3 \cdot 5$

   $= \log_b 3 + \log_b 5$ (P1)

   $= 1.099 + 1.609$

   $= 2.708$

19. $\log_b \dfrac{1}{5} = \log_b 1 - \log_b 5$ (P2)

   $= 0 - \log_b 5$ (P6)

   $= -\log_b 5$

   $= -1.609$

21. $\log_b 5b = \log_b 5 + \log_b b$ (P1)

   $= 1.609 + 1$ (P4)

   $= 2.609$

23. $\ln 20 = \ln 4 \cdot 5$

   $= \ln 4 + \ln 5$ (P1)

   $= 1.3863 + 1.6094$

   $= 2.9957$

25. $\ln \dfrac{1}{4} = \ln 1 - \ln 4$ (P2)

   $= 0 - 1.3863$ (P6)

   $= -1.3863$

27. $\ln \sqrt{e^8} = \ln e^{8/2}$

   $= \ln e^4$

   $= 4$ (P5)

29. $\ln 5894 = 8.681690$ Using a calculator and
   rounding to six decimal
   places

31. $\ln 0.0182 = -4.006334$

33. $\ln 8100 = 8.999619$

35. $e^t = 100$

   $\ln e^t = \ln 100$ Taking the natural logarithm
   on both sides

   $t = \ln 100$ (P5)

   $t = 4.605170$ Using a calculator

   $t \approx 4.6$

37. $e^t = 60$

   $\ln e^t = \ln 60$ Taking the natural logarithm
   on both sides

   $t = \ln 60$ (P5)

   $t = 4.094345$ Using a calculator

   $t \approx 4.1$

39. $e^{-t} = 0.1$

   $\ln e^{-t} = \ln 0.1$ Taking the natural logarithm
   on both sides

   $-t = \ln 0.1$ (P5)

   $t = -\ln 0.1$

   $t = -(-2.302585)$ Using a calculator

   $t = 2.302585$

   $t \approx 2.3$

41. $e^{-0.02t} = 0.06$

   $\ln e^{-0.02t} = \ln 0.06$ Taking the natural
   logarithm on both sides

   $-0.02t = \ln 0.06$ (P5)

   $t = \dfrac{\ln 0.06}{-0.02}$

   $t = \dfrac{-2.813411}{-0.02}$ Using a calculator

   $t \approx 141$

43. $y = -6 \ln x$

   $\dfrac{dy}{dx} = -6 \cdot \dfrac{1}{x}$ $\left[\dfrac{d}{dx}[c \cdot f(x)] = c \cdot f'(x)\right]$

   $= -\dfrac{6}{x}$

45. $y = x^4 \ln x - \dfrac{1}{2}x^2$

   $\dfrac{dy}{dx} = \underbrace{x^4 \cdot \dfrac{1}{x} + 4x^3 \cdot \ln x}_{} - \dfrac{1}{2} \cdot 2x$

   ⟶ Product Rule on $x^4 \ln x$

   $= x^3 + 4x^3 \ln x - x$

   $= x^3(1 + 4 \ln x) - x$

47. $y = \dfrac{\ln x}{x^4}$

$\dfrac{dy}{dx} = \dfrac{x^4 \cdot \dfrac{1}{x} - 4x^3 \cdot \ln x}{x^8}$ Using the Quotient Rule

$= \dfrac{x^3 - 4x^3 \ln x}{x^8}$

$= \dfrac{x^3(1 - 4\ln x)}{x^3 \cdot x^5}$ Factoring both numerator and denominator

$= \dfrac{x^3}{x^3} \cdot \dfrac{1 - 4\ln x}{x^5}$

$= \dfrac{1 - 4\ln x}{x^5}$ Simplifying

49. $y = \ln \dfrac{x}{4}$

$y = \ln x - \ln 4$ (P2)

$\dfrac{dy}{dx} = \dfrac{1}{x} - 0$

$= \dfrac{1}{x}$

51. $f(x) = \ln(5x^2 - 7)$

$f'(x) = 10x \cdot \dfrac{1}{5x^2 - 7}$ $\left[\dfrac{d}{dx} \ln g(x) = g'(x) \cdot \dfrac{1}{g(x)}\right]$

                        $\left[\dfrac{d}{dx}(5x^2 - 7) = 10x\right]$

$= \dfrac{10x}{5x^2 - 7}$

53. $f(x) = \ln(\ln 4x)$

$f'(x) = 4 \cdot \dfrac{1}{4x} \cdot \dfrac{1}{\ln 4x}$ $\left[\dfrac{d}{dx} \ln g(x) = g'(x) \cdot \dfrac{1}{g(x)}\right]$

                    $\left[\dfrac{d}{dx} \ln 4x = 4 \cdot \dfrac{1}{4x}\right]$

$= \dfrac{1}{x} \cdot \dfrac{1}{\ln 4x}$

$= \dfrac{1}{x \ln 4x}$

55. $f(x) = \ln\left(\dfrac{x^2 - 7}{x}\right)$

$f'(x) = \dfrac{x \cdot 2x - 1 \cdot (x^2 - 7)}{x^2} \cdot \dfrac{1}{\dfrac{x^2 - 7}{x}}$

$\left[\dfrac{d}{dx} \ln g(x) = g'(x) \cdot \dfrac{1}{g(x)}\right]$

$\left(\text{Using Quotient Rule to find}\right.$

$\left.\dfrac{d}{dx} \dfrac{x^2 - 7}{x}\right)$

$= \dfrac{2x^2 - x^2 + 7}{x^2} \cdot \dfrac{x}{x^2 - 7}$ $\left[\dfrac{1}{\dfrac{x^2-7}{x}} = \dfrac{x}{x^2-7}\right]$

$= \dfrac{x^2 + 7}{x^2} \cdot \dfrac{x}{x^2 - 7}$

$= \dfrac{x}{x} \cdot \dfrac{x^2 + 7}{x(x^2 - 7)}$

$= \dfrac{x^2 + 7}{x(x^2 - 7)}$

We could also have done this another way using P2:

$f(x) = \ln\left(\dfrac{x^2 - 7}{x}\right)$

$= \ln(x^2 - 7) - \ln x$

$f'(x) = 2x \cdot \dfrac{1}{x^2 - 7} - \dfrac{1}{x}$

$= \dfrac{2x}{x^2 - 7} \cdot \dfrac{x}{x} - \dfrac{1}{x} \cdot \dfrac{x^2 - 7}{x^2 - 7}$ Multiplying by forms of 1

$= \dfrac{2x^2}{x(x^2 - 7)} - \dfrac{x^2 - 7}{x(x^2 - 7)}$

$= \dfrac{2x^2 - x^2 + 7}{x(x^2 - 7)}$

$= \dfrac{x^2 + 7}{x(x^2 - 7)}$

57. $f(x) = e^x \ln x$

$f'(x) = e^x \cdot \dfrac{1}{x} + e^x \cdot \ln x$ Using the Product Rule

$= e^x\left(\dfrac{1}{x} + \ln x\right)$

59. $f(x) = \ln(e^x + 1)$

$f'(x) = e^x \cdot \dfrac{1}{e^x + 1}$ $\left[\dfrac{d}{dx}(e^x + 1) = e^x\right]$

$= \dfrac{e^x}{e^x + 1}$

61. $f(x) = (\ln x)^2$

$f'(x) = 2(\ln x)^1 \cdot \dfrac{1}{x}$ Extended Power Rule

$= \dfrac{2\ln x}{x}$

63. a) $N(a) = 1000 + 200\ln a$, $a \geq 1$

$N(1) = 1000 + 200\ln 1$ Substituting 1 for a

$\qquad = 1000 + 200 \cdot 0$

$\qquad = 1000 + 0$

$\qquad = 1000$

Thus 1000 units were sold after spending $1000 on advertising.

b) $N(a) = 1000 + 200\ln a$, $a \geq 1$

$N'(a) = 0 + 200 \cdot \dfrac{1}{a}$

$\qquad = \dfrac{200}{a}$

$N'(10) = \dfrac{200}{10} = 20$

c) $N'(a) > 0$ for all $a \geq 1$. Thus $N(a)$ is an increasing function and has a minimum value of 1000 when $a = 1$. There is no maximum.

d) $\boxed{tw}$

65. a) Find $R(t)$.

$R(t) = \begin{matrix}\text{Price} \\ \text{per} \\ \text{unit}\end{matrix} \cdot \begin{matrix}\text{Target} \\ \text{market}\end{matrix} \cdot \begin{matrix}\text{Percentage} \\ \text{buying}\end{matrix}$

$R(t) = 0.5(1,000,000)(1 - e^{-0.04t})$

$\qquad = 500,000 - 500,000e^{-0.04t}$

b) Find $C(t)$.

$C(t) = \begin{matrix}\text{Advertising costs} \\ \text{per day}\end{matrix} \cdot \begin{matrix}\text{Number of} \\ \text{days}\end{matrix}$

$C(t) = 2000t$

c) Find $P(t)$ and take its derivative.

$P(t) = R(t) - C(t)$

$P(t) = 500,000 - 500,000e^{-0.04t} - 2000t$

$P'(t) = 20,000e^{-0.04t} - 2000$

d) Set the first derivative equal to 0 and solve.

$20,000e^{-0.04t} - 2000 = 0$

$20,000e^{-0.04t} = 2000$

$e^{-0.04t} = \dfrac{1}{10} = 0.1$

$\ln e^{-0.04t} = \ln 0.1$

$-0.04t = -2.302585$

$t = \dfrac{-2.302585}{-0.04}$

$t \approx 58$

e) We have only one critical point. So we can use the second derivative to determine whether we have a maximum.

$P''(t) = -800e^{-0.04t}$

Now since exponential functions are positive, $e^{-0.04t} > 0$ for all numbers t. Thus, since $-800e^{-0.04t} < 0$ for all numbers t, $P''(t)$ is less than 0 for $t = 58$ and we have a maximum.

The length of the advertising campaign must be 58 days to result in maximum profit.

67. $V(t) = \$58(1 - e^{-1.1t}) + \$20$

V is the value of the stock after time t, in months.

a) $V(1) = 58(1 - e^{-1.1(1)}) + 20$

$\qquad = 58(1 - 0.332871) + 20$

$\qquad = 58(0.667129) + 20$

$\qquad = 38.693482 + 20$

$\qquad = 58.693482$

$\qquad \approx \$58.69$

$V(12) = 58(1 - e^{-1.1(12)}) + 20$

$\qquad = 58(1 - e^{-13.2}) + 20$

$\qquad = 58(1 - 0.000002) + 20$

$\qquad = 58(0.999998) + 20$

$\qquad = 57.999884 + 20$

$\qquad = 77.999884$

$\qquad \approx \$78.00$

b) $V'(t) = 58(1.1e^{-1.1t}) + 0$

$\qquad = \$63.80e^{-1.1t}$

c) $\qquad V(t) = 58(1 - e^{-1.1t} + 20$

$\qquad 75 = 58(1 - e^{-1.1t}) + 20$

$\qquad\qquad$ Substituting 75 for $V(t)$

$\qquad 55 = 58(1 - e^{-1.1t})$

$\qquad \dfrac{55}{58} = 1 - e^{-1.1t}$

$\qquad 0.948276 = 1 - e^{-1.1t}$

$\qquad -0.051724 = -e^{-1.1t}$

$\qquad 0.051724 = e^{-1.1t}$

$\ln 0.051724 = \ln e^{-1.1t}$ Taking the natural log on both sides

$\ln 0.051724 = -1.1t$

$\dfrac{\ln 0.051724}{-1.1} = t$

$2.7 \text{ months} \approx t$

d) $\boxed{tw}$

69. a) $P(t) = 100(1 - e^{-0.2t})$

$P(1) = 100(1 - e^{-0.2 \cdot 1})$ Substituting 1 for t

$\qquad = 100(1 - e^{-0.2})$

$\qquad = 100(0.181269)$ Using a calculator

$\qquad \approx 18.1\%$

$P(6) = 100(1 - e^{-0.2 \cdot 6})$

$\qquad = 100(1 - e^{-1.2})$

$\qquad = 100(0.698806)$

$\qquad \approx 69.9\%$

b) $P(t) = 100(1 - e^{-0.2t})$

$P'(t) = 100[-(-0.2)e^{-0.2t}]$

$\qquad = 20e^{-0.2t}$

c)
$$P(t) = 100(1 - e^{-0.2t})$$
$$90 = 100(1 - e^{-0.2t}) \quad \text{Replacing } P(t) \text{ by } 90$$
$$0.9 = 1 - e^{-0.2t}$$
$$-0.1 = -e^{-0.2t} \quad \text{Adding } -1$$
$$0.1 = e^{-0.2t} \quad \text{Multiplying by } -1$$
$$\ln 0.1 = \ln e^{-0.2t} \quad \begin{array}{l}\text{Taking the natural} \\ \text{logarithm on both sides}\end{array}$$
$$\ln 0.1 = -0.2t$$
$$\frac{\ln 0.1}{-0.2} = t$$
$$\frac{-2.302585}{-0.2} = t \quad \text{Using a calculator}$$
$$11.5 \approx t$$

Thus it will take approximately 11.5 months for 90% of the doctors to become aware of the new medicine.

d) $\boxed{tw}$

71. a) $\quad S(t) = 68 - 20\ln(t+1), \; t \geq 0$
$$S(0) = 68 - 20\ln(0+1) \quad \text{Substituting 0 for } t$$
$$= 68 - 20\ln 1$$
$$= 68 - 20 \cdot 0$$
$$= 68 - 0$$
$$= 68$$

Thus the average score when they initially took the test was 68%.

b) $\quad S(4) = 68 - 20\ln(4+1) \quad \text{Substituting 4 for } t$
$$= 68 - 20\ln 5$$
$$= 68 - 20(1.609438) \quad \text{Using a calculator}$$
$$= 68 - 32.18876$$
$$\approx 36\%$$

c) $\quad S(24) = 68 - 20\ln(24+1) \quad \text{Substituting 24 for } t$
$$= 68 - 20\ln 25$$
$$= 68 - 20(3.218876) \quad \text{Using a calculator}$$
$$= 68 - 64.37752$$
$$\approx 3.6\%$$

d) First we reword the question:

3.6 (the average score after 24 months) is what percent of 68 (the average score when $t = 0$).

Then we translate and solve:
$$3.6 = x \cdot 68$$
$$\frac{3.6}{68} = x$$
$$0.052941 = x$$
$$5\% \approx x$$

e) $\quad S(t) = 68 - 20\ln(t+1), \; t \geq 0$
$$S'(t) = 0 - 20 \cdot 1 \cdot \frac{1}{t+1}$$
$$= -\frac{20}{t+1}$$

f) $S'(t) < 0$ for all $t \geq 0$. Thus $S(t)$ is a decreasing function and has a maximum value of 68% when $t = 0$.

g) $\boxed{tw}$

73. $\quad v(p) = 0.37\ln p + 0.05 \quad \begin{array}{l} p \text{ in thousands,} \\ v \text{ in ft per sec}\end{array}$

a) $\quad v(531) = 0.37\ln 531 + 0.05 \quad \begin{array}{l}\text{Substituting} \\ 531 \text{ for } p\end{array}$
$$= 0.37(6.274762) + 0.05$$
$$= 2.321662 + 0.05$$
$$= 2.371662$$
$$\approx 2.37 \text{ ft/sec}$$

b) $\quad v(7900) = 0.37\ln 7900 + 0.05$
$$= 0.37(8.974618) + 0.05$$
$$= 3.320609 + 0.05$$
$$= 3.370609$$
$$\approx 3.37 \text{ ft/sec}$$

c) $\quad v'(p) = 0.37 \cdot \frac{1}{p} + 0$
$$= \frac{0.37}{p}$$

d) $\boxed{tw}$

75. $\quad y = (\ln x)^{-4}$
$$\frac{dy}{dx} = -4(\ln x)^{-5} \cdot \frac{1}{x}$$
$$= \frac{-4(\ln x)^{-5}}{x}$$

77. $\quad f(t) = \ln(t^3+1)^5$
$$f'(t) = 5(t^3+1)^4 \cdot 3t^2 \cdot \frac{1}{(t^3+1)^5}$$
$$= \frac{15t^2}{t^3+1} \quad \text{Simplifying}$$

79. $\quad f(x) = [\ln(x+5)]^4$
$$f'(x) = 4[\ln(x+5)]^3 \cdot 1 \cdot \frac{1}{x+5}$$
$$= \frac{4[\ln(x+5)]^3}{x+5}$$

81. $\quad f(t) = \ln[(t^3+3)(t^2-1)]$
$$f'(t) = [(t^3+3)(2t)+(3t^2)(t^2-1)] \cdot \frac{1}{(t^3+3)(t^2-1)}$$
$$= \frac{2t^4 + 6t + 3t^4 - 3t^2}{(t^3+3)(t^2-1)}$$
$$= \frac{5t^4 - 3t^2 + 6t}{(t^3+3)(t^2-1)}$$

83.　$y = \ln \dfrac{x^5}{(8x+5)^2}$

$\qquad y = \ln x^5 - \ln(8x+5)^2$　　　　Property 2

$\qquad \dfrac{dy}{dx} = 5x^4 \cdot \dfrac{1}{x^5} - 2(8x+5) \cdot 8 \cdot \dfrac{1}{(8x+5)^2}$

$\qquad\qquad = \dfrac{5}{x} - \dfrac{16}{8x+5}$

$\qquad\qquad = \dfrac{5}{x} \cdot \dfrac{8x+5}{8x+5} - \dfrac{16}{8x+5} \cdot \dfrac{x}{x}$　Multiplying by forms of 1

$\qquad\qquad = \dfrac{40x + 25 - 16x}{x(8x+5)}$

$\qquad\qquad = \dfrac{24x + 25}{8x^2 + 5x}$

85.　$f(t) = \dfrac{\ln t^2}{t^2} = t^{-2} \ln t^2$

$\qquad f'(t) = t^{-2} \cdot 2t \cdot \dfrac{1}{t^2} + (-2)t^{-3} \ln t^2$

$\qquad\qquad = \dfrac{2t}{t^4} - \dfrac{2 \ln t^2}{t^3}$

$\qquad\qquad = \dfrac{2}{t^3} - \dfrac{2 \ln t^2}{t^3}$

$\qquad\qquad = \dfrac{2 - 2 \ln t^2}{t^3}, \text{ or } \dfrac{2(1 - \ln t^2)}{t^3}$

87.　$y = \dfrac{x^{n+1}}{n+1}\left(\ln x - \dfrac{1}{n+1} \right)$

$\qquad \dfrac{dy}{dx} = \dfrac{x^{n+1}}{n+1}\left(\dfrac{1}{x} - 0 \right) + x^n \left(\ln x - \dfrac{1}{n+1} \right)$

$\qquad\qquad \left[\dfrac{d}{dx} \dfrac{x^{n+1}}{n+1} = \dfrac{1}{n+1} \cdot (n+1)x^n = x^n \right]$

$\qquad\qquad = \dfrac{x^n}{n+1} + x^n \ln x - \dfrac{x^n}{n+1}$

$\qquad\qquad = x^n \ln x$

89.　$y = \ln\left(t + \sqrt{1+t^2} \right) = \ln[t + (1+t^2)^{1/2}]$

$\qquad \dfrac{dy}{dx} = \left[1 + \dfrac{1}{2}(1+t^2)^{-1/2} \cdot 2t \right] \cdot \dfrac{1}{t + (1+t^2)^{1/2}}$

$\qquad\qquad = \left(1 + \dfrac{t}{(1+t^2)^{1/2}} \right) \cdot \dfrac{1}{t + (1+t^2)^{1/2}}$

$\qquad\qquad = \left(\dfrac{(1+t^2)^{1/2}}{(1+t^2)^{1/2}} + \dfrac{t}{(1+t^2)^{1/2}} \right) \cdot \dfrac{1}{t+(1+t^2)^{1/2}}$

$\qquad\qquad = \dfrac{t + (1+t^2)^{1/2}}{(1+t^2)^{1/2}} \cdot \dfrac{1}{t+(1+t^2)^{1/2}}$

$\qquad\qquad = \dfrac{1}{(1+t^2)^{1/2}}$

$\qquad\qquad = \dfrac{1}{\sqrt{1+t^2}}$

91.　$f(x) = \ln[\ln x]^3$

$\qquad f'(x) = 3[\ln x]^2 \cdot \dfrac{1}{x} \cdot \dfrac{1}{[\ln x]^3}$

$\qquad\qquad = \dfrac{3}{x \ln x}$

93.　$\lim\limits_{h \to 0} \dfrac{\ln(1+h)}{h}$

$\ln(1+h)$ exists only for $1 + h > 0$, or $h > -1$.

The function $\dfrac{\ln(1+h)}{h}$ is not continuous at $h = 0$. We use input-output tables.

h	$f(h)$	
-0.9	2.56	h approaches 0
-0.7	1.72	from the left
-0.5	1.39	
-0.3	1.19	
-0.1	1.05	
-0.01	1.01	
-0.001	1.001	

h	$f(h)$	
0.9	0.71	h approaches 0
0.7	0.76	from the right
0.5	0.81	
0.3	0.87	
0.1	0.95	
0.01	0.995	
0.001	0.9995	

From the tables we observe that

$\lim\limits_{h \to 0^-} \dfrac{\ln(1+h)}{h} = 1$

and

$\lim\limits_{h \to 0^+} \dfrac{\ln(1+h)}{h} = 1$

Thus, $\lim\limits_{h \to 0} \dfrac{\ln(1+h)}{h} = 1.$

95.　$\qquad P = P_0 e^{kt}$

$\qquad\qquad \dfrac{P}{P_0} = e^{kt}$

$\qquad\qquad \ln \dfrac{P}{P_0} = \ln e^{kt}$

$\qquad\quad \ln P - \ln P_0 = kt$

$\qquad\quad \dfrac{\ln P - \ln P_0}{k} = t$

97. Let $a = \ln x$; then $e^a = x$.

$\qquad\qquad \log x = \log e^a$

$\qquad\qquad \log x = a \log e$　(P3)

$\qquad\qquad \dfrac{\log x}{\log e} = a$　　　Multiplying by $\dfrac{1}{\log e}$

$\qquad\qquad \dfrac{\log x}{\log e} = \ln x$　　Substituting $\ln x$ for a

$\qquad\qquad \dfrac{\log x}{0.4343} \approx \ln x$

$\qquad\quad 2.3026 \log x \approx \ln x \quad \left(\dfrac{1}{0.4343} \approx 2.3026 \right)$

Thus, $\ln x = \dfrac{\log x}{\log e} \approx 2.3026 \log x.$

99.　[tw]

101. $\sqrt[e]{e} = e^{1/e} \approx 1.444667861$

$e \approx 2.71828183$

For comparison select values of x $(x > 0)$ less than e and greater than e and compute $\sqrt[x]{x}$.

For

$x = 0.5$, $\quad \sqrt[0.5]{0.5} = 0.25$

$x = 0.8$, $\quad \sqrt[0.8]{0.8} \approx 0.756593$

$x = 1.5$, $\quad \sqrt[1.5]{1.5} \approx 1.310371$

$x = 2.0$, $\quad \sqrt{2} \approx 1.414214$

$x = 2.5$, $\quad \sqrt[2.5]{2.5} \approx 1.442700$

$x = 2.7$, $\quad \sqrt[2.7]{2.7} \approx 1.444656$

$x = 2.71$, $\sqrt[2.71]{2.71} \approx 1.44466538$

$x = 2.72$, $\sqrt[2.72]{2.72} \approx 1.44466775$

$x = 3.0$, $\quad \sqrt[3]{3} \approx 1.442250$

$x = 3.5$, $\quad \sqrt[3.5]{3.5} \approx 1.430369$

For any $x > 0$ such that $x \neq e$, $\sqrt[e]{e} > \sqrt[x]{x}$.

103. Find $\lim\limits_{x \to \infty} \ln x$.

x	$\ln x$	
1	0	$\ln x$ exists only
10	2.3	for $x > 0$
100	4.6	
1000	6.9	
10,000	9.2	
100,000	11.5	
1,000,000	13.8	

$\lim\limits_{x \to \infty} \ln x = \infty$

105. See the answer section in the text.

107. See the answer section in the text.

109. $f(x) = x^2 \ln x$, $x > 0$

$f'(x) = x^2 \cdot \dfrac{1}{x} + 2x \cdot \ln x$

$\qquad = x + 2x \ln x$

Solve $f'(x) = 0$.

$x + 2x \ln x = 0$

$x(1 + 2\ln x) = 0$

$x = 0$ or $1 + 2\ln x = 0$

$x = 0$ or $\qquad 2\ln x = -1$

$x = 0$ or $\qquad \ln x = -\dfrac{1}{2}$

$x = 0$ or $\qquad x = e^{-1/2}$

Only $e^{-1/2}$ is in the interval $(0, \infty)$.

We use Max-Min Principle 2.

$f''(x) = 1 + 2x \cdot \dfrac{1}{x} + 2\ln x$

$\qquad = 1 + 2 + 2\ln x$

$\qquad = 3 + 2\ln x$

Since $f''(x) > 0$ for all x in $(0, \infty)$, there is a minimum value at $x = e^{-1/2}$.

$f(x) = x^2 \ln x$

$f(e^{-1/2}) = (e^{-1/2})^2 \ln e^{-1/2}$

$\qquad = e^{-1} \cdot \left(-\dfrac{1}{2}\right)$

$\qquad = -\dfrac{1}{2e} \approx -0.184$

Exercise Set 4.3

1. The solution of $\dfrac{dQ}{dt} = kQ$ is $Q(t) = ce^{kt}$, where t is the time. At $t = 0$, we have some "initial" population $Q(0)$ that we will represent by Q_0. Thus $Q_0 = Q(0) = ce^{k \cdot 0} = ce^0 = c \cdot 1 = c$.

Thus, $Q_0 = c$, so we can express $Q(t)$ as $Q(t) = Q_0 e^{kt}$.

3. It is estimated that the number of franchises N will increase at a rate of 10% per year. That is $\dfrac{dN}{dt} = 0.10N$.

a) $N(t) = N_0 e^{kt}$

$N(t) = 50 e^{0.10t}$ $\quad$ Substituting 50 for N_0 and 0.10 for k

b) $N(20) = 50 e^{0.10 \cdot 20}$ $\quad$ Substituting 20 for t

$\qquad = 50 e^2$

$\qquad = 50(7.389056)$ $\quad$ Using a calculator

$\qquad \approx 369$

Thus in 20 years there will be 369 franchises.

c) $T = \dfrac{\ln 2}{k}$

$\quad = \dfrac{0.693147}{0.10}$ $\quad$ Substituting 0.693147 for $\ln 2$ and 0.10 for k

$\quad \approx 6.9$

The initial number of franchises will double in 6.9 years.

5. The balance grows at the rate given by

$\dfrac{dP}{dt} = 0.09P$.

a) $P(t) = P_0 e^{0.065t}$

b) $P(1) = 1000e^{0.065 \cdot 1}$ $\quad$ Substituting 1000 for P_0 and 1 for t

$\qquad = 1000e^{0.065}$

$\qquad = 1000(1.067159)$

$\qquad \approx 1067.16$

The balance after 1 year is $1067.16.

$P(2) = 1000e^{0.065 \cdot 2}$ $\quad$ Substituting 1000 for P_0 and 2 for t

$\qquad = 1000e^{0.13}$

$\qquad = 1000(1.138828)$

$\qquad \approx 1138.83$

The balance after 2 years is $1138.83.

c) $T = \dfrac{\ln 2}{k}$

$= \dfrac{0.693147}{0.065}$ Substituting 0.693147 for $\ln 2$ and 0.065 for k

≈ 10.7

An investment of \$1000 will double itself in 10.7 years.

7. $k = \dfrac{\ln 2}{T}$

$= \dfrac{0.693147}{10}$ Substituting 0.693147 for $\ln 2$ and 10 for T

$= 0.0693147$

$\approx 6.9\%$

The annual interest rate is 6.9%.

9. $T = \dfrac{\ln 2}{k}$

$= \dfrac{0.693147}{0.10}$ Substituting 0.693147 for $\ln 2$ and 0.10 for k

≈ 6.9

The demand for oil in the U.S. will be double that of 2000 6.9 years after 2000.

11. Find the doubling time:

$T = \dfrac{\ln 2}{k} = \dfrac{\ln 2}{0.062} \approx 11.2 \text{ yr}$

Find the amount after 5 years:

$P(t) = P_0 e^{0.062t}$

$P(5) = 75,000 e^{0.062(5)}$

$= 75,000 e^{0.31}$

$\approx \$102,256.88$

13. Find the initial investment:

$P(t) = P_0 e^{0.084t}$

$11,414.71 = P_0 e^{0.084(5)}$

$11,414.71 = P_0 e^{0.42}$

$\dfrac{11,414.71}{e^{0.42}} = P_0$

$\$7500 \approx P_0$

Find the doubling time:

$T = \dfrac{\ln 2}{k} = \dfrac{\ln 2}{0.084} \approx 8.3 \text{ yr}$

15. a) The exponential growth function is $V(t) = V_0 e^{kt}$. We will express $V(t)$ in thousands of dollars and t as the number of years after 1947. Since $V_0 = 84$ thousand, we have

$V(t) = 84 e^{kt}.$

In 1987 ($t = 40$), we know that $V(t) = 53,900$ thousand. We substitute and solve for k.

$53,900 = 84 e^{k(40)}$

$\dfrac{1925}{3} = e^{40k}$

$\ln \dfrac{1925}{3} = \ln e^{40k}$

$\ln \dfrac{1925}{3} = 40k$

$\dfrac{\ln \dfrac{1925}{3}}{40} = k$

$\dfrac{6.4641}{40} \approx k$

$0.161602 \approx k$

The exponential growth rate is about 0.161602, or 16.1602%. The exponential growth function is $V(t) = 84 e^{0.161602t}$, where V is in thousands of dollars and t is in the number of years after 1947.

b) In 1997, $t = 50$. We find $V(50)$.

$V(50) = 84 e^{0.161602(50)} = 84 e^{8.0801}$

$\approx \$271,283 \text{ thousand, or}$

$\$271,283,000$

(Answers may vary slightly due to rounding.)

c) We will use the expression relating growth rate k and doubling time T. (We could also set $V(t) = 168$ and solve for t.)

$T = \dfrac{\ln 2}{k} \approx \dfrac{0.6931}{0.161602} \approx 4.3 \text{ years}$

d) \$1 billion = \$1,000,000 thousand. We set $V(t) = 1,000,000$ and solve for t.

$1,000,000 = 84 e^{0.161602t}$

$\dfrac{250,000}{21} = e^{0.161602t}$

$\ln \dfrac{250,000}{21} = \ln e^{0.161602t}$

$\ln \dfrac{250,000}{21} = 0.161602t$

$\dfrac{\ln \dfrac{250,000}{21}}{0.161602} = t$

$\dfrac{9.3847}{0.161602} \approx t$

$58 \approx t$

The value of the painting will be \$1 billion about 58 years after 1947.

17. a)
$$P(t) = P_0 e^{kt}$$
$$0.573 = 0.386 e^{k \cdot 2}$$
$$0.573 = 0.386 e^{2k}$$
$$\frac{0.573}{0.386} = e^{2k}$$
$$\ln\left(\frac{0.573}{0.386}\right) = \ln e^{2k}$$
$$\ln\left(\frac{0.573}{0.386}\right) = 2k$$
$$\frac{\ln\left(\frac{0.573}{0.386}\right)}{2} = k$$
$$0.198 \approx k$$

Thus, we have $P(t) = 0.386 e^{0.198t}$.

b) In 2000, $t = 2000 - 1996 = 4$.
$$P(4) = 0.386 e^{0.198(4)} \approx \$0.852 \text{ billion}$$

In 2001, $t = 2001 - 1996 = 5$.
$$P(5) = 0.386 e^{0.198(5)} \approx \$1.039 \text{ billion}$$

c) Substitute 2.0 for $P(t)$ and solve for t.
$$2.0 = 0.386 e^{0.198t}$$
$$\frac{2.0}{0.386} = e^{0.198t}$$
$$\ln\left(\frac{2.0}{0.386}\right) = \ln e^{0.198t}$$
$$\ln\left(\frac{2.0}{0.386}\right) = 0.198t$$
$$\frac{\ln\left(\frac{2.0}{0.386}\right)}{0.198} = t$$
$$8.31 \approx t$$

Sales will be \$2.0 billion after about 8.31 yr.

d) $T = \dfrac{\ln 2}{0.198} \approx 3.5$ yr

e) $\boxed{tw}$

19. a) $y = 0.1097854248(1.442743433)^x$, where y is in millions and x is the number of years after 1990.

Using $b^x = e^{x(\ln b)}$, we have:
$$1.442743433^x = e^{x(\ln 1.442743433)}$$
$$= e^{0.3665464629x}$$

Rounding, we have $P(t) = 0.1 e^{0.367t}$ and $k \approx 0.367$, or 36.7%.

b) In 2004, $t = 2004 - 1990 = 14$.
$$P(14) = 0.1 e^{0.367(14)} \approx 17.0 \text{ million}$$

In 2010, $t = 2010 - 1990 = 20$.
$$P(20) = 0.1 e^{0.367(20)} \approx 154.1 \text{ million}$$

c) Substitute 200 for $P(t)$ and solve for t.
$$200 = 0.1 e^{0.367t}$$
$$\frac{200}{0.1} = e^{0.367t}$$
$$2000 = e^{0.367t}$$
$$\ln 2000 = \ln e^{0.367t}$$
$$\ln 2000 = 0.367t$$
$$\frac{\ln 2000}{0.367} = t$$
$$20.7 \approx t$$

Sales will be 200 million about 20.7 yr after 1990.

d) $T = \dfrac{\ln 2}{0.367} \approx 1.9$ yr

21. $V(t) = V_0 e^{kt}$
$$V(t) = 24 e^{0.08(374)} \qquad \begin{array}{l}\text{Substituting 24 for } P_0, \\ \text{0.08 for } k, \text{ and 374 for } t \\ (2000 - 1626 = 374)\end{array}$$
$$= 24 e^{29.92}$$
$$\approx 2.37 \times 10^{14}$$
$$= 237{,}000{,}000{,}000{,}000$$

Manhattan Island will be worth \$237,000,000,000,000 in 2000.

23. $S(t) = S_0 e^{kt}$

Find the growth rate k. Express salaries in millions of dollars.
$$2.24 = 0.029303 e^{k \cdot 28} \qquad (1998 - 1970 = 28)$$
$$\frac{2.24}{0.029303} = e^{28k}$$
$$\ln\left(\frac{2.24}{0.029303}\right) = \ln e^{28k}$$
$$\ln\left(\frac{2.24}{0.029303}\right) = 28k$$
$$\frac{\ln\left(\frac{2.24}{0.029303}\right)}{28} = k$$
$$0.154876 \approx k$$

The growth rate was 0.154876, or 15.4876%.

Find the average salary in 2000:
$$S(30) = 0.029303 e^{0.154876(30)}6 \qquad (2000 - 1970 = 30)$$
$$= 0.029303 e^{4.64628}$$
$$\approx 3.053274$$

The average salary in 2000 will be about \$3.053274 million, or \$3,053,274.

Find the average salary in 2008:
$$S(38) = 0.029303 e^{0.154876(38)} \qquad (2008 - 1970 = 38)$$
$$= 0.029303 e^{5.885288}$$
$$\approx 10.540475$$

The average salary in 2008 will be about \$10.540475 million, or \$10,540,475.

25. $P(t) = \dfrac{100\%}{1 + 49\, e^{-0.13t}}$

a) $P(0) = \dfrac{100\%}{1 + 49\, e^{-0.13(0)}}$

$= \dfrac{100\%}{1 + 49\, e^{0}}$

$- \dfrac{100\%}{1 + 49 \cdot 1}$

$= \dfrac{100\%}{50}$

$= 2\%$

b) $P(5) = \dfrac{100\%}{1 + 49\, e^{-0.13(5)}}$

$= \dfrac{100\%}{1 + 49\, e^{-0.65}}$

$\approx 3.8\%$

$P(10) = \dfrac{100\%}{1 + 49\, e^{-0.13(10)}}$

$= \dfrac{100\%}{1 + 49\, e^{-1.3}}$

$\approx 7.0\%$

$P(20) = \dfrac{100\%}{1 + 49\, e^{-0.13(20)}}$

$= \dfrac{100\%}{1 + 49\, e^{-2.6}}$

$\approx 21.6\%$

$P(30) = \dfrac{100\%}{1 + 49\, e^{-0.13(30)}}$

$= \dfrac{100\%}{1 + 49\, e^{-3.9}}$

$\approx 50.2\%$

$P(50) = \dfrac{100\%}{1 + 49\, e^{-0.13(50)}}$

$= \dfrac{100\%}{1 + 49\, e^{-6.5}}$

$\approx 93.1\%$

$P(60) = \dfrac{100\%}{1 + 49\, e^{-0.13(60)}}$

$= \dfrac{100\%}{1 + 49\, e^{-7.8}}$

$\approx 98.0\%$

c) $P'(t) =$

$\dfrac{(1 + 49\, e^{-0.13t})(0) - 49(-0.13)e^{-0.13t} \cdot 100\%}{(1 + 49\, e^{-0.13t})^2}$

$= \dfrac{637\% \cdot e^{-0.13t}}{(1 + 49\, e^{-0.13t})^2},\ \text{or}\ \dfrac{6.37\, e^{-0.13t}}{(1 + 49\, e^{-0.13t})^2}$

d) The derivative $P'(t)$ exists for all real numbers. The equation $P'(t) = 0$ has no solution. Thus, the function has no critical points and hence no relative extrema. $P'(t) > 0$ for all real numbers, so $P(t)$ is increasing on $[0, \infty)$. The second derivative can be used to show that the graph has an inflection point at $(29.9, 50\%)$, or $(29.9, 0.5)$. The function is concave up on $(0, 29.9)$ and concave down on $(29.9, \infty)$.

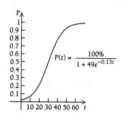

27. $T = \dfrac{\ln 2}{k} = \dfrac{\ln 2}{0.035} \approx 19.8\ \text{yr}$

29. $k = \dfrac{\ln 2}{T} = \dfrac{\ln 2}{6.931} \approx 0.10 \approx 10\%$

31. $T = \dfrac{\ln 2}{k} = \dfrac{\ln 2}{0.02794} \approx 24.8\ \text{yr}$

33. It was estimated that the population P was growing exponentially at a rate of 0.9% per year. That is,

$\dfrac{dP}{dt} = 0.009P.$

a) $P(t) = P_0\, e^{kt}$

$P(t) = 241\, e^{0.009t}$

b) $P(14) = 241\, e^{0.009(14)}$

$= 241\, e^{0.126}$

≈ 273

In 2000 the population of the United States will be approximately 273 million.

c) $T = \dfrac{\ln 2}{k}$

$= \dfrac{0.693147}{0.009}$ Substituting 0.693147 for $\ln 2$ and 0.009 for k

≈ 77

The population will be double that of 2000 in approximately 77 years.

35. $R(b) = e^{21.4b}$ See Example 7

$80 = e^{21.4b}$ Substituting 80 for $R(b)$

$\ln 80 = \ln e^{21.4b}$

$\ln 80 = 21.4b$

$\dfrac{\ln 80}{21.4} = b$

$0.20 \approx b$ Rounding to the nearest hundredth

Thus when the blood alcohol level is 0.20%, the risk of having an accident is 80%.

37. a)
$$P(t) = P_0\, e^{kt}$$
$$216{,}000{,}000 = 2{,}508{,}000\, e^{k\cdot 200}$$
$$\frac{216{,}000{,}000}{2{,}508{,}000} = e^{200k}$$
$$\ln\frac{216{,}000}{2508} = \ln e^{200k}$$
$$\frac{\ln\dfrac{216{,}000}{2508}}{200} = k$$
$$0.022 \approx k$$

The growth rate was approximately 2.2%.

b) $\boxed{tw}$

39. $P(t) = \dfrac{2500}{1 + 5.25\, e^{-0.32t}}$

a) $P(0) = \dfrac{2500}{1 + 5.25\, e^{-0.32(0)}}$
$$= \frac{2500}{1 + 5.25\, e^0}$$
$$= \frac{2500}{1 + 5.25(1)}$$
$$= 400$$

$P(1) = \dfrac{2500}{1 + 5.25\, e^{-0.32(1)}}$
$$= \frac{2500}{1 + 5.25\, e^{-0.32}}$$
$$\approx 520$$

$P(5) = \dfrac{2500}{1 + 5.25\, e^{-0.32(5)}}$
$$= \frac{2500}{1 + 5.25\, e^{-1.6}}$$
$$\approx 1214$$

$P(10) = \dfrac{2500}{1 + 5.25\, e^{-0.32(10)}}$
$$= \frac{2500}{1 + 5.25\, e^{-3.2}}$$
$$\approx 2059$$

$P(15) = \dfrac{2500}{1 + 5.25\, e^{-0.32(15)}}$
$$= \frac{2500}{1 + 5.25\, e^{-4.8}}$$
$$\approx 2396$$

$P(20) = \dfrac{2500}{1 + 5.25\, e^{-0.32(20)}}$
$$= \frac{2500}{1 + 5.25\, e^{-6.4}}$$
$$\approx 2478$$

b) $P'(t) =$
$$\frac{(1 + 5.25e^{-0.32t})(0) - 5.25(-0.32)e^{-0.32t}(2500)}{(1 + 5.25\, e^{-0.32t})^2}$$
$$\text{Quotient Rule}$$
$$= \frac{4200\, e^{-0.32t}}{(1 + 5.25\, e^{-0.32t})^2}$$

c) The derivative $P'(t)$ exists for all real numbers. The equation $P'(t) = 0$ has no solution. Thus, the function has no critical points and hence no relative extrema. $P'(t) > 0$ for all real numbers, so $P(t)$ is increasing on $[0,\infty)$. The second derivative can be used to show that the graph has an inflection point at $(5.18, 1250)$. The function is concave up on $(0, 5.18)$ and concave down on $(5.18, \infty)$.

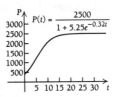

41. $P(t) = 100\%(1 - e^{-0.4t})$

a) $P(0) = 100\%(1 - e^{-0.4(0)}) = 100\%(1 - e^0) =$
$100\%(1 - 1) = 100\%(0) = 0\%$

$P(1) = 100\%(1 - e^{-0.4(1)}) = 100\%(1 - e^{-0.4}) \approx$ 33.0%

$P(2) = 100\%(1 - e^{-0.4(2)}) = 100\%(1 - e^{-0.8}) \approx$ 55.1%

$P(3) = 100\%(1 - e^{-0.4(3)}) = 100\%(1 - e^{-1.2}) \approx$ 69.9%

$P(5) = 100\%(1 - e^{-0.4(5)}) = 100\%(1 - e^{-2}) \approx$ 86.5%

$P(12) = 100\%(1 - e^{-0.4(12)}) =$
$100\%(1 - e^{-4.8}) \approx 99.2\%$

$P(16) = 100\%(1 - e^{-0.4(16)}) =$
$100\%(1 - e^{-6.4}) \approx 99.8\%$

b) $P'(t) = 100\%[-(-0.4)\, e^{-0.4t}]$
$$= 100\%(0.4)\, e^{-0.4t}$$
$$= 0.4\, e^{-0.4t} \qquad (100\% = 1)$$

c) The derivative $P'(t)$ exists for all real numbers. The equation $P'(t) = 0$ has no solution. Thus, the function has no critical points and hence no relative extrema. $P'(t) > 0$ for all real numbers, so $P(t)$ is increasing on $[0,\infty)$. $P''(t) = -0.16\, e^{-0.4t}$, so $P''(t) < 0$ for all real numbers and hence is concave down on $[0,\infty)$.

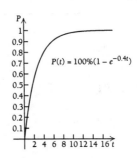

43. $i = e^k - 1$

$\quad = e^{0.073} - 1 \qquad$ Substituting 0.073 for k

$\quad = 1.075731 - 1$

$\quad = 0.075731$

$\quad \approx 7.573\%$

The effective annual yield was 7.573%.

45. $\qquad i = e^k - 1$

$\quad 9.42\% = e^k - 1 \quad$ Substituting 9.42% for i

$\quad 0.0942 = e^k - 1$

$\quad 1.0942 = e^k$

$\quad \ln 1.0942 = \ln e^k$

$\quad 0.090024 = k$

$\quad 9.0\% \approx k$

The investment was invested at 9%.

47. We can find a general expression relating the growth rate k and the tripling time T_3 by solving the following equation.

$\quad 3 P_0 = P_0\, e^{kT_3}$

$\quad\quad 3 = e^{kT_3}$

$\quad \ln 3 = \ln e^{kT_3}$

$\quad \ln 3 = kT_3$

$\quad \dfrac{\ln 3}{k} = T_3$

49. Answers depend on particular data.

51. If an amount grows at a rate of 100% per day, this means it doubles each day. The doubling time T is 1 day, or 24 hours.

The growth rate k and the doubling time T are related by

$\quad k = \dfrac{\ln 2}{T}$

$\quad\quad = \dfrac{\ln 2}{24} \qquad$ Substituting for T

$\quad\quad \approx 0.028881$

A growth rate of 100% per day corresponds to a growth rate of approximately 2.9% per hour.

53. $\boxed{tw}$

55. $\boxed{tw}$

Exercise Set 4.4

1. $P_0 = P\, e^{-kt}$

$\quad P_0 = \$50,000\, e^{-0.06(20)}$

$\quad\quad = \$50,000\, e^{-1.2}$

$\quad\quad \approx \$50,000(0.301194)$

$\quad\quad \approx \$15,059.71$

3. $P_0 = P\, e^{-kt}$

$\quad P_0 = \$60,000\, e^{-0.064(8)}$

$\quad\quad = \$60,000\, e^{-0.512}$

$\quad\quad \approx \$60,000(0.599296)$

$\quad\quad \approx \$35,957.75$

5. $\ln p = \ln \dfrac{163,000}{p}$

$\quad p = \dfrac{163,000}{p}$

$\quad p^2 = 163,000$

$\quad p = 403.73 \qquad (p > 0)$

$S(403.73) = \ln(403.73) \approx 6$

The equilibrium point is ($403.73, 6).

7. $P_0 = P\, e^{-kt}$

$\quad = \$80,000\, e^{-0.083(13)}$

$\quad = \$80,000\, e^{-1.079}$

$\quad \approx \$80,000(0.339935)$

$\quad \approx \$27,194.82$

9. a) $y = 34,001.78697(0.6702977719)^x$

Using $b^x = e^{x(\ln b)}$, we have:

$\quad 0.6702977719^x = e^{x(\ln 0.6702977719)}$

$\quad\quad\quad = e^{-0.4t}$

Thus, we have $V(t) = 34,001.78697 e^{-0.4t}$.

b) $\quad V(7) = 34,001.78697 e^{-0.4(7)} \approx \$2068$

$\quad V(10) = 34,001.78697 e^{-0.4(10)} \approx \$623$

c) $\quad\quad 1000 = 34,001.78697 e^{-0.4t}$

$\quad \dfrac{1000}{34,001.78697} = e^{-0.4t}$

$\quad \ln\left(\dfrac{1000}{34,001.78697}\right) = \ln e^{-0.4t}$

$\quad \ln\left(\dfrac{1000}{34,001.78697}\right) = -0.4t$

$\quad \dfrac{\ln\left(\frac{1000}{34,001.78697}\right)}{-0.4} = t$

$\quad\quad 8.8 \approx t$

The salvage value will be \$1000 after about 8.8 yr.

d) $T = \dfrac{\ln 2}{k} = \dfrac{\ln 2}{0.4} \approx 1.7$ yr

e) $\boxed{tw}$

11. (f)

13. (b)

15. (c)

17. $k = \dfrac{\ln 2}{T} = \dfrac{\ln 2}{22} \approx 0.032 \approx 3.2\%$ per yr

19. $k = \dfrac{\ln 2}{T} = \dfrac{\ln 2}{25} \approx 0.028 \approx 2.8\%$ per year

21. $k = \dfrac{\ln 2}{T} = \dfrac{\ln 2}{23,105} \approx 0.00003 \approx 0.003\%$ per yr

23. $P(t) = P_0 \, e^{-kt}$

$P(20) = 1000 \, e^{-0.231(20)}$ Substituting 1000 for P_0,
                                                                0.231 for k and 20 for t

$\qquad = 1000 \, e^{-4.62}$

$\qquad \approx 9.9$

Thus 9.9 grams of polonium will remain after 20 minutes.

25. $N(t) = N_0 \, e^{-0.0001205t}$ See Example 3(b).

If a piece of wood has lost 90% of its carbon-14 from an initial amount P_0, then 10% P_0 is the amount present. To find the age of the wood, we solve the following equation for t:

$10\% \, P_0 = P_0 \, e^{-0.0001205t}$

$\qquad\qquad$ Substituting 10% P_0 for $P(t)$

$0.1 = e^{-0.0001205t}$

$\ln 0.1 = \ln e^{-0.0001205t}$

$\ln 0.1 = -0.0001205t$

$-2.302585 = -0.0001205t$ Using a calculator

$\dfrac{-2.302585}{-0.0001205} = t$

$19,109 \approx t$

Thus, the piece of wood is about 19,109 years old.

27. If an artifact has lost 60% of its carbon-14 from an initial amount P_0, then 40% P_0 is the amount present. To find the age t we solve the following equation for t.

$40\% P_0 = P_0 \, e^{-0.0001205t}$

$\qquad\qquad$ Substituting 40% P_0 for $P(t)$
$\qquad\qquad$ (See Example 3(b).)

$0.4 = e^{-0.0001205t}$

$\ln 0.4 = \ln e^{-0.0001205t}$

$\ln 0.4 = -0.0001205t$

$-0.916291 = -0.0001205t$ Using a calculator

$\dfrac{-0.916291}{-0.0001205} = t$

$7604 \approx t$

The artifact is about 7604 years old.

29. a) When A decomposes at a rate proportional to the amount of A present, we know that

$\dfrac{dA}{dt} = -kA.$

The solution of this equation is $A = A_0 \, e^{-kt}$.

b) We first find k. The half-life of A is 3 hr.

$k = \dfrac{\ln 2}{T}$

$k = \dfrac{0.693147}{3}$ Substituting 0.693147 for
                                                $\ln 2$ and 3 for T

$\quad \approx 0.23$, or 23%

We now substitute 8 for A_0, 1 for A, and 0.23 for k and solve for t.

$A = A_0 \, e^{-kt}$

$1 = 8 \, e^{-0.23t}$ Substituting

$\dfrac{1}{8} = e^{-0.23t}$

$0.125 = e^{-0.23t}$

$\ln 0.125 = \ln e^{-0.23t}$

$-2.079442 = -0.23t$ Using a calculator

$\dfrac{-2.079442}{-0.23} = t$

$9 \approx t$

After 9 hr there will be 1 gram left.

31. a) $W = W_0 \, e^{-0.008t}$

$k = 0.008$, or 0.8%

The starving animal loses 0.8% of its weight each day.

b) $W = W_0 \, e^{-0.008t}$

$W = W_0 \, e^{-0.008(30)}$ Substituting 30 for t

$W = W_0 \, e^{-0.24}$

$W = 0.786628 \, W_0$ Using a calculator

$W \approx 78.7\% \, W_0$

Thus, after 30 days, 78.7% of the initial weight remains.

33. a) $I = I_0 \, e^{-\mu x}$

$I = I_0 \, e^{-1.4(1)}$ Substituting 1.4 for μ
                                        and 1 for x

$I = I_0 \, e^{-1.4}$

$I = I_0(0.246597)$ Using a calculator

$I \approx 25\% \, I_0$

$I = I_0 \, e^{-1.4(2)}$ Substituting 1.4 for μ
                                        and 2 for x

$I = I_0 \, e^{-2.8}$

$I = I_0(0.060810)$ Using a calculator

$I \approx 6.1\% \, I_0$

$I = I_0 \, e^{-1.4(3)}$ Substituting 1.4 for μ
                                        and 3 for x

$I = I_0 \, e^{-4.2}$

$I = I_0(0.014996)$ Using a calculator

$I \approx 1.5\% \, I_0$

b) $I = I_0 \, e^{-\mu x}$

$I = I_0 \, e^{-1.4(10)}$ Substituting 1.4 for μ
                                          and 10 for x

$I = I_0 \, e^{-14}$

$I = I_0(0.00000083)$ Using a calculator

$I \approx 0.00008\% \, I_0$

35. a) $T(t) = ae^{-kt} + C$ Newton's law of Cooling

At $t = 0$, $T = 100°$. We solve the following equation for a.

$100 = ae^{-k \cdot 0} + 75$ Substituting 100 for T, 0 for t, and 75 for C

$25 = ae^0$

$25 = a$ $(e^0 = 1)$

Thus, $T(t) = 25\,e^{-kt} + 75$.

b) Now we find k using the fact that at $t = 10$, $T = 90°$.

$T(t) = 25\,e^{-kt} + 75$

$90 = 25\,e^{-k \cdot 10} + 75$ Substituting 90 for T and 10 for t

$15 = 25\,e^{-10k}$

$\dfrac{15}{25} = e^{-10k}$

$0.6 = e^{-10k}$

$\ln 0.6 = e^{-10k}$

$\ln 0.6 = -10k$

$\dfrac{\ln 0.6}{-10} = k$

$\dfrac{-0.510826}{-10} = k$

$0.05 \approx k$

Thus, $T(t) = 25\,e^{-0.05t} + 75$.

c) $T(t) = 25\,e^{-0.05t} + 75$

$T(20) = 25\,e^{-0.05(20)} + 75$ Substituting 20 for t

$= 25\,e^{-1} + 75$

$= 25(0.367879) + 75$ Using a calculator

$= 9.196975 + 75$

≈ 84.2

The temperature after 20 minutes is $84.2°$.

d) $T(t) = 25\,e^{-0.05t} + 75$

$80 = 25\,e^{-0.05t} + 75$ Substituting 80 for T

$5 = 25\,e^{-0.05t}$

$\dfrac{5}{25} = e^{-0.05t}$

$0.2 = e^{-0.05t}$

$\ln 0.2 = \ln e^{-0.05t}$

$\ln 0.2 = -0.05t$

$\dfrac{\ln 0.02}{-0.05} = t$

$\dfrac{-1.609438}{-0.05} = t$

$32 \approx t$

It takes 32 minutes for the liquid to cool to $80°$.

e) $\boxed{tw}$

37. We first find a in the equation $T(t) = ae^{-kt} + C$.

Assuming the temperature of the body was normal when the murder occurred, we have $T = 98.6°$ at $t = 0$. Thus

$98.6° = ae^{-k \cdot 0} + 60°$ (Room temperature is $60°$.)

so

$a = 38.6°$.

Thus T is given by $T(t) = 38.6\,e^{-kt} + 60$.

We want to find the number of hours N since the murder was committed. To find N we must first determine k. From the two temperature readings, we have

$85.9 = 38.6\,e^{-kN} + 60$, or $25.9 = 38.6\,e^{-kN}$

$83.4 = 38.6\,e^{-k(N+1)} + 60$, or $23.4 = 38.6\,e^{-k(N+1)}$

Dividing the first equation by the second, we get

$\dfrac{25.9}{23.4} = \dfrac{38.6\,e^{-kN}}{38.6\,e^{-k(N+1)}} = e^{-kN + k(N+1)} = e^k$.

We solve this equation for k:

$\ln \dfrac{25.9}{23.4} = \ln e^k$

$\ln 1.106838 = k$

$0.10 \approx k$

Now we substitute 0.10 for k in the equation $25.9 = 38.6\,e^{-kN}$ and solve for N.

$25.9 = 38.6\,e^{-0.10N}$

$\dfrac{25.9}{38.6} = e^{-0.10N}$

$\ln \dfrac{25.9}{38.6} = \ln e^{-0.10N}$

$\ln 0.670984 = -0.10N$

$\dfrac{-0.399009}{-0.10} = N$

$4\ \text{hr} \approx N$

The coroner arrived at 11 P.M., so the murder was committed about 7 P.M.

39. a) $P(t) = P_0\,e^{-kt}$

$385,000 = 453,000\,e^{-k \cdot 10}$

        Substituting 385,000 for P, 453,000 for P_0, and 10 for t $(1980 - 1970 = 10)$

$\dfrac{385,000}{453,000} = e^{-10k}$

$\dfrac{385}{453} = e^{-10k}$

$\ln \dfrac{385}{453} = \ln e^{-10k}$

$\ln 0.849890 = -10k$

$\dfrac{\ln 0.849890}{-10} = k$

$\dfrac{-0.162649}{-10} = k$

$0.0162649 \approx k$

Thus, $P(t) = 453,000\,e^{-0.0162649t}$, where $t = $ years since 1970.

b)
$$P(t) = 453,000\, e^{-0.0162649t}$$
$$P(30) = 453,000\, e^{-0.0162649(38)}$$
Substituting 38 for t
$$= 453,000\, e^{-0.6180662}$$
$$\approx 244,161$$

Thus, the population of Cincinnati in 2000 will be approximately 244,161.

c)
$$P(t) = 453,000\, e^{-0.0162649t}$$
$$1 = 453,000\, e^{-0.0162649t}$$
Substituting 1 for $P(t)$
$$\frac{1}{453,000} = e^{-0.0162649t}$$
$$\ln \frac{1}{453,000} = \ln e^{-0.0162649t}$$
$$\ln 1 - \ln 453,000 = -0.0162649t$$
$$-\ln 453,000 = -0.0162649t$$
$$\frac{\ln 453,000}{0.0162649} = t$$
$$\frac{13.023647}{0.0162649} = t$$
$$801 \approx t$$

Thus, the population of Cincinnati will be only 1 person 801 years after 1970.

41. a)
$$P(t) = 50\, e^{-0.004t}$$
$$P(375) = 50\, e^{-0.004(375)}$$
Substituting 375 for t
$$= 50\, e^{-1.5}$$
$$= 50(0.223130)$$
$$\approx 11$$

After 375 days, 11 watts will be available.

b)
$$T = \frac{\ln 2}{k}$$
$$= \frac{0.693147}{0.004}$$
Substituting 0.693147 for $\ln 2$ and 0.004 for k
$$\approx 173$$

The half-life of the power supply is 173 days.

c)
$$P(t) = 50\, e^{-0.004t}$$
$$10 = 50\, e^{-0.004t}$$
Substituting 10 for $P(t)$
$$\frac{10}{50} = e^{-0.004t}$$
$$0.2 = e^{-0.004t}$$
$$\ln 0.2 = \ln e^{-0.004t}$$
$$\ln 0.2 = -0.004t$$
$$\frac{\ln 0.2}{-0.004} = t$$
$$\frac{-1.609438}{-0.004} = t$$
Using a calculator
$$402 \approx t$$

The satellite can stay in operation 402 days.

d) When $t = 0$,
$$P = 50\, e^{-0.004(0)}$$
Substituting 0 for t
$$= 50\, e^0$$
$$= 50 \cdot 1$$
$$= 50$$

At the beginning the power output was 50 watts.

e) $\boxed{tw}$

43. a) $y = 84.94353992 - 0.5412834098 \ln x$

b) When $x = 8$, $y = 83.8\%$;
when $x = 10$, $y = 83.7\%$;
when $x = 24$, $y = 83.2\%$;
when $x = 36$, $y = 83.0\%$

c)
$$82 = 84.94353992 - 0.5412834098 \ln x$$
$$-2.94353992 = -0.5412834098 \ln x$$
$$5.438075261 \approx \ln x$$
$$x \approx e^{5.438075261}$$
$$x \approx 230$$

The test scores will fall below 82% after about 230 months.

d) $\boxed{tw}$

45. We solve $D(p) = S(p)$.
$$480\, e^{-0.003p} = 150\, e^{0.004p}$$
$$\frac{480}{150} = \frac{e^{0.004p}}{e^{-0.003p}}$$
$$3.2 = e^{0.007p}$$
$$\ln 3.2 = \ln e^{0.007p}$$
$$\ln 3.2 = 0.007p$$
$$\frac{\ln 3.2}{0.007} = p$$
$$166.16 \approx p$$

Now find $D(p)$ or $S(p)$ for $p = 166.16$. We will find $S(p)$.
$$S(166.16) = 150\, e^{0.004(166.16)}$$
$$= 150\, e^{0.66464}$$
$$\approx 150(1.943791)$$
$$\approx 292$$

The equilibrium point is ($166.16, 292$).

47. $\boxed{tw}$

Exercise Set 4.5

1. $5^4 = e^{4 \cdot \ln 5}$ Theorem 13: $a^x = e^{x \cdot \ln a}$

$\approx e^{4(1.609438)}$ Using a calculator

$\approx e^{6.4378}$

3. $3.4^{10} = e^{10 \cdot \ln 3.4}$ Theorem 13: $a^x = e^{x \cdot \ln a}$

$\approx e^{10(1.223775)}$ Using a calculator

$\approx e^{12.238}$

5. $4^k = e^{k \cdot \ln 4}$ Theorem 13: $a^x = e^{x \cdot \ln a}$

7. $8^{kT} = e^{kT \cdot \ln 8}$ Theorem 13: $a^x = e^{x \cdot \ln a}$

9. $y = 6^x$

$\frac{dy}{dx} = (\ln 6)6^x$ Theorem 14: $\frac{dy}{dx}a^x = (\ln a)a^x$

11. $f(x) = 10^x$

$f'(x) = (\ln 10)10^x$ Theorem 14

13. $f(x) = x(6.2)^x$

$f'(x) = x\left[\frac{d}{dx}(6.2)^x\right] + \left[\frac{d}{dx}x\right](6.2)^x$ Product Rule

$= x \cdot \underbrace{(\ln 6.2)(6.2)^x}_{\text{Theorem 14}} + 1 \cdot (6.2)^x$

$= (6.2)^x[x \ln 6.2 + 1]$

15. $y = x^3 10^x$

$\frac{dy}{dx} = x^3 \cdot \underbrace{(\ln 10)10^x}_{\text{Theorem 14}} + 3x^2 \cdot 10^x$ Product Rule

$= 10^x x^2(x \ln 10 + 3)$

17. $y = \log_4 x$

$\frac{dy}{dx} = \frac{1}{\ln 4} \cdot \frac{1}{x}$ Theorem 16: $\frac{d}{dx}\log_a x = \frac{1}{\ln a} \cdot \frac{1}{x}$

19. $f(x) = 2 \log x$

$f'(x) = 2 \cdot \frac{d}{dx}\log x$

$= 2 \cdot \frac{1}{\ln 10} \cdot \frac{1}{x}$ Theorem 16 ($\log x = \log_{10} x$)

$= \frac{2}{\ln 10} \cdot \frac{1}{x}$

21. $f(x) = \log \frac{x}{3}$

$f(x) = \log x - \log 3$ (P2)

$f'(x) = \frac{1}{\ln 10} \cdot \frac{1}{x} - 0$ Theorem 16 ($\log x = \log_{10} x$)

$= \frac{1}{\ln 10} \cdot \frac{1}{x}$

23. $y = x^3 \log_8 x$

$\frac{dy}{dx} = x^3\underbrace{\left(\frac{1}{\ln 8} \cdot \frac{1}{x}\right)}_{\text{Theorem 16}} + 3x^2 \cdot \log_8 x$ Product Rule

$= x^2 \cdot \frac{1}{\ln 8} + 3x^2 \cdot \log_8 x$

$= x^2\left(\frac{1}{\ln 8} + 3\log_8 x\right)$

25. a) $N(t) = 250,000\left(\frac{1}{4}\right)^t$

$N'(t) = 250,000\frac{d}{dx}\left(\frac{1}{4}\right)^t$

$= 250,000 \cdot \left(\ln \frac{1}{4}\right)\left(\frac{1}{4}\right)^t$ Theorem 14

$= 250,000(\ln 1 - \ln 4)\left(\frac{1}{4}\right)^t$ (P2)

$= -250,000(\ln 4)\left(\frac{1}{4}\right)^t$ ($\ln 1 = 0$)

b) $\boxed{tw}$

27. $R = \log \frac{I}{I_0}$

$R = \log \frac{10^5 \cdot I_0}{I_0}$ Substituting $10^5 \cdot I_0$ for I

$= \log 10^5$

$= 5$ (P5)

The magnitude on the Richter scale is 5.

29. a) $I = I_0 \cdot 10^R$

$I = I_0 \cdot 10^7$ Substituting 7 for R

$= 10^7 \cdot I_0$

b) $I = I_0 \cdot 10^R$

$I = I_0 \cdot 10^8$ Substituting 8 for R

$= 10^8 \cdot I_0$

c) The intensity in (b) is 10 times that in (a).

$10^8 I_0 = 10 \cdot 10^7 I_0$

d) $I = I_0 10^R$

$\frac{dI}{dR} = I_0 \cdot \frac{d}{dR} 10^R$ I_0 is a constant

$= I_0 \cdot (\ln 10)10^R$ Theorem 14

$= (I_0 \cdot \ln 10)10^R$

e) $\boxed{tw}$

31. a) $R = \log \frac{I}{I_0}$

$R = \log I - \log I_0$ (P2)

$\frac{dR}{dI} = \frac{1}{\ln 10} \cdot \frac{1}{I} - 0$ Theorem 16; I_0 is a constant.

$= \frac{1}{\ln 10} \cdot \frac{1}{I}$

b) $\boxed{tw}$

33. a) $y = m \log x + b$

$\frac{dy}{dx} = m \cdot \frac{d}{dx}\log x + 0$ m and b are constants

$= m\left(\frac{1}{\ln 10} \cdot \frac{1}{x}\right)$ Theorem 16

$= \frac{m}{\ln 10} \cdot \frac{1}{x}$

b) $\boxed{tw}$

35. $f(x) = 3^{2x} = (3^x)^2$

$f'(x) = 2(3^x) \cdot (\ln\ 3)3^x$

$\qquad = 2(\ln\ 3)(3^x)^2$

$\qquad = 2(\ln\ 3)(3^{2x})$

37. $y = x^x,\ x > 0$

$y = e^{x\ \ln\ x}$ Theorem 13: $a^x = e^{x\ \ln\ a}$

$\dfrac{dy}{dx} = \left(x \cdot \dfrac{1}{x} + 1 \cdot \ln\ x\right)e^{x\ \ln\ x}$

$\qquad = (1 + \ln\ x)x^x$ Substituting x^x for $e^{x\ \ln\ x}$

39. $f(x) = x^{e^x},\ x > 0$

$f(x) = e^{e^x\ \ln\ x}$ Theorem 13: $a^x = e^{x\ \ln\ a}$

$f'(x) = \left(e^x \cdot \dfrac{1}{x} + e^x\ \ln\ x\right)e^{e^x\ \ln\ x}$

$\qquad = e^x\left(\dfrac{1}{x} + \ln\ x\right)x^{e^x}$ Substituting x^{e^x} for $e^{e^x\ \ln\ x}$

$\qquad = e^x\ x^{e^x}\left(\ln\ x + \dfrac{1}{x}\right)$

41. $y = \log_a f(x),\ f(x) > 0$

$a^y = f(x)$ Exponential equation

$e^{y\ \ln\ a} = f(x)$ Theorem 13

Differential implicitly to find dy/dx.

$\dfrac{d}{dx}e^{y\ \ln\ a} = \dfrac{d}{dx}f(x)$

$\dfrac{dy}{dx} \cdot \ln\ a \cdot e^{y\ \ln\ a} = f'(x)$

$\dfrac{dy}{dx} \cdot \ln\ a \cdot f(x) = f'(x)$ Substituting $f(x)$ for $e^{y\ \ln\ a}$

$\dfrac{dy}{dx} = \dfrac{1}{\ln\ a} \cdot \dfrac{f'(x)}{f(x)}$

43. $\boxed{tw}$

Exercise Set 4.6

1. $x = D(p) = 400 - p;\ p = \$125$

a) To find the elasticity, we first find $D'(p)$:

$D'(p) = -1$

Then we substitute -1 for $D'(p)$ and $400 - p$ for $D(p)$ in the expression for elasticity:

$E(p) = -\dfrac{pD'(p)}{D(p)} = -\dfrac{p \cdot (-1)}{400 - p} = \dfrac{p}{400 - p}$

b) $E(125) = \dfrac{125}{400 - 125} = \dfrac{125}{275} = \dfrac{5}{11}$

Since $E(125) < 1$, the demand is inelastic.

c) Total revenue is a maximum at the value(s) of p for which $E(p) = 1$. We solve $E(p) = 1$.

$\dfrac{p}{400 - p} = 1$

$p = 400 - p$

$2p = 400$

$p = \$200$

3. $x = D(p) = 200 - 4p;\ p = \$46$

a) $D'(p) = -4$

$E(p) = -\dfrac{pD'(p)}{D(p)} = -\dfrac{p \cdot (-4)}{200 - 4p} = \dfrac{4p}{200 - 4p} = \dfrac{p}{50 - p}$

b) $E(46) = \dfrac{46}{50 - 46} = \dfrac{46}{4} = \dfrac{23}{2}$

Since $E(46) > 1$, the demand is elastic.

c) Solve $E(p) = 1$.

$\dfrac{p}{50 - p} = 1$

$p = 50 - p$

$2p = 50$

$p = \$25$

5. $x = D(p) = \dfrac{400}{p} = 400p^{-1};\ p = \$50$

a) $D'(p) = -400p^{-2} = -\dfrac{400}{p^2}$

$E(p) = -\dfrac{pD'(p)}{D(p)} = -\dfrac{p\left(-\dfrac{400}{p^2}\right)}{\dfrac{400}{p}} = \dfrac{\dfrac{400}{p}}{\dfrac{400}{p}} = 1$

b) $E(50) = 1$

Since $E(50) = 1$, demand has unit elasticity.

c) $E(p) = 1$ for all $p > 0$, so total revenue is a maximum for all $p > 0$.

7. $x = D(p) = \sqrt{500 - p} = (500 - p)^{1/2};\ p = \$400$

a) $D'(p) = \dfrac{1}{2}(500 - p)^{1/2}(-1) = -\dfrac{1}{2\sqrt{500 - p}}$

$E(p) = -\dfrac{pD'(p)}{D(p)} = -\dfrac{p\left(-\dfrac{1}{2\sqrt{500 - p}}\right)}{\sqrt{500 - p}} = \dfrac{\dfrac{p}{2\sqrt{500 - p}}}{\sqrt{500 - p}} = \dfrac{p}{2(500 - p)} = \dfrac{p}{1000 - 2p}$

b) $E(400) = \dfrac{400}{1000 - 2 \cdot 400} = \dfrac{400}{200} = 2$

Since $E(400) > 1$, the demand is elastic.

c) Solve $E(v) = 1$.

$\dfrac{p}{1000 - 2p} = 1$

$p = 1000 - 2p$

$3p = 1000$

$p = \dfrac{\$1000}{3} \approx \$333.33$

9. $x = D(p) = 100\ e^{-0.25p};\ p = \$10$

a) $D'(p) = 100(-0.25)\ e^{-0.25p} = -25\ e^{-0.25p}$

$E(p) = -\dfrac{pD'(p)}{D(p)} = -\dfrac{p(-25\ e^{-0.25p})}{100\ e^{-0.25p}} = \dfrac{25p}{100} = \dfrac{p}{4}$

b) $E(10) = \dfrac{10}{4} = \dfrac{5}{2}$

Since $E(10) > 1$, the demand is elastic.

c) Solve $E(p) = 1$.

$$\frac{p}{4} = 1$$

$$p = \$4$$

11. $x = D(p) = \dfrac{100}{(p+3)^2} = 100(p+3)^{-2};\ p = \$1$

a) $D'(p) = 100(-2)(p+3)^{-3} = -\dfrac{200}{(p+3)^3}$

$$E(p) = -\frac{pD'(p)}{D(p)} = -\frac{p\left[-\dfrac{200}{(p+3)^3}\right]}{\dfrac{100}{(p+3)^2}} =$$

$$\frac{\dfrac{200p}{(p+3)^3}}{\dfrac{100}{(p+3)^2}} = \frac{2p}{p+3}$$

b) $E(1) = \dfrac{2 \cdot 1}{1+3} = \dfrac{2}{4} = \dfrac{1}{2}$

Since $E(1) < 1$, the demand is inelastic.

c) Solve $E(p) = 1$.

$$\frac{2p}{p+3} = 1$$

$$2p = p + 3$$

$$p = \$3$$

13. $x = D(p) = 967 - 25p$

a) $D'(p) = -25$

$$E(p) = -\frac{pD'(p)}{D(p)} = -\frac{p \cdot (-25)}{967 - 25p} = \frac{25p}{967 - 25p}$$

b) We set $E(p) = 1$ and solve for p

$$\frac{25p}{967 - 25v} = 1$$

$$25p = 967 - 25p$$

$$50p = 967$$

$$p = 19.34 \cent$$

c) Demand is elastic when $E(p) > 1$. Test a value on either side of 19.34.

$$E(19) = \frac{25 \cdot 19}{967 - 25 \cdot 19} \approx 0.97 < 1$$

$$E(20) = \frac{25 \cdot 20}{967 - 25 \cdot 20} \approx 1.07 > 1$$

Thus, demand is elastic for $p > 19.34\cent$.

d) Demand is inelastic when $E(p) < 1$. Using the calculations in part (c), we see that demand is inelastic for $p < 19.34\cent$.

e) Since $E(19.34\cent) = 1$, total revenue is a maximum when $p = 19.34\cent$.

f) At a price of $20\cent$ per cookie the demand is elastic. (See part (c).) Thus, a small increase in price will cause the total revenue to decrease.

15. $x = D(p) = \sqrt{200 - p^3}$, or $(200 - p^3)^{1/2}$

a) $D'(p) = \dfrac{1}{2}(200 - p^3)^{-1/2}(-3p^2)$

$$= -\frac{3p^2}{2\sqrt{200 - p^3}}$$

$$E(p) = -\frac{pD'(p)}{D(p)} = -\frac{p\left(-\dfrac{3p^2}{2\sqrt{200 - p^3}}\right)}{\sqrt{200 - p^3}} =$$

$$\frac{3p^3}{2(200 - p^3)}$$

b) $E(3) = \dfrac{3(3)^3}{2(200 - 3^3)} = \dfrac{81}{2 \cdot 173} = \dfrac{81}{346} \approx 0.234$

c) $E(3) < 1$, so the demand is inelastic at $p = \$3$. Therefore, a small price increase will cause total revenue to increase.

17. a) $x = D(p) = \dfrac{k}{p^n} = kp^{-n}$

$$D'(p) = -nkp^{-n-1} = -\frac{nk}{p^{n+1}}$$

$$E(p) = -\frac{pD'(p)}{D(p)} = -\frac{p\left(-\dfrac{nk}{p^{n+1}}\right)}{\dfrac{k}{p^n}} = \frac{\dfrac{nk}{p^n}}{\dfrac{k}{p^n}} = n$$

b) No: $E(p)$ is a constant, n.

c) Solve $E(p) = 1$.

$$n = 1$$

Thus, total revenue is a maximum for $p = \$1$.

19. We will use implicit differentiation.

$$L(p) = \ln D(p)$$

$$L'(p) = D'(p) \cdot \frac{1}{D(p)}$$

$$L'(p)D(p) = D'(p)$$

$$E(p) = -\frac{pD'(p)}{D(p)} = -\frac{pL'(p)D(p)}{D(p)} = -pL'(p)$$

21. $\boxed{tw}$

Chapter 5

Integration

1. $\int x^6\, dx$

$= \dfrac{x^{6+1}}{6+1} + C \quad \left(\int x^r\, dx = \dfrac{x^{r+1}}{r+1} + C\right)$

$= \dfrac{x^7}{7} + C$

3. $\int 2\, dx$

$= 2x + C \quad$ (For k a constant, $\int k\, dx = kx + C$.)

5. $\int x^{1/4}\, dx$

$= \dfrac{x^{1/4+1}}{\frac{1}{4}+1} + C \quad \left(\int x^r\, dx = \dfrac{x^{r+1}}{r+1} + C\right)$

$= \dfrac{x^{5/4}}{\frac{5}{4}} + C$

$= \dfrac{4}{5}x^{5/4} + C$

7. $\int (x^2 + x - 1)\, dx$

$= \int x^2\, dx + \int x\, dx - \int 1\, dx$

$\qquad$ The integral of a sum is the
$\qquad$ sum of the integrals.

$= \dfrac{x^3}{3} + \dfrac{x^2}{2} - x + C \quad \leftarrow$ DON'T FORGET THE C!

$\qquad \left[\int x^r\, dx = \dfrac{x^{r+1}}{r+1} + C\right]$

$\qquad$ (For k a constant, $\int k\, dx = kx + C$.)

9. $\int (t^2 - 2t + 3)\, dt$

$= \int t^2\, dt - \int 2t\, dt + \int 3\, dt$

$\qquad$ The integral of a sum is the
$\qquad$ sum of the integrals.

$= \dfrac{t^3}{3} - 2 \cdot \dfrac{t^2}{2} + 3t + C$

$\qquad \left[\int x^r\, dx = \dfrac{x^{r+1}}{r+1} + C\right]$

$\qquad$ (For k a constant, $\int k\, dx = kx + C$.)

$= \dfrac{t^3}{3} - t^2 + 3t + C$

11. $\int 5\, e^{8x}\, dx$

$= \dfrac{5}{8}\, e^{8x} + C \quad \left(\int b\, e^{ax}\, dx = \dfrac{b}{a}\, e^{ax} + C\right)$

13. $\int (x^3 - x^{8/7})\, dx$

$= \int x^3\, dx - \int x^{8/7}\, dx$

$\qquad$ The integral of a sum is the
$\qquad$ sum of the integrals.

$= \dfrac{x^4}{4} - \dfrac{x^{8/7+1}}{\frac{8}{7}+1} + C \quad \left[\int x^r\, dx = \dfrac{x^{r+1}}{r+1} + C\right]$

$= \dfrac{x^4}{4} - \dfrac{x^{15/7}}{\frac{15}{7}} + C$

$= \dfrac{x^4}{4} - \dfrac{7}{15}x^{15/7} + C$

15. $\displaystyle\int \dfrac{1000}{x}\, dx$

$= 1000 \displaystyle\int \dfrac{1}{x}\, dx \quad$ The integral of a constant
$\qquad\qquad\qquad$ times a function is the
$\qquad\qquad\qquad$ constant times the integral.

$= 1000 \ln x + C \quad \int \dfrac{1}{x}\, dx = \ln x + C,\ x > 0;$
$\qquad\qquad\qquad$ we generally consider $x > 0$

17. $\displaystyle\int \dfrac{dx}{x^2} = \int \dfrac{1}{x^2}\, dx = \int x^{-2}\, dx$

$= \dfrac{x^{-2+1}}{-2+1} + C \quad \left[\int x^r\, dx = \dfrac{x^{r+1}}{r+1} + C\right]$

$= \dfrac{x^{-1}}{-1} + C$

$= -x^{-1} + C,\ $ or $\ -\dfrac{1}{x} + C$

19. $\int \sqrt{x}\, dx = \int x^{1/2}\, dx$

$= \dfrac{x^{1/2+1}}{\frac{1}{2}+1} + C$

$= \dfrac{x^{3/2}}{\frac{3}{2}} + C$

$= \dfrac{2}{3}x^{3/2} + C$

21. $\int \dfrac{-6}{\sqrt[3]{x^2}}\, dx = \int \dfrac{-6}{x^{2/3}}\, dx = \int -6x^{-2/3}\, dx$

$$= -6 \int x^{-2/3}\, dx$$

$$= -6 \cdot \dfrac{x^{-2/3+1}}{-\dfrac{2}{3}+1} + C$$

$$= -6 \cdot \dfrac{x^{1/3}}{\dfrac{1}{3}} + C$$

$$= -6 \cdot 3x^{1/3} + C$$

$$= -18x^{1/3} + C$$

23. $\int 8\, e^{-2x}\, dx = \dfrac{8}{-2}\, e^{-2x} + C$

$$= -4\, e^{-2x} + C$$

25. $\int \left(x^2 - \dfrac{3}{2}\sqrt{x} + x^{-4/3} \right) dx$

$$= \int x^2\, dx - \dfrac{3}{2} \int x^{1/2}\, dx + \int x^{-4/3}\, dx$$

$$= \dfrac{x^{2+1}}{2+1} - \dfrac{3}{2} \cdot \dfrac{x^{1/2+1}}{\dfrac{1}{2}+1} + \dfrac{x^{-4/3+1}}{-\dfrac{4}{3}+1} + C$$

$$= \dfrac{x^3}{3} - \dfrac{3}{2} \cdot \dfrac{x^{3/2}}{\dfrac{3}{2}} + \dfrac{x^{-1/3}}{-\dfrac{1}{3}} + C$$

$$= \dfrac{x^3}{3} - x^{3/2} - 3x^{-1/3} + C$$

27. Find the function f such that
$f'(x) = x - 3$ and $f(2) = 9$.

We first find $f(x)$ by integrating.

$f(x) = \int (x - 3)\, dx$

$\quad = \int x\, dx - \int 3\, dx$

$\quad = \dfrac{x^2}{2} - 3x + C$

The condition $f(2) = 9$ allows us to find C.

$f(x) = \dfrac{x^2}{2} - 3x + C$

$f(2) = \dfrac{2^2}{2} - 3 \cdot 2 + C = 9 \quad$ Substituting 2 for x and 9 for $f(2)$

$\qquad\qquad 2 - 6 + C = 9$

$\qquad\qquad\qquad C = 13$

Thus, $f(x) = \dfrac{x^2}{2} - 3x + 13$.

29. Find the function f such that
$f'(x) = x^2 - 4$ and $f(0) = 7$.

We first find $f(x)$ by integrating.

$f(x) = \int (x^2 - 4)\, dx$

$\quad = \int x^2\, dx - \int 4\, dx$

$\quad = \dfrac{x^3}{3} - 4x + C$

The condition $f(0) = 7$ allows us to find C.

$$f(x) = \dfrac{x^3}{3} - 4x + C$$

$$f(0) = \dfrac{0^3}{3} - 4 \cdot 0 + C = 7 \quad \begin{array}{l}\text{Substituting 0 for } x \\ \text{and 7 for } f(0)\end{array}$$

Solving for C we get $C = 7$.

Thus, $f(x) = \dfrac{x^3}{3} - 4x + 7$.

31. $C'(x) = x^3 - 2x$

We integrate to find $C(x)$ using K for the integration constant to avoid confusion with the cost function C.

$C(x) = \int C'(x)\, dx$

$\quad = \int (x^3 - 2x)\, dx$

$\quad = \dfrac{x^4}{4} - x^2 + K$

Fixed costs are \$100. This means $C(0) = 100$. This allows us to determine the value of K.

$$C(0) = \dfrac{0^4}{4} - 0^2 + K = 100$$

$$\qquad\qquad \text{Substituting 0 for } x \text{ and 100 for } C(0)$$

Solving for K we get $K = 100$.

Thus, $C(x) = \dfrac{x^4}{4} - x^2 + 100$.

33. $R'(x) = x^2 - 3$

a) We integrate to find $R(x)$.

$R(x) = \int R'(x)\, dx$

$\quad = \int (x^2 - 3)\, dx$

$\quad = \dfrac{x^3}{3} - 3x + C$

The condition $R(0) = 0$ allows us to find C.

$$R(0) = \dfrac{0^3}{3} - 3 \cdot 0 + C = 0$$

$$\qquad\qquad \text{Substituting 0 for } x \text{ and 0 for } R(0)$$

Solving for C we get $C = 0$.

Thus, $R(x) = \dfrac{x^3}{3} - 3x$.

b) $\boxed{tw}$

35. $D'(p) = -\dfrac{4000}{p^2} = -4000p^{-2}$

We integrate to find $D(p)$.

$D(p) = \int -4000p^{-2}\, dp$

$\quad = -4000 \int p^{-2}\, dp$

$\quad = -4000 \cdot \dfrac{p^{-1}}{-1} + C$

$\quad = \dfrac{4000}{p} + C$

When $p = \$4$ per unit, $D(p) = 1003$ units, or $D(4) = 1003$. We use this to find C.

$$D(4) = \frac{4000}{4} + C = 1003$$
$$1000 + C = 1003$$
$$C = 3$$
$$D(p) = \frac{4000}{p} + 3$$

37. $\frac{dE}{dt} = 30 - 10t$, where t = the number of hours the operator has been at work.

a) We find $E(t)$ by integrating $E'(t)$.
$$E(t) = \int E'(t)\, dt$$
$$= \int (30 - 10t)\, dt$$
$$= 30t - 5t^2 + C$$

The condition $E(2) = 72$ allows us to find C.
$$E(2) = 30 \cdot 2 - 5 \cdot 2^2 + C = 72$$

Substituting 2 for t and 72 for $E(2)$

Solving for C we get:
$$60 - 20 + C = 72$$
$$40 + C = 72$$
$$C = 32$$

Thus, $E(t) = 30t - 5t^2 + 32$.

b) $E(t) = 30t - 5t^2 + 32$
$$E(3) = 30 \cdot 3 - 5 \cdot 3^2 + 32 \quad \text{Substituting 3 for } t$$
$$= 90 - 45 + 32$$
$$= 77$$

The operator's efficiency after 3 hours is 77%.
$$E(5) = 30 \cdot 5 - 5 \cdot 5^2 + 32 \quad \text{Substituting 5 for } t$$
$$= 150 - 125 + 32$$
$$= 57$$

The operator's efficiency after 5 hours is 57%.

39. $v(t) = 3t^2$, $\quad s(0) = 4$

We find $s(t)$ by integrating $v(t)$.
$$s(t) = \int v(t)\, dt$$
$$= \int 3t^2\, dt$$
$$= 3 \cdot \frac{t^3}{3} + C$$
$$= t^3 + C$$

The condition $s(0) = 4$ allows us to find C.
$$s(0) = 0^3 + C = 4 \quad \text{Substituting 0 for } t \text{ and}$$
$$4 \text{ for } s(0)$$

Solving for C, we get $C = 4$.

Thus, $s(t) = t^3 + 4$.

41. $a(t) = 4t$, $\quad v(0) = 20$

We find $v(t)$ by integrating $a(t)$.
$$v(t) = \int a(t)\, dt$$
$$= \int 4t\, dt$$
$$= 4 \cdot \frac{t^2}{2} + C$$
$$= 2t^2 + C$$

The condition $v(0) = 20$ allows us to find C.
$$v(0) = 2 \cdot 0^2 + C = 20 \quad \text{Substituting 0 for } t$$
$$\text{and 20 for } v(0)$$

Solving for C, we get $C = 20$.

Thus, $v(t) = 2t^2 + 20$.

43. $a(t) = -2t + 6$, $\quad v(0) = 6 \quad$ and $s(0) = 10$

We find $v(t)$ by integrating $a(t)$.
$$v(t) = \int a(t)\, dt$$
$$= \int (-2t + 6)\, dt$$
$$= -t^2 + 6t + C_1$$

The condition $v(0) = 6$ allows us to find C_1.
$$v(0) = -0^2 + 6 \cdot 0 + C_1 = 6 \quad \text{Substituting 0 for } t$$
$$\text{and 6 for } v(0)$$

Solving for C_1, we get $C_1 = 6$.

Thus, $v(t) = -t^2 + 6t + 6$.

We find $s(t)$ by integrating $v(t)$.
$$s(t) = \int v(t)\, dt$$
$$= \int (-t^2 + 6t + 6)\, dt$$
$$= -\frac{t^3}{3} + 3t^2 + 6t + C_2$$

The condition $s(0) = 10$ allows us to find C_2.
$$s(0) = -\frac{0^3}{3} + 3 \cdot 0^2 + 6 \cdot 0 + C_2 \quad \text{Substituting 0 for } t$$
$$\text{and 10 for } s(0)$$

Solving for C_2, we get $C_2 = 10$.

Thus, $s(t) = -\frac{1}{3}t^3 + 3t^2 + 6t + 10$.

45. $a(t) = -32 \text{ ft/sec}^2$
$v(0) = \text{initial velocity} = v_0$
$s(0) = \text{initial height} = s_0$

We find $v(t)$ by integrating $a(t)$.
$$v(t) = \int a(t)\, dt$$
$$= \int (-32)\, dt$$
$$= -32t + C_1$$

The condition $v(0) = v_0$ allows us to find C_1.
$$v(0) = -32 \cdot 0 + C_1 = v_0 \quad \text{Substituting 0 for } t$$
$$\text{and } v_0 \text{ for } v(0)$$
$$C_1 = v_0$$

Thus, $v(t) = -32t + v_0$.

We find $s(t)$ by integrating $v(t)$.
$$s(t) = \int v(t)\, dt$$
$$= \int (-32t + v_0)\, dt$$
$$= -16t^2 + v_0 t + C_2 \quad v_0 \text{ is constant}$$

The condition $s(0) = s_0$ allows us to find C_2.
$$s(0) = -16 \cdot 0^2 + v_0 \cdot 0 + C_2 = s_0 \quad \text{Substituting 0 for } t$$
$$\text{and } s_0 \text{ for } s(0)$$
$$C_2 = s_0$$

Thus, $s(t) = -16t^2 + v_0 t + s_0$.

47. $a(t) = k$ Constant acceleration

$v(t) = \int a(t)\, dt = \int k\, dt = kt$ $(v(0) = 0;$ thus $C = 0.)$

$$s(t) = \int v(t)\, dt = \int kt\, dt = k \cdot \frac{t^2}{2} = \frac{1}{2}kt^2$$

$$(s(0) = 0;\ \text{thus}\ C = 0.)$$

We know that

$$a(t) = k = \frac{60\ \text{mph}}{\dfrac{1}{2}\ \text{min}}$$

and that

$$t = \frac{1}{2}\ \text{min.}$$

Thus

$$s(t) = \frac{1}{2}kt^2$$

$$s\!\left(\frac{1}{2}\ \text{min}\right) = \frac{1}{2} \cdot \frac{60\ \text{mph}}{\dfrac{1}{2}\ \text{min}} \cdot \left(\frac{1}{2}\ \text{min}\right)^2$$

$$= \frac{1}{2} \cdot \frac{60\ \text{mi}}{\text{hr}} \cdot \frac{1}{2}\ \text{min}$$

$$= \frac{1}{2} \cdot \frac{60\ \text{mi}}{\text{hr}} \cdot \frac{1}{120}\ \text{hr}$$

$$= \frac{60}{240}\ \text{mi}$$

$$= \frac{1}{4}\ \text{mi}$$

The car travels $\dfrac{1}{4}$ mi during that time.

49. $M'(t) = 0.2t - 0.003t^2$

a) We integrate to find $M(t)$.

$$M(t) = \int (0.2t - 0.003t^2)\, dt$$
$$= 0.1t^2 - 0.001t^3 + C$$

We use $M(0) = 0$ to find C.

$$M(0) = 0.1(0)^2 - 0.001(0)^3 + C = 0$$
$$C = 0$$
$$M(t) = 0.1t^2 - 0.001t^3$$

b) $M(8) = 0.1(8)^2 - 0.001(8)^3$

$$= 6.4 - 0.512$$
$$= 5.888$$
$$\approx 6\ \text{words}$$

51. Find the function f such that

$f'(t) = t^{\sqrt{3}}$ and $f(0) = 8$.

We first find $f'(x)$ by integrating.

$$f(t) = \int f'(t)\, dt$$
$$= \int t^{\sqrt{3}}\, dt$$
$$= \frac{t^{\sqrt{3}+1}}{\sqrt{3}+1} + C$$

The condition $f(0) = 8$ allows us to find C.

$$f(t) = \frac{t^{\sqrt{3}+1}}{\sqrt{3}+1} + C$$

$$f(0) = \frac{0^{\sqrt{3}+1}}{\sqrt{3}+1} + C = 8 \quad \begin{array}{l}\text{Substituting 0 for } t\\ \text{and 8 for } f(t)\end{array}$$

Solving for C we get $C = 8$.

Thus, $f(t) = \dfrac{t^{\sqrt{3}+1}}{\sqrt{3}+1} + 8.$

53. $\int (x-1)^2\, x^3\, dx$

$$= \int (x^2 - 2x + 1)x^3\, dx$$
$$= \int (x^5 - 2x^4 + x^3)\, dx$$
$$= \int x^5\, dx - \int 2x^4\, dx + \int x^3\, dx$$
$$= \frac{x^6}{6} - 2 \cdot \frac{x^5}{5} + \frac{x^4}{4} + C$$
$$= \frac{1}{6}x^6 - \frac{2}{5}x^5 + \frac{1}{4}x^4 + C$$

55. $\displaystyle\int \frac{(t+3)^2}{\sqrt{t}}\, dt$

$$= \int \frac{t^2 + 6t + 9}{t^{1/2}}\, dt$$
$$= \int (t^{3/2} + 6t^{1/2} + 9t^{-1/2})\, dt$$
$$= \int t^{3/2}\, dt + \int 6t^{1/2}\, dt + \int 9t^{-1/2}\, dt$$
$$= \frac{t^{5/2}}{5/2} + 6 \cdot \frac{t^{3/2}}{3/2} + 9 \cdot \frac{t^{1/2}}{1/2} + C$$
$$= \frac{2}{5}t^{5/2} + 4t^{3/2} + 18t^{1/2} + C$$

57. $\int (t+1)^3\, dt$

$$= \int (t^3 + 3t^2 + 3t + 1)\, dt$$
$$= \int t^3\, dt + \int 3t^2\, dt + \int 3t\, dt + \int dt$$
$$= \frac{t^4}{4} + 3 \cdot \frac{t^3}{3} + 3 \cdot \frac{t^2}{2} + t + C$$
$$= \frac{1}{4}t^4 + t^3 + \frac{3}{2}t^2 + t + C,\ \text{or}\ \frac{(t+1)^4}{4} + C$$

59. $\displaystyle\int be^{ax}\, dx = \frac{b}{a}e^{ax} + C$ See Theorem 2.

61. $\int \sqrt[3]{64x^4}\, dx$

$$= \int 4\sqrt[3]{x^4}\, dx$$
$$= 4\int x^{4/3}\, dx$$
$$= 4 \cdot \frac{x^{7/3}}{7/3} + C$$
$$= \frac{12}{7}x^{7/3} + C$$

63. $\displaystyle\int \frac{t^3 + 8}{t + 2}\, dt$

$\displaystyle = \int \frac{(t+2)(t^2 - 2t + 4)}{(t+2)}\, dt$

$\displaystyle = \int (t^2 - 2t + 4)\, dt$

$\displaystyle = \int t^2\, dt - \int 2t\, dt + \int 4\, dt$

$\displaystyle = \frac{t^3}{3} - 2 \cdot \frac{t^2}{2} + 4t + C$

$\displaystyle = \frac{1}{3}t^3 - t^2 + 4t + C$

65. $\boxed{tw}$

Exercise Set 5.2

1. Find the area under the curve $y = 4$ on the interval $[1, 3]$.

$A(x) = \int 4\, dx$

$\quad = 4x + C$

Since we know that $A(1) = 0$ (there is no area above the number 1), we can substitute for x and $A(x)$ to determine C.

$A(1) = 4 \cdot 1 + C = 0$ Substituting 1 for x and 0 for $A(1)$

Solving for C we get:

$\quad 4 + C = 0$

$\quad\quad C = -4$

Thus, $A(x) = 4x - 4$.

Then the area on the interval $[1, 3]$ is $A(3)$.

$A(3) = 4 \cdot 3 - 4$ Substituting 3 for x

$\quad = 12 - 4$

$\quad = 8$

3. Find the area under the curve $y = 2x$ on the interval $[1, 3]$.

$A(x) = \int 2x\, dx$

$\quad = x^2 + C$

Since we know that $A(1) = 0$ (there is no area above the number 1), we can substitute for x and $A(x)$ to determine C.

$A(1) = 1^2 + C = 0$ Substituting 1 for x and 0 for $A(1)$

Solving for C we get:

$\quad 1 + C = 0$

$\quad\quad C = -1$

Thus, $A(x) = x^2 - 1$.

Then the area on the interval $[1, 3]$ is $A(3)$.

$A(3) = 3^2 - 1$ Substituting 3 for x

$\quad = 9 - 1$

$\quad = 8$

5. Find the area under the curve $y = x^2$ on the interval $[0, 5]$.

$A(x) = \int x^2\, dx$

$\quad = \dfrac{x^3}{3} + C$

Since we know that $A(0) = 0$ (there is no area above the number 0), we can substitute for x and $A(x)$ to determine C.

$A(0) = \dfrac{0^3}{3} + C = 0$ Substituting 0 for x and 0 for $A(0)$

Solving for C, we get $C = 0$:

Thus, $A(x) = \dfrac{x^3}{3}$.

Then the area on the interval $[0, 5]$ is $A(5)$.

$A(5) = \dfrac{5^3}{3}$ Substituting 5 for x

$\quad = \dfrac{125}{3}$, or $41\dfrac{2}{3}$

7. Find the area under the curve $y = x^3$ on the interval $[0, 1]$.

$A(x) = \int x^3\, dx$

$\quad = \dfrac{x^4}{4} + C$

Since we know that $A(0) = 0$, we can substitute for x and $A(x)$ to determine C.

$A(0) = \dfrac{0^4}{4} + C = 0$ Substituting 0 for x and 0 for $A(0)$

Solving for C, we get $C = 0$.

Thus, $A(x) = \dfrac{x^4}{4}$.

Then the area on the interval $[0, 1]$ is $A(1)$.

$A(1) = \dfrac{1^4}{4}$ Substituting 1 for x

$\quad = \dfrac{1}{4}$

9. Find the area under the curve $y = 4 - x^2$ on the interval $[-2, 2]$.

$A(x) = \int (4 - x^2)\, dx$

$\quad = 4x - \dfrac{x^3}{3} + C$

Since we know that $A(-2) = 0$, (there is no area above the number -2), we can substitute for x and $A(x)$ to determine C.

$A(-2) = 4(-2) - \dfrac{(-2)^3}{3} + C = 0$

Substituting -2 for x and 0 for $A(-2)$

Solving for C, we get:

$-8 + \dfrac{8}{3} + C = 0$

$-\dfrac{24}{8} + \dfrac{8}{3} + C = 0$

$-\dfrac{16}{3} + C = 0$

$C = \dfrac{16}{3}$

Thus, $A(x) = 4x - \dfrac{x^3}{3} + \dfrac{16}{3}$.

The area on the interval $[-2, 2]$ is $A(2)$.

$$A(2) = 4 \cdot 2 - \frac{2^3}{3} + \frac{16}{3} \quad \text{Substituting 2 for } x$$

$$= 8 - \frac{8}{3} + \frac{16}{3}$$

$$= \frac{24}{3} - \frac{8}{3} + \frac{16}{3}$$

$$= \frac{32}{3}, \text{ or } 10\frac{2}{3}$$

11. Find the area under the curve $y = e^x$ on the interval $[0, 3]$.

$$A(x) = \int e^x \, dx$$
$$= e^x + C$$

Since we know that $A(0) = 0$, (there is no area above the number 0), we can substitute for x and $A(x)$ to determine C.

$$A(0) = e^0 + C = 0 \quad \text{Substituting 0 for } x \text{ and}$$
$$\qquad\qquad\qquad\qquad 0 \text{ for } A(0)$$
$$1 + C = 0 \quad (e^0 = 1)$$
$$C = -1$$

Thus, $A(x) = e^x - 1$.

The area on the interval $[0, 3]$ is $A(3)$.

$$A(3) = e^3 - 1$$
$$= 20.085537 - 1 \quad \text{Using a calculator}$$
$$\approx 19.086$$

13. Find the area under the curve $y = \frac{1}{x}$ on the interval $[1, 3]$.

$$A(x) = \int \frac{1}{x} \, dx$$
$$= \ln x + C$$

Since we know that $A(1) = 0$, (there is no area above the number 1), we can substitute for x and $A(x)$ to determine C.

$$A(1) = \ln 1 + C = 0 \quad \text{Substituting 1 for } x$$
$$\qquad\qquad\qquad\qquad \text{and 0 for } A(1)$$
$$0 + C = 0 \qquad \ln 1 = 0$$
$$C = 0$$

Thus, $A(x) = \ln x$.

The area on the interval $[1, 3]$ is $A(3)$.

$$A(3) = \ln 3 \quad \text{Substituting 3 for } x$$
$$\approx 1.0986 \quad \text{Using a calculator}$$

15.
$$\int_0^{1.5} (x - x^2) \, dx$$
$$= \left[\frac{x^2}{2} - \frac{x^3}{3} \right]_0^{1.5}$$
$$= \left(\frac{(1.5)^2}{2} - \frac{(1.5)^3}{3} \right) - \left(\frac{0^2}{2} - \frac{0^3}{3} \right)$$

Substituting 0 for x
Substituting 1.5 for x

$$= \left(\frac{2.25}{2} - \frac{3.375}{3} \right) - (0 - 0)$$
$$= 1.125 - 1.125$$
$$= 0$$

The area above the x-axis is equal to the area below the x-axis.

17.
$$\int_{-1}^1 (x^4 - x^2) \, dx$$
$$= \left[\frac{x^5}{5} - \frac{x^3}{3} \right]_1^1$$
$$= \left[\frac{1^5}{5} - \frac{1^3}{3} \right] - \left[\frac{(-1)^5}{5} - \frac{(-1)^3}{3} \right]$$
$$= \left(\frac{1}{5} - \frac{1}{3} \right) - \left(-\frac{1}{5} + \frac{1}{3} \right)$$
$$= \frac{1}{5} - \frac{1}{3} + \frac{1}{5} - \frac{1}{3}$$
$$= \frac{2}{5} - \frac{2}{3}$$
$$= \frac{6}{15} - \frac{10}{15}$$
$$= -\frac{4}{15}$$

The area is below the x-axis.

19. - 35.

37.
$$\int_a^b e^t \, dt$$
$$= [e^t]_a^b$$
$$= e^b - e^a$$

39.
$$\int_a^b 3t^2 \, dt$$
$$= \left[3 \cdot \frac{t^3}{3} \right]_a^b$$
$$= [t^3]_a^b$$
$$= b^3 - a^3$$

41.
$$\int_1^e \left(x + \frac{1}{x} \right) dx$$
$$= \left[\frac{x^2}{2} + \ln x \right]_1^e$$
$$= \left(\frac{e^2}{2} + \ln e \right) - \left(\frac{1^2}{2} + \ln 1 \right)$$
$$= \frac{e^2}{2} + 1 - \frac{1}{2} \quad (\ln e = 1, \ln 1 = 0)$$
$$= \frac{e^2}{2} + \frac{1}{2}$$

43.
$$\int_0^1 \sqrt{x} \, dx$$
$$= \int_0^1 x^{1/2} \, dx$$
$$= \left[\frac{x^{3/2}}{3/2} \right]_0^1$$
$$= \left[\frac{2}{3} x^{3/2} \right]_0^1$$
$$= \frac{2}{3} \cdot 1^{3/2} - \frac{2}{3} \cdot 0^{3/2}$$

$$= \frac{2}{3} \cdot 1 - \frac{2}{3} \cdot 0$$

$$= \frac{2}{3} - 0$$

$$= \frac{2}{3}$$

45. $\displaystyle\int_{-4}^{1} \frac{10}{17} t^3 \, dt$

$$= \frac{10}{17} \int_{-4}^{1} t^3 \, dt$$

$$= \frac{10}{17} \left[\frac{t^4}{4} \right]_{-4}^{1}$$

$$= \frac{10}{17} \left(\frac{1^4}{4} - \frac{(-4)^4}{4} \right)$$

$$= \frac{10}{17} \left(\frac{1}{4} - 64 \right)$$

$$= \frac{10}{17} \cdot \left(-\frac{255}{4} \right)$$

$$= -\frac{2550}{68}$$

$$= -\frac{1275}{34}$$

47. Find the area under $y = x^3$ on $[0, 2]$.

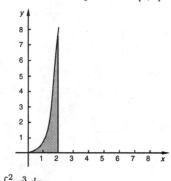

$\displaystyle\int_0^2 x^3 \, dx$

$$= \left[\frac{x^4}{4} \right]_0^2$$

$$= \frac{2^4}{4} - \frac{0^4}{4}$$

$$= 4 - 0$$

$$= 4$$

49. Find the area under $y = x^2 + x + 1$ on $[2, 3]$.

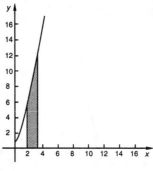

$\displaystyle\int_2^3 (x^2 + x + 1) \, dx$

$$= \left[\frac{x^3}{3} + \frac{x^2}{2} + x \right]_2^3$$

$$= \left(\frac{3^3}{3} + \frac{3^2}{2} + 3 \right) - \left(\frac{2^3}{3} + \frac{2^2}{2} + 2 \right)$$

$$= \left(9 + \frac{9}{2} + 3 \right) - \left(\frac{8}{3} + 2 + 2 \right)$$

$$= 12 + \frac{9}{2} - \frac{8}{3} - 4$$

$$= 8 + \frac{27}{6} - \frac{16}{6}$$

$$= 8 + \frac{11}{6}$$

$$= 8 + 1\frac{5}{6}$$

$$= 9\frac{5}{6}$$

51. Find the area under $y = 5 - x^2$ on $[-1, 2]$.

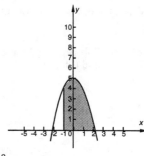

$\displaystyle\int_{-1}^{2} (5 - x^2) \, dx$

$$= \left[5x - \frac{x^3}{3} \right]_{-1}^{2}$$

$$= \left(5 \cdot 2 - \frac{2^3}{3} \right) - \left[5(-1) - \frac{(-1)^3}{3} \right]$$

$$= \left(10 - \frac{8}{3} \right) - \left(-5 + \frac{1}{3} \right)$$

$$= \left(\frac{30}{3} - \frac{8}{3} \right) - \left(-\frac{15}{3} + \frac{1}{3} \right)$$

$$= \frac{22}{3} - \left(-\frac{14}{3} \right)$$

$$= \frac{22}{3} + \frac{14}{3}$$

$$= \frac{36}{3}$$

$$= 12$$

53. Find the area under $y = e^x$ on $[-1, 5]$.

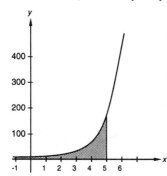

$$\int_{-1}^{5} e^x \, dx$$
$$= [e^x]_{-1}^{5}$$
$$= e^5 - e^{-1}$$
$$= e^5 - \frac{1}{e}$$

55. $\boxed{tw}$

57. Find the area under the curve $y = \dfrac{x^2 - 1}{x - 1}$ on the interval $[2, 3]$.

$$A(x) = \int \frac{x^2 - 1}{x - 1} \, dx$$
$$= \int \frac{(x + 1)(x - 1)}{x - 1} \, dx$$
$$= \int (x + 1) \, dx$$
$$= \frac{x^2}{2} + x + C$$

Since we know that $A(2) = 0$, we can substitute for x and $A(x)$ to determine C.

$$A(2) = \frac{2^2}{2} + 2 + C = 0 \quad \begin{array}{l}\text{Substituting 2 for } x \\ \text{and 0 for } A(2)\end{array}$$

Solve for C we get:

$$2 + 2 + C = 0$$
$$C = -4$$

Thus, $A(x) = \dfrac{x^2}{2} + x - 4$.

The area on the interval $[2, 3]$ is $A(3)$.

$$A(3) = \frac{3^2}{2} + 3 - 4 \quad \text{Substituting 3 for } x$$
$$= \frac{9}{2} + \frac{6}{2} - \frac{8}{2}$$
$$= \frac{7}{2}, \text{ or } 3\frac{1}{2}$$

59. Find the area under the curve $y = (x-1)\sqrt{x}$ on the interval $[4, 16]$.

$$A(x) = \int (x - 1)\sqrt{x} \, dx$$
$$= \int (x - 1)x^{1/2} \, dx$$
$$= \int (x^{3/2} - x^{1/2}) \, dx$$
$$= \frac{x^{5/2}}{5/2} - \frac{x^{3/2}}{3/2} + C$$
$$= \frac{2}{5}x^{5/2} - \frac{2}{3}x^{3/2} + C$$

Since we know that $A(4) = 0$, we can substitute for x and $A(x)$ to determine C.

$$A(4) = \frac{2}{5} \cdot 4^{5/2} - \frac{2}{3} \cdot 4^{3/2} + C = 0 \quad \text{Substituting}$$

$$\frac{2}{5} \cdot 32 - \frac{2}{3} \cdot 8 + C = 0$$

$$(4^{5/2} = (2^2)^{5/2} = 2^5 = 32)$$
$$(4^{3/2} = (2^2)^{3/2} = 2^3 = 8)$$

$$\frac{64}{5} - \frac{16}{3} + C = 0$$

$$\frac{192}{15} - \frac{80}{15} + C = 0$$

$$\frac{112}{15} + C = 0$$

$$C = -\frac{112}{15}$$

Thus, $A(x) = \dfrac{2}{5}x^{5/2} - \dfrac{2}{3}x^{3/2} - \dfrac{112}{15}$.

The area on the interval $[4, 16]$ is $A(16)$.

$$A(16) = \frac{2}{5} \cdot 16^{5/2} - \frac{2}{3} \cdot 16^{3/2} - \frac{112}{15}.$$
$$= \frac{2}{5} \cdot 2^{10} - \frac{2}{3} \cdot 2^6 - \frac{112}{15}$$
$$= \frac{2^{11}}{5} - \frac{2^7}{3} - \frac{112}{15}$$
$$= \frac{3 \cdot 2^{11}}{15} - \frac{5 \cdot 2^7}{15} - \frac{112}{15}$$
$$= \frac{6144}{15} - \frac{640}{15} - \frac{112}{15}$$
$$= \frac{5392}{15}$$
$$= 359\frac{7}{15}$$

61. Find the area under the curve $y = \dfrac{\sqrt[3]{x^2} - 1}{\sqrt[3]{x}}$ on the interval $[1, 8]$.

$$A(x) = \int \frac{\sqrt[3]{x^2} - 1}{\sqrt[3]{x}}\, dx$$

$$= \int \frac{x^{2/3} - 1}{x^{1/3}}\, dx$$

$$= \int (x^{1/3} - x^{-1/3})\, dx$$

$$= \int x^{1/3}\, dx - \int x^{-1/3}\, dx$$

$$= \frac{x^{4/3}}{4/3} - \frac{x^{2/3}}{2/3} + C$$

$$= \frac{3}{4} x^{4/3} - \frac{3}{2} x^{2/3} + C$$

Since we know that $A(1) = 0$, we can substitute for x and $A(x)$ to determine C.

$$A(1) = \frac{3}{4} \cdot 1^{4/3} - \frac{3}{2} \cdot 1^{2/3} + C = 0 \quad \text{Substituting}$$

$$\frac{3}{4} - \frac{3}{2} + C = 0$$

$$C = -\frac{3}{4} + \frac{6}{4}$$

$$C = \frac{3}{4}$$

Thus, $A(x) = \dfrac{3}{4} x^{4/3} - \dfrac{3}{2} x^{2/3} + \dfrac{3}{4}$.

The area on the interval $[1, 8]$ is $A(8)$.

$$A(8) = \frac{3}{4} \cdot 8^{4/3} - \frac{3}{2} \cdot 8^{2/3} + \frac{3}{4}$$

$$= \frac{3}{4} \cdot 2^4 - \frac{3}{2} \cdot 2^2 + \frac{3}{4}$$

$$= 12 - 6 + \frac{3}{4}$$

$$= 6\frac{3}{4}$$

63. $\int_1^2 (4x + 3)(5x - 2)\, dx$

$$= \int_1^2 (20x^2 + 7x - 6)\, dx$$

$$= \left[20 \cdot \frac{x^3}{3} + 7 \cdot \frac{x^2}{2} - 6x \right]_1^2$$

$$= \left(20 \cdot \frac{2^3}{3} + 7 \cdot \frac{2^2}{2} - 6 \cdot 2 \right) - \left(20 \cdot \frac{1^3}{3} + 7 \cdot \frac{1^2}{2} - 6 \cdot 1 \right)$$

$$= \left(\frac{160}{3} + 14 - 12 \right) - \left(\frac{20}{3} + \frac{7}{2} - 6 \right)$$

$$= \left(53\frac{1}{3} + 14 - 12 \right) - \left(\frac{40}{6} + \frac{21}{6} - \frac{36}{6} \right)$$

$$= 55\frac{1}{3} - 4\frac{1}{6}$$

$$= 51\frac{1}{6}, \text{ or } \frac{307}{6}$$

65. $\int_0^1 (t + 1)^3\, dt$

$$= \int_0^1 (t^3 + 3t^2 + 3t + 1)\, dt$$

$$= \left[\frac{t^4}{4} + t^3 + \frac{3}{2} t^2 + t \right]_0^1$$

$$= \left(\frac{1^4}{4} + 1^3 + \frac{3}{2} \cdot 1^2 + 1 \right) - \left(\frac{0^4}{4} + 0^3 + \frac{3}{2} \cdot 0^2 + 0 \right)$$

$$= \left(\frac{1}{4} + 1 + \frac{3}{2} + 1 \right) - 0$$

$$= 3\frac{3}{4}, \text{ or } \frac{15}{4}$$

67. $\displaystyle\int_1^3 \frac{t^5 - t}{t^3}\, dt$

$$= \int_1^3 (t^2 - t^{-2})\, dt$$

$$= \left[\frac{t^3}{3} - \frac{t^{-1}}{-1} \right]_1^3$$

$$= \left[\frac{t^3}{3} + \frac{1}{t} \right]_1^3$$

$$= \left(\frac{3^3}{3} + \frac{1}{3} \right) - \left(\frac{1^3}{3} + \frac{1}{1} \right)$$

$$= 9\frac{1}{3} - 1\frac{1}{3}$$

$$= 8$$

69. $\displaystyle\int_3^5 \frac{x^2 - 4}{x - 2}\, dx$

$$= \int_3^5 \frac{(x - 2)(x + 2)}{x - 2}\, dx$$

$$= \int_3^5 (x + 2)\, dx$$

$$= \left[\frac{x^2}{2} + 2x \right]_3^5$$

$$= \left(\frac{5^2}{2} + 2 \cdot 5 \right) - \left(\frac{3^2}{2} + 2 \cdot 3 \right)$$

$$= \left(\frac{25}{2} + 10 \right) - \left(\frac{9}{2} + 6 \right)$$

$$= \frac{45}{2} - \frac{21}{2}$$

$$= \frac{24}{2}$$

$$= 12$$

71. $\boxed{tw}$

73. Using the fnInt feature on a grapher, we get
$$\int_{-2}^3 (x^3 - 4x)\, dx = 6.25.$$

75. Using the fnInt feature on a grapher, we get
$$\int_{-1.2}^{6.3} (x^3 - 9x^2 + 27x + 50)\, dx \approx 529.36.$$

77. Using the fnInt feature on a grapher, we get
$$\int_{-1}^{20} \frac{4x}{x^2 + 1}\, dx \approx 10.60.$$

79. Using the fnInt feature on a grapher, we get
$\int_{-2}^{2} \sqrt{4 - x^2}\, dx \approx 6.28$.

81. Using the fnInt feature on a grapher, we get
$\int_{0}^{8} x(x - 5)^4\, dx \approx 885.33$.

83. Using the fnInt feature on a grapher, we get
$\int_{2}^{4} \dfrac{x^2 - 4}{x^2 - 3}\, dx \approx 1.51$.

Exercise Set 5.3

1. a) $f(x) = \dfrac{1}{x^2}$

In the drawing in the text the interval $[1, 7]$ has been divided into 6 subintervals, each having width $1 \left(\Delta x = \dfrac{7 - 1}{6} = 1 \right)$.

The heights of the rectangles shown are

$f(1) = \dfrac{1}{1^2} = 1$

$f(2) = \dfrac{1}{2^2} = \dfrac{1}{4} = 0.2500$

$f(3) = \dfrac{1}{3^2} = \dfrac{1}{9} \approx 0.1111$

$f(4) = \dfrac{1}{4^2} = \dfrac{1}{16} = 0.0625$

$f(5) = \dfrac{1}{5^2} = \dfrac{1}{25} = 0.0400$

$f(6) = \dfrac{1}{6^2} = \dfrac{1}{36} \approx 0.0278$

The area of the region under the curve over $[1, 7]$ is approximately the sum of the areas of the 6 rectangles.

Area of each rectangle:

1st rectangle: $1 \cdot 1 = 1$
$\quad [f(1) = 1 \text{ and } \Delta x = 1]$
2nd rectangle: $0.2500 \cdot 1 = 0.2500$
$\quad [f(2) = 0.2500 \text{ and } \Delta x = 1]$
3rd rectangle: $0.1111 \cdot 1 = 0.1111$
4th rectangle: $0.0625 \cdot 1 = 0.0625$
5th rectangle: $0.0400 \cdot 1 = 0.0400$
6th rectangle: $0.0278 \cdot 1 = 0.0278$

The total area is $1 + 0.2500 + 0.1111 + 0.0625 + 0.0400 + 0.0278$, or 1.4914.

b) $f(x) = \dfrac{1}{x^2}$

The interval $[1, 7]$ has been divided into 12 subintervals, each having width 0.5 $\left(\Delta x = \dfrac{7 - 1}{12} = \dfrac{6}{12} = 0.5 \right)$. The heights of six of the rectangles were computed in part (a). The others are computed below:

$f(1.5) = \dfrac{1}{(1.5)^2} = \dfrac{1}{2.25} \approx 0.4444$

$f(2.5) = \dfrac{1}{(2.5)^2} = \dfrac{1}{6.25} \approx 0.1600$

$f(3.5) = \dfrac{1}{(3.5)^2} = \dfrac{1}{12.25} \approx 0.0816$

$f(4.5) = \dfrac{1}{(4.5)^2} = \dfrac{1}{20.25} \approx 0.0494$

$f(5.5) = \dfrac{1}{(5.5)^2} = \dfrac{1}{30.25} \approx 0.0331$

$f(6.5) = \dfrac{1}{(6.5)^2} = \dfrac{1}{42.25} \approx 0.0237$

The area of the region under the curve over $[1, 7]$ is approximately the sum of the areas of the 12 rectangles.

Area of each rectangle:

1st rectangle: $1(0.5) = 0.5$
$\quad [f(1) = 1 \text{ and } \Delta x = 0.5]$
2nd rectangle: $0.4444(0.5) = 0.2222$
$\quad [f(1.5) \approx 0.4444 \text{ and } \Delta x = 0.5]$
3rd rectangle: $0.2500(0.5) = 0.1250$
4th rectangle: $0.1600(0.5) = 0.0800$
5th rectangle: $0.1111(0.5) \approx 0.0556$
6th rectangle: $0.0816(0.5) = 0.0408$
7th rectangle: $0.0625(0.5) \approx 0.0313$
8th rectangle: $0.0494(0.5) = 0.0247$
9th rectangle: $0.0400(0.5) = 0.0200$
10th rectangle: $0.0331(0.5) \approx 0.0166$
11th rectangle: $0.0278(0.5) = 0.0139$
12th rectangle: $0.0237(0.5) \approx 0.0119$

The total area is $0.5 + 0.2222 + 0.1250 + 0.0800 + 0.0556 + 0.0408 + 0.0313 + 0.0247 + 0.0200 + 0.0166 + 0.0139 + 0.0119$, or 1.1420. (Answers may vary slightly depending on when rounding was done.)

c) $\int_1^7 (dx/x^2)$

$= \int_1^7 \frac{1}{x^2}\, dx$

$= \int_1^7 x^{-2}\, dx$

$= \left[\frac{x^{-1}}{-1}\right]_1^7$

$= \left[-\frac{1}{x}\right]_1^7$

$= -\frac{1}{7} - \left(-\frac{1}{1}\right)$

$= -\frac{1}{7} + 1$

$= \frac{6}{7}$

≈ 0.8571

3. The shaded region represents an antiderivative. It also represents velocity, the antiderivative of acceleration.

5. The shaded region represents an antiderivative. It also represents total energy used in time t.

7. The shaded region represents an antiderivative. It also represents total revenue.

9. The shaded region represents an antiderivative. It also represents the amount of the drug in the blood.

11. The shaded region represents an antiderivative. It also represents the number of words memorized in time t.

13. $y = 2x^3;\ [-1, 1]$

$y_{av} = \frac{1}{b-a}\int_a^b f(x)\, dx$

$y_{av} = \frac{1}{1-(-1)}\int_{-1}^1 2x^3\, dx$ $\quad [a=-1,\ b=1,\ \text{and}\ f(x)=2x^3]$

$= \frac{1}{2}\cdot 2 \int_{-1}^1 x^3\, dx$

$= \int_{-1}^1 x^3\, dx$

$= \left[\frac{x^4}{4}\right]_{-1}^1$

$= \frac{1^4}{4} - \frac{(-1)^4}{4}$

$= \frac{1}{4} - \frac{1}{4}$

$= 0$

15. $y = e^x;\ [0, 1]$

$y_{av} = \frac{1}{b-a}\int_a^b f(x)\, dx$

$y_{av} = \frac{1}{1-0}\int_0^1 e^x\, dx$ $\quad [a=0,\ b=1,\ \text{and}\ f(x)=e^x]$

$= \int_0^1 e^x\, dx$

$= [e^x]_0^1$

$= e^1 - e^0$

$= e - 1$

17. $y = x^2 - x + 1;\ [0, 2]$

$y_{av} = \frac{1}{b-a}\int_a^b f(x)\, dx$

$y_{av} = \frac{1}{2-0}\int_0^2 (x^2 - x + 1)\, dx$ $\quad [a=0,\ b=2,\ \text{and}\ f(x)=x^2-x+1]$

$= \frac{1}{2}\left[\frac{x^3}{3} - \frac{x^2}{2} + x\right]_0^2$

$= \frac{1}{2}\left[\left(\frac{2^3}{3} - \frac{2^2}{2} + 2\right) - \left(\frac{0^3}{3} - \frac{0^2}{2} + 0\right)\right]$

$= \frac{1}{2}\left(\frac{8}{3} - 2 + 2\right)$

$= \frac{1}{2}\cdot\frac{8}{3}$

$= \frac{4}{3}$

19. $y = 3x + 1;\ [2, 6]$

$y_{av} = \frac{1}{b-a}\int_a^b f(x)\, dx$

$y_{av} = \frac{1}{6-2}\int_2^6 (3x + 1)\, dx$ $\quad [a=2,\ b=6,\ \text{and}\ f(x)=3x+1]$

$= \frac{1}{4}\left[\frac{3}{2}x^2 + x\right]_2^6$

$= \frac{1}{4}\left[\left(\frac{3}{2}\cdot 6^2 + 6\right) - \left(\frac{3}{2}\cdot 2^2 + 2\right)\right]$

$= \frac{1}{4}\left[(54 + 6) - (6 + 2)\right]$

$= \frac{1}{4}(60 - 8)$

$= \frac{1}{4}\cdot 52$

$= 13$

21. $y = x^n$; $[0, 1]$

$$y_{av} = \frac{1}{b - a} \int_a^b f(x)\, dx$$

$$y_{av} = \frac{1}{1 - 0} \int_0^1 x^n\, dx \qquad [a = 0,\ b = 1,\ \text{and}\\ \hspace{3.5cm} f(x) = x^n]$$

$$= \int_0^1 x^n\, dx$$

$$= \left[\frac{x^{n+1}}{n+1}\right]_0^1$$

$$= \frac{1^{n+1}}{n+1} - \frac{0^{n+1}}{n+1}$$

$$= \frac{1}{n+1} - 0$$

$$= \frac{1}{n+1}$$

23. a) Accumulated sales through day 5 are

$$\int_0^5 S'(t)\, dt = \int_0^5 20e^t\, dt$$

$$= 20 \int_0^5 e^t\, dt$$

$$= 20[e^t]_0^5$$

$$= 20(e^5 - e^0)$$

$$= 20(148.413159 - 1)$$

$$= 20(147.413159)$$

$$\approx \$2948.26$$

b) Accumulated sales from the 2nd through the 5th day are

$$\int_1^5 S'(t)\, dt = \int_1^5 20e^t\, dt$$

$$= 20 \int_1^5 e^t\, dt$$

$$= 20[e^t]_1^5$$

$$= 20(e^5 - e^1)$$

$$= 20(148.413159 - 2.718282)$$

$$= 20(145.694877)$$

$$\approx \$2913.90$$

c) Accumulated sales through day k are

$$\int_0^k S'(t)\, dt = \int_0^k 20e^t\, dt$$

$$= 20 \int_0^k e^t\, dt$$

$$= 20[e^t]_0^k$$

$$= 20(e^k - e^0)$$

$$= 20(e^k - 1)$$

We set this equal to \$20,000 and solve for k.

$$20(e^k - 1) = 20,000$$

$$e^k - 1 = 1000$$

$$e^k = 1001$$

$$\ln e^k = \ln 1001$$

$$k = \ln 1001$$

$$k = 6.908755$$

Thus, accumulated sales will exceed \$20,000 on the 7th day.

25. $\int_{100}^{400} (0.0003x^2 - 0.2x + 50)\, dx$

$$= [0.0001x^3 - 0.1x^2 + 50x]_{100}^{400}$$

$$= [0.0001(400)^3 - 0.1(400)^2 + 50 \cdot 400] -$$

$$\qquad [0.0001(100)^3 - 0.1(100)^2 + 50 \cdot 100]$$

$$= \$6300$$

27. $R'(x) = 200x - 1080$

$$R(x) = \int_{1000}^{1300} (200x - 1080)\, dx$$

$$= [100x^2 - 1080x]_{1000}^{1300}$$

$$= [100(1300)^2 - 1080 \cdot 1300] -$$

$$\qquad [100(1000)^2 - 1080 \cdot 1000]$$

$$= (169,000,000 - 1,404,000) -$$

$$\qquad (100,000,000 - 1,080,000)$$

$$= 167,596,000 - 98,920,000$$

$$= \$68,676,000$$

29. $W(t) = -6t^2 + 12t + 90$

a) $W(0) = -6 \cdot 0^2 + 12 \cdot 0 + 90$

$$= 90 \text{ words per minute}$$

b) We first find $W'(t)$.

$$W'(t) = -12t + 12$$

Then we find the critical points. Since $W'(t)$ exists for all values of t in $[0, 5]$, we solve $W'(t) = 0$.

$$-12t + 12 = 0$$

$$-12t = -12$$

$$t = 1 \qquad \text{Critical point}$$

Find the function values at 0, 1, and 5.

$$W(0) = 90$$

$$W(1) = -6 \cdot 1^2 + 12 \cdot 1 + 90$$

$$= 96 \hspace{3cm} \text{Maximum}$$

$$W(5) = -6 \cdot 5^2 + 12 \cdot 5 + 90$$

$$= 0$$

The maximum speed is 96 words per minute at $t = 1$ minute.

c) $W_{av} = \dfrac{1}{b - a} \displaystyle\int_a^b W(t)\, dt$

$$= \frac{1}{5 - 0} \int_0^5 (-6t^2 + 12t + 90)\, dt$$

$$= \frac{1}{5}\left[-2t^3 + 6t^2 + 90t\right]_0^5$$

$$= \frac{1}{5}\big[(-2 \cdot 5^3 + 6 \cdot 5^2 + 90 \cdot 5 -$$

$$\qquad (-2 \cdot 0^3 + 6 \cdot 0^2 + 90 \cdot 0)\big]$$

$$= \frac{1}{5}\big[-250 + 150 + 450 - (0 + 0 + 0)\big]$$

$$= 70 \text{ words per minute}$$

31. $s(t) = \int v(t)\, dt$

To find the distance traveled (s) from the first through the third hour we integrate. We find the area under $v(t) =$

$4t^3 + 2t$ on $[0, 3]$. Note that the first hour begins at $t = 0$ and the third hour ends at $t = 3$.

$$\int_0^3 v(t)\, dt$$
$$= \int_0^3 (4t^3 + 2t)\, dt$$
$$= [t^4 + t^2]_0^3$$
$$= (3^4 + 3^2) - (0^4 + 0^2)$$
$$= 81 + 9$$
$$= 90$$

33. a) $P'(t) = 1200e^{0.32t}$

$$P(t) = \int_0^{20} 1200e^{0.32t}\, dt$$
$$= \left[\frac{1200}{0.32}e^{0.32t}\right]_0^{20}$$
$$= [3750e^{0.32t}]_0^{20}$$
$$= 3750e^{0.32(20)} - 3750e^{0.32(0)}$$
$$= 3750e^{6.4} - 3750$$
$$\approx 2,253,169$$

b) The population through day k is

$$\int_0^k 1200e^{0.32t}\, dt$$
$$= \left[\frac{1200}{0.32}e^{0.32t}\right]_0^k$$
$$= [3750e^{0.32t}]_0^k$$
$$= 3750e^{0.32k} - 3750e^{0.32(0)}$$
$$= 3750e^{0.32k} - 3750$$

We set this equal to 4,000,000 and solve for k.

$$3750e^{0.32k} - 3750 = 4,000,000$$
$$3750e^{0.32k} = 4,003,750$$
$$e^{0.32k} = \frac{4,003,750}{3750}$$
$$\ln e^{0.32k} = \ln \frac{4,003,750}{3750}$$
$$0.32k = \ln \frac{4,003,750}{3750}$$
$$k = \frac{\ln \frac{4,003,750}{3750}}{0.32}$$
$$k \approx 21.8 \text{ days}$$

35. $f(t) = -t^2 + 5t + 40,\ 0 \le t \le 10$

a) $f_{av} = \dfrac{1}{10-0}\displaystyle\int_0^{10}(-t^2 + 5t + 40)\, dt$

$$= \frac{1}{10}\left[-\frac{t^3}{3} + \frac{5t^2}{2} + 40t\right]_0^{10}$$
$$= \frac{1}{10}\left[-\frac{10^3}{3} + \frac{5 \cdot 10^2}{2} + 40 \cdot 10 -\right.$$
$$\left.\left(-\frac{0^3}{3} + \frac{5 \cdot 0^2}{2} + 40 \cdot 0\right)\right]$$
$$= \frac{95}{3} \approx 31.7°$$

b) $f'(t) = -2t + 5$

$f'(t)$ exists for all values of t in $[0, 10]$, so we solve $f'(t) = 0$.

$$-2t + 5 = 0$$
$$t = \frac{5}{2}$$

Evaluate $f(t)$ at 0, $\dfrac{5}{2}$, and 10.

$$f(0) = -0^2 + 5 \cdot 0 + 40 = 40$$
$$f\left(\frac{5}{2}\right) = -\left(\frac{5}{2}\right)^2 + 5 \cdot \frac{5}{2} + 40 = 46.25$$
$$f(10) = -10^2 + 5 \cdot 10 + 40 = -10$$

The minimum temperature is $-10°$.

c) From part (b), we see that the maximum temperature is $46.25°$.

37. $M'(t) = -0.003t^2 + 0.2t$

$$\int_0^{10} M'(t)\, dt$$
$$= \int_0^{10}(-0.003t^2 + 0.2t)\, dt$$
$$= [-0.001t^3 + 0.1t^2]_0^{10}$$
$$= [-0.001(10)^3 + 0.1(10)^2] -$$
$$[-0.001(0)^3 + 0.1(0)^2]$$
$$= [-1 + 10] - [0 + 0]$$
$$= 9$$

Thus, 9 words are memorized in the first 10 minutes.

39. $P(t) = 241e^{0.009t}$

$$P_{av} = \frac{1}{b-a}\int_a^b P(t)\, dt$$
$$P_{av} = \frac{1}{14-0}\int_0^{14} 241e^{0.009t}\, dt \quad (a = 0,\ b = 14;$$
$$2000 - 1986 = 14)$$
$$= \frac{1}{14} \cdot 241 \int_0^{14} e^{0.009t}\, dt$$
$$= \frac{241}{14}\left[\frac{e^{0.009t}}{0.009}\right]_0^{14} \quad \text{Using Formula 5 in Table 1}$$
$$= \frac{241}{0.126}[e^{0.009t}]_0^{14}$$
$$= \frac{241}{0.126}(e^{0.009(14)} - e^{0.009(0)})$$
$$= \frac{241}{0.126}(e^{0.126} - 1)$$
$$= 1912.70(1.134282 - 1)$$
$$= 1912.70(0.134282)$$
$$\approx 256.8$$

The average value of the population from 1986 to 2000 is approximately 256.8 million.

41. a) $S(t) = t^2$, t in $[0, 10]$

We first find $S'(t)$.

$S'(t) = 2t$

Then we find the critical points.

Since $S'(t)$ exists for all values of t in $[0, 10]$, we solve $S'(t) = 0$.

$2t = 0$

$t = 0$ Critical point

Find the function values at 0 and 10.

$S(t) = t^2$

$S(0) = 0^2 = 0$ Minimum

$S(10) = 10^2 = 100$ Maximum

Thus, the maximum score is 100 after the student studies 10 hours.

b) $S(t)_{av} = \dfrac{1}{b-a} \displaystyle\int_a^b S(t)\, dt$

$S_{av} = \dfrac{1}{10-0} \displaystyle\int_0^{10} t^2\, dt$

$= \dfrac{1}{10}\left[\dfrac{t^3}{3}\right]_0^{10}$

$= \dfrac{1}{10}\left(\dfrac{10^3}{3} - \dfrac{0^3}{3}\right)$

$= \dfrac{1}{10}\cdot\dfrac{10^3}{3}$

$= \dfrac{10^2}{3}$

$= \dfrac{100}{3}$, or $33\dfrac{1}{3}$

The average score over the 10-hr interval is $33\dfrac{1}{3}$.

43. $\Delta x = 1$

$\displaystyle\int_0^5 (x^2 + 1)\, dx$

$\approx 1\cdot\left[\dfrac{f(0)}{2} + f(1) + f(2) + f(3) + f(4) + \dfrac{f(5)}{2}\right]$

$\approx \dfrac{1}{2} + 2 + 5 + 10 + 17 + \dfrac{26}{2}$

$\approx 47\dfrac{1}{2}$

45. Using the fnInt feature on a grapher, we find

$\displaystyle\int_0^4 \sqrt{x}\, dx \approx 5.333$.

47. Using the fnInt feature on a grapher, we find

$\displaystyle\int_2^4 \ln x\, dx \approx 2.159$.

Exercise Set 5.4

1. First graph the system of equations and shade the region bounded by the graphs.

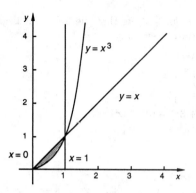

Here the boundaries are easily determined by looking at the graph. Note which is the upper graph. Here it is $x \geq x^3$ over the interval $[0, 1]$.

Compute the area as follows:

$\displaystyle\int_0^1 (x - x^3)\, dx$

$= \left[\dfrac{x^2}{2} - \dfrac{x^4}{4}\right]_0^1$

$= \left(\dfrac{1^2}{2} - \dfrac{1^4}{4}\right) - \left(\dfrac{0^2}{2} - \dfrac{0^4}{4}\right)$

$= \dfrac{1}{2} - \dfrac{1}{4}$

$= \dfrac{1}{4}$

3. First graph the system of equations and shade the region bounded by the graphs.

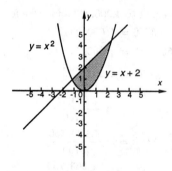

Here the boundaries are easily determined by the graph. Note which is the upper graph. Here it is $(x+2) \geq x^2$ over the interval $[-1, 2]$.

Compute the area as follows:

$\displaystyle\int_{-1}^2 [(x+2) - x^2]\, dx$

$= \displaystyle\int_{-1}^2 (-x^2 + x + 2)\, dx$

$= \left[-\dfrac{x^3}{3} + \dfrac{x^2}{2} + 2x\right]_{-1}^2$

$= \left[-\dfrac{2^3}{3} + \dfrac{2^2}{2} + 2\cdot 2\right] - \left[-\dfrac{(-1)^3}{3} + \dfrac{(-1)^2}{2} + 2(-1)\right]$

$= \left(-\dfrac{8}{3} + 2 + 4\right) - \left(\dfrac{1}{3} + \dfrac{1}{2} - 2\right)$

$= -\dfrac{8}{3} + 2 + 4 - \dfrac{1}{3} - \dfrac{1}{2} + 2$

$= -\dfrac{9}{3} - \dfrac{1}{2} + 2 + 4 + 2 = 4\dfrac{1}{2}$, or $\dfrac{9}{2}$

5. First graph the system of equations and shade the region bounded by the graphs.

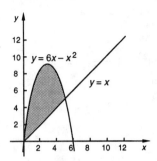

Here the boundaries are easily determined by the graph. Note which is the upper graph. Here it is $(6x - x^2) \geq x$ over the interval $[0, 5]$.

Compute the area as follows:

$\int_0^5 [(6x - x^2) - x]\, dx$

$= \int_0^5 (-x^2 + 5x)\, dx$

$= \left[-\dfrac{x^3}{3} + 5 \cdot \dfrac{x^2}{2} \right]_0^5$

$= \left(-\dfrac{5^3}{3} + 5 \cdot \dfrac{5^2}{2} \right) - \left(-\dfrac{0^3}{3} + 5 \cdot \dfrac{0^2}{2} \right)$

$= \left(-\dfrac{125}{3} + \dfrac{125}{2} \right) - 0$

$= -\dfrac{250}{6} + \dfrac{375}{6}$

$= \dfrac{125}{6}$

7. First graph the system of equations and shade the region bounded by the graphs.

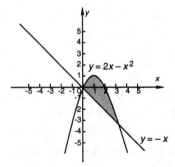

The boundaries are easily determined by looking at the graph. Note which is the upper graph over the shaded region. Here it is $(2x - x^2) \geq -x$ over the interval $[0, 3]$.

Compute the area as follows:

$\int_0^3 [(2x - x^2) - (-x)]\, dx$

$= \int_0^3 (3x - x^2)\, dx$

$= \left[\dfrac{3}{2}x^2 - \dfrac{x^3}{3} \right]_0^3$

$= \left(\dfrac{3}{2} \cdot 3^2 - \dfrac{3^3}{3} \right) - \left(\dfrac{3}{2} \cdot 0^2 - \dfrac{0^3}{3} \right)$

$= \dfrac{27}{2} - \dfrac{27}{3}$

$= \dfrac{81}{6} - \dfrac{54}{6}$

$= \dfrac{27}{6}$

$= \dfrac{9}{2}$

9. First graph the system of equations and shade the region bounded by the graphs.

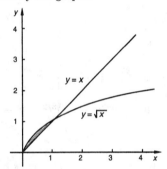

The boundaries are easily determined by looking at the graph. Note which is the upper graph over the shaded region. Here it is $\sqrt{x} \geq x$ over the interval $[0, 1]$.

Compute the area as follows:

$\int_0^1 (\sqrt{x} - x)\, dx$

$= \int_0^1 (x^{1/2} - x)\, dx$

$= \left[\dfrac{2}{3}x^{3/2} - \dfrac{1}{2}x^2 \right]_0^1$

$= \left(\dfrac{2}{3} \cdot 1^{3/2} - \dfrac{1}{2} \cdot 1^2 \right) - \left(\dfrac{2}{3} \cdot 0^{3/2} - \dfrac{1}{2} \cdot 0^2 \right)$

$= \dfrac{2}{3} - \dfrac{1}{2} - 0$

$= \dfrac{4}{6} - \dfrac{3}{6}$

$= \dfrac{1}{6}$

11. Graph the system of equations and shade the region bounded by the graphs.

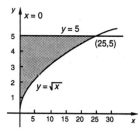

The boundaries are easily determined by looking at the graph. Here $5 \geq \sqrt{x}$ over the interval $[0, 25]$.

Compute the area as follows:

$$\int_0^{25} (5 - \sqrt{x}) \, dx$$
$$= \int_0^{25} (5 - x^{1/2}) \, dx$$
$$= \left[5x - \frac{x^{3/2}}{3/2} \right]_0^{25}$$
$$= \left[5x - \frac{2}{3} x^{3/2} \right]_0^{25}$$
$$= \left(5 \cdot 25 - \frac{2}{3} \cdot 25^{3/2} \right) - \left(5 \cdot 0 - \frac{2}{3} \cdot 0^{3/2} \right)$$
$$= 125 - \frac{250}{3} - 0 \qquad [25^{3/2} = (5^2)^{3/2} = 5^3 = 125]$$
$$= \frac{375}{3} - \frac{250}{3}$$
$$= \frac{125}{3}, \text{ or } 41\frac{2}{3}$$

13. First graph the system of equations and shade the region bounded by the graphs.

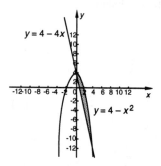

Then determine the first coordinates of possible points of intersection by solving a system of equations as follows. At the points of intersection, $y = 4 - x^2$ and $y = 4 - 4x$, so

$$4 - x^2 = 4 - 4x$$
$$0 = x^2 - 4x$$
$$0 = x(x - 4)$$
$$x = 0 \text{ or } x = 4$$

Thus the interval with which we are concerned is $[0, 4]$. Note that $4 - x^2 \geq 4 - 4x$ over the interval $[0, 4]$.

Compute the area as follows:

$$\int_0^4 [(4 - x^2) - (4 - 4x)] \, dx$$
$$= \int_0^4 (-x^2 + 4x) \, dx$$
$$= \left[-\frac{x^3}{3} + 2x^2 \right]_0^4$$
$$= \left(-\frac{4^3}{3} + 2 \cdot 4^2 \right) - \left(-\frac{0^3}{3} + 2 \cdot 0^2 \right)$$
$$= -\frac{64}{3} + 32$$
$$= -\frac{64}{3} + \frac{96}{3}$$
$$= \frac{32}{3}$$

15. First graph the system of equations and shade the region bounded by the graphs.

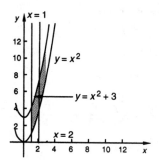

From the graph we can easily determine the interval with which we are concerned. Here $x^2 + 3 \geq x^2$ over the interval $[1, 2]$.

Compute the area as follows:

$$\int_1^2 [(x^2 + 3) - x^2] \, dx$$
$$= \int_1^2 3 \, dx$$
$$= [3x]_1^2$$
$$= 3 \cdot 2 - 3 \cdot 1$$
$$= 6 - 3$$
$$= 3$$

17. $f(x) \geq g(x)$ on $[-5, -1]$, and $g(x) \geq f(x)$ on $[-1, 3]$. We use two integrals to find the total area.

$$\int_{-5}^{-1} [(x^3 + 3x^2 - 9x - 12) - (4x + 3)] \, dx +$$
$$\int_{-1}^{3} [(4x + 3) - (x^3 + 3x^2 - 9x - 12)] \, dx$$
$$= \int_{-5}^{-1} (x^3 + 3x^2 - 13x - 15) \, dx +$$
$$\int_{-1}^{3} (-x^3 - 3x^2 + 13x + 15) \, dx$$
$$= \left[\frac{x^4}{4} + x^3 - \frac{13x^2}{2} - 15x \right]_{-5}^{-1} +$$
$$\left[-\frac{x^4}{4} - x^3 + \frac{13x^2}{2} + 15x \right]_{-1}^{3}$$

$$= \left[\frac{(-1)^4}{4} + (-1)^3 - \frac{13(-1)^2}{2} - 15(-1) \right] -$$

$$\left[\frac{(-5)^4}{4} + (-5)^3 - \frac{13(-5)^2}{2} - 15(-5) \right] +$$

$$\left[-\frac{3^4}{4} - 3^3 + \frac{13(3)^2}{2} + 15 \cdot 3 \right] -$$

$$\left[-\frac{(-1)^4}{4} - (-1)^3 + \frac{13(-1)^2}{2} + 15(-1) \right]$$

$$= \left(\frac{1}{4} - 1 - \frac{13}{2} + 15 \right) - \left(\frac{625}{4} - 125 - \frac{325}{2} + 75 \right) +$$

$$\left(-\frac{81}{4} - 27 + \frac{117}{2} + 45 \right) - \left(-\frac{1}{4} + 1 + \frac{13}{2} - 15 \right)$$

$$= \frac{31}{4} + \frac{225}{4} + \frac{225}{4} + \frac{31}{4}$$

$$= 128$$

19. $f(x) \geq g(x)$ on $[1,4]$. We find the area.

$$\int_1^4 [(4x - x^2) - (x^2 - 6x + 8)] \, dx$$

$$= \int_1^4 (-2x^2 + 10x - 8) \, dx$$

$$= \left[-\frac{2x^3}{3} + 5x^2 - 8x \right]_1^4$$

$$= \left(-\frac{2 \cdot 4^3}{3} + 5 \cdot 4^2 - 8 \cdot 4 \right) - \left(-\frac{2 \cdot 1^3}{3} + 5 \cdot 1^2 - 8 \cdot 1 \right)$$

$$= \left(-\frac{128}{3} + 80 - 32 \right) - \left(-\frac{2}{3} + 5 - 8 \right)$$

$$= \frac{16}{3} + \frac{11}{3} = \frac{27}{3}$$

$$= 9$$

21. Find the area under

$$f(x) = \begin{cases} 4 - x^2, & \text{if } x < 0 \\ 4, & \text{if } x \geq 0 \end{cases} \quad \text{on } [-2, 3]$$

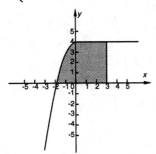

$$\int_{-2}^3 f(x) \, dx$$

$$= \int_{-2}^0 f(x) \, dx + \int_0^3 f(x) \, dx$$

$$= \int_{-2}^0 (4 - x^2) \, dx + \int_0^3 4 \, dx$$

$$= \left[4x - \frac{x^3}{3} \right]_{-2}^0 + [4x]_0^3$$

$$= \left\{ \left[4 \cdot 0 - \frac{0^3}{3} \right] - \left[4(-2) - \frac{(-2)^3}{3} \right] \right\} + (4 \cdot 3 - 4 \cdot 0)$$

$$= (0 - 0) - \left(-8 + \frac{8}{3} \right) + (12 - 0)$$

$$= -\left(-\frac{24}{3} + \frac{8}{3} \right) + 12$$

$$= -\left(-\frac{16}{3} \right) + 12$$

$$= \frac{16}{3} + \frac{36}{3}$$

$$= \frac{52}{3}, \text{ or } 17\frac{1}{3}$$

23. a) $-0.003t^2 \geq -0.009t^2$

Thus, $M'(t) \geq m'(t)$, so subject B has the higher rate of memorization.

b) $\int_0^{10} [(-0.003t^2 + 0.2t) - (-0.009t^2 + 0.2t)] \, dt$

$$= \int_0^{10} 0.006t^2 \, dt$$

$$= [0.002t^3]_0^{10}$$

$$= 0.002 \cdot 10^3 - 0.002 \cdot 0^3$$

$$= 2 - 0$$

$$= 2$$

Subject B memorizes 2 more words then subject A during the first 10 minutes.

25. First graph the system of equations and shade the region bounded by the graph.

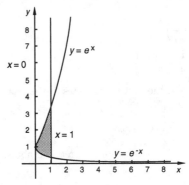

From the graph we can easily determine the interval with which we are concerned. Here $e^x \geq e^{-x}$ over the interval $[0, 1]$.

Compute the area as follows:

$$\int_0^1 (e^x - e^{-x})\, dx$$
$$= \left[e^x + e^{-x} \right]_0^1$$
$$= (e^1 + e^{-1}) - (e^0 + e^{-0})$$
$$= \left(e + \frac{1}{e} \right) - (1 + 1)$$
$$= e + \frac{1}{e} - 2$$
$$= \frac{e^2 - 2e + 1}{e}$$
$$= \frac{(e-1)^2}{e} \approx 1.086$$

27. First graph the system of equations and shade the region bounded by the graph.

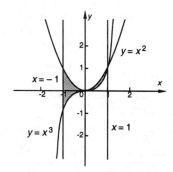

From the graph we can easily determine the interval with which we are concerned. Here $x^2 \geq x^3$ over the interval $[-1, 1]$.

Compute the area as follows:
$$\int_{-1}^1 (x^2 - x^3)\, dx$$
$$= \left[\frac{x^3}{3} - \frac{x^4}{4} \right]_{-1}^1$$
$$= \left(\frac{1^3}{3} - \frac{1^4}{4} \right) - \left(\frac{(-1)^3}{3} - \frac{(-1)^4}{4} \right)$$
$$= \left(\frac{1}{3} - \frac{1}{4} \right) - \left(-\frac{1}{3} - \frac{1}{4} \right)$$
$$= \frac{1}{3} - \frac{1}{4} + \frac{1}{3} + \frac{1}{4} = \frac{2}{3}$$

29. First we find the first coordinates of the relative extrema.
$$y = 3x^5 - 20x^3$$
$$\frac{dy}{dx} = 15x^4 - 60x^2$$

The derivative exists for all real numbers. We solve $\frac{dy}{dx} = 0$.

$$15x^4 - 60x^2 = 0$$
$$15x^2(x^2 - 4) = 0$$
$$15x^2(x+2)(x-2) = 0$$
$$x^2 = 0 \text{ or } x+2 = 0 \quad \text{or } x-2 = 0$$
$$x = 0 \text{ or } \quad x = -2 \text{ or } \quad x = 2$$

We use the Second Derivative Test.
$$\frac{d^2y}{dx^2} = 60x^3 - 120x$$

When $x = -2$, $\frac{d^2y}{dx^2} = 60(-2)^3 - 120(-2) = -240 < 0$, so there is a relative maximum at $x = -2$. When $x = 2$, $\frac{d^2y}{dx^2} = 60 \cdot 2^3 - 120 \cdot 2 = 240 > 0$, so there is a relative minimum at $x = 2$.

When $x = 0$, $\frac{d^2y}{dx^2} = 60 \cdot 0^3 - 120 \cdot 0 = 0$, so the test fails. We use the first derivative and test a value in $(-2, 0)$ and a value in $(0, 2)$.

Test -1, $15 \cdot 1^4 - 60 \cdot 1^2 = -45 < 0$

Test 1, $15 \cdot 1^4 - 60 \cdot 1^2 = -45 < 0$

Thus, the function is decreasing on both intervals, so there is no relative extremum at $x = 0$.

We graph the region. We use different scales on the x- and y-axes. Note that the equation of the x-axis is $y = 0$.

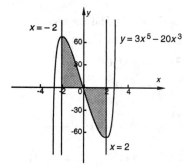

On the interval $[-2, 0]$, $3x^5 - 20x^3 \geq 0$, and on the interval $[0, 2]$, $0 \geq 3x^5 - 20x^3$. We evaluate two integrals to find the total area.

$$\int_{-2}^0 [(3x^5 - 20x^3) - 0]\, dx + \int_0^2 [0 - (3x^5 - 20x^3)\, dx$$
$$= \int_{-2}^0 (3x^5 - 20x^3)\, dx + \int_0^2 (-3x^5 + 20x^3)\, dx$$
$$= \left[\frac{x^6}{2} - 5x^4 \right]_{-2}^0 + \left[-\frac{x^6}{2} + 5x^4 \right]_0^2$$
$$= \left(\frac{0^6}{2} - 5 \cdot 0^4 \right) - \left(\frac{(-2)^6}{2} - 5(-2)^4 \right) + \left(-\frac{2^6}{2} + 5 \cdot 2^4 \right) - \left(-\frac{0^6}{2} + 5 \cdot 0^4 \right)$$
$$= (0 - 0) - (32 - 80) + (-32 + 80) - (-0 + 0)$$
$$= 0 + 48 + 48 - 0$$
$$= 96$$

31. $Q = \int_0^R 2\pi \cdot \frac{p}{4\,Lv}(R^2 - r^2) \cdot r \cdot dr$ Substituting

$$= \frac{\pi p}{2\,Lv} \int_0^R (R^2 r - r^3)\, dr$$
$$= \frac{\pi p}{2\,Lv} \left[\frac{R^2 r^2}{2} - \frac{r^4}{4} \right]_0^R$$
$$= \frac{\pi p}{2\,Lv} \left[\left(\frac{R^2 \cdot R^2}{2} - \frac{R^4}{4} \right) - \left(\frac{R^2 \cdot 0^2}{2} - \frac{0^4}{4} \right) \right]$$
$$= \frac{\pi p}{2\,Lv} \cdot \frac{R^4}{4}$$
$$= \frac{\pi p\, R^4}{8\,Lv}$$

33. From the graph we find that the interval we are concerned with is approximately $[-4, 0.8]$. Then we use the fnInt feature to find that $\int_{-4}^{0.8}(\sqrt{16-x^2}-(x^2+4x))\,dx \approx 24.961$.

35. From the graph we find that the interval we are concerned with is approximately $[-1.55, 1.35]$. Then we use the fnInt feature to find that
$$\int_{-1.55}^{1.35}(1-x+8x^2-4x^4-(2x^2+x-4))\,dx \approx 16.7040.$$
(Answers may vary slightly due to the interval used.)

37. a)

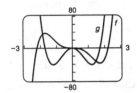

b) From the graph we estimate that the first coordinates of the three points of intersection of the graphs are $a \approx -1.8623$, $b = 0$, and $c \approx 1.4594$.

c) Using the fnInt feature on a grapher, we find that $\int_{-1.8623}^{0}(3.8x^5-18.6x^3-(19x^4-55.8x^2))\,dx \approx 64.5239$.

d) Using the fnInt feature on a grapher, we find that $\int_{0}^{1.4594}(3.8x^5-18.6x^3-(19x^4-55.8x^2))\,dx \approx 17.6830$.

Exercise Set 5.5

1. $\displaystyle \int \frac{3x^2\,dx}{7+x^3}$

Let $u = 7+x^3$, then $du = 3x^2\,dx$.

$$= \int \frac{du}{u} \qquad \text{Substituting } u \text{ for } 7+x^3 \text{ and } du \text{ for } 3x^2\,dx$$
$$= \int \frac{1}{u}\,du$$
$$= \ln u + C \qquad \text{Using Formula C}$$
$$= \ln(7+x^3) + C$$

3. $\int e^{4x}\,dx$

Let $u = 4x$, then $du = 4\,dx$.

We do not have $4\,dx$. We only have dx and need to supply a 4. We do this by multiplying by $\frac{1}{4}\cdot 4$ as follows.

$$\frac{1}{4}\cdot 4\int e^{4x}\,dx \qquad \text{Multiplying by 1}$$
$$= \frac{1}{4}\int 4e^{4x}\,dx$$
$$= \frac{1}{4}\int e^{4x}(4\,dx)$$
$$= \frac{1}{4}\int e^u\,du \qquad \begin{array}{l}\text{Substituting } u \text{ for } 4x \text{ and} \\ du \text{ for } 4\,dx\end{array}$$
$$= \frac{1}{4}e^u + C \qquad \text{Using Formula B}$$
$$= \frac{1}{4}e^{4x} + C$$

5. $\int e^{x/2}\,dx = \int e^{(1/2)x}\,dx$

Let $u = \frac{1}{2}x$, then $du = \frac{1}{2}\,dx$.

We do not have $\frac{1}{2}\,dx$. We only have dx and need to supply a $\frac{1}{2}$ by multiplying by $2\cdot\frac{1}{2}$ as follows.

$$2\cdot\frac{1}{2}\int e^{x/2}\,dx \qquad \text{Multiplying by 1}$$
$$= 2\int \frac{1}{2}e^{x/2}\,dx$$
$$= 2\int e^{x/2}\left(\frac{1}{2}\,dx\right)$$
$$= 2\int e^u\,du \qquad \text{Substituting } u \text{ for } x/2 \text{ and } du \text{ for } \frac{1}{2}\,dx$$
$$= 2\,e^u + C \qquad \text{Using Formula B}$$
$$= 2\,e^{x/2} + C$$

7. $\int x^3 e^{x^4}\,dx$

Let $u = x^4$, then $du = 4x^3\,dx$.

We do not have $4x^3\,dx$. We only have $x^3\,dx$ and need to supply a 4. We do this by multiplying by $\frac{1}{4}\cdot 4$ as follows.

$$\frac{1}{4}\cdot 4\int x^3 e^{x^4}\,dx \qquad \text{Multiplying by 1}$$
$$= \frac{1}{4}\int 4x^3 e^{x^4}\,dx$$
$$= \frac{1}{4}\int e^{x^4}(4x^3\,dx)$$
$$= \frac{1}{4}\int e^u\,du \qquad \begin{array}{l}\text{Substituting } u \text{ for } x^4 \text{ and} \\ du \text{ for } 4x^3\,dx\end{array}$$
$$= \frac{1}{4}\cdot e^u + C \qquad \text{Using Formula B}$$
$$= \frac{1}{4}e^{x^4} + C$$

9. $\int t^2 e^{-t^3}\,dt$

Let $u = -t^3$, then $du = -3t^2\,dt$.

We do not have $-3t^2\,dt$. We only have $t^2\,dt$. We need to supply a -3 by multiplying by $-\frac{1}{3}\cdot(-3)$ as follows.

$$-\frac{1}{3}\cdot(-3)\int t^2 e^{-t^3}\,dt \qquad \text{Multiplying by 1}$$
$$= -\frac{1}{3}\int -3t^2 e^{-t^3}\,dt$$
$$= -\frac{1}{3}\int e^{-t^3}(-3t^2\,dt)$$
$$= -\frac{1}{3}\int e^u\,du \qquad \begin{array}{l}\text{Substituting } u \text{ for } -t^3 \text{ and} \\ du \text{ for } -3t^2\,dt\end{array}$$
$$= -\frac{1}{3}e^u + C \qquad \text{Using Formula B}$$
$$= -\frac{1}{3}e^{-t^3} + C$$

11. $\int \dfrac{\ln 4x \, dx}{x}$

Let $u = \ln 4x$, then $du = \dfrac{1}{x} \, dx$.

$= \int \ln 4x \left(\dfrac{1}{x} \, dx \right)$

$= \int u \, du \qquad$ Substituting u for $\ln 4x$ and du for $\dfrac{1}{x} \, dx$

$= \dfrac{u^2}{2} + C \qquad$ Using Formula A

$= \dfrac{(\ln 4x)^2}{2} + C$

13. $\int \dfrac{dx}{1+x}$

Let $u = 1 + x$, then $du = dx$.

$= \int \dfrac{du}{u} \qquad$ Substituting u for $1+x$ and du for dx

$= \int \dfrac{1}{u} \, du$

$= \ln u + C \qquad$ Using Formula C

$= \ln (1+x) + C$

15. $\int \dfrac{dx}{4-x}$

Let $u = 4 - x$, then $du = -dx$.

We do not have $-dx$. We only have dx and need to supply a -1 by multiplying by $-1 \cdot (-1)$ as follows.

$-1 \cdot (-1) \int \dfrac{dx}{4-x} \qquad$ Multiplying by 1

$= -1 \int -1 \cdot \dfrac{dx}{4-x}$

$= -\int \dfrac{1}{4-x}(-dx)$

$= -\int \dfrac{1}{u} \, du \qquad$ Substituting u for $4-x$ and du for $-dx$

$= -\ln u + C \qquad$ Using Formula C

$= -\ln (4-x) + C$

17. $\int t^2 (t^3 - 1)^7 \, dt$

Let $u = t^3 - 1$, then $du = 3t^2 \, dt$.

We do not have $3t^2 \, dt$. We only have $t^2 \, dt$. We need to supply a 3 by multiplying by $\dfrac{1}{3} \cdot 3$ as follows.

$\dfrac{1}{3} \cdot 3 \int t^2 (t^3 - 1)^7 \, dt \qquad$ Multiplying by 1

$= \dfrac{1}{3} \int 3t^2 (t^3 - 1)^7 \, dt$

$= \dfrac{1}{3} \int (t^3 - 1)^7 \, 3t^2 \, dt$

$= \dfrac{1}{3} \int u^7 \, du \qquad$ Substituting u for $t^3 - 1$ and du for $3t^2 \, dt$

$= \dfrac{1}{3} \cdot \dfrac{u^8}{8} + C \qquad$ Using Formula A

$= \dfrac{1}{24} (t^3 - 1)^8 + C$

19. $\int (x^4 + x^3 + x^2)^7 (4x^3 + 3x^2 + 2x) \, dx$

Let $u = x^4 + x^3 + x^2$, then $du = (4x^3 + 3x^2 + 2x) \, dx$.

$= \int u^7 \, du \qquad$ Substituting u for $x^4 + x^3 + x^2$ and du for $(4x^3 + 3x^2 + 2x) \, dx$

$= \dfrac{u^8}{8} + C \qquad$ Using Formula A

$= \dfrac{1}{8} (x^4 + x^3 + x^2)^8 + C$

21. $\int \dfrac{e^x \, dx}{4 + e^x}$

Let $u = 4 + e^x$, then $du = e^x \, dx$.

$= \int \dfrac{du}{u} \qquad$ Substituting u for $4 + e^x$ and du for $e^x \, dx$

$= \int \dfrac{1}{u} \, du$

$= \ln u + C \qquad$ Using Formula C

$= \ln (4 + e^x) + C$

23. $\int \dfrac{\ln x^2}{x} \, dx$

Let $u = \ln x^2$, then $du = \left(2x \cdot \dfrac{1}{x^2} \right) dx = \dfrac{2}{x} \, dx$.

We do not have $\dfrac{2}{x} \, dx$. We only have $\dfrac{1}{x} \, dx$ and need to supply a 2 by multiplying by $\dfrac{1}{2} \cdot 2$ as follows.

$\dfrac{1}{2} \cdot 2 \int \dfrac{\ln x^2}{x} \, dx \quad$ Multiplying by 1

$= \dfrac{1}{2} \int 2 \cdot \dfrac{\ln x^2}{x} \, dx$

$= \dfrac{1}{2} \int \ln x^2 \cdot \dfrac{2}{x} \, dx$

$= \dfrac{1}{2} \int u \, du \qquad$ Substituting u for $\ln x^2$ and du for $\dfrac{2}{x} \, dx$

$= \dfrac{1}{2} \cdot \dfrac{u^2}{2} + C \qquad$ Using Formula A

$$= \frac{u^2}{4} + C$$

$$= \frac{1}{4}(\ln x^2)^2 + C,$$

$$\text{or } \frac{1}{4}(2\ln x)^2 + C = \frac{1}{4} \cdot 4(\ln x)^2 + C$$

$$= (\ln x)^2 + C$$

25. $\displaystyle\int \frac{dx}{x \ln x}$

Let $u = \ln x$, then $du = \frac{1}{x} dx$.

$$= \int \frac{1}{\ln x}\left(\frac{1}{x} dx\right)$$

$$= \int \frac{1}{u} du \qquad \text{Substituting } u \text{ for } \ln x \text{ and}$$
$$\qquad du \text{ for } \frac{1}{x} dx$$

$$= \ln u + C \qquad \text{Using Formula C}$$

$$= \ln (\ln x) + C$$

27. $\int \sqrt{ax+b}\, dx$, or $\int (ax+b)^{1/2}\, dx$

Let $u = ax + b$, then $du = a\, dx$.

We do not have $a\, dx$. We only have dx and need to supply an a by multiplying by $\frac{1}{a} \cdot a$ as follows.

$$\frac{1}{a} \cdot a \int \sqrt{ax+b}\, dx \qquad \text{Multiplying by 1}$$

$$= \frac{1}{a} \int a\sqrt{ax+b}\, dx$$

$$= \frac{1}{a} \int \sqrt{ax+b}\, (a\, dx)$$

$$= \frac{1}{a} \int \sqrt{u}\, du \qquad \text{Substituting } u \text{ for } ax+b \text{ and}$$
$$\qquad du \text{ for } a\, dx$$

$$= \frac{1}{a} \int u^{1/2}\, du$$

$$= \frac{1}{a} \cdot \frac{u^{3/2}}{\frac{3}{2}} + C \qquad \text{Using Formula A}$$

$$= \frac{2}{3a} \cdot u^{3/2} + C$$

$$= \frac{2}{3a}(ax+b)^{3/2} + C$$

29. $\int b\, e^{ax}\, dx$

$$= b \int e^{ax}\, dx$$

Let $u = ax$, then $du = a\, dx$.

We do not have $a\, dx$. We only have dx and need to supply an a by multiplying by $\frac{1}{a} \cdot a$ as follows.

$$= b \cdot \frac{1}{a} \cdot a \int e^{ax}\, dx$$

$$= \frac{b}{a} \int a\, e^{ax}\, dx$$

$$= \frac{b}{a} \int e^{ax}\, (a\, dx)$$

$$= \frac{b}{a} \int e^u\, du \qquad \text{Substituting } u \text{ for } ax \text{ and}$$
$$\qquad du \text{ for } a\, dx$$

$$= \frac{b}{a} e^u + C \qquad \text{Using Formula B}$$

$$= \frac{b}{a} e^{ax} + C$$

31. $\displaystyle\int \frac{3x^2\, dx}{(1+x^3)^5}$

Let $u = 1 + x^3$, then $du = 3x^2\, dx$.

$$= \int \frac{1}{(1+x^3)^5} \cdot 3x^2\, dx$$

$$= \int \frac{1}{u^5}\, du \qquad \text{Substituting } u \text{ for } 1+x^3$$
$$\qquad \text{and } du \text{ for } 3x^2\, dx$$

$$= \int u^{-5}\, du$$

$$= \frac{u^{-4}}{-4} + C \qquad \text{Using Formula A}$$

$$= -\frac{1}{4u^4} + C$$

$$= -\frac{1}{4(1+x^3)^4} + C$$

33. $\int 7x\sqrt[3]{4-x^2}\, dx$

$$= 7 \int x\sqrt[3]{4-x^2}\, dx$$

Let $u = 4 - x^2$, then $du = -2x\, dx$.

We supply a -2 by multiplying by $-\frac{1}{2}(-2)$.

$$= 7\left(-\frac{1}{2}\right)(-2) \int x\sqrt[3]{4-x^2}\, dx$$

$$= 7\left(-\frac{1}{2}\right) \int -2x\sqrt[3]{4-x^2}\, dx$$

$$= -\frac{7}{2} \int \sqrt[3]{4-x^2}(-2x\, dx)$$

$$= -\frac{7}{2} \int \sqrt[3]{u}\, du \qquad \text{Substituting } u \text{ for } 4-x^2 \text{ and}$$
$$\qquad du \text{ for } -2x\, dx$$

$$= -\frac{7}{2} \int u^{1/3}\, du$$

$$= -\frac{7}{2} \cdot \frac{u^{4/3}}{\frac{4}{3}} + C \qquad \text{Using Formula A}$$

$$= -\frac{21}{8}(4-x^2)^{4/3} + C$$

35. $\int_0^1 2x\, e^{x^2}\, dx$

First find the indefinite integral.

$$\int 2x\, e^{x^2}\, dx$$

Let $u = x^2$, then $du = 2x\, dx$.

$$= \int e^{x^2}(2x\, dx)$$

$$= \int e^u\, du \qquad \text{Substituting } u \text{ for } x^2 \text{ and } du \text{ for } 2x\, dx$$

$$= e^u + C$$

$$= e^{x^2} + C$$

Then evaluate the definite integral on $[0, 1]$.

$$\int_0^1 2x\, e^{x^2}\, dx$$

$$= \left[e^{x^2} \right]_0^1$$

$$= e^{1^2} - e^{0^2}$$

$$= e - 1$$

37. $\int_0^1 x(x^2 + 1)^5\, dx$

First find the indefinite integral.

$$\int x(x^2 + 1)^5\, dx$$

Let $u = x^2 + 1$, then $du = 2x\, dx$.

We only have $x\, dx$ and need to supply a 2 by multiplying by $\frac{1}{2} \cdot 2$.

$$\frac{1}{2} \cdot 2 \int x(x^2 + 1)^5\, dx \qquad \text{Multiplying by 1}$$

$$= \frac{1}{2} \int 2x(x^2 + 1)^5\, dx$$

$$= \frac{1}{2} \int (x^2 + 1)^5 \cdot 2x\, dx$$

$$= \frac{1}{2} \int u^5\, du \qquad \text{Substituting } u \text{ for } x^2 + 1 \text{ and } du \text{ for } 2x\, dx$$

$$= \frac{1}{2} \cdot \frac{u^6}{6} + C \qquad \text{Using Formula A}$$

$$= \frac{(x^2 + 1)^6}{12} + C$$

Then evaluate the definite integral on $[0, 1]$.

$$\int_0^1 x(x^2 + 1)^5\, dx$$

$$= \left[\frac{(x^2 + 1)^6}{12} \right]_0^1$$

$$= \frac{(1^2 + 1)^6}{12} - \frac{(0^2 + 1)^6}{12}$$

$$= \frac{64}{12} - \frac{1}{12}$$

$$= \frac{63}{12}$$

$$= \frac{21}{4}$$

39. $\int_1^3 \frac{dt}{1 + t}$

First find the indefinite integral.

$$\int \frac{dt}{1 + t}$$

Let $u = 1 + t$, then $du = dt$.

$$= \int \frac{du}{u} \qquad \text{Substituting } u \text{ for } 1 + t \text{ and } du \text{ for } dt$$

$$= \int \frac{1}{u}\, du$$

$$= \ln u + C \qquad \text{Using Formula C}$$

$$= \ln(1 + t) + C$$

Then evaluate the definite integral on $[1, 3]$.

$$\int_1^3 \frac{dt}{1 + t}$$

$$= \left[\ln(1 + t) \right]_1^3$$

$$= \ln(1 + 3) - \ln(1 + 1)$$

$$= \ln 4 - \ln 2$$

$$= \ln \frac{4}{2}$$

$$= \ln 2$$

41. $\int_1^4 \frac{2x + 1}{x^2 + x - 1}\, dx$

First find the indefinite integral.

$$\int \frac{2x + 1}{x^2 + x - 1}\, dx$$

Let $u = x^2 + x = 1$, then $du = (2x + 1)\, dx$.

$$= \int \frac{1}{x^2 + x - 1}(2x + 1)\, dx$$

$$= \int \frac{1}{u}\, du \qquad \text{Substituting } u \text{ for } x^2 + x - 1 \text{ and } du \text{ for } (2x + 1)\, dx$$

$$= \ln u + C$$

$$= \ln(x^2 + x - 1) + C$$

Then evaluate the definite integral on $[1, 4]$.

$$\int_1^4 \frac{2x + 1}{x^2 + x - 1}\, dx$$

$$= \left[\ln(x^2 + x - 1) \right]_1^4$$

$$= \ln(4^4 + 4 - 1) - \ln(1^2 + 1 - 1)$$

$$= \ln 19 - \ln 1$$

$$= \ln 19 \qquad (\ln 1 = 0)$$

43. $\int_0^b e^{-x}\, dx$

First find the indefinite integral.

$$\int e^{-x}\, dx$$

Let $u = -x$, then $du = -dx$.

We only have dx and need to supply a -1 by multiplying by $-1 \cdot (-1)$.

$$-1 \cdot (-1) \int e^{-x}\, dx$$

$$= -\int -e^{-x}\, dx$$

$$= -\int e^{-x}(-dx)$$

$$= -\int e^u\, du \qquad \text{Substituting } u \text{ for } -x \text{ and } du \text{ for } -dx$$

$$= -e^u + C \qquad \text{Using Formula B}$$

$$= -e^{-x} + C$$

Then evaluate the definite integral on $[0, b]$.

$$\int_0^b e^{-x}\, dx$$
$$= \left[-e^{-x}\right]_0^b$$
$$= (-e^{-b}) - (-e^{-0})$$
$$= -e^{-b} + e^0$$
$$= -e^{-b} + 1$$
$$= 1 - \frac{1}{e^b}$$

45. $\int_0^b m\, e^{-mx}\, dx$

First find the indefinite integral.

$\int m\, e^{-mx}\, dx$ m is a constant

Let $u = -mx$, then $du = -m\, dx$.

We only have $m\, dx$ and need to supply a -1 by multiplying by $-1 \cdot (-1)$.

$$-1 \cdot (-1) \int m\, e^{-mx}\, dx$$
$$= -\int -m\, e^{-mx}\, dx$$
$$= -\int e^{-mx}\,(-m\, dx)$$
$$= -\int e^u\, du \qquad \text{Substituting } u \text{ for } -mx \text{ and } du \text{ for } -m\, dx$$
$$= -e^u + C \qquad \text{Using Formula B}$$
$$= -e^{-mx} + C$$

Then evaluate the definite integral on $[0, b]$.

$$\int_0^b m e^{-mx}\, dx$$
$$= \left[-e^{-mx}\right]_0^b$$
$$= (-e^{-mb}) - (-e^{-m \cdot 0})$$
$$= -e^{-mb} + 1 \qquad (e^{-m \cdot 0} = e^0 = 1)$$
$$= 1 - e^{-mb}$$
$$= 1 - \frac{1}{e^{mb}}$$

47. $\int_0^4 (x-6)^2\, dx$

First find the indefinite integral.

$$\int (x-6)^2\, dx$$

Let $u = x - 6$, then $du = dx$.

$$= \int u^2\, du \qquad \text{Substituting } u \text{ for } x-6 \text{ and } du \text{ for } dx$$
$$= \frac{u^3}{3} + C$$
$$= \frac{(x-6)^3}{3} + C$$

Then evaluate the definite integral on $[0, 4]$.

$$\int_0^4 (x-6)^2\, dx$$
$$= \left[\frac{(x-6)^3}{3}\right]_0^4$$
$$= \frac{(4-6)^3}{3} - \frac{(0-6)^3}{3}$$
$$= -\frac{8}{3} - \left(-\frac{216}{3}\right)$$
$$= -\frac{8}{3} + \frac{216}{3}$$
$$= \frac{208}{3}$$

49. $\int_0^2 \frac{3x^2\, dx}{(1+x^3)^5}$

From Exercise 31 we know that the indefinite integral is

$$\int \frac{3x^2\, dx}{(1+x^3)^5} = -\frac{1}{4(1+x^3)^4} + C.$$

Now we evaluate the definite integral on $[0, 2]$.

$$\int_0^2 \frac{3x^2\, dx}{(1+x^3)^5}$$
$$= \left[-\frac{1}{4(1+x^3)^4}\right]_0^2$$
$$= \left[-\frac{1}{4(1+2^3)^4}\right] - \left[-\frac{1}{4(1+0^3)^4}\right]$$
$$= -\frac{1}{26,244} + \frac{1}{4}$$
$$= \frac{6560}{26,244}$$
$$= \frac{1640}{6561}$$

51. $\int_0^{\sqrt{7}} 7x\sqrt[3]{1+x^2}\, dx = 7\int_0^{\sqrt{7}} x\sqrt[3]{1+x^2}\, dx$

First find the indefinite integral.

$$7\int x\sqrt[3]{1+x^2}\, dx$$

Let $u = 1 + x^2$, then $du = 2x\, dx$.

$$= 7 \cdot \frac{1}{2} \cdot 2 \int x\sqrt[3]{1+x^2}\, dx$$
$$= 7 \cdot \frac{1}{2} \int 2x\sqrt[3]{1+x^2}\, dx$$
$$= \frac{7}{2} \int \sqrt[3]{1+x^2}\,(2x\, dx)$$
$$= \frac{7}{2} \int \sqrt[3]{u}\, du$$
$$= \frac{7}{2} \int u^{1/3}\, du$$
$$= \frac{7}{2} \cdot \frac{u^{4/3}}{4/3} + C$$
$$= \frac{21}{8} u^{4/3} + C$$
$$= \frac{21}{8}(1+x^2)^{4/3} + C$$

Then evaluate the definite integral on $[0, \sqrt{7}]$.

$$7 \int_0^{\sqrt{7}} x \sqrt[3]{1 + x^2} \, dx$$

$$= \left[\frac{21}{8} (1 + x^2)^{4/3} \right]_0^{\sqrt{7}}$$

$$= \frac{21}{8} (1 + (\sqrt{7})^2)^{4/3} - \frac{21}{8} (1 + 0^2)^{4/3}$$

$$= \frac{21}{8} \cdot 8^{4/3} - \frac{21}{8} \cdot 1^{4/3}$$

$$= \frac{21}{8} \cdot 16 - \frac{21}{8} \cdot 1$$

$$= 42 - \frac{21}{8}$$

$$= \frac{315}{8}$$

53.

55. a) We will evaluate each integral separately and then do the operations to find $P(T)$.

Find $\int_0^T R'(t) \, dt$: First find the indefinite integral.

$$\int 4000t \, dt = \frac{4000t^2}{2} + C = 2000t^2 + C$$

Then evaluate the definite integral on $[0, T]$.

$$\left[2000t^2 \right]_0^T = 2000T^2 - 2000 \cdot 0^2$$
$$= 2000T^2$$

Find $\int_0^T V'(t) \, dt$: First find the indefinite integral.

$$\int 25,000 \, e^{-0.1t} \, dt$$

Let $u = -0.1t$, then $du = -0.1 \, dt$.

$$= \int 25,000(-0.1)\left(\frac{1}{-0.1}\right) e^{-0.1t} \, dt$$

$$= \int 25,000\left(\frac{1}{-0.1}\right) e^{-0.1t} (-0.1) \, dt$$

$$= 25,000\left(\frac{1}{-0.1}\right) \int e^{-0.1t} (-0.1) \, dt$$

$$= 25,000(-10) \int e^u \, du \qquad \text{Substituting;}$$
$$\frac{1}{-0.1} = -10$$

$$= -250,000 \, e^u + C$$

$$= -250,000 \, e^{-0.1t} + C$$

Then evaluate the definite integral on $[0, T]$.

$$-250,000 \left[e^{-0.1t} \right]_0^T$$
$$= -250,000(e^{-0.1T} - e^{-0.1(0)})$$
$$= -250,000(e^{-0.1T} - 1) \qquad (e^0 = 1)$$

$$P(T) = 2000T^2 + [-250,000(e^{-0.1T} - 1)] - 250,000$$

$$= 2000T^2 - 250,000 \, e^{-0.1T} + 250,000 - 250,000$$

$$= 2000T^2 - 250,000 \, e^{-0.1T}$$

b) $P(10) = 2000(10)^2 - 250,000 \, e^{-0.1(10)}$

$$= 2000 \cdot 100 - 250,000 \, e^{-1}$$

$$= 200,000 - 250,000 \, e^{-1}$$

$$\approx \$108,030$$

57. a) First find the indefinite integral.

$$\int 100,000 \, e^{0.025t} \, dt$$

Let $u = 0.025t$, then $du = 0.025 \, dt$.

$$= 100,000\left(\frac{1}{0.025}\right) \int e^{0.025t} (0.025) \, dt$$

$$= 4,000,000 \int e^u \, du$$

$$= 4,000,000 \, e^u + C$$

$$= 4,000,000 \, e^{0.025t} + C$$

Then evaluate the definite integral on $[0, 99]$.

$$4,000,000 \left[e^{0.025t} \right]_0^{99}$$
$$= 4,000,000(e^{0.025(99)} - e^{0.025(0)})$$
$$= 4,000,000(e^{2.475} - e^0)$$
$$= 4,000,000(e^{2.475} - 1)$$
$$\approx 43,526,828$$

b) From part a) we know that

$$\int D(t) \, dt = 4,000,000 \, e^{0.025t} + C$$

Now evaluate the definite integral on $[80, 99]$.

$$4,000,000 \left[e^{0.025t} \right]_{80}^{99}$$
$$= 4,000,000(e^{0.025(99)} - e^{0.025(80)})$$
$$= 4,000,000(e^{2.475} - e^2)$$
$$\approx 17,970,604$$

59. The area of the shaded region is the area between the curves $y = 0$ (the x-axis) and $y = x\sqrt{16 - x^2}$. On $[-4, 0]$, $x(16 - x^2) \leq 0$, and on $[0, 4]$, $x(16 - x^2) \geq 0$. We will calculate each portion of the area separately and then add them. On $[-4, 0]$: First find the indefinite integral.

$$\int (0 - x\sqrt{16 - x^2}) \, dx$$
$$= \int -x\sqrt{16 - x^2} \, dx$$

Let $u = 16 - x^2$, then $du = -2x \, dx$.

$$= \frac{1}{2}(2) \int -x\sqrt{16 - x^2} \, dx$$

$$= \frac{1}{2} \int \sqrt{16 - x^2} \, (-2x) \, dx$$

$$= \frac{1}{2} \int \sqrt{u} \, du$$

$$= \frac{1}{2} \int u^{1/2} \, du$$

$$= \frac{1}{2} \cdot \frac{u^{3/2}}{3/2} + C$$

$$= \frac{1}{3} u^{3/2} + C$$

$$= \frac{1}{3} (16 - x^2)^{3/2} + C$$

Then find the definite integral on $[-4, 0]$.

$$\frac{1}{3}\left[(16 - x^2)^{3/2}\right]_{-4}^{0}$$

$$= \frac{1}{3}\left[(16 - 0^2)^{3/2} - (16 - (-4)^2)^{3/2}\right]$$

$$= \frac{1}{3}(16^{3/2} - 0^{3/2})$$

$$= \frac{1}{3} \cdot 64$$

$$= \frac{64}{3}$$

On $[0, 4]$: First find the indefinite integral.

$$\int (x\sqrt{16 - x^2} - 0)\, dx$$

$$= \int x\sqrt{16 - x^2}\, dx$$

Let $u = 16 - x^2$, then $du = -2x\, dx$.

$$= -\frac{1}{2}(-2)\int x\sqrt{16 - x^2}\, dx$$

$$= -\frac{1}{2}\int \sqrt{16 - x^2}(-2x)\, dx$$

$$= -\frac{1}{2}\int \sqrt{u}\, dx$$

$$= -\frac{1}{2}\int u^{1/2}\, du$$

$$= -\frac{1}{2} \cdot \frac{u^{3/2}}{3/2} + C$$

$$= -\frac{1}{3}u^{3/2} + C$$

$$= -\frac{1}{3}(16 - x^2)^{3/2} + C$$

Then find the definite integral on $[0, 4]$.

$$-\frac{1}{3}\left[(16 - x^2)^{3/2}\right]_{0}^{4}$$

$$= -\frac{1}{3}\left[(16 - 4^2)^{3/2} - (16 - 0^2)^{3/2}\right]$$

$$= -\frac{1}{3}(0^{3/2} - 16^{3/2})$$

$$= -\frac{1}{3}(-64)$$

$$= \frac{64}{3}$$

The total area is $\frac{64}{3} + \frac{64}{3} = \frac{128}{3}$.

61. $\int 5x\sqrt{1 - 4x^2}\, dx$, or $5\int x(1 - 4x^2)^{1/2}\, dx$

Let $u = 1 - 4x^2$, then $du = -8x\, dx$.

We do not have $-8x\, dx$. We only have $x\, dx$ and need to supply a -8 by multiplying by $-\frac{1}{8} \cdot (-8)$ as follows.

$$-\frac{1}{8} \cdot (-8) \cdot 5\int x(1 - 4x^2)^{1/2}\, dx \quad \begin{array}{l}\text{Multiplying}\\\text{by 1}\end{array}$$

$$= -\frac{5}{8}\int (1 - 4x^2)^{1/2}(-8x\, dx)$$

$$= -\frac{5}{8}\int u^{1/2}\, du \quad \begin{array}{l}\text{Substituting } u \text{ for } 1 - 4x^2\\\text{and } du \text{ for } -8x\, dx\end{array}$$

$$= -\frac{5}{8} \cdot \frac{u^{3/2}}{3/2} + C \quad \text{Using Formula A}$$

$$= -\frac{10}{24}u^{3/2} + C$$

$$= -\frac{5}{12}u^{3/2} + C$$

$$= -\frac{5}{12}(1 - 4x^2)^{3/2} + C$$

63. $\int \frac{x^2}{e^{x^3}}\, dx$, or $\int x^2 e^{-x^3}\, dx$

Let $u = -x^3$, then $du = -3x^2\, dx$.

We do not have $-3x^2\, dx$. We only have $x^2\, dx$ and need to supply a -3 by multiplying by $-\frac{1}{3} \cdot (-3)$ as follows.

$$-\frac{1}{3} \cdot (-3) \cdot \int x^2 e^{-x^3}\, dx \quad \text{Multiplying by 1}$$

$$= -\frac{1}{3}\int e^{-x^3}(-3x^2\, dx)$$

$$= -\frac{1}{3}\int e^u\, du \quad \begin{array}{l}\text{Substituting } u \text{ for } -x^3 \text{ and}\\ du \text{ for } -3x^2\, dx\end{array}$$

$$= -\frac{1}{3} \cdot e^u + C \quad \text{Using Formula B}$$

$$= -\frac{1}{3}e^{-x^3} + C$$

65. $\int \frac{e^{1/t}}{t^2}\, dt$, or $\int e^{t^{-1}} t^{-2}\, dt$

Let $u = t^{-1}$, then $du = -t^{-2}\, dt$.

We do not have $-t^{-2}\, dt$. We only have $t^{-2}\, dt$ and need to supply a -1 by multiplying by $-1 \cdot (-1)$ as follows.

$$-1 \cdot (-1) \cdot \int e^{t^{-1}} t^{-2}\, dt \quad \text{Multiplying by 1}$$

$$= -\int e^{t^{-1}}(-t^{-2}\, dt)$$

$$= -\int e^u\, du \quad \begin{array}{l}\text{Substituting } u \text{ for } t^{-1} \text{ and}\\ du \text{ for } -t^{-2}\, dt\end{array}$$

$$= -e^u + C$$

$$= -e^{t^{-1}} + C$$

$$= -e^{1/t} + C$$

67. $\int \dfrac{dx}{x(\ln x)^4}$, or $\int \dfrac{1}{x}(\ln x)^{-4}\, dx$

Let $u = \ln x$, then $du = \dfrac{1}{x}\, dx$.

$= \int u^{-4}\, du$ Substituting u for $\ln x$ and du for $\dfrac{1}{x}\, dx$

$= \dfrac{u^{-3}}{-3} + C$ Using Formula A

$= -\dfrac{1}{3}u^{-3} + C$

$= -\dfrac{1}{3}(\ln x)^{-3} + C$

69. $\int x^2 \sqrt{x^3 + 1}\, dx$, or $\int x^2 (x^3 + 1)^{1/2}\, dx$

Let $u = x^3 + 1$, then $du = 3x^2\, dx$.

We do not have $3x^2\, dx$. We only have $x^2\, dx$ and need to supply a 3 by multiplying by $\dfrac{1}{3} \cdot 3$ as follows.

$\dfrac{1}{3} \cdot 3 \int x^2 (x^3 + 1)^{1/2}\, dx$ Multiplying by 1

$= \dfrac{1}{3} \int (x^3 + 1)^{1/2}\, 3x^2\, dx$

$= \dfrac{1}{3} \int u^{1/2}\, du$ Substituting u for $x^3 + 1$ and du for $3x^2\, dx$

$= \dfrac{1}{3} \cdot \dfrac{u^{3/2}}{3/2} + C$ Using Formula A

$= \dfrac{2}{9}u^{3/2} + C$

$= \dfrac{2}{9}(x^3 + 1)^{3/2} + C$

71. $\int \dfrac{x - 3}{(x^2 - 6x)^{1/3}}\, dx$, or $\int (x - 3)(x^2 - 6x)^{-1/3}\, dx$

Let $u = x^2 - 6x$, then $du = (2x - 6)\, dx$, or $2(x - 3)\, dx$.

We do not have $2(x - 3)\, dx$. We only have $(x - 3)\, dx$ and need to supply a 2 by multiplying by $\dfrac{1}{2} \cdot 2$.

$\dfrac{1}{2} \cdot 2 \int (x - 3)(x^2 - 6x)^{-1/3}\, dx$ Multiplying by 1

$= \dfrac{1}{2} \int (x^2 - 6x)^{-1/3}\, 2(x - 3)\, dx$

$= \dfrac{1}{2} \int u^{-1/3}\, du$ Substituting u for $x^2 - 6x$ and du for $2(x - 3)\, dx$

$= \dfrac{1}{2} \cdot \dfrac{u^{2/3}}{2/3} + C$

$= \dfrac{3}{4}u^{2/3} + C$

$= \dfrac{3}{4}(x^2 - 6x)^{2/3} + C$

73. $\int \dfrac{t^2 + 2t}{(t + 1)^2}\, dt$

$= \int \left[1 - \dfrac{1}{(t + 1)^2} \right] dt$ See the hint.

Let $u = t + 1$, then $du = dt$.

$= \int \left[1 - \dfrac{1}{u^2} \right] du$

$= \int (1 - u^{-2})\, du$

$= u - \dfrac{u^{-1}}{-1} + C$

$= u + \dfrac{1}{u} + C$

$= t + 1 + \dfrac{1}{t + 1} + C$

$= t + \dfrac{1}{t + 1} + K$ where $K = 1 + C$

75. $\int \dfrac{x + 3}{x + 1}\, dx$

$= \int \left(1 + \dfrac{2}{x + 1} \right) dx$ See the hint.

Let $u = x + 1$, then $du = dx$.

$= \int \left(1 + \dfrac{2}{u} \right) du$

$= u + 2 \ln u + C$

$= x + 1 + 2 \ln (x + 1) + C$

$= x + 2 \ln (x + 1) + K$ where $K = 1 + C$

77. $\int \dfrac{dx}{x(\ln x)^n}$

Let $u = \ln x$, then $du = \dfrac{1}{x}\, dx$.

$= \int \dfrac{du}{u^n}$

$= \int u^{-n}\, du$

$= \dfrac{u^{-n+1}}{-n + 1} + C$

$= \dfrac{(\ln x)^{-n+1}}{-n + 1} + C$, or

$-\dfrac{(\ln x)^{-n+1}}{n - 1}$

79. $\int \dfrac{e^x - e^{-x}}{e^x + e^{-x}}\, dx$

Let $u = e^x + e^{-x}$, then $du = (e^x - e^{-x})\, dx$.

$= \int \dfrac{du}{u}$

$= \ln u + C$

$= \ln (e^x + e^{-x}) + C$

81. $\int \dfrac{dx}{x \ln x[\ln (\ln x)]}$

Let $u = \ln (\ln x)$, then $du = \dfrac{1}{x \ln x}\, dx$.

$= \int \dfrac{1}{u}\, du$ Substituting

$= \ln u + C$

$= \ln [\ln (\ln x)] + C$

83. $\int 9x(7x^2+9)^n\,dx$

$= 9\int x(7x^2+9)^n\,dx$

Let $u = 7x^2+9$, then $du = 14x\,dx$.

We only have $x\,dx$ and need to supply a 14 by multiplying by $\frac{1}{14}\cdot 14$ as follows.

$\frac{1}{14}\cdot 14\cdot 9\int x(7x^2+9)^n\,dx$

$= \frac{9}{14}\int (7x^2+9)^n\, 14x\,dx$

$= \frac{9}{14}\int u^n\,du$ \qquad Substituting

$= \frac{9}{14}\cdot\frac{u^{n+1}}{n+1}+C$ \qquad Using Formula A

$= \frac{9(7x^2+9)^{n+1}}{14(n+1)}+C$

85. $\boxed{tw}$

Exercise Set 5.6

1. $\int 5x\,e^{5x}\,dx = \int x(5e^{5x}\,dx)$

Let

$u = x$ and $dv = 5e^{5x}\,dx$.

Then $du = dx$ and $v = e^{5x}$.

$\begin{matrix} u & dv & u & v & v & du \end{matrix}$

$\int x(5e^{5x}\,dx) = x\cdot e^{5x} - \int e^{5x}\cdot dx$

Using Theorem 7:
$\int u\,dv = uv - \int v\,du$

$= xe^{5x} - \frac{1}{5}e^{5x}+C$

3. $\int x^3(3x^2\,dx)$

Let

$u = x^3$ and $dv = 3x^2\,dx$.

Then $du = 3x^2\,dx$ and $v = x^3$.

$\begin{matrix} u & dv & u & v & v & du \end{matrix}$

$\int x^3(3x^2\,dx) = x^3\cdot x^3 - \int x^3\cdot 3x^2\,dx$

Integration by Parts

$= x^6 - \int 3x^5\,dx$

$= x^6 - 3\int x^5\,dx$

$= x^6 - 3\cdot\frac{x^6}{6}+C$

$= x^6 - \frac{1}{2}x^6+C$

$= \frac{1}{2}x^6+C$

This problem can also be worked with substitution or with Formula A.

$\int x^3(3x^2\,dx)$

Let $u = x^3$, then $du = 3x^2\,dx$.

$= \int u\,du$ \qquad Substituting u for x^3 and du for $3x^2\,dx$

$= \frac{u^2}{2}+C$

$= \frac{(x^3)^2}{2}+C$

$= \frac{1}{2}x^6+C$

$\int x^3(3x^2\,dx)$

$= \int 3x^5\,dx$

$= 3\int x^5\,dx$

$= 3\cdot\frac{x^6}{6}+C$ \qquad Using Formula A

$= \frac{1}{2}x^6+C$

5. $\int xe^{2x}\,dx$

Let

$u = x$ and $dv = e^{2x}\,dx$.

Then $du = dx$ and $v = \frac{1}{2}e^{2x}$. $\left(\int be^{ax}\,dx = \frac{b}{a}e^{ax}+C\right)$

$\begin{matrix} u & dv & u & v & v & du \end{matrix}$

$\int x\,e^{2x}\,dx = x\cdot\frac{1}{2}e^{2x} - \int\frac{1}{2}e^{2x}\,dx$

Integration by Parts

$= \frac{1}{2}xe^{2x} - \frac{\frac{1}{2}}{2}e^{2x}+C$

$\left(\int be^{ax}\,dx = \frac{b}{a}e^{ax}+C\right)$

$= \frac{1}{2}xe^{2x} - \frac{1}{4}e^{2x}+C$

7. $\int xe^{-2x}\,dx$

Let

$u = x$ and $dv = e^{-2x}\,dx$.

Then $du = dx$ and $v = -\frac{1}{2}e^{-2x}$. $\left(\int be^{ax}\,dx = \frac{b}{a}e^{ax}+C\right)$

$\begin{matrix} u & dv & u & v & v & du \end{matrix}$

$\int x\,e^{-2x}\,dx = x\cdot\left(-\frac{1}{2}e^{-2x}\right) - \int\left(-\frac{1}{2}e^{-2x}\right)dx$

Integration by Parts

$= -\frac{1}{2}xe^{-2x} - \frac{-\frac{1}{2}}{-2}e^{-2x}+C$

$\left(\int be^{ax}\,dx = \frac{b}{a}e^{ax}+C\right)$

$= -\frac{1}{2}xe^{-2x} - \frac{1}{4}e^{-2x}+C$

9. $\int x^2\ln x\,dx = \int(\ln x)x^2\,dx$

Let

$u = \ln x$ and $dv = x^2\,dx$.

Then $du = \dfrac{1}{x}\,dx$ and $v = \dfrac{x^3}{3}$.

$$\quad\;\; u \qquad dv \qquad u \quad v \qquad v \quad du$$

$$\int (\ln x)\, x^2\, dx = \ln x \cdot \frac{x^3}{3} - \int \frac{x^3}{3} \cdot \frac{1}{x}\, dx$$

$$\text{Integration by Parts}$$

$$= \frac{x^3}{3} \ln x - \frac{1}{3} \int x^2\, dx$$

$$= \frac{x^3}{3} \ln x - \frac{1}{3} \cdot \frac{x^3}{3} + C$$

$$= \frac{x^3}{3} \ln x - \frac{x^3}{9} + C$$

11. $\int x \ln x^2\, dx = \int (\ln x^2)\, x\, dx$

Let

$u = \ln x^2$ and $dv = x\, dx$.

Then $du = 2x \cdot \dfrac{1}{x^2}\, dx = \dfrac{2}{x}\, dx$ and $v = \dfrac{x^2}{2}$.

$$\quad\;\; u \qquad dv \qquad u \quad v \qquad v \quad du$$

$$\int (\ln x^2)\, x\, dx = (\ln x^2) \cdot \frac{x^2}{2} - \int \frac{x^2}{2} \cdot \frac{2}{x}\, dx$$

$$\text{Integration by Parts}$$

$$= \frac{x^2}{2} \ln x^2 - \int x\, dx$$

$$= \frac{x^2}{2} \ln x^2 - \frac{x^2}{2} + C, \text{ or}$$

$$x^2 \ln x - \frac{x^2}{2} + C$$

13. $\int \ln (x+3)\, dx$

Let

$u = \ln (x+3)$ and $dv = dx$.

Then $du = \dfrac{1}{x+3}\, dx$ and $v = x + 3$.

Choosing $x+3$ as an antiderivative of dv

$$\quad u \qquad dv \qquad u \qquad v$$

$$\int \ln (x+3)\, dx = [\ln (x+3)] \cdot [x+3] -$$

$$\qquad\qquad v \qquad du$$

$$\qquad\quad \int (x+3) \cdot \frac{1}{x+3}\, dx$$

$$\text{Integration by Parts}$$

$$= (x+3) \ln (x+3) - \int dx$$

$$= (x+3) \ln (x+3) - x + C$$

15. $\int (x+2) \ln x\, dx = \int (\ln x)(x+2)\, dx$

Let

$u = \ln x$ and $dv = (x+2)\, dx$.

Then

$$du = \frac{1}{x}\, dx \text{ and } v = \frac{(x+2)^2}{2}.$$

$$\quad u \qquad dv$$

$$\int (\ln x)\,(x+2)\, dx$$

$$\quad\;\; u \qquad v \qquad\qquad v \qquad du$$

$$= (\ln x) \cdot \frac{(x+2)^2}{2} - \int \frac{(x+2)^2}{2} \cdot \frac{1}{x}\, dx$$

$$\text{Integration by Parts}$$

$$= \frac{x^2+4x+4}{2} \ln x - \int \frac{x^2+4x+4}{2x}\, dx$$

$$= \frac{x^2+4x+4}{2} \ln x - \int \left(\frac{x}{2} + 2 + \frac{2}{x}\right) dx$$

$$= \frac{x^2+4x+4}{2} \ln x - \frac{1}{2}\int x\, dx - 2\int dx - 2\int \frac{1}{x}\, dx$$

$$= \frac{x^2+4x+4}{2} \ln x - \frac{1}{2} \cdot \frac{x^2}{2} - 2 \cdot x - 2 \cdot \ln x + C$$

$$= \frac{x^2+4x+4}{2} \ln x - \frac{1}{2} \cdot \frac{x^2}{2} - 2 \cdot x - 2 \cdot \ln x + C$$

$$= \left(\frac{x^2+4x+4}{2} - 2\right) \ln x - \frac{x^2}{4} - 2x + C$$

$$= \left(\frac{x^2}{2} + 2x\right) \ln x - \frac{x^2}{4} - 2x + C$$

17. $\int (x-1) \ln x\, dx = \int (\ln x)(x-1)\, dx$

Let

$u = \ln x$ and $dv = (x-1)\, dx$.

Then

$$du = \frac{1}{x}\, dx \text{ and } v = \frac{(x-1)^2}{2}.$$

$$\quad u \qquad dv$$

$$\int (\ln x)\,(x-1)\, dx$$

$$\quad\;\; u \qquad v \qquad\qquad v \qquad du$$

$$= (\ln x) \cdot \frac{(x-1)^2}{2} - \int \frac{(x-1)^2}{2} \cdot \frac{1}{x}\, dx$$

$$\text{Integration by Parts}$$

$$= \frac{x^2-2x+1}{2} \ln x - \int \frac{x^2-2x+1}{2x}\, dx$$

$$= \frac{x^2-2x+1}{2} \ln x - \int \left(\frac{x}{2} - 1 + \frac{1}{2x}\right) dx$$

$$= \frac{x^2-2x+1}{2} \ln x - \frac{1}{2}\int x\, dx + \int dx - \frac{1}{2}\int \frac{1}{x}\, dx$$

$$= \frac{x^2-2x+1}{2} \ln x - \frac{1}{2} \cdot \frac{x^2}{2} + x - \frac{1}{2} \ln x + C$$

$$= \left(\frac{x^2-2x+1}{2} - \frac{1}{2}\right) \ln x - \frac{x^2}{4} + x + C$$

$$= \left(\frac{x^2}{2} - x\right) \ln x - \frac{x^2}{4} + x + C$$

19. $\int x\sqrt{x+2}\,dx$

Let
$$u = x \text{ and } dv = \sqrt{x+2}\,dx = (x+2)^{1/2}\,dx.$$

Then
$$du = dx \text{ and } v = \frac{(x+2)^{3/2}}{3/2} = \frac{2}{3}(x+2)^{3/2}.$$

$$\overset{u}{} \quad \overset{dv}{}$$
$$\int x \; \sqrt{x+2}\,dx$$

$$\overset{u}{} \quad \overset{v}{} \qquad \overset{v}{} \quad \overset{du}{}$$
$$= x \cdot \frac{2}{3}(x+2)^{3/2} - \int \frac{2}{3}(x+2)^{3/2}\,dx$$

Integration by Parts

$$= \frac{2}{3}x(x+2)^{3/2} - \frac{2}{3}\int (x+2)^{3/2}\,dx$$

$$= \frac{2}{3}x(x+2)^{3/2} - \frac{2}{3}\cdot\frac{(x+2)^{5/2}}{5/2} + C$$

$$= \frac{2}{3}x(x+2)^{3/2} - \frac{4}{15}(x+2)^{5/2} + C$$

21. $\int x^3 \ln 2x\,dx = \int (\ln 2x)(x^3\,dx)$

Let
$$u = \ln 2x \text{ and } dv = x^3\,dx.$$

Then $du = \frac{1}{x}\,dx$ and $v = \frac{x^4}{4}$.

$$\overset{u}{} \quad \overset{dv}{} \qquad \overset{u}{} \quad \overset{v}{} \qquad \overset{v}{} \quad \overset{du}{}$$
$$\int (\ln 2x)\,(x^3\,dx) = (\ln 2x)\cdot\frac{x^4}{4} - \int \frac{x^4}{4}\cdot\frac{1}{x}\,dx$$

Integration by Parts

$$= \frac{x^4}{4}\ln 2x - \frac{1}{4}\int x^3\,dx$$

$$= \frac{x^4}{4}\ln 2x - \frac{1}{4}\cdot\frac{x^4}{4} + C$$

$$= \frac{x^4}{4}\ln 2x - \frac{x^4}{16} + C$$

23. $\int x^2 e^x\,dx$

Let
$$u = x^2 \text{ and } dv = e^x\,dx.$$

Then $du = 2x\,dx$ and $v = e^x$.

$$\overset{u}{} \quad \overset{dv}{} \qquad \overset{u}{} \overset{v}{} \qquad \overset{v}{} \quad \overset{du}{}$$
$$\int x^2 e^x\,dx = x^2 e^x - \int e^x \cdot 2x\,dx$$

Integration by Parts

$$= x^2 e^x - \int 2xe^x\,dx$$

We evaluate $\int 2xe^x\,dx$ using the Integration by Parts formula.

$\int 2xe^x\,dx$

Let
$$u = 2x \text{ and } dv = e^x\,dx.$$

Then
$$du = 2\,dx \text{ and } v = e^x.$$

$$\overset{u}{} \quad \overset{dv}{} \qquad \overset{u}{} \quad \overset{v}{} \qquad \overset{v}{} \,\, \overset{du}{}$$
$$\int 2x\,e^x\,dx = 2x\cdot e^x - \int 2e^x\,dx$$

$$= 2xe^x - 2e^x + K$$

Thus,
$$\int x^2 e^x\,dx = x^2 e^x - (2xe^x - 2e^x + K)$$

$$= x^2 e^x - 2xe^x + 2e^x + C \quad (C = -K)$$

Since we have an integral $\int f(x)g(x)\,dx$ where $f(x)$, or x^2, can be differentiated repeatedly to a derivative that is eventually 0 and $g(x)$, or e^x, can be integrated repeatedly easily, we can use tabular integration.

$f(x)$ and repeated derivatives		$g(x)$ and repeated integrals
x^2	$+$	e^x
$2x$	$-$	e^x
2	$+$	e^x
0		e^x

We add the products along the arrows, making the alternate sign changes.

$$\int x^2 e^x\,dx = x^2 e^x - 2xe^x + 2e^x + C$$

25. $\int x^2 e^{2x}\,dx$

Let
$$u = x^2 \text{ and } dv = e^{2x}\,dx.$$

Then $du = 2x\,dx$ and $v = \frac{1}{2}e^{2x}$.

$$\overset{u}{} \quad \overset{dv}{} \qquad \overset{u}{} \quad \overset{v}{} \qquad \overset{v}{} \quad \overset{du}{}$$
$$\int x^2 e^{2x}\,dx = x^2 \cdot \frac{1}{2}e^{2x} - \int \frac{1}{2}e^{2x}\cdot 2x\,dx$$

Integration by Parts

$$= \frac{1}{2}x^2 e^{2x} - \int xe^{2x}\,dx$$

We integrate $xe^{2x}\,dx$ using the Integration by Parts formula.

$\int xe^{2x}\,dx$

Let
$$u = x \text{ and } dv = e^{2x}\,dx.$$

Then
$$du = dx \text{ and } v = \frac{1}{2}e^{2x}.$$

$$\overset{u}{} \quad \overset{dv}{} \qquad \overset{u}{} \quad \overset{v}{} \qquad \overset{v}{} \quad \overset{du}{}$$
$$\int x\,e^{2x}\,dx = x\cdot\frac{1}{2}e^{2x} - \int \frac{1}{2}e^{2x}\cdot dx$$

$$= \frac{1}{2}xe^{2x} - \frac{1}{2}\cdot\frac{1}{2}e^{2x} + K$$

$$= \frac{1}{2}xe^{2x} - \frac{1}{4}e^{2x} + K$$

Thus,
$$\int x^2 e^{2x}\,dx = \frac{1}{2}x^2 e^{2x} - \frac{1}{2}xe^{2x} + \frac{1}{4}e^{2x} + C \quad (C = -K)$$

We could also use tabular integration.

$f(x)$ and repeated derivatives	$g(x)$ and repeated integrals
x^2	e^{2x}
$2x$	$\frac{1}{2}e^{2x}$
2	$\frac{1}{4}e^{2x}$
0	$\frac{1}{8}e^{2x}$

(with alternating signs $+$, $-$, $+$)

$$\int x^2\, e^{2x}\, dx$$

$$= x^2 \cdot \frac{1}{2}\, e^{2x} - 2x \cdot \frac{1}{4}e^{2x} + 2 \cdot \frac{1}{8}e^{2x} + C$$

$$= \frac{1}{2}x^2\, e^{2x} - \frac{1}{2}xe^{2x} + \frac{1}{4}e^{2x} + C$$

27. $\int x^3\, e^{-2x}\, dx$

We will use tabular integration.

$f(x)$ and repeated derivatives	$g(x)$ and repeated integrals
x^3	e^{-2x}
$3x^2$	$-\frac{1}{2}e^{-2x}$
$6x$	$\frac{1}{4}e^{-2x}$
6	$-\frac{1}{8}e^{-2x}$
0	$\frac{1}{16}e^{-2x}$

(with alternating signs $+$, $-$, $+$, $-$)

$$\int x^3\, e^{-2x}\, dx$$

$$= x^3\left(-\frac{1}{2}e^{-2x}\right) - 3x^2\left(\frac{1}{4}e^{-2x}\right) + 6x\left(-\frac{1}{8}e^{-2x}\right) -$$

$$6\left(\frac{1}{16}e^{-2x}\right) + C$$

$$= -\frac{1}{2}x^3\, e^{-2x} - \frac{3}{4}x^2\, e^{-2x} - \frac{3}{4}xe^{-2x} - \frac{3}{8}e^{-2x} + C$$

$$= e^{-2x}\left(-\frac{1}{2}x^3 - \frac{3}{4}x^2 - \frac{3}{4}x - \frac{3}{8}\right) + C$$

29. $\int (x^4 + 1)\, e^{3x}\, dx$

We will use tabular integration.

$f(x)$ and repeated derivatives	$g(x)$ and repeated integrals
$x^4 + 1$	e^{3x}
$4x^3$	$\frac{1}{3}e^{3x}$
$12x^2$	$\frac{1}{9}e^{3x}$
$24x$	$\frac{1}{27}e^{3x}$
24	$\frac{1}{81}e^{3x}$
0	$\frac{1}{243}e^{3x}$

(with alternating signs $+$, $-$, $+$, $-$, $+$)

$$\int (x^4 + 1)\, e^{3x}\, dx = (x^4 + 1) \cdot \frac{1}{3}e^{3x} -$$

$$4x^3 \cdot \frac{1}{9}e^{3x} + 12x^2 \cdot \frac{1}{27}e^{3x} - 24x \cdot \frac{1}{81}e^{3x} +$$

$$24 \cdot \frac{1}{243}e^{3x} + C$$

$$= e^{3x}\left[\frac{1}{3}(x^4 + 1) - \frac{4}{9}x^3 + \frac{4}{9}x^2 - \frac{8}{27}x + \frac{8}{81}\right] + C,$$

or $e^{3x}\left[\frac{1}{3}x^4 - \frac{4}{9}x^3 + \frac{4}{9}x^2 - \frac{8}{27}x + \frac{35}{81}\right] + C$

Adding $\frac{1}{3}$ and $\frac{8}{81}$

31. $\int_1^2 x^2\, \ln x\, dx$

In Exercise 9 above we found the indefinite integral.

$$\int x^2\, \ln x\, dx = \frac{x^3}{3}\ln x - \frac{x^3}{9} + C$$

Evaluate the definite integral.

$$\int_1^2 x^2\, \ln x\, dx = \left[\frac{x^3}{3}\ln x - \frac{x^3}{9}\right]_1^2$$

$$= \left(\frac{2^3}{3}\ln 2 - \frac{2^3}{9}\right) - \left(\frac{1^3}{3}\ln 1 - \frac{1^3}{9}\right)$$

$$= \left(\frac{8}{3}\ln 2 - \frac{8}{9}\right) - \left(\frac{1}{3}\ln 1 - \frac{1}{9}\right)$$

$$= \frac{8}{3}\ln 2 - \frac{8}{9} + \frac{1}{9} \qquad (\ln 1 = 0)$$

$$= \frac{8}{3}\ln 2 - \frac{7}{9}$$

33. $\int_2^6 \ln(x + 3)\, dx$

In Exercise 13 above we found the indefinite integral.

$$\int \ln (x + 3)\, dx = (x + 3)\ln (x + 3) - x + C$$

Evaluate the definite integral.

$\int_2^6 \ln x(x+3)\, dx$

$= \left[(x+3)\ln(x+3) - x \right]_2^6$

$= \left[(6+3)\ln(6+3) - 6 \right] - \left[(2+3)\ln(2+3) - 2 \right]$

$= (9\ln 9 - 6) - (5\ln 5 - 2)$

$= 9\ln 9 - 6 - 5\ln 5 + 2$

$= 9\ln 9 - 5\ln 5 - 4$

35. a) We first find the indefinite integral.

$\int xe^x\, dx$

Let

$u = x$ and $dv = e^x\, dx$.

Then

$du = dx$ and $v = e^x$.

$\int xe^x\, dx = xe^x - \int e^x\, dx$

$\qquad\qquad = xe^x - e^x + C$

b) Evaluate the definite integral.

$\int_0^1 xe^x\, dx = \left[xe^x - e^x \right]_0^1$

$\qquad\qquad = (1\cdot e^1 - e^1) - (0\cdot e^0 - e^0)$

$\qquad\qquad = (e - e) - (0 - 1)$

$\qquad\qquad = 0 - (-1)$

$\qquad\qquad = 1$

37. $C(x) = \int 4x\sqrt{x+3}\, dx$

Let

$u = 4x$ and $dv = \sqrt{x+3}\, dx$.

Then

$du = 4\, dx$ and $v = \dfrac{2}{3}(x+3)^{3/2}$.

$C(x) = 4x\cdot\dfrac{2}{3}(x+3)^{3/2} - \int \dfrac{2}{3}(x+3)^{3/2}\cdot 4\, dx$

$\quad = \dfrac{8}{3}x(x+3)^{3/2} - \dfrac{8}{3}\int (x+3)^{3/2}\, dx$

$\quad = \dfrac{8}{3}x(x+3)^{3/2} - \dfrac{8}{3}\cdot\dfrac{2}{5}(x+3)^{5/2} + C$

$\quad = \dfrac{8}{3}x(x+3)^{3/2} - \dfrac{16}{15}(x+3)^{5/2} + C$

Use $C(13) = \$1126.40$ to find C.

$C(13) = \dfrac{8}{3}\cdot 13(13+3)^{3/2} - \dfrac{16}{15}(13+3)^{5/2} + C = 1126.40$

$\dfrac{104}{3}(16)^{3/2} - \dfrac{16}{15}(16)^{5/2} + C = 1126.40$

$\dfrac{104}{3}\cdot 64 - \dfrac{16}{15}\cdot 1024 + C = 1126.40$

$\dfrac{6656}{3} - \dfrac{16,384}{15} + C = 1126.40$

$\dfrac{16,896}{15} + C = 1126.40$

$1126.40 + C = 1126.40$

$C = 0$

$C(x) = \dfrac{8}{3}x(x+3)^{3/2} - \dfrac{16}{15}(x+3)^{5/2}$

39. a) We first find the indefinite integral.

$\int 10te^{-t}\, dt = 10\int te^{-t}\, dt$

Let

$u = t$ and $dv = e^{-t}\, dt$.

Then

$du = dt$ and $v = -e^{-t}$.

$10\int te^{-t}\, dt = 10\left[t(-e^{-t}) - \int -e^{-t}\, dt \right]$

$\qquad\qquad = 10(-te^{-t} + \int e^{-t}\, dt)$

$\qquad\qquad = 10(-te^{-t} - e^{-t} + K)$

$\qquad\qquad = -10te^{-t} - 10e^{-t} + C \quad (C = 10K)$

Then evaluate the definite integral.

$\int_0^T te^{-t}\, dt = \left[-10te^{-t} - 10e^{-t} \right]_0^T$

$\qquad\qquad = (-10Te^{-T} - 10e^{-T}) -$

$\qquad\qquad\qquad (-10\cdot 0\cdot e^{-0} - 10e^{-0})$

$\qquad\qquad = (-10Te^{-T} - 10e^{-T}) - (0 - 10)$

$\qquad\qquad = -10Te^{-T} - 10e^{-T} + 10$

$\qquad\qquad = -10\left[e^{-T}(T+1) - 1 \right]$, or

$\qquad\qquad 10\left[e^{-T}(-T-1) + 1 \right]$

b) Substitute 4 for T.

$\int_0^4 te^{-t}\, dt = -10\left[e^{-4}(4+1) - 1 \right]$

$\qquad\qquad = -50e^{-4} + 10$

$\qquad\qquad \approx -50(0.018316) + 10$

$\qquad\qquad \approx -0.915800 + 10$

$\qquad\qquad \approx 9.084$

41. $\int \sqrt{x}\ln x\, dx = \int (\ln x)x^{1/2}\, dx$

Let

$u = \ln x$ and $dv = x^{1/2}\, dx$.

Then

$du = \dfrac{1}{x}\, dx$ and $v = \dfrac{2}{3}x^{3/2}$.

$\int (\ln x)x^{1/2}\, dx = (\ln x)\left(\dfrac{2}{3}x^{3/2}\right) - \int \dfrac{2}{3}x^{3/2}\cdot\dfrac{1}{x}\, dx$

$\qquad\qquad = \dfrac{2}{3}x^{3/2}\ln x - \dfrac{2}{3}\int x^{1/2}\, dx$

$\qquad\qquad = \dfrac{2}{3}x^{3/2}\ln x - \dfrac{2}{3}\cdot\dfrac{x^{3/2}}{3/2} + C$

$\qquad\qquad = \dfrac{2}{3}x^{3/2}\ln x - \dfrac{4}{9}x^{3/2} + C$, or

$\qquad\qquad = \dfrac{2}{9}x^{3/2}(3\ln x - 2) + C$ Factoring

43. $\int \dfrac{te^t}{(t+1)^2}\, dt = \int te^t(t+1)^{-2}\, dt$

Let

$u = te^t$ $\qquad\qquad$ and $dv = (t+1)^{-2}\, dt$.

Then

$du = (te^t + e^t)\, dt$ and $\quad v = \dfrac{(t+1)^{-1}}{-1}$

$\qquad = e^t(t+1)\, dt$ $\qquad\qquad = -\dfrac{1}{t+1}$.

$\int te^t(t+1)^{-2}\,dt$

$= te^t\left(-\dfrac{1}{t+1}\right) - \int\left(-\dfrac{1}{t+1}\right)e^t(t+1)\,dt$

$= -\dfrac{te^t}{t+1} + \int e^t\,dt$

$= -\dfrac{te^t}{t+1} + e^t + C$

$= e^t\left(1 - \dfrac{t}{t+1}\right) + C$

$= \dfrac{e^t}{t+1} + C$

45. $\displaystyle\int \dfrac{\ln x}{\sqrt{x}}\,dx = \int (\ln x)x^{-1/2}\,dx$

Let

$u = \ln x$ and $dv = x^{-1/2}\,dx.$

Then

$du = \dfrac{1}{x}\,dx$ and $v = 2x^{1/2}.$

$\displaystyle\int (\ln x)x^{-1/2}\,dx = (\ln x)2x^{1/2} - \int 2x^{1/2}\cdot\dfrac{1}{x}\,dx$

$= 2x^{1/2}\ln x - 2\int x^{-1/2}\,dx$

$= 2x^{1/2}\ln x - 2\cdot\dfrac{x^{1/2}}{1/2} + C$

$= 2\sqrt{x}\,\ln x - 4\sqrt{x} + C$

47. $\displaystyle\int \dfrac{13t^2 - 48}{\sqrt[5]{4t+7}}\,dt = \int (13t^2-48)(4t+7)^{-1/5}\,dt$

Use tabular integration.

$f(x)$ and repeated derivatives		$g(x)$ and repeated integrals
$13t^2 - 48$	$+$	$(4t+7)^{-1/5}$
$26t$	$-$	$\dfrac{5}{16}(4t+7)^{4/5}$
26	$+$	$\dfrac{25}{576}(4t+7)^{9/5}$
0		$\dfrac{125}{32,256}(4t+7)^{14/5}$

$\displaystyle\int \dfrac{13t^2-48}{\sqrt[5]{4t+7}}\,dt = (13t^2-48)\left[\dfrac{5}{16}(4t+7)^{4/5}\right] -$

$26t\left[\dfrac{25}{576}(4t+7)^{9/5}\right] + 26\left[\dfrac{125}{32,256}(4t+7)^{14/5}\right] + C$

$= (13t^2-48)\left[\dfrac{5}{16}(4t+7)^{4/5}\right] -$

$\dfrac{325}{288}t(4t+7)^{9/5} + \dfrac{1625}{16,128}(4t+7)^{14/5} + C$

49. $\int x^n\,e^x\,dx$

Let

$u = x^n$ and $dv = e^x\,dx.$

Then

$du = nx^{n-1}\,dx$ and $v = e^x.$

$\int x^n\,e^x\,dx = x^n\,e^x - \int e^x(nx^{n-1})\,dx$

$= x^n\,e^x - n\int x^{n-1}\,e^x\,dx$

51. $\boxed{tw}$

53. Using the fnInt feature on a grapher, we find that $\int_1^{10} x^5\,\ln x\,dx \approx 355,986.$

Exercise Set 5.7

1. $\int xe^{-3x}\,dx$

This integral fits Formula 6 in Table 1.

$\displaystyle\int xe^{ax}\,dx = \dfrac{1}{a^2}\cdot e^{ax}(ax-1) + C$

In our integral $a = -3$, so we have, by the formula,

$\int xe^{-3x}\,dx = \dfrac{1}{(-3)^2}\cdot e^{-3x}(-3x-1) + C$

$= \dfrac{1}{9}e^{-3x}(-3x-1) + C$

or $-\dfrac{1}{9}e^{-3x}(3x+1) + C$

3. $\int 5^x\,dx$

This integral fits Formula 11 in Table 1.

$\displaystyle\int a^x\,dx = \dfrac{a^x}{\ln a} + C,\ a > 0,\ a\ne 1$

In our integral $a = 5$, so we have, by the formula,

$\displaystyle\int 5^x\,dx = \dfrac{5^x}{\ln 5} + C$

5. $\displaystyle\int \dfrac{1}{16 - x^2}\,dx$

This integral fits Formula 15 in Table 1.

$\displaystyle\int \dfrac{1}{a^2 - x^2}\,dx = \dfrac{1}{2a}\ln\left(\dfrac{a+x}{a-x}\right) + C$

In our integral $a = 4$, so we have, by the formula,

$\displaystyle\int \dfrac{1}{16 - x^2}\,dx = \int \dfrac{1}{4^2 - x^2}\,dx$

$= \dfrac{1}{2\cdot 4}\ln\dfrac{4+x}{4-x} + C$

$= \dfrac{1}{8}\ln\dfrac{4+x}{4-x} + C$

7. $\displaystyle\int \dfrac{x}{5-x}\,dx$

This integral fits Formula 18 in Table 1.

$\displaystyle\int \dfrac{x}{ax+b}\,dx = \dfrac{b}{a^2} + \dfrac{x}{a} - \dfrac{b}{a^2}\ln(ax+b) + C$

In our integral $a = -1$ and $b = 5$, so we have, by the formula,

$\displaystyle\int \dfrac{x}{5-x}\,dx = \dfrac{5}{(-1)^2} + \dfrac{x}{(-1)} - \dfrac{5}{(-1)^2}\ln(-1\cdot x+5) + C$

$= 5 - x - 5\ln(5-x) + C$

9. $\int \dfrac{1}{x(5-x)^2}\,dx$

This integral fits Formula 21 in Table 1.

$$\int \frac{1}{x(ax+b)^2}\,dx = \frac{1}{b(ax+b)} + \frac{1}{b^2}\ln\left(\frac{x}{ax+b}\right) + C$$

In our integral $a = -1$ and $b = 5$, so we have, by the formula,

$$\int \frac{1}{x(5-x)^2}\,dx = \int \frac{1}{x(-x+5)^2}\,dx$$
$$= \frac{1}{5(-x+5)} + \frac{1}{5^2}\ln\left(\frac{x}{-x+5}\right) + C$$
$$= \frac{1}{5(5-x)} + \frac{1}{25}\ln\left(\frac{x}{5-x}\right) + C$$

11. $\int \ln 3x\,dx$

$= \int (\ln 3 + \ln x)\,dx$

$= \int \ln 3\,dx + \int \ln x\,dx$

$= (\ln 3)\,x + \int \ln x\,dx$

The integral in the second term fits Formula 8 in Table 1.

$\int \ln x\,dx = x\ln x - x + C$

$\int \ln 3x\,dx = (\ln 3)x + \int \ln x\,dx$

$= (\ln 3)x + x\ln x - x + C$

13. $\int x^4 e^{5x}\,dx$

This integral first Formula 7 in Table 1.

$$\int x^n e^{ax}\,dx = \frac{x^n e^{ax}}{a} - \frac{n}{a}\int x^{n-1} e^{ax}\,dx$$

In our integral $n = 4$ and $a = 5$, so we have, by the formula,

$\int x^4 e^{5x}\,dx$

$= \dfrac{x^4 e^{5x}}{5} - \dfrac{4}{5}\int x^3 e^{5x}\,dx$

In the integral in the second term where $n = 3$ and $a = 5$, we again apply Formula 7.

$= \dfrac{x^4 e^{5x}}{5} - \dfrac{4}{5}\left[\dfrac{x^3 e^{5x}}{5} - \dfrac{3}{5}\int x^2 e^{5x}\,dx\right]$

We continue to apply Formula 7.

$= \dfrac{x^4 e^{5x}}{5} - \dfrac{4}{25}x^3 e^{5x} + \dfrac{12}{25}\left[\dfrac{x^2 e^{5x}}{5} - \dfrac{2}{5}\int x e^{5x}\,dx\right]$

$= \dfrac{x^4 e^{5x}}{5} - \dfrac{4}{25}x^3 e^{5x} + \dfrac{12}{125}x^2 e^{5x} -$

$\qquad \dfrac{24}{125}\left[\dfrac{x e^{5x}}{5} - \dfrac{1}{5}\int x^0 e^{5x}\,dx\right]$

$= \dfrac{x^4 e^{5x}}{5} - \dfrac{4}{25}x^3 e^{5x} + \dfrac{12}{125}x^2 e^{5x} - \dfrac{24}{625}x e^{5x} +$

$\qquad \dfrac{24}{625}\int e^{5x}\,dx$

We now apply Formula 5, $\int e^{ax}\,dx = \dfrac{1}{a}\cdot e^{ax} + C.$

$= \dfrac{x^4 e^{5x}}{5} - \dfrac{4}{25}x^3 e^{5x} + \dfrac{12}{125}x^2 e^{5x} - \dfrac{24}{625}x e^{5x} +$

$\qquad \dfrac{24}{3125}e^{5x} + C$

15. $\int x^3 \ln x\,dx$

This integral fits Formula 10 in Table 1.

$$\int x^n \ln x\,dx = x^{n+1}\left[\frac{\ln x}{n+1} - \frac{1}{(n+1)^2}\right] + C,\ n \neq 1$$

In our integral $n = 3$, so we have, by the formula,

$\int x^3 \ln x\,dx = x^{3+1}\left[\dfrac{\ln x}{3+1} - \dfrac{1}{(3+1)^2}\right] + C$

$= x^4\left[\dfrac{\ln x}{4} - \dfrac{1}{16}\right] + C$

17. $\int \dfrac{dx}{\sqrt{x^2+7}}$

This integral fits Formula 12 in Table 1.

$$\int \frac{1}{\sqrt{x^2+a^2}}\,dx = \ln(x + \sqrt{x^2+a^2}) + C$$

In our integral $a^2 = 7$, so we have, by the formula,

$$\int \frac{dx}{\sqrt{x^2+7}} = \ln\left(x + \sqrt{x^2+7}\right) + C$$

19. $\int \dfrac{10\,dx}{x(5-7x)^2} = 10\int \dfrac{1}{x(-7x+5)^2}\,dx$

This integral fits Formula 21 in Table 1.

$$\int \frac{1}{x(ax+b)^2}\,dx = \frac{1}{b(ax+b)} + \frac{1}{b^2}\ln\left(\frac{x}{ax+b}\right) + C$$

In our integral $a = -7$ and $b = 5$, so we have, by the formula,

$= 10\int \dfrac{1}{x(-7x+5)^2}\,dx$

$= 10\left[\dfrac{1}{5(-7x+5)} + \dfrac{1}{5^2}\ln\left(\dfrac{x}{-7x+5}\right)\right] + C$

$= \dfrac{2}{5-7x} + \dfrac{2}{5}\ln\left(\dfrac{x}{5-7x}\right) + C$

21. $\int \dfrac{-5}{4x^2-1}\,dx = -5\int \dfrac{1}{4x^2-1}\,dx$

This integral almost fits Formula 14 in Table 1.

$$\int \frac{1}{x^2-a^2}\,dx = \frac{1}{2a}\ln\left(\frac{x-a}{x+a}\right) + C$$

But the x^2 coefficient needs to be 1. We factor out 4 as follows. Then we apply Formula 14.

$-5\int \dfrac{1}{4x^2-1}\,dx = -5\int \dfrac{1}{4\left(x^2-\frac{1}{4}\right)}\,dx$

$= -\dfrac{5}{4}\int \dfrac{1}{x^2-\frac{1}{4}}\,dx \quad \left(a^2 = \dfrac{1}{4},\ a = \dfrac{1}{2}\right)$

$= -\dfrac{5}{4}\left[\dfrac{1}{2\cdot\frac{1}{2}}\ln\left(\dfrac{x-\frac{1}{2}}{x+\frac{1}{2}}\right)\right] + C$

$= -\dfrac{5}{4}\ln\left(\dfrac{x-1/2}{x+1/2}\right) + C$

23. $\int \sqrt{4m^2+16}\,dm$

This integral almost fits Formula 22 in Table 1.

$\int \sqrt{x^2+a^2}\,dx$

$= \dfrac{1}{2}\left[x\sqrt{x^2+a^2} + a^2\ln\left(x + \sqrt{x^2+a^2}\right)\right] + C$

But the x^2 coefficient needs to be 1. We factor out 4 as follows. Then we apply Formula 22.

$$\int \sqrt{4m^2 + 16}\, dm$$
$$= \int \sqrt{4(m^2 + 4)}\, dm$$
$$= 2 \int \sqrt{m^2 + 4}\, dm$$
$$= 2 \cdot \frac{1}{2}\left[m\sqrt{m^2 + 4} + 4 \ln\left(m + \sqrt{m^2 + 4}\right)\right] + C$$
$$= m\sqrt{m^2 + 4} + 4 \ln\left(m + \sqrt{m^2 + 4}\right) + C$$

25. $\int \dfrac{-5 \ln x}{x^3}\, dx = -5 \int x^{-3} \ln x\, dx$

This integral fits Formula 10 in Table 1.

$$\int x^n \ln x\, dx = x^{n+1}\left[\frac{\ln x}{n+1} - \frac{1}{(n+1)^2}\right] + C, \; n \neq -1$$

In our integral $n = -3$, so we have, by the formula,

$$-5 \int x^{-3} \ln x\, dx$$
$$= -5\left[x^{-3+1}\left(\frac{\ln x}{-3+1} - \frac{1}{(-3+1)^2}\right)\right] + C$$
$$= -5\left[x^{-2}\left(\frac{\ln x}{-2} - \frac{1}{4}\right)\right] + C$$
$$= \frac{5 \ln x}{2x^2} + \frac{5}{4x^2} + C$$

27. $\int \dfrac{e^x}{x^{-3}}\, dx = \int x^3\, e^x\, dx$

This integral fits Formula 7 in Table 1.

$$\int x^n\, e^{ax}\, dx = \frac{x^n\, e^{ax}}{a} - \frac{n}{a} \int x^{n-1}\, e^{ax}\, dx$$

In our integral $n = 3$ and $a = 1$, so we have, by the formula,

$$\int x^3\, e^x\, dx$$
$$= x^3\, e^x - 3 \int x^2\, e^x\, dx$$

We continue to apply Formula 7.

$$= x^3\, e^x - 3\left(x^2\, e^x - 2 \int x\, e^x\, dx\right) \quad (n = 2,\; a = 1)$$
$$= x^3\, e^x - 3x^2\, e^x + 6 \int x\, e^x\, dx$$
$$= x^3\, e^x - 3x^2\, e^x + 6\left(x\, e^x - \int x^0\, e^x\, dx\right) \quad (n = 1,$$
$$\hspace{9cm} a = 1)$$
$$= x^3\, e^x - 3x^2\, e^x + 6x\, e^x - 6 \int e^x\, dx$$
$$= x^3\, e^x - 3x^2\, e^x + 6x\, e^x - 6\, e^x + C$$

29. $S(p) = \int \dfrac{100p}{(20-p)^2}\, dp, \; 0 \leq p \leq 19$

$$= 100 \int \frac{p}{(20-p)^2}\, dp$$

This integral fits formula 19 in Table 1.

$$\int \frac{x}{(ax+b)^2}\, dx = \frac{b}{a^2(ax+b)} + \frac{1}{a^2}\, \ln(ax+b) + C$$

In our integral $a = -1$ and $b = 20$.

$$100 \int \frac{p}{(20-p)^2}$$
$$= 100\left[\frac{20}{(-1)^2(-1 \cdot p + 20)} + \frac{1}{(-1)^2}\, \ln(-1 \cdot p + 20)\right] + C$$
$$= 100\left[\frac{20}{20-p} + \ln(20-p)\right] + C$$

Use $S(19) = 2000$ to find C.

$$S(19) = 100\left[\frac{20}{20-19} + \ln(20-19)\right] + C = 2000$$
$$100\left(\frac{20}{1} + \ln 1\right) + C = 2000$$
$$100(20 + 0) + C = 2000$$
$$2000 + C = 2000$$
$$C = 0$$

$$S(p) = 100\left[\frac{20}{20-p} + \ln(20-p)\right]$$

31. $\int \dfrac{8}{3x^2 - 2x}\, dx = 8 \int \dfrac{1}{x(3x-2)}\, dx$

This integral fits Formula 20 in Table 1.

$$\int \frac{1}{x(ax+b)}\, dx = \frac{1}{a}\, \ln\left(\frac{x}{ax+b}\right) + C$$

In our integral $a = 3$ and $b = -2$, so we have, by the formula,

$$8 \int \frac{1}{x(3x-2)}\, dx = 8\left[\frac{1}{-2}\, \ln\left(\frac{x}{3x-2}\right)\right] + C$$
$$= -4 \ln\left(\frac{x}{3x-2}\right) + C$$

33. $\int \dfrac{dx}{x^3 - 4x^2 + 4x} = \int \dfrac{1}{x(x^2 - 4x + 4)}\, dx =$

$$\int \frac{1}{x(x-2)^2}\, dx$$

This integral fits Formula 21 in Table 1.

$$\int \frac{1}{x(ax+b)^2}\, dx = \frac{1}{b(ax+b)} + \frac{1}{b^2}\, \ln\left(\frac{x}{ax+b}\right) + C$$

In our integral $a = 1$ and $b = -2$, so we have, by the formula,

$$\int \frac{1}{x(x-2)^2}\, dx = \frac{1}{-2(x-2)} + \frac{1}{(-2)^2}\, \ln\left(\frac{x}{x-2}\right) + C$$
$$= -\frac{1}{2(x-2)} + \frac{1}{4}\, \ln\left(\frac{x}{x-2}\right) + C$$

35. $\int \dfrac{-e^{-2x}\, dx}{9 - 6e^{-x} + e^{-2x}} = \int \dfrac{e^{-x}(-e^{-x})\, dx}{(e^{-x} - 3)^2}$

This integral fits formula 19 in Table 1.

$$\int \frac{x}{(ax+b)^2}\, dx = \frac{b}{a^2(ax+b)} + \frac{1}{a^2}\, \ln(ax+b) + C$$

In our integral x is represented by e^{-x}, dx is represented by $-e^{-x}\, dx$, $a = 1$, and $b = -3$.

$$\int \frac{e^{-x}(-e^{-x})\, dx}{(e^{-x} - 3)^2}$$
$$= \frac{-3}{1^2(1 \cdot e^{-x} - 3)} + \frac{1}{1^2}\, \ln(1 \cdot e^{-x} - 3) + C$$
$$= \frac{-3}{e^{-x} - 3} + \ln(e^{-x} - 3) + C$$

Chapter 6

Applications of Integration

1. $D(x) = -\dfrac{5}{6}x + 10$

$S(x) = \dfrac{1}{2}x + 2$

a) To find the equilibrium point we set $D(x) = S(x)$ and solve.

$$-\dfrac{5}{6}x + 10 = \dfrac{1}{2}x + 2$$

$$10 - 2 = \dfrac{1}{2}x + \dfrac{5}{6}x$$

$$8 = \dfrac{4}{3}x \qquad \left(\dfrac{1}{2} + \dfrac{5}{6} = \dfrac{8}{6} = \dfrac{4}{3}\right)$$

$$\dfrac{3}{4} \cdot 8 = \dfrac{3}{4} \cdot \dfrac{4}{3}x \qquad \text{Multiplying by } \dfrac{3}{4}$$

$$6 = x$$

Thus $x_E = 6$ units. To find p_E we substitute x_E into $D(x)$ or $S(x)$. Here we use $D(x)$.

$$p_E = D(x_E) = D(6) = -\dfrac{5}{6} \cdot 6 + 10$$

$$= -5 + 10$$

$$= \$5 \text{ per unit}$$

Thus the equilibrium point is $(6, \$5)$.

b) The consumer's surplus is

$$\int_0^{x_E} D(x)\, dx - x_E\, p_E,$$

or

$$\int_0^6 \left(-\dfrac{5}{6}x + 10\right) dx - 6 \cdot 5$$

$$\qquad \text{Substituting } \left(-\dfrac{5}{6}x + 10\right) \text{ for } D(x),$$
$$\qquad 6 \text{ for } x_E, \text{ and } 5 \text{ for } p_E$$

$$= \left[-\dfrac{5x^2}{12} + 10x\right]_0^6 - 30$$

$$= \left[\left(-\dfrac{5 \cdot 6^2}{12} + 10 \cdot 6\right) - \left(-\dfrac{5 \cdot 0^2}{12} + 10 \cdot 0\right)\right] - 30$$

$$= (-15 + 60) - 30$$

$$= \$15$$

c) The producer's surplus is

$$x_E\, p_E - \int_0^{x_E} S(x)\, dx,$$

or

$$6 \cdot 5 - \int_0^6 \left(\dfrac{1}{2}x + 2\right) dx$$

$$\qquad \text{Substituting } \left(\dfrac{1}{2}x + 2\right) \text{ for } S(x),$$
$$\qquad 6 \text{ for } x_E, \text{ and } 5 \text{ for } p_E$$

$$= 30 - \left[\dfrac{x^2}{4} + 2x\right]_0^6$$

$$= 30 - \left[\left(\dfrac{6^2}{4} + 2 \cdot 6\right) - \left(\dfrac{0^2}{4} + 2 \cdot 0\right)\right]$$

$$= 30 - (9 + 12)$$

$$= 30 - 21$$

$$= \$9$$

3. $D(x) = (x - 4)^2$

$S(x) = x^2 + 2x + 6$

a) To find the equilibrium point we set $D(x) = S(x)$ and solve.

$$(x - 4)^2 = x^2 + 2x + 6$$

$$x^2 - 8x + 16 = x^2 + 2x + 6$$

$$-8x + 16 = 2x + 6$$

$$10 = 10x$$

$$1 = x$$

Thus $x_E = 1$ unit. To find p_E we substitute x_E into $D(x)$ or $S(x)$. Here we use $D(x)$.

$$p_E = D(x_E) = D(1) = (1 - 4)^2 = (-3)^2$$

$$= \$9 \text{ per unit}$$

Thus the equilibrium point is $(1, \$9)$.

b) The consumer's surplus is

$$\int_0^{x_E} D(x)\, dx - x_E\, p_E,$$

or

$$= \int_0^1 (x - 4)^2\, dx - 1 \cdot 9 \quad \text{Substituting } (x-4)^2$$
$$\qquad\qquad\qquad\qquad \text{for } D(x), 1 \text{ for } x_E$$
$$\qquad\qquad\qquad\qquad \text{and } 9 \text{ for } p_E$$

$$= \int_0^1 (x^2 - 8x + 16)\, dx - 9$$

$$= \left[\dfrac{x^3}{3} - 4x^2 + 16x\right]_0^1 - 9$$

$$= \left[\left(\dfrac{1}{3} - 4 + 16\right) - (0 - 0 + 0)\right] - 9$$

$$= 12\dfrac{1}{3} - 9$$

$$= 3\dfrac{1}{3}$$

$$= \$3.33$$

c) The producer's surplus is

$$x_E\, p_E - \int_0^{x_E} S(x)\, dx$$

or

$$= 1 \cdot 9 - \int_0^1 (x^2 + 2x + 6)\, dx$$
$$\text{Substituting } x^2 + 2x + 6 \text{ for } S(x),$$
$$1 \text{ for } x_E \text{ and } 9 \text{ for } p_E$$

$$= 9 - \left[\frac{x^3}{3} + x^2 + 6x\right]_0^1$$

$$= 9 - \left[\left(\frac{1}{3} + 1 + 6\right) - (0 + 0 + 0)\right]$$

$$= 9 - 7\frac{1}{3}$$

$$= 1\frac{2}{3}$$

$$= \$1.67$$

5. $D(x) = (x - 6)^2$

$S(x) = x^2$

a) To find the equilibrium point we set $D(x) = S(x)$ and solve.

$$(x - 6)^2 = x^2$$
$$x^2 - 12x + 36 = x^2$$
$$-12x + 36 = 0$$
$$36 = 12x$$
$$3 = x$$

Thus $x_E = 3$ units. To find p_E we substitute x_E into $D(x)$ or $S(x)$. Here we use $S(x)$.

$$p_E = S(x_E) = S(3) = 3^2 = \$9 \text{ per unit}$$

Thus the equilibrium point is $(3, \$9)$.

b) The consumer's surplus is

$$\int_0^{x_E} D(x)\, dx - x_E\, p_E,$$

or

$$= \int_0^3 (x - 6)^2\, dx - 3 \cdot 9$$
$$\text{Substituting } (x - 6)^2 \text{ for } D(x),$$
$$3 \text{ for } x_E, \text{ and } 9 \text{ for } p_E$$

$$= \int_0^3 (x^2 - 12x + 36)\, dx - 27$$

$$= \left[\frac{x^3}{3} - 6x^2 + 36x\right]_0^3 - 27$$

$$= \left[\left(\frac{3^3}{3} - 6 \cdot 3^2 + 36 \cdot 3\right) - \left(\frac{0^3}{3} - 6 \cdot 0^2 + 36 \cdot 0\right)\right] - 27$$

$$= 9 - 54 + 108 - 27$$

$$= \$36$$

c) The producer's surplus is

$$x_E\, p_E - \int_0^{x_E} S(x)\, dx$$

or

$$3 \cdot 9 - \int_0^3 x^2\, dx \quad \text{Substituting } x^2 \text{ for } S(x),$$
$$3 \text{ for } x_E, \text{ and } 9 \text{ for } p_E$$

$$= 27 - \left[\frac{x^3}{3}\right]_0^3$$

$$= 27 - \left(\frac{3^3}{3} - \frac{0^3}{3}\right)$$

$$= 27 - 9$$

$$= \$18$$

7. $D(x) = 1000 - 10x$

$S(x) = 250 + 5x$

a) To find the equilibrium point we set $D(x) = S(x)$ and solve.

$$1000 - 10x = 250 + 5x$$
$$750 = 15x$$
$$50 = x$$

Thus $x_E = 50$ units. To find p_E we substitute x_E into $D(x)$ or $S(x)$. Here we use $D(x)$.

$$p_E = D(x_E) = 1000 - 10 \cdot 50$$
$$= 1000 - 500$$
$$= \$500 \text{ per unit}$$

Thus the equilibrium point is $(50, \$500)$.

b) The consumer's surplus is

$$\int_0^{x_E} D(x)\, dx - x_E\, p_E$$

or

$$\int_0^{50} (1000 - 10x)\, dx - 50 \cdot 500$$

$$= \left[1000x - 5x^2\right]_0^{50} - 25,000$$

$$= [(1000 \cdot 50 - 5(50)^2) - (1000 \cdot 0 - 5(0)^2)] - 25,000$$

$$= 50,000 - 12,500 - 0 - 25,000$$

$$= \$12,500$$

c) The producer's surplus is

$$x_E\, p_E - \int_0^{x_E} S(x)\, dx$$

or

$$= 50 \cdot 500 - \int_0^{50} (250 + 5x)\, dx$$

$$= 25,000 - \left[250x + \frac{5x^2}{2}\right]_0^{50}$$

$$= 25,000 - \left[\left(250 \cdot 50 + \frac{5(50)^2}{2}\right) - \left(250 \cdot 0 + \frac{5(0)^2}{2}\right)\right]$$

$$= 25,000 - 12,500 - 6250 + 0$$

$$= \$6250$$

9. $D(x) = 5 - x$, $0 \le x \le 5$

$S(x) = \sqrt{x + 7}$

a) To find the equilibrium point we set $D(x) = S(x)$ and solve.

$$5 - x = \sqrt{x + 7}$$

$$(5 - x)^2 = (\sqrt{x + 7})^2$$

$$25 - 10x + x^2 = x + 7$$

$$x^2 - 11x + 18 = 0$$

$$(x - 2)(x - 9) = 0$$

$$x - 2 = 0 \text{ or } x - 9 = 0$$

$$x = 2 \text{ or } \qquad x = 9$$

Only 2 is in the domain of $D(x)$, so $x_E = 2$ units. To find p_E we substitute x_E into $D(x)$ or $S(x)$. Here we use $D(x)$.

$$p_E = D(x_E) = 5 - 2 = \$3$$

Thus the equilibrium point is $(2, \$3)$.

b) The consumer's surplus is

$$\int_0^{x_E} D(x)\, dx - x_E\, p_E$$

or

$$\int_0^2 (5 - x)\, dx - 2 \cdot 3$$

$$= \left[5x - \frac{x^2}{2} \right]_0^2 - 6$$

$$= \left[\left(5 \cdot 2 - \frac{2^2}{2} \right) - \left(5 \cdot 0 - \frac{0^2}{2} \right) \right] - 6$$

$$= 8 - 0 - 6$$

$$= \$2$$

c) The producer's surplus is

$$x_E\, p_E - \int_0^{x_E} S(x)\, dx$$

or

$$2 \cdot 3 - \int_0^2 \sqrt{x + 7}\, dx$$

$$= 6 - \left[\frac{2}{3} (x + 7)^{3/2} \right]_0^2$$

$$= 6 - \left[\frac{2}{3} (2 + 7)^{3/2} - \frac{2}{3} (0 + 7)^{3/2} \right]$$

$$= 6 - \frac{2}{3} \cdot 27 + \frac{2}{3} \cdot 7^{3/2}$$

$$= 6 - 18 + \frac{2}{3} \cdot 7^{3/2}$$

$$\approx \$0.35 \qquad \text{Using a calculator}$$

11. $D(x) = e^{-x+4.5}$

$S(x) = e^{x-5.5}$

a) To find the equilibrium point we set $D(x) = S(x)$ and solve.

$$e^{-x+4.5} = e^{x-5.5}$$

$$\ln e^{-x+4.5} = \ln e^{x-5.5}$$

$$-x + 4.5 = x - 5.5$$

$$10 = 2x$$

$$5 = x$$

Thus $x_E = 5$ units. To find p_E we substitute x_E into $D(x)$ or $S(x)$. Here we use $S(x)$.

$$p_E = S(x_E) = S(5) = e^{5-5.5} = e^{-0.5} \approx \$0.61$$

Thus the equilibrium point is $(5, \$0.61)$.

b) The consumer's surplus is

$$\int_0^{x_E} D(x)\, dx - x_E\, p_E$$

or

$$\int_0^5 (e^{-x+4.5})\, dx - 5(0.61)$$

$$= \left[-e^{-x+4.5} \right]_0^5 - 3.05$$

$$= \left[(-e^{-5+4.5}) - (-e^{-0+4.5}) \right] - 3.05$$

$$= (-e^{-0.5} + e^{4.5}) - 3.05$$

$$\approx -0.606531 + 90.017131 - 3.05$$

$$\approx \$86.36$$

c) The producer's surplus is

$$x_E\, p_E - \int_0^{x_E} S(x)\, dx$$

or

$$5(0.61) - \int_0^5 e^{x-5.5}\, dx$$

$$= 3.05 - \left[e^{x-5.5} \right]_0^5$$

$$= 3.05 - (e^{5-5.5} - e^{0-5.5})$$

$$= 3.05 - (e^{-0.5} - e^{-5.5})$$

$$\approx 3.05 - (0.606531 - 0.004087)$$

$$\approx 3.05 - 0.602444$$

$$\approx \$2.45$$

13. $\boxed{tw}$

15. a) $(1, \$9)$

b)

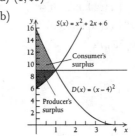

c) $\int_0^1 (x - 4)^2\, dx - 1 \cdot 9 \approx 12.33 - 9 \approx \$3.33$

d) $9 \cdot 1 - \int_0^1 (x^2 + 2x + 6)\, dx \approx 9 - 7.33 \approx \$1.67$

17. a) A scatterplot shows that a linear function appears to fit the data.

b) $y = -2.5x + 22.5$

c) $\int_0^6 (-2.5x + 22.5)\, dx - 6(7.50) = \$45$

d) First find the value of x when $y = \$11.50$.

$$11.5 = -2.5x + 22.5$$
$$-11 = -2.5x$$
$$4.4 = x$$

Now find the consumer surplus.

$$\int_0^{4.4} (-2.5x + 22.5)\, dx - 4.4(11.50) = \$24.20$$

Exercise Set 6.2

1. $P(t) = P_0\, e^{kt}$

$P(3) = 100\, e^{0.09(3)}$ Substituting 3 for t, 100 for P_0, and 0.09 for k

$= 100\, e^{0.27}$

$\approx 100(1.309964)$

$\approx \$131.00$

3. $\int_0^T P_0\, e^{kt}\, dt = \dfrac{P_0}{k}(e^{kT} - 1)$

$\int_0^{20} 100\, e^{0.09t}\, dt = \dfrac{100}{0.09}(e^{0.09(20)} - 1)$

     Substituting 100 for P_0, 20 for T, and 0.09 for k

$= 1111.111(e^{1.8} - 1)$

$\approx 1111.111(6.049647 - 1)$

$\approx 1111.111(5.049647)$

$\approx \$5610.72$

5. $\int_0^T P_0\, e^{kt}\, dt = \dfrac{P_0}{k}(e^{kT} - 1)$

$\int_0^{40} 1000\, e^{0.085t}\, dt = \dfrac{1000}{0.085}(e^{0.085(40)} - 1)$

     Substituting 1000 for P_0, 40 for T, and 0.085 for k

$= 11,764.71(e^{3.4} - 1)$

$\approx 11,764.71(29.964100 - 1)$

$\approx 11,764.71(28.964100)$

$\approx \$340,754.12$

7. $50,000 = \int_0^{20} P_0\, e^{0.085t}\, dt$

$50,000 = \dfrac{P_0}{0.085}(e^{0.085(20)} - 1)$

$4250 = P_0(e^{1.7} - 1)$

$4250 \approx P_0(5.473547 - 1)$

$4250 \approx P_0(4.473947)$

$\dfrac{4250}{4.473947} \approx P_0$

$\$949.94 \approx P_0$

9. $40,000 = \int_0^{30} P_0\, e^{0.09t}\, dt$

$40,000 = \dfrac{P_0}{0.09}(e^{0.09(30)} - 1)$

$3600 = P_0(e^{2.7} - 1)$

$3600 = P_0(14.879732 - 1)$

$3600 \approx P_0(13.879732)$

$\dfrac{3600}{13.879732} \approx P_0$

$\$259.37 \approx P_0$

11. $P_0 = Pe^{-kt}$

$P_0 = 50,000\, e^{-0.09(20)}$ Substituting 50,000 for P, 0.09 for k and 20 for t

$P_0 = 50,000\, e^{-1.8}$

$P_0 \approx 50,000(0.165299)$

$P_0 \approx \$8264.94$

13. $P_0 = Pe^{-kt}$

$P_0 = 60,000\, e^{-0.088(8)}$ Substituting 60,000 for P, 0.088 for k and 8 for t

$P_0 = 60,000\, e^{-0.704}$

$P_0 \approx \$29,676.18$

15. $\begin{aligned} \text{Accumulated} \\ \text{present} \\ \text{value} \end{aligned} = \int_0^T P\, e^{-kt}\, dt = \dfrac{P}{k}(1 - e^{-kT})$

$\int_0^{10} 2700\, e^{-0.09t}\, dt = \dfrac{2700}{0.09}\left(1 - e^{-0.09(10)}\right)$

$= 30,000(1 - e^{-0.9})$

$\approx 30,000(1 - 0.406570)$

     Using a calculator

$\approx 30,000(0.593430)$

$\approx \$17,802.90$

Thus the accumulated present value is \$17,802.90. Note that if we had waited to round until after we had multiplied by 30,000, the answer would have been \$17,802.91.

17. $\int_0^T P\, e^{-kt}\, dt = \dfrac{P}{k}(1 - e^{-kt})$

$\int_0^{30} 85,000\, e^{-0.08t}\, dt = \dfrac{85,000}{0.08}(1 - e^{-0.08(30)})$

$= 1,062,500(1 - e^{-2.4})$

$\approx 1,062,500(1 - 0.090718)$

     Using a calculator

$\approx 1,062,500(0.909282)$

$\approx \$966,112.13$

(Answers may vary slightly due to rounding differences.)

19.
$$\int_0^T P_0 \, e^{kt} \, dt = \frac{P_0}{k}(e^{kT} - 1)$$

$$\int_0^{14} 101,000,000 \, e^{0.12t} \, dt = \frac{101,000,000}{0.12}(e^{0.12(14)} - 1)$$

$$= 841,666,666.7(e^{1.68} - 1)$$

$$= 841,666,666.7(5.365556 - 1)$$

$$= 841,666,666.7(4.365556)$$

$$\approx 3,674,343,000$$

Thus from 1990 to 2004 the world will use approximately 3,674,343,000 tons of aluminum ore.

21.
$$75,000,000,000 = \frac{101,000,000}{0.12}(e^{0.12T} - 1)$$

$$75,000,000,000 \approx 841,666,666.7(e^{0.12T} - 1)$$

$$\frac{75,000,000,000}{841,666,666.7} \approx e^{0.12T} - 1$$

$$89.108911 \approx e^{0.12T} - 1$$

$$90.108911 \approx e^{0.12T}$$

$$\ln 90.108911 \approx \ln e^{0.12T}$$

$$\ln 90.108911 \approx 0.12T$$

$$\frac{\ln 90.108911}{0.12} \approx T$$

$$\frac{4.501019}{0.12} \approx T$$

$$37.5 \approx T$$

$$38 \approx T$$

Thus in 38 years from 1990 the world reserves of aluminum ore will be exhausted.

23.
$$\int_0^T P \, e^{-kt} \, dt = \frac{P}{k}(1 - e^{-kT})$$

$$\int_0^{20} 1 \cdot e^{-0.00003t} \, dt = \frac{1}{0.00003}\left(1 - e^{-0.00003(20)}\right)$$

$$= 33,333.33(1 - e^{-0.0006})$$

$$\approx 33,333.33(1 - 0.999400)$$

$$\approx 33,333.33(0.000600)$$

$$\approx 19.994$$

The total amount of radioactive buildup is about 19.994 lb.

25.
$$\int_0^T P_0 \, e^{kt} \, dt = \frac{P_0}{k}(e^{kT} - 1)$$

$$\int_0^{10} 6570 \, e^{0.1t} \, dt = \frac{6570}{0.1}(e^{0.1(10)} - 1)$$

$$= 65,700(e - 1)$$

$$= 65,700(2.718282 - 1)$$

$$= 65,700(1.718282)$$

$$\approx 112,891.1$$

The world will use approximately 112,891.1 million barrels of oil from 1990 to 2000.

27.
$$\int_0^T R(t) \, e^{k(T-t)} \, dt$$

$$= \int_0^{30} (2000t + 7) \, e^{0.08(30-t)} \, dt$$

$$= \int_0^{30} 2000t \, e^{0.08(30-t)} \, dt + \int_0^{30} 7 \, e^{0.08(30-t)} \, dt$$

$$= 2000 \int_0^{30} t \, e^{0.08(30-t)} \, dt + 7 \int_0^{30} e^{0.08(30-t)} \, dt$$

We will use integration by parts to evaluate the first integral.

$$\int t \, e^{0.08(30-t)} \, dt$$

Let

$$u = t \text{ and } dv = e^{0.08(30-t)}.$$

Then

$$du = dt \text{ and } v = -\frac{1}{0.08} e^{0.08(30-t)}.$$

$$\int t \, e^{0.08(30-t)} \, dt$$

$$= -\frac{1}{0.08} t \, e^{0.08(30-t)} - \int -\frac{1}{0.08} \, e^{0.08(30-t)} \, dt$$

$$= -\frac{1}{0.08} t \, e^{0.08(30-t)} - \frac{1}{0.08} \cdot \frac{1}{0.08} \, e^{0.08(30-t)} + C$$

$$= -12.5t \, e^{0.08(30-t)} - 156.25 \, e^{0.08(30-t)} + C$$

Find the definite integral.

$$\int_0^{30} t \, e^{0.08(30-t)} \, dt$$

$$= \left[-12.5t \, e^{0.08(30-t)} - 156.25 \, e^{0.08(30-t)} \right]_0^{30}$$

$$= [(-12.5(30) \, e^{0.08(30-30)} - 156.25 \, e^{0.08(30-30)} -$$

$$(-12.5(0) \, e^{0.08(30-0)} - 156.25 \, e^{0.08(30-0)}]$$

$$= (-375 \, e^0 - 156.25 \, e^0) - (0 - 156.25 \, e^{2.4})$$

$$\approx -375 - 156.25 - 0 + 1722.37$$

$$\approx 1191.12$$

From the process of integration by parts we know

$$\int e^{0.08(30-t)} \, dt = -\frac{1}{0.08} e^{0.08(30-t)} + C$$

$$= -12.5e^{0.08(30-t)} + C$$

Find the definite integral.

$$\int_0^{30} e^{0.08(30-t)} \, dt$$

$$= \left[-12.5e^{0.08(30-t)} \right]_0^{30}$$

$$= [(-12.5 \, e^{0.08(30-30)} - (-12.5 \, e^{0.08(30-0)}]$$

$$= (12.5 \, e^0 + 12.5 \, e^{2.4})$$

$$\approx -12.5 + 137.79$$

$$\approx 125.29$$

Then

$$2000 \int_0^{30} t \, e^{0.08(30-t)} \, dt + 7 \int_0^{30} e^{0.08(30-t)} \, dt$$

$$\approx 2000(1191.12) + 7(125.29)$$

$$\approx \$2,383,117$$

If we had done all the calculations before rounding, the result would have been \$2,383,120.

29.
$$\int_0^T R(t) \, e^{-k(T-t)} \, dt$$

$$= \int_0^{20} t \, e^{-0.08(20-t)} \, dt$$

Use integration by parts.

Let

$$u = t \text{ and } dv = e^{-0.08(20-t)} \, dt.$$

Then

$$du = dt \text{ and } v = \frac{1}{0.08} e^{-0.08(20-t)}.$$

$$= \frac{1}{0.08} t\, e^{-0.08(20-t)} - \int \frac{1}{0.08}\, e^{-0.08(20-t)}\, dt$$

$$= \frac{1}{0.08} t\, e^{-0.08(20-t)} - \frac{1}{0.08} \cdot \frac{1}{0.08}\, e^{-0.08(20-t)} + C$$

$$= 12.5t\, e^{-0.08(20-t)} - 156.25\, e^{-0.08(20-t)} + C$$

Find the definite integral.

$$\int_0^{20} t\, c^{-0.08(20-t)}\, dt$$

$$= \left[12.5t\, e^{-0.08(20-t)} - 156.25\, e^{-0.08(20-t)} \right]_0^{20}$$

$$= [(12.5(20)\, e^{-0.08(20-20)} - 156.25\, e^{-0.08(20-20)}) -$$
$$(12.5(0)\, e^{-0.08(20-0)} - 156.25\, e^{-0.08(20-0)})]$$

$$= (250\, e^0 - 156.25\, e^0) - (0 - 156.25\, e^{-1.6})$$

$$\approx 250 - 156.25 - 0 + 31.55$$

$$\approx \$125.30$$

31. $\boxed{tw}$

33. The area and the amount of the continuous money flow are the same. Both are 5610.72.

Exercise Set 6.3

1. $\displaystyle \int_3^\infty \frac{dx}{x^2}$

$$= \lim_{b \to \infty} \int_3^b x^{-2}\, dx$$

$$= \lim_{b \to \infty} \left[\frac{x^{-1}}{-1} \right]_3^b$$

$$= \lim_{b \to \infty} \left[-\frac{1}{x} \right]_3^b$$

$$= \lim_{b \to \infty} \left(-\frac{1}{b} - \left(-\frac{1}{3} \right) \right)$$

$$= \frac{1}{3} \quad \left(\text{As } b \to \infty,\ -\frac{1}{b} \to 0 \text{ and } -\frac{1}{b} + \frac{1}{3} \to \frac{1}{3}. \right)$$

The limit does exist. Thus the improper integral is convergent.

3. $\displaystyle \int_3^\infty \frac{dx}{x} = \lim_{b \to \infty} \int_3^b \frac{1}{x}\, dx$

$$= \lim_{b \to \infty} \left[\ln x \right]_3^b$$

$$= \lim_{b \to \infty} (\ln b - \ln 3)$$

Note that $\ln b$ increases indefinitely as b increases. Therefore, the limit does not exist. If the limit does not exist, we say the improper integral is divergent.

5. $\displaystyle \int_0^\infty e^{-3x}\, dx$

$$= \lim_{b \to \infty} \int_0^b 3\, e^{-3x}\, dx$$

$$= \lim_{b \to \infty} \left[\frac{3}{-3} e^{-3x} \right]_0^b$$

$$= \lim_{b \to \infty} \left[-e^{-3x} \right]_0^b$$

$$= \lim_{b \to \infty} [-e^{-3b} - (-e^{-3 \cdot 0})]$$

$$= \lim_{b \to \infty} (-e^{-3b} + 1)$$

$$= \lim_{b \to \infty} \left(1 - \frac{1}{e^{3b}} \right)$$

$$= 1 \quad \left[\text{As } b \to \infty,\ e^{3b} \to \infty, \text{ so } \frac{1}{e^{3b}} \to 0 \text{ and} \right.$$
$$\left. \left(1 - \frac{1}{e^{3b}} \right) \to 1. \right]$$

The limit does exist. Thus the improper integral is convergent.

7. $\displaystyle \int_1^\infty \frac{dx}{x^3}$

$$= \lim_{b \to \infty} \int_1^b x^{-3}\, dx$$

$$= \lim_{b \to \infty} \left[\frac{x^{-2}}{-2} \right]_1^b$$

$$= \lim_{b \to \infty} \left[-\frac{1}{2x^2} \right]_1^b$$

$$= \lim_{b \to \infty} \left[-\frac{1}{2b^2} - \left(-\frac{1}{2 \cdot 1^2} \right) \right]$$

$$= \lim_{b \to \infty} \left(-\frac{1}{2b^2} + \frac{1}{2} \right)$$

$$= \frac{1}{2} \quad \left(\text{As } b \to \infty,\ 2b^2 \to \infty, \text{ so } -\frac{1}{2b^2} \to 0. \right)$$

The limit does exist. Thus the improper integral is convergent.

9. $\displaystyle \int_0^\infty \frac{dx}{1+x} = \lim_{b \to \infty} \int_0^b \frac{1}{1+x}\, dx$

$$= \lim_{b \to \infty} \left[\ln (1+x) \right]_0^b$$

$$= \lim_{b \to \infty} [\ln (1+b) - \ln (1+0)]$$

$$= \lim_{b \to \infty} [\ln (1+b) - \ln 1]$$

$$= \lim_{b \to \infty} \ln (1+b)$$

Note that $\ln (1+b)$ increases indefinitely as b increases. Therefore, the limit does not exist. Thus, the improper integral is divergent.

11. $\int_1^\infty 5x^{-2}\,dx$

$= \lim_{b\to\infty} \int_1^b 5x^{-2}\,dx$

$= \lim_{b\to\infty} \left[5\cdot\frac{x^{-1}}{-1} \right]_1^b$

$= \lim_{b\to\infty} \left[-\frac{5}{x} \right]_1^b$

$= \lim_{b\to\infty} \left[-\frac{5}{b} - \left(-\frac{5}{1}\right) \right]$

$= \lim_{b\to\infty} \left(-\frac{5}{b} + 5 \right)$

$= 5 \qquad \left[\text{As } b\to\infty, -\dfrac{5}{b}\to 0, \text{ so} \right.$

$\left. \left(-\dfrac{5}{b}+5\right)\to 5. \right]$

The limit does exist. Thus the improper integral is convergent.

13. $\int_0^\infty e^x\,dx = \lim_{b\to\infty} \int_0^b e^x\,dx$

$= \lim_{b\to\infty} \left[e^x \right]_0^b$

$= \lim_{b\to\infty} (e^b - e^0)$

$= \lim_{b\to\infty} (e^b - 1)$

As $b\to\infty$, $e^b\to\infty$. Thus the limit does not exist. The improper integral is divergent.

15. $\int_3^\infty x^2\,dx = \lim_{b\to\infty} \int_3^b x^2\,dx$

$= \lim_{b\to\infty} \left[\frac{x^3}{3} \right]_3^b$

$= \lim_{b\to\infty} \left(\frac{b^3}{3} - \frac{3^3}{3} \right)$

$= \lim_{b\to\infty} \frac{b^3}{3} - 9$

Since b^3 increases indefinitely as b increases, the limit does not exist. The improper integral is divergent.

17. $\int_0^\infty x\,e^x\,dx = \lim_{b\to\infty} \int_0^b x\,e^x\,dx$

$= \lim_{b\to\infty} \left[e^x(x-1) \right]_0^b$

Using Integration by Parts or Formula 6

$= \lim_{b\to\infty} \left[e^b(b-1) - e^0(0-1) \right]$

$= \lim_{b\to\infty} \left[e^b(b-1) + 1 \right]$

Since $e^b(b-1)$ increases indefinitely as b increases, the limit does not exist. The improper integral is divergent.

19. $\int_0^\infty m\,e^{-mx}\,dx,\ m>0$

$= \lim_{b\to\infty} \int_0^b m\,e^{-mx}\,dx$

$= \lim_{b\to\infty} \left[\frac{m}{-m} e^{-mx} \right]_0^b$

$= \lim_{b\to\infty} \left[-e^{-mx} \right]_0^b$

$= \lim_{b\to\infty} \left[-e^{-mb} - (-e^{-m\cdot 0}) \right]$

$= \lim_{b\to\infty} \left(1 - \frac{1}{e^{mb}} \right)$

$= 1 \qquad \left[\text{As } b\to\infty, e^{mb}\to\infty. \text{ Thus } \dfrac{1}{e^{mb}}\to 0 \right.$

$\left. \text{and } \left(1 - \dfrac{1}{e^{mb}}\right)\to 1. \right]$

The limit does exist. Thus the improper integral is convergent.

21. The area is given by

$\int_2^\infty \frac{1}{x^2}\,dx = \lim_{b\to\infty} \int_2^b x^{-2}\,dx$

$= \lim_{b\to\infty} \left[-x^{-1} \right]_2^b$

$= \lim_{b\to\infty} \left[-\frac{1}{x} \right]_2^b$

$= \lim_{b\to\infty} \left(-\frac{1}{b} - \left(-\frac{1}{2}\right) \right)$

$= \frac{1}{2}$

The area of the region is $\dfrac{1}{2}$.

23. The area is given by

$\int_0^\infty 2x\,e^{-x^2}\,dx$

$= \lim_{b\to\infty} \int_0^b 2x\,e^{-x^2}\,dx$

(We use the substitution $u = -x^2$ to integrate.)

$= \lim_{b\to\infty} \left[-e^{-x^2} \right]_0^b$

$= \lim_{b\to\infty} \left[-e^{-b^2} - (-e^{-0^2}) \right]$

$= \lim_{b\to\infty} \left(-\frac{1}{e^{b^2}} + 1 \right)$

$= 1 \qquad \left(\text{As } b\to\infty, -\dfrac{1}{e^{b^2}}\to 0 \text{ and } -\dfrac{1}{e^{b^2}}+1\to 1. \right)$

The area of the region is 1.

25. $\dfrac{P}{k} = \dfrac{3600}{0.08}$ Substituting 3600 for P and 0.08 for k

$= 45,000$

The accumulated present value is \$45,000.

27. $P(x) = \int_0^\infty 200\, e^{-0.032x}\, dx$

$$= \lim_{b\to\infty} \int_0^b 200\, e^{-0.032x}\, dx$$

$$= \lim_{b\to\infty} \left[\frac{200}{-0.032} e^{-0.032x}\right]_0^b$$

$$= \lim_{b\to\infty} [-6250\, e^{-0.032b} - (-6250\, e^{-0.032(0)})]$$

$$= \lim_{b\to\infty} \left[-\frac{6250}{e^{0.032b}} + 6250\right]$$

$$= 6250 \qquad \left(\text{As } b\to\infty, -\frac{6250}{e^{0.032b}} \to 0.\right)$$

The total profit would be $6250.

29. $C(x) = \int_1^\infty 3600x^{-1.8}\, dx$

$$= \lim_{b\to\infty} \int_1^b 3600x^{-1.8}\, dx$$

$$= \lim_{b\to\infty} \left[\frac{3600}{-0.8} x^{-0.8}\right]_1^b$$

$$= \lim_{b\to\infty} [-4500b^{-0.8} - (-4500)1^{-0.8}]$$

$$= \lim_{b\to\infty} \left[-\frac{4500}{b^{0.8}} + 4500\right]$$

$$= 4500 \qquad \left(\text{As } b\to\infty, -\frac{4500}{b^{0.8}} \to 0.\right)$$

The total cost would be $4500.

31. $\int_0^T P\, e^{-kt}\, dt = \frac{P}{k}(1 - e^{-kT})$

$$= \frac{P}{k}\left(1 - \frac{1}{e^{kT}}\right)$$

As $T \to \infty$, the buildup of radioactive material approaches a limiting value P/k.

$$\frac{P}{k} = \frac{1}{0.00003} \quad \begin{array}{l}\text{Substituting 1 for } P \text{ and } 0.00003 \\ \text{for } k \quad (0.003\% = 0.00003)\end{array}$$

$$\approx 33,333 \text{ lb}$$

33. $\int_0^\infty \frac{dx}{x^{2/3}} = \lim_{b\to\infty} \int_0^b x^{-2/3}\, dx$

$$= \lim_{b\to\infty} \left[\frac{x^{1/3}}{1/3}\right]_0^b$$

$$= \lim_{b\to\infty} [3x^{1/3}]_0^b$$

$$= \lim_{b\to\infty} (3b^{1/3} - 3\cdot 0^{1/3})$$

$$= \lim_{b\to\infty} 3\sqrt[3]{b}$$

Now, as $b\to\infty$, we know that $\sqrt[3]{b} \to \infty$. Thus, the limit does not exist. The improper integral is divergent.

35. $\int_0^\infty \frac{dx}{(x+1)^{3/2}}$

$$= \lim_{b\to\infty} \int_0^b (x+1)^{-3/2}\, dx$$

$$= \lim_{b\to\infty} \left[\frac{(x+1)^{-1/2}}{-1/2}\right]_0^b$$

$$= \lim_{b\to\infty} \left[-\frac{2}{\sqrt{x+1}}\right]_0^b$$

$$= \lim_{b\to\infty} \left[-\frac{2}{\sqrt{b+1}} - \left(-\frac{2}{\sqrt{0+1}}\right)\right]$$

$$= \lim_{b\to\infty} \left[-\frac{2}{\sqrt{b+1}} + 2\right]$$

$$= 2 \qquad \left(\text{As } b\to\infty, \sqrt{b+1}\to\infty, \text{ so } -\frac{2}{\sqrt{b+1}} \to 0.\right)$$

The limit does exist. The improper integral is convergent.

37. $\int_0^\infty x\, e^{-x^2}\, dx$

$$= \int_0^\infty -\frac{1}{2}\cdot(-2)x\, e^{-x^2}\, dx \quad \text{Multiplying by 1}$$

$$= -\frac{1}{2} \int_0^\infty e^{-x^2}(-2x)\, dx$$

$$= -\frac{1}{2} \lim_{b\to 0} \left[e^{-x^2}\right]_0^b \quad \begin{array}{l}\text{Using substitution where}\\ u=-x^2 \text{ and } du=-2x\,dx\end{array}$$

$$= -\frac{1}{2} \lim_{b\to\infty} \left(e^{-b^2} - e^{-0^2}\right)$$

$$= -\frac{1}{2} \lim_{b\to\infty} \left(\frac{1}{e^{b^2}} - 1\right)$$

$$= -\frac{1}{2}(-1) \qquad \left[\begin{array}{l}\text{As } b\to\infty, e^{b^2}\to\infty, \text{ so}\\ \frac{1}{e^{b^2}} \to 0 \text{ and } \left(\frac{1}{e^{b^2}}-1\right)\to -1.\end{array}\right]$$

$$= \frac{1}{2}$$

The limit does exist. The improper integral is convergent.

39. $\int_0^\infty E(t)\, dt$

$$= \int_0^\infty t\, e^{-kt}\, dt$$

$$= \lim_{b\to\infty} \int_0^b t\, e^{-kt}\, dt$$

$$= \lim_{b\to\infty} \left[\frac{1}{(-k)^2}\cdot e^{-kt}(-kt-1)\right]_0^b$$

$$= \lim_{b\to\infty} \left[-\frac{kt+1}{k^2\, e^{kt}}\right]_0^b$$

$$= \lim_{b\to\infty} \left[-\frac{kb+1}{k^2\, e^{kb}} - \left(-\frac{k\cdot 0+1}{k^2\, e^{k\cdot 0}}\right)\right]$$

$$= \lim_{b\to\infty} \left[-\frac{kb+1}{k^2\, e^{kb}} + \frac{1}{k^2}\right]$$

$$= \frac{1}{k^2} \qquad \left(\begin{array}{l}\text{Input-output tables show that}\\ \text{as } b\to\infty, -\frac{kb+1}{k^2\, e^{kb}} \to 0.\end{array}\right)$$

The integral represents the total dose of the drug.

41. $\boxed{tw}$

43. See the answer section in the text.

45. Using the fnInt feature on a grapher with a large value for the upper limit, we find

$$\int_1^\infty \frac{6}{5+e^x}\, dx \approx 1.25.$$

Exercise Set 6.4

1. $f(x) = 2x$, $[0,1]$

$$\begin{aligned}
P([0,1]) &= \int_0^1 f(x)\, dx \\
&= \int_0^1 2x\, dx \\
&= \left[x^2\right]_0^1 \\
&= 1^2 - 0^2 \\
&= 1
\end{aligned}$$

3. $f(x) = \frac{1}{3}$, $[4,7]$

$$\begin{aligned}
P([4,7]) &= \int_4^7 f(x)\, dx \\
&= \int_4^7 \frac{1}{3}\, dx \\
&= \left[\frac{1}{3}x\right]_4^7 \\
&= \frac{1}{3}\cdot 7 - \frac{1}{3}\cdot 4 \\
&= \frac{7}{3} - \frac{4}{3} \\
&= \frac{3}{3} \\
&= 1
\end{aligned}$$

5. $f(x) = \frac{3}{26}x^2$, $[1,3]$

$$\begin{aligned}
P([1,3]) &= \int_1^3 f(x)\, dx \\
&= \int_1^3 \frac{3}{26}x^2\, dx \\
&= \left[\frac{3}{26}\cdot\frac{x^3}{3}\right]_1^3 \\
&= \left[\frac{x^3}{26}\right]_1^3 \\
&= \frac{3^3}{26} - \frac{1^3}{26} \\
&= \frac{27}{26} - \frac{1}{26} \\
&= \frac{26}{26} \\
&= 1
\end{aligned}$$

7. $f(x) = \frac{1}{x}$, $[1,e]$

$$\begin{aligned}
P([1,e]) &= \int_1^e f(x)\, dx \\
&= \int_1^e \frac{1}{x}\, dx \\
&= \left[\ln x\right]_1^e \\
&= \ln e - \ln 1 \\
&= 1 - 0 \\
&= 1
\end{aligned}$$

9. $f(x) = \frac{3}{2}x^2$, $[-1,1]$

$$\begin{aligned}
P([-1,1]) &= \int_{-1}^1 f(x)\, dx \\
&= \int_{-1}^1 \frac{3}{2}x^2\, dx \\
&= \left[\frac{3}{2}\cdot\frac{x^3}{3}\right]_{-1}^1 \\
&= \left[\frac{x^3}{2}\right]_{-1}^1 \\
&= \frac{1^3}{2} - \frac{(-1)^3}{2} \\
&= \frac{1}{2} + \frac{1}{2} \\
&= 1
\end{aligned}$$

11. $f(x) = 3\,e^{-3x}$, $[0,\infty]$

$$\begin{aligned}
P([0,\infty)) &= \int_0^\infty f(x)\, dx \\
&= \int_0^\infty 3\,e^{-3x}\, dx \\
&= \lim_{b\to\infty}\int_0^b 3\,e^{-3x}\, dx \\
&= \lim_{b\to\infty}\left[\frac{3}{-3}e^{-3x}\right]_0^b \\
&= \lim_{b\to\infty}\left[-e^{-3x}\right]_0^b \\
&= \lim_{b\to\infty}\left[-e^{-3b} - (-e^{-3\cdot 0})\right] \\
&= \lim_{b\to\infty}\left[-e^{-3b} + 1\right] \\
&= \lim_{b\to\infty}\left[-\frac{1}{e^{3b}} + 1\right] \\
&= 1 \qquad \left[\text{As } b\to\infty,\ e^{3b}\to\infty,\text{ so}\right. \\
&\qquad\qquad \left. -\frac{1}{e^{3b}}\to 0 \text{ and } \left(-\frac{1}{e^{3b}} + 1\right)\to 1.\right]
\end{aligned}$$

13. $f(x) = kx$, $[1,3]$

Find k such that $\int_1^3 kx\, dx = 1$.

$$\begin{aligned}
\int_1^3 x\, dx &= \left[\frac{x^2}{2}\right]_1^3 \\
&= \frac{3^2}{2} - \frac{1^2}{2} = \frac{9}{2} - \frac{1}{2} = \frac{8}{2} = 4
\end{aligned}$$

Thus $k = \frac{1}{4}$ and $f(x) = \frac{1}{4}x$.

15. $f(x) = kx^2$, $[-1, 1]$

Find k such that $\int_{-1}^{3} kx^2\, dx = 1$.

$$\int_{-1}^{1} x^2\, dx = \left[\frac{x^3}{3}\right]_{-1}^{1}$$

$$= \frac{1^3}{3} - \frac{(-1)^3}{3} = \frac{1}{3} + \frac{1}{3} = \frac{2}{3}$$

Thus $k = \frac{1}{2/3} = \frac{3}{2}$ and $f(x) = \frac{3}{2}x^2$.

17. $f(x) = k$, $[2, 7]$

Find k such that $\int_{2}^{7} k\, dx = 1$.

$\int_{2}^{7} dx = [x]_{2}^{7} = 7 - 2 = 5$

Thus $k = \frac{1}{5}$, and $f(x) = \frac{1}{5}$.

19. $f(x) = k(2 - x)$, $[0, 2]$

Find k such that $\int_{0}^{2} k(2 - x)\, dx = 1$.

$$\int_{0}^{2} (2 - x)\, dx = \left[2x - \frac{x^2}{2}\right]_{0}^{2}$$

$$= \left(2 \cdot 2 - \frac{2^2}{2}\right) - \left(2 \cdot 0 - \frac{0^2}{2}\right)$$

$$= 4 - 2 - 0$$

$$= 2$$

Thus $k = \frac{1}{2}$, and $f(x) = \frac{1}{2}(2 - x)$, or $\frac{2 - x}{2}$.

21. $f(x) = \dfrac{k}{x}$, $[1, 3]$

Find k such that $\displaystyle\int_{1}^{3} k \cdot \frac{1}{x}\, dx = 1$.

$$\int_{1}^{3} \frac{1}{x}\, dx = \left[\ln x\right]_{1}^{3}$$

$$= \ln 3 - \ln 1$$

$$= \ln 3 \qquad (\ln 1 = 0)$$

Thus $k = \dfrac{1}{\ln 3}$, and $f(x) = \dfrac{1}{\ln 3} \cdot \dfrac{1}{x} = \dfrac{1}{x \ln 3}$.

23. $f(x) = k\, e^x$, $[0, 3]$

Find k such that $\int_{0}^{3} k\, e^x\, dx = 1$.

$\int_{0}^{3} e^x\, dx = \left[e^x\right]_{0}^{3} = e^3 - e^0 = e^3 - 1$

Thus $k = \dfrac{1}{e^3 - 1}$, and $f(x) = \dfrac{1}{e^3 - 1} \cdot e^x$, or $\dfrac{e^x}{e^3 - 1}$.

25. a) $f(x) = \dfrac{1}{50}x$, for $0 \le x \le 10$

$$P(2 \le x \le 6) = \int_{2}^{6} \frac{1}{50}x\, dx$$

$$= \left[\frac{1}{50} \cdot \frac{x^2}{2}\right]_{2}^{6}$$

$$= \left[\frac{x^2}{100}\right]_{2}^{6}$$

$$= \frac{6^2}{100} - \frac{2^2}{100}$$

$$= \frac{36 - 4}{100}$$

$$= \frac{32}{100}$$

$$= \frac{8}{25}, \text{ or } 0.32$$

The probability that the dart lands in $[2, 6]$ is 0.32.

b) $\boxed{tw}$

27. $f(x) = \dfrac{1}{16}$, for $4 \le x \le 20$

$$P(9 \le x \le 17) = \int_{9}^{17} \frac{1}{16}\, dx$$

$$= \frac{1}{16}\left[x\right]_{9}^{17}$$

$$= \frac{1}{16}(17 - 9)$$

$$= \frac{8}{16}$$

$$= \frac{1}{2}, \text{ or } 0.5$$

The probability that a number selected is in the subinterval $[9, 17]$ is 0.5.

29. $f(x) = k\, e^{-kx}$, for $0 \le x < \infty$, where $k = 1/a$ and $a = $ the average distance between successive cars over some period of time.

We first determine k:

$k = \dfrac{1}{100} = 0.01$.

The probability density function for x is
$f(x) = 0.01\, e^{-0.01x}$, for $0 \le x < \infty$.

The probability that the distance between cars is 40 ft or less is

$$P(0 \le x \le 40) = \int_{0}^{40} 0.01\, e^{-0.01x}\, dx$$

$$= \left[\frac{0.01}{-0.01} e^{-0.01x}\right]_{0}^{40}$$

$$= \left[-e^{-0.01x}\right]_{0}^{40}$$

$$= \left(-e^{-0.01(40)}\right) - \left(-e^{-0.01(0)}\right)$$

$$= -e^{-0.4} + 1$$

$$= 1 - e^{-0.4}$$

$$\approx 1 - 0.670320$$

$$\approx 0.329680$$

$$\approx 0.3297$$

31. $f(t) = 2\,e^{-2t},\ 0 \le t < \infty$

$P([0,5]) = \int_0^5 2\,e^{-2t}\,dt$

$\qquad = \left[\dfrac{2}{-2}e^{-2t}\right]_0^5$

$\qquad = \left[-e^{-2t}\right]_0^5$

$\qquad = (-e^{-2\cdot 5}) - (-e^{-2\cdot 0})$

$\qquad = -e^{-10} + 1$

$\qquad = 1 - e^{-10}$

$\qquad \approx 1 - 0.0000454$

$\qquad \approx 0.9999546$

$\qquad \approx 0.99995$

The probability that a phone call will last no more than 5 minutes is 0.99995.

33. $f(x) = k\,e^{-kt},\ 0 \le t < \infty$, where $k = 1/a$ and $a =$ the average time that will pass before a failure occurs.

We first find k:

$k = \dfrac{1}{100} = 0.01.$

The probability density function for x is
$f(x) = 0.01\,e^{-0.01x}$, for $0 \le x < \infty$.

The probability that a failure will occur in 50 hours or less is

$P(0 \le x \le 50) = \int_0^{50} 0.01\,e^{-0.01x}\,dx$

$\qquad = \left[\dfrac{0.01}{-0.01}e^{-0.01x}\right]_0^{50}$

$\qquad = \left[-e^{-0.01x}\right]_0^{50}$

$\qquad = \left(-e^{-0.01(50)}\right) - \left(-e^{-0.01(0)}\right)$

$\qquad = -e^{-0.5} + 1$

$\qquad = 1 - e^{-0.5}$

$\qquad \approx 1 - 0.606531$

$\qquad \approx 0.393469$

$\qquad \approx 0.3935$

The probability that a failure will occur in 50 hours or less is 0.3935.

35. $f(t) = 0.02\,e^{-0.02t},\ 0 \le t < \infty$

$P(0 \le t \le 150) = \int_0^{150} 0.02\,e^{-0.02t}\,dt$

$\qquad = \left[\dfrac{0.02}{-0.02}e^{-0.02t}\right]_0^{150}$

$\qquad = \left[-e^{-0.02t}\right]_0^{150}$

$\qquad = \left(-e^{-0.02(150)}\right) - \left(-e^{-0.02(0)}\right)$

$\qquad = -e^{-3} + 1$

$\qquad \approx -0.049787 + 1$

$\qquad \approx 0.950213$

$\qquad \approx 0.9502$

The probability that a rat will learn its way through a maze in 150 seconds, or less, is 0.9502.

37. $f(x) = x^3$ is a probability function on $[0, b]$. Thus,

$\int_0^b f(x)\,dx = 1$

$\int_0^b x^3\,dx = 1$

$\left[\dfrac{x^4}{4}\right]_0^b = 1$

$\dfrac{b^4}{4} - \dfrac{0^4}{4} = 1$

$\dfrac{b^4}{4} = 1$

$b^4 = 4$

$b = \sqrt[4]{4}$

$b = \sqrt[4]{2^2}$

$b = \sqrt{2}$

39. $\boxed{tw}$

41. - 51.

Exercise Set 6.5

---

1. $f(x) = \dfrac{1}{3},\ [2, 5]$

$E(x) = \int_a^b x\,f(x)\,dx \qquad$ Expected value of x

$E(x) = \displaystyle\int_2^5 x \cdot \dfrac{1}{3}\,dx$

$\qquad = \dfrac{1}{3}\left[\dfrac{x^2}{2}\right]_2^5$

$\qquad = \dfrac{1}{3}\left(\dfrac{5^2}{2} - \dfrac{2^2}{2}\right)$

$\qquad = \dfrac{1}{3}\left(\dfrac{25}{2} - \dfrac{4}{2}\right)$

$\qquad = \dfrac{1}{3} \cdot \dfrac{21}{2}$

$\qquad = \dfrac{7}{2}$

$E(x^2) = \int_a^b x^2\,f(x)\,dx \qquad$ Expected value of x^2

$E(x^2) = \displaystyle\int_2^5 x^2 \cdot \dfrac{1}{3}\,dx$

$\qquad = \dfrac{1}{3}\left[\dfrac{x^3}{3}\right]_2^5$

$\qquad = \dfrac{1}{3}\left(\dfrac{5^3}{3} - \dfrac{2^3}{3}\right)$

$\qquad = \dfrac{1}{3}\left(\dfrac{125}{3} - \dfrac{8}{3}\right)$

$\qquad = \dfrac{1}{3} \cdot \dfrac{117}{3}$

$\qquad = \dfrac{117}{9}$

$\qquad = 13$

$\mu = E(x) = \dfrac{7}{2}$ Mean

$\sigma^2 = E(x^2) - \big[E(x)\big]^2$

$\quad = 13 - \left(\dfrac{7}{2}\right)^2$ Substituting 13 for $E(x^2)$
$\qquad\qquad\qquad$ and $\dfrac{7}{2}$ for $E(x)$

$\quad = \dfrac{52}{4} - \dfrac{49}{4}$

$\quad = \dfrac{3}{4}$ Variance

$\sigma = \sqrt{\text{variance}}$

$\quad = \sqrt{\dfrac{3}{4}}$ Substituting $\dfrac{3}{4}$ for the variance

$\quad = \dfrac{1}{2}\sqrt{3}$ Standard deviation

3. $f(x) = \dfrac{2}{9}x$, $[0, 3]$

$E(x) = \int_a^b x\, f(x)\, dx$ Expected value of x

$E(x) = \int_0^3 x \cdot \dfrac{2}{9}x\, dx$

$\quad = \int_0^3 \dfrac{2}{9}x^2\, dx$

$\quad = \dfrac{2}{9}\left[\dfrac{x^3}{3}\right]_0^3$

$\quad = \dfrac{2}{9}\left(\dfrac{3^3}{3} - \dfrac{0^3}{3}\right)$

$\quad = \dfrac{2}{9} \cdot 9$

$\quad = 2$

$E(x^2) = \int_a^b x^2\, f(x)\, dx$ Expected value of x^2

$E(x^2) = \int_0^3 x^2 \cdot \dfrac{2}{9}x\, dx$

$\quad = \int_0^3 \dfrac{2}{9}x^3\, dx$

$\quad = \dfrac{2}{9}\left[\dfrac{x^4}{4}\right]_0^3$

$\quad = \dfrac{2}{9}\left(\dfrac{3^4}{4} - \dfrac{0^4}{4}\right)$

$\quad = \dfrac{2}{9} \cdot \dfrac{81}{4}$

$\quad = \dfrac{9}{2}$

$\mu = E(x) = 2$ Mean

$\sigma^2 = E(x^2) - \big[E(x)\big]^2$

$\quad = \dfrac{9}{2} - 2^2$ Substituting $\dfrac{9}{2}$ for $E(x^2)$ and
$\qquad\qquad\qquad$ 2 for $E(x)$

$\quad = \dfrac{9}{2} - \dfrac{8}{2}$

$\quad = \dfrac{1}{2}$ Variance

$\sigma = \sqrt{\text{variance}}$

$\quad = \sqrt{\dfrac{1}{2}}$ Substituting $\dfrac{1}{2}$ for the variance;
$\qquad\qquad\qquad$ standard deviation

5. $f(x) = \dfrac{2}{3}x$, $[1, 2]$

$E(x) = \int_a^b x\, f(x)\, dx$ Expected value of x

$E(x) = \int_1^2 x \cdot \dfrac{2}{3}x\, dx$

$\quad = \int_1^2 \dfrac{2}{3}x^2\, dx$

$\quad = \dfrac{2}{3}\left[\dfrac{x^3}{3}\right]_1^2$

$\quad = \dfrac{2}{3}\left(\dfrac{2^3}{3} - \dfrac{1^3}{3}\right)$

$\quad = \dfrac{2}{3}\left(\dfrac{8}{3} - \dfrac{1}{3}\right)$

$\quad = \dfrac{2}{3} \cdot \dfrac{7}{3}$

$\quad = \dfrac{14}{9}$

$E(x^2) = \int_a^b x^2\, f(x)\, dx$ Expected value of x^2

$E(x^2) = \int_1^2 x^2 \cdot \dfrac{2}{3}x\, dx$

$\quad = \int_1^2 \dfrac{2}{3}x^3\, dx$

$\quad = \dfrac{2}{3}\left[\dfrac{x^4}{4}\right]_1^2$

$\quad = \dfrac{2}{3}\left(\dfrac{2^4}{4} - \dfrac{1^4}{4}\right)$

$\quad = \dfrac{2}{3}\left(\dfrac{16}{4} - \dfrac{1}{4}\right)$

$\quad = \dfrac{2}{3} \cdot \dfrac{15}{4}$

$\quad = \dfrac{5}{2}$

$\mu = E(x) = \dfrac{14}{9}$ Mean

$\sigma^2 = E(x^2) - \big[E(x)\big]^2$

$\quad = \dfrac{5}{2} - \left(\dfrac{14}{9}\right)^2$ Substituting $\dfrac{5}{2}$ for $E(x^2)$ and
$\qquad\qquad\qquad$ $\dfrac{14}{9}$ for $E(x)$

$\quad = \dfrac{5}{2} - \dfrac{196}{81}$

$\quad = \dfrac{405}{162} - \dfrac{392}{162}$

$\quad = \dfrac{13}{162}$ Variance

$\sigma = \sqrt{\text{variance}}$

$= \sqrt{\dfrac{13}{162}}$ Substituting $\dfrac{13}{162}$ for the variance

$= \dfrac{1}{9}\sqrt{\dfrac{13}{2}}$ Standard deviation

7. $f(x) = \dfrac{1}{3}x^2,\ [-2, 1]$

$E(x) = \int_a^b x\, f(x)\, dx$ Expected value of x

$E(x) = \int_{-2}^1 x \cdot \dfrac{1}{3}x^2\, dx$

$= \int_{-2}^1 \dfrac{1}{3}x^3\, dx$

$= \dfrac{1}{3}\left[\dfrac{x^4}{4}\right]_{-2}^1$

$= \dfrac{1}{3}\left(\dfrac{1^4}{4} - \dfrac{(-2)^4}{4}\right)$

$= \dfrac{1}{3}\left(\dfrac{1}{4} - \dfrac{16}{4}\right)$

$= \dfrac{1}{3}\left(-\dfrac{15}{4}\right)$

$= -\dfrac{5}{4}$

$E(x^2) = \int_a^b x^2\, f(x)\, dx$ Expected value of x^2

$E(x^2) = \int_{-2}^1 x^2 \cdot \dfrac{1}{3}x^2\, dx$

$= \int_{-2}^1 \dfrac{1}{3}x^4\, dx$

$= \dfrac{1}{3}\left[\dfrac{x^5}{5}\right]_{-2}^1$

$= \dfrac{1}{3}\left[\dfrac{1^5}{5} - \dfrac{(-2)^5}{5}\right]$

$= \dfrac{1}{3}\left(\dfrac{1}{5} + \dfrac{32}{5}\right)$

$= \dfrac{1}{3}\left(\dfrac{33}{5}\right)$

$= \dfrac{11}{5}$

$\mu = E(x) = -\dfrac{5}{4}$ Mean

$\sigma^2 = E(x^2) - \left[E(x)\right]^2$

$= \dfrac{11}{5} - \left(-\dfrac{5}{4}\right)^2$ Substituting $\dfrac{11}{5}$ for $E(x^2)$ and $-\dfrac{5}{4}$ for $E(x)$

$= \dfrac{11}{5} - \dfrac{25}{16}$

$= \dfrac{176}{80} - \dfrac{125}{80}$

$= \dfrac{51}{80}$ Variance

$\sigma = \sqrt{\text{variance}}$

$= \sqrt{\dfrac{51}{80}}$ Substituting $\dfrac{51}{80}$ for the variance

$= \dfrac{1}{4}\sqrt{\dfrac{51}{5}}$ Standard deviation

9. $f(x) = \dfrac{1}{\ln 3} \cdot \dfrac{1}{x},\ [1, 3]$

$E(x) = \int_a^b x\, f(x)\, dx$ Expected value of x

$E(x) = \int_1^3 x\left(\dfrac{1}{\ln 3} \cdot \dfrac{1}{x}\right) dx$

$= \int_1^3 \dfrac{1}{\ln 3}\, dx$

$= \dfrac{1}{\ln 3}\int_1^3 dx$

$= \dfrac{1}{\ln 3}\left[x\right]_1^3$

$= \dfrac{1}{\ln 3}(3 - 1)$

$= \dfrac{2}{\ln 3}$

$E(x^2) = \int_a^b x^2\, f(x)\, dx$ Expected value of x^2

$E(x^2) = \int_1^3 x^2\left(\dfrac{1}{\ln 3} \cdot \dfrac{1}{x}\right) dx$

$= \int_1^3 \dfrac{1}{\ln 3} \cdot x\, dx$

$= \dfrac{1}{\ln 3}\left[\dfrac{x^2}{2}\right]_1^3$

$= \dfrac{1}{\ln 3}\left(\dfrac{3^2}{2} - \dfrac{1^2}{2}\right)$

$= \dfrac{1}{\ln 3}\left(\dfrac{9}{2} - \dfrac{1}{2}\right)$

$= \dfrac{4}{\ln 3}$

$\mu = E(x) = \dfrac{2}{\ln 3}$ Mean

$\sigma^2 = E(x^2) - \left[E(x)\right]^2$

$= \dfrac{4}{\ln 3} - \left(\dfrac{2}{\ln 3}\right)^2$ Substituting $\dfrac{4}{\ln 3}$ for $E(x^2)$ and $\dfrac{2}{\ln 3}$ for $E(x)$

$= \dfrac{4\ln 3}{(\ln 3)^2} - \dfrac{4}{(\ln 3)^2}$

$= \dfrac{4\ln 3 - 4}{(\ln 3)^2}$

$= \dfrac{4(\ln 3 - 1)}{(\ln 3)^2}$ Variance

$\sigma = \sqrt{\text{variance}}$

$= \sqrt{\dfrac{4(\ln 3 - 1)}{(\ln 3)^2}}$ Substituting $\dfrac{4(\ln 3 - 1)}{(\ln 3)^2}$ for the variance

$= \dfrac{2}{\ln 3}\sqrt{\ln 3 - 1}$ Standard deviation

11.

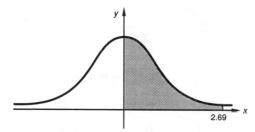

$P(0 \leq x \leq 2.69)$

$= 0.4964$ Using Table 2

13.

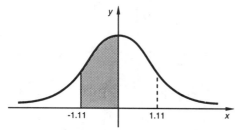

$P(-1.11 \leq x \leq 0)$

$= P(0 \leq x \leq 1.11)$ Symmetry of the graph

$= 0.3665$ Using Table 2

15.

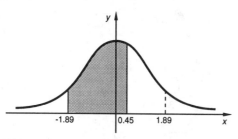

$P(-1.89 \leq x \leq 0.45)$

$= P(-1.89 \leq x \leq 0) + P(0 \leq x \leq 0.45)$

$= P(0 \leq x \leq 1.89) + P(0 \leq x \leq 0.45)$

                      Symmetry of the graph

$= 0.4706 + 0.1736$ Using Table 2

$= 0.6442$

17.

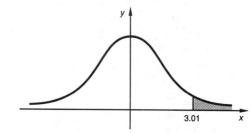

$P(1.76 \leq x \leq 1.86)$

$= P(0 \leq x \leq 1.86) - P(0 \leq x \leq 1.76)$

$= 0.4686 - 0.4608$ Using Table 2

$= 0.0078$

19.

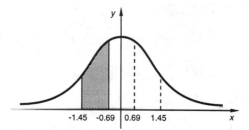

$P(-1.45 \leq x \leq -0.69)$

$= P(0.69 \leq x \leq 1.45)$ Symmetry of the graph

$= P(0 \leq x \leq 1.45) - P(0 \leq x \leq 0.69)$

$= 0.4265 - 0.2549$ Using Table 2

$= 0.1716$

21.

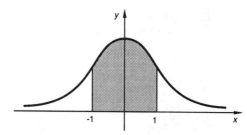

$P(x \geq 3.01)$

$= P(x \geq 0) - P(0 \leq x \leq 3.01)$

$= 0.5000 - 0.4987$ Using Table 2

      Half of the area is on each side of the
      line $x = 0$. Since the entire area is 1,
      $P(x \geq 0) = 0.5000$

$= 0.0013$

23. a)

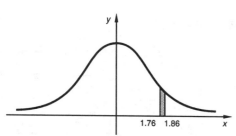

$P(-1 \leq x \leq 1)$

$= P(-1 \leq x \leq 0) + P(0 \leq x \leq 1)$

$= P(0 \leq x \leq 1) + P(0 \leq x \leq 1)$ Symmetry of
                                the graph

$= 2[P(0 \leq x \leq 1)]$

$= 2(0.3413)$ Using Table 2

$= 0.6826$

b) $0.6826 = 68.26\%$

25. $P(24 \leq x \leq 30)$

Mean $\mu = 22$

Standard deviation $\sigma = 5$

We first standardize the numbers 24 and 30.

30 is standardized to $\dfrac{b - \mu}{\sigma} = \dfrac{30 - 22}{5} = \dfrac{8}{5} = 1.6$

24 is standardized to $\dfrac{a - \mu}{\sigma} = \dfrac{24 - 22}{5} = \dfrac{2}{5} = 0.4$

Then

$P(24 \leq x \leq 30)$

$= P(0.4 \leq X \leq 1.6)$

$= P(0 \leq X \leq 1.6) - P(0 \leq X \leq 0.4)$

$= 0.4452 - 0.1554$ Using Table 2

$= 0.2898$

27. $P(19 \leq x \leq 25)$

Mean $\mu = 22$

Standard deviation $\sigma = 5$

We first standardize the numbers 19 and 25.

25 is standardized to $\dfrac{b - \mu}{\sigma} = \dfrac{25 - 22}{5} = \dfrac{3}{5} = 0.6$

19 is standardized to $\dfrac{a - \mu}{\sigma} = \dfrac{19 - 22}{5} = \dfrac{-3}{5} = -0.6$

Then

$P(19 \leq x \leq 25)$

$= P(-0.6 \leq X \leq 0.6)$

$= P(-0.6 \leq X \leq 0) + P(0 \leq X \leq 0.6)$

$= P(0 \leq X \leq 0.6) + P(0 \leq X \leq 0.6)$

                 Symmetry of the graph

$= 2[P(0 \leq X \leq 0.6)]$

$= 2(0.2257)$ Using Table 2

$= 0.4514$

29. - 45.

47. We first standardize the number 300.

300 is standardized to

$\dfrac{b - \mu}{\sigma} = \dfrac{300 - 250}{20}$ Substituting 300 for b, 250 for μ, and 20 for σ

$= \dfrac{50}{20}$

$= \dfrac{5}{2}$, or 2.5

$P(x \geq 300)$

$= P(X \geq 2.5)$

$= P(X \geq 0) - P(0 \leq X \leq 2.5)$

$= 0.5000 - 0.4938$ Entire area is 1; $P(X \geq 0) = 0.5000$; using Table 2

$= 0.0062$

$= 0.62\%$

Thus the company will have to hire extra help or pay overtime 0.62% of the days.

49. $\mu = 65$, $\sigma = 20$

We first standardize 80 and 89.

80 is standardized to $\dfrac{a - \mu}{\sigma} = \dfrac{80 - 65}{20} = \dfrac{15}{20} = 0.75$

89 is standardized to $\dfrac{b - \mu}{\sigma} = \dfrac{89 - 65}{20} = \dfrac{24}{20} = 1.2$

Then $P(80 \leq S \leq 89)$

$= P(0.75 \leq X \leq 1.2)$

$= P(0 \leq X \leq 1.2) - P(0 \leq X \leq 0.75)$

$= 0.3849 - 0.2734$

$= 0.1115$

51. $f(x) = \dfrac{1}{b - a}$, $[a, b]$

$E(x) = \int_a^b x\, f(x)\, dx$ Expected value of x

$E(x) = \int_a^b x \cdot \dfrac{1}{b - a}\, dx$

$= \int_a^b \dfrac{1}{b - a} \cdot x\, dx$

$= \dfrac{1}{b - a}\left[\dfrac{x^2}{2}\right]_a^b$

$= \dfrac{1}{b - a}\left(\dfrac{b^2}{2} - \dfrac{a^2}{2}\right)$

$= \dfrac{1}{b - a}\left(\dfrac{b^2 - a^2}{2}\right)$

$= \dfrac{1}{b - a} \cdot \dfrac{(b - a)(b + a)}{2}$

$= \dfrac{b + a}{2}$

$E(x^2) = \int_a^b x^2\, f(x)\, dx$ Expected value of x^2

$E(x^2) = \int_a^b x^2 \cdot \dfrac{1}{b - a}\, dx$

$= \int_a^b \dfrac{1}{b - a} \cdot x^2\, dx$

$= \dfrac{1}{b - a}\left[\dfrac{x^3}{3}\right]_a^b$

$= \dfrac{1}{b - a}\left(\dfrac{b^3}{3} - \dfrac{a^3}{3}\right)$

$= \dfrac{1}{b - a}\left(\dfrac{b^3 - a^3}{3}\right)$

$= \dfrac{1}{b - a} \cdot \dfrac{(b - a)(b^2 + ba + a^2)}{3}$

$= \dfrac{b^2 + ba + a^2}{3}$

$$\mu = E(x) = \frac{b+a}{2} \qquad \text{Mean}$$

$$\sigma^2 = E(x^2) - \left[E(x)\right]^2$$
$$= \frac{b^2 + ba + a^2}{3} - \left(\frac{b+a}{2}\right)^2 \qquad \text{Substituting}$$
$$= \frac{b^2 + ba + a^2}{3} - \frac{b^2 + 2ba + a^2}{4}$$
$$= \frac{4b^2 + 4ba + 4a^2}{12} - \frac{3b^2 + 6ba + 3a^2}{12}$$
$$= \frac{4b^2 + 4ba + 4a^2 - 3b^2 - 6ba - 3a^2}{12}$$
$$= \frac{b^2 - 2ba + a^2}{12}$$
$$= \frac{(b-a)^2}{12} \qquad \text{Variance}$$

$$\sigma = \sqrt{\text{variance}}$$
$$= \sqrt{\frac{(b-a)^2}{12}} \qquad \text{Substituting}$$
$$= \frac{b-a}{2\sqrt{3}} \qquad \text{Standard deviation}$$

53. $f(x) = \dfrac{1}{2}x, \ [0,2]$

$$\int_0^m f(x)\, dx = \frac{1}{2}$$
$$\int_0^m \frac{1}{2}x\, dx = \frac{1}{2}$$
$$\frac{1}{2}\int_0^m x\, dx = \frac{1}{2}$$
$$\int_0^m x\, dx = 1$$
$$\left[\frac{x^2}{2}\right]_0^m = 1$$
$$\frac{m^2}{2} - \frac{0^2}{2} = 1$$
$$\frac{m^2}{2} = 1$$
$$m^2 = 2$$
$$m = \sqrt{2}$$

55. $f(x) = k\,e^{-kx}, \ [0,\infty)$

$$\int_0^m f(x)\, dx = \frac{1}{2}$$
$$\int_0^m k\,e^{-kx} = \frac{1}{2}$$
$$\left[\frac{k}{-k}e^{-kx}\right]_0^m = \frac{1}{2}$$
$$\left[-e^{-kx}\right]_0^m = \frac{1}{2}$$
$$(-e^{-km}) - (-e^{-k\cdot 0}) = \frac{1}{2}$$
$$-e^{-km} + e^0 = \frac{1}{2}$$
$$-e^{-km} + 1 = \frac{1}{2}$$
$$\frac{1}{2} = e^{-km}$$
$$\ln\frac{1}{2} = \ln e^{-km}$$
$$\ln 1 - \ln 2 = -km$$
$$-\ln 2 = -km \qquad (\ln 1 = 0)$$
$$\frac{\ln 2}{k} = m$$

57. Standardize 6.5.
$$X = \frac{6.5 - \mu}{\sigma} = \frac{6.5 - \mu}{0.3}$$
We are looking for a value of t for which
$$P(x > t) = \frac{1}{100} = 0.01. \text{ Now if } t \geq 0, \text{ then}$$
$$P(X > t) = P(X \geq 0) - P(0 \leq X \leq t)$$
$$= 0.5000 - P(0 \leq X \leq t)$$
Then $P(0 \leq X \leq t) = 0.5000 - P(X > t)$
$$= 0.5000 - 0.01$$
$$= 0.4900$$

Looking in the body of the table for the number closest to 0.4900, we find 0.4901. The value of t that corresponds to 0.4901 is 2.33. Set X equal to 2.33 and solve for μ.
$$2.33 = \frac{6.5 - \mu}{0.3}$$
$$0.699 = 6.5 - \mu$$
$$\mu = 5.801$$

The setting of μ should be 5.801 oz.

59. $\boxed{tw}$

Exercise Set 6.6

---

1. Find the volume of the solid of revolution generated by rotating about the x-axis the region under the graph of

$$y = x$$

from $x = 0$ to $x = 1$.

$V = \int_a^b \pi\left[f(x)\right]^2 dx$ Volume of a solid of revolution

$V = \int_0^1 \pi\, x^2 \, dx$ Substituting 0 for a, 1 for b, and x for $f(x)$

$= \left[\pi \cdot \dfrac{x^3}{3}\right]_0^1$

$= \dfrac{\pi}{3}\left[x^3\right]_0^1$

$= \dfrac{\pi}{3}(1^3 - 0^3)$

$= \dfrac{\pi}{3}\cdot 1$

$= \dfrac{\pi}{3}$

3. Find the volume of the solid of revolution generated by rotating about the x-axis the region under the graph of

$$y = x$$

from $x = 1$ to $x = 2$.

$V = \int_a^b \pi\left[f(x)\right]^2 dx$ Volume of a solid of revolution

$V = \int_1^2 \pi\left[x\right]^2 dx$ Substituting 1 for a, 2 for b, and x for $f(x)$

$= \int_1^2 \pi\, x^2 \, dx$

$= \left[\pi \cdot \dfrac{x^3}{3}\right]_1^2$

$= \dfrac{\pi}{3}\left[x^3\right]_1^2$

$= \dfrac{\pi}{3}(2^3 - 1^3)$

$= \dfrac{\pi}{3}(8 - 1)$

$= \dfrac{7}{3}\pi$

5. Find the volume of the solid of revolution generated by rotating about the x-axis the region under the graph of

$$y = e^x$$

from $x = -2$ to $x = 5$.

$V = \int_a^b \pi\left[f(x)\right]^2 dx$ Volume of a solid of revolution

$V = \int_{-2}^5 \pi\left[e^x\right]^2 dx$ Substituting -2 for a, 5 for b, and e^x for $f(x)$

$= \int_{-2}^5 \pi\, e^{2x} \, dx$

$= \left[\pi \cdot \dfrac{1}{2}e^{2x}\right]_{-2}^5$

$= \dfrac{\pi}{2}\left[e^{2x}\right]_{-2}^5$

$= \dfrac{\pi}{2}\left(e^{2\cdot 5} - e^{2(-2)}\right)$

$= \dfrac{\pi}{2}\left(e^{10} - e^{-4}\right)$

7. Find the volume of the solid of revolution generated by rotating about the x-axis the region under the graph of

$$y = \dfrac{1}{x}$$

from $x = 1$ to $x = 3$.

$V = \int_a^b \pi\left[f(x)\right]^2 dx$ Volume of a solid of revolution

$V = \int_1^3 \pi\left[\dfrac{1}{x}\right]^2 dx$ Substituting 1 for a, 3 for b, and $\dfrac{1}{x}$ for $f(x)$

$= \int_1^3 \pi \cdot \dfrac{1}{x^2} \, dx$

$= \int_1^3 \pi\, x^{-2} \, dx$

$= \left[\pi\dfrac{x^{-1}}{-1}\right]_1^3$

$= -\pi\left[\dfrac{1}{x}\right]_1^3$

$= -\pi\left(\dfrac{1}{3} - \dfrac{1}{1}\right)$

$= -\pi \cdot \left(-\dfrac{2}{3}\right)$

$= \dfrac{2}{3}\pi$

9. Find the volume of the solid of revolution generated by rotating about the x-axis the region under the graph of

$$y = \dfrac{1}{\sqrt{x}}$$

from $x = 1$ to $x = 3$.

$V = \int_a^b \pi\left[f(x)\right]^2 dx$ Volume of a solid of revolution

$V = \int_1^3 \pi\left[\dfrac{1}{\sqrt{x}}\right]^2 dx$ Substituting 1 for a, 3 for b, and $\dfrac{1}{\sqrt{x}}$ for $f(x)$

$= \int_1^3 \pi \cdot \dfrac{1}{x} \, dx$

$= \pi\left[\ln x\right]_1^3$

$= \pi(\ln 3 - \ln 1)$

$= \pi \ln 3$ $(\ln 1 = 0)$

11. Find the volume of the solid of revolution generated by rotating about the x-axis the region under the graph of

$$y = 4$$

from $x = 1$ to $x = 3$.

$V = \int_a^b \pi\left[f(x)\right]^2 dx$ Volume of a solid of revolution

$V = \int_1^3 \pi\left[4\right]^2 dx$ Substituting 1 for a, 3 for b, and 4 for $f(x)$

$= \int_1^3 16\pi \, dx$

$= 16\pi\left[x\right]_1^3$

$= 16\pi(3 - 1)$

$= 32\pi$

13. Find the volume of the solid of revolution generated by rotating about the x-axis the region under the graph of

$$y = x^2$$

from $x = 0$ to $x = 2$.

$V = \int_a^b \pi [f(x)]^2 \, dx$ Volume of a solid of revolution

$V = \int_0^2 \pi \left[x^2\right]^2 dx$ Substituting 0 for a,
$\qquad\qquad\qquad\qquad$ 2 for b, and x^2 for $f(x)$

$= \int_0^2 \pi \, x^4 \, dx$

$= \left[\pi \cdot \dfrac{x^5}{5}\right]_0^2$

$= \dfrac{\pi}{5}(2^5 - 0^5)$

$= \dfrac{32}{5}\pi$

15. Find the volume of the solid of revolution generated by rotating about the x-axis the region under the graph of

$$y = \sqrt{1+x}$$

from $x = 2$ to $x = 10$.

$V = \int_a^b \pi [f(x)]^2 \, dx$ Volume of a solid of revolution

$V = \int_2^{10} \pi \left(\sqrt{1+x}\right)^2 dx$ Substituting 2 for a, 10 for b, and $\sqrt{1+x}$ for $f(x)$

$= \int_2^{10} \pi \, (1+x) \, dx$

$= \left[\pi \cdot \dfrac{(1+x)^2}{2}\right]_2^{10}$

$= \dfrac{\pi}{2}\left[(1+x)^2\right]_2^{10}$

$= \dfrac{\pi}{2}[(1+10)^2 - (1+2)^2]$

$= \dfrac{\pi}{2}(11^2 - 3^2)$

$= \dfrac{\pi}{2}(121 - 9)$

$= \dfrac{\pi}{2} \cdot 112$

$= 56\pi$

17. Find the volume of the solid of revolution generated by rotating about the x-axis the region under the graph of

$$y = \sqrt{4 - x^2}$$

from $x = -2$ to $x = 2$.

$V = \int_a^b \pi [f(x)]^2 \, dx$ Volume of a solid of revolution

$V = \int_{-2}^2 \pi \left(\sqrt{4-x^2}\right)^2 dx$ Substituting -2 for a, 2 for b, and $\sqrt{4-x^2}$ for $f(x)$

$= \int_{-2}^2 \pi (4 - x^2) \, dx$

$= \pi \left[4x - \dfrac{x^3}{3}\right]_{-2}^2$

$= \pi \left[\left(4 \cdot 2 - \dfrac{2^3}{3}\right) - \left(4 \cdot (-2) - \dfrac{(-2)^3}{3}\right)\right]$

$= \pi \left(8 - \dfrac{8}{3} + 8 - \dfrac{8}{3}\right)$

$= \pi \left(16 - \dfrac{16}{3}\right)$

$= \pi \left(\dfrac{48}{3} - \dfrac{16}{3}\right)$

$= \dfrac{32}{3}\pi$

19. Find the volume of the solid of revolution generated by rotating about the x-axis the region under the graph of

$$y = \sqrt{\ln x}$$

from $x = e$ to $x = e^3$.

$V = \int_a^b \pi [f(x)]^2 \, dx$ Volume of a solid of revolution

$V = \int_e^{e^3} \pi \left[\sqrt{\ln x}\right]^2 dx$ Substituting

$= \int_e^{e^3} \pi \ln x \, dx$

$= \pi \left[x \ln x - x\right]_e^{e^3}$ Using Formula 8

$= \pi[(e^3 \ln e^3 - e^3) - (e \ln e - e)]$

$= \pi[(3e^3 - e^3) - (e - e)]$

$= 2\pi e^3$

21. $V = \displaystyle\int_1^\infty \pi \left(\dfrac{1}{x}\right)^2 dx$

$= \displaystyle\int_1^\infty \pi \cdot \dfrac{1}{x^2} \, dx$

$= \displaystyle\int_1^\infty \pi \, x^{-2} \, dx$

$= \lim_{b \to \infty} \displaystyle\int_1^b \pi \, x^{-2} \, dx$

$= \lim_{b \to \infty} \left[\pi \dfrac{x^{-1}}{-1}\right]_1^b$

$= \lim_{b \to \infty} \left[-\dfrac{\pi}{x}\right]_1^b$

$= \lim_{b \to \infty} \left[-\dfrac{\pi}{b} - \left(-\dfrac{\pi}{1}\right)\right]$

$= \lim_{b \to \infty} \left(-\dfrac{\pi}{b} + \pi\right)$

$= \pi$ $\left(\text{As } b \to \infty, \dfrac{\pi}{b} \to 0.\right)$

Exercise Set 6.7

1. Solve: $y' = 4x^3$.
$$y = \int 4x^3 \, dx$$
Solution: $y = x^4 + C$

This solution is called a general solution because taking all values of C gives all the solutions. Taking specific values of C gives particular solutions. The following are particular solutions of $y' = 4x^3$. Answers may vary.

$$y = x^4 - 13$$
$$y = x^4$$
$$y = x^4 + \frac{7}{8}$$

3. Solve: $y' = e^{2x} + x$.
$$y = \int (e^{2x} + x) \, dx$$
Solution: $y = \frac{1}{2}e^{2x} + \frac{1}{2}x^2 + C$

This solution is called a general solution because taking all values of C gives all the solutions. Taking specific values of C gives particular solutions. The following are particular solutions of $y' = e^{2x} + x$. Answers may vary.

$$y = \frac{1}{2}e^{2x} + \frac{1}{2}x^2 + \frac{5}{8}$$
$$y = \frac{1}{2}e^{2x} + \frac{1}{2}x^2 - 7$$
$$y = \frac{1}{2}e^{2x} + \frac{1}{2}x^2 + 83$$

5. Solve: $y' = \dfrac{3}{x} - x^2 + x^5$
$$y = \int \left(\frac{3}{x} - x^2 + x^5 \right) dx$$
Solution: $y = 3 \ln x - \dfrac{1}{3}x^3 + \dfrac{1}{6}x^6 + C$

This solution is called a general solution because taking all values of C gives all the solutions. Taking specific values of C gives particular solutions. The following are particular solutions of $y' = \dfrac{3}{x} - x^2 + x^5$. Answers may vary.

$$y = 3 \ln x - \frac{1}{3}x^3 + \frac{1}{6}x^6$$
$$y = 3 \ln x - \frac{1}{3}x^3 + \frac{1}{6}x^6 - 10$$
$$y = 3 \ln x - \frac{1}{3}x^3 + \frac{1}{6}x^6 + \frac{11}{16}$$

7. Solve $y' = x^2 + 2x - 3$, given that $y = 4$ when $x = 0$.
First find the general solution.
$$y = \int y' \, dx$$
$$y = \frac{1}{3}x^3 + x^2 - 3x + C$$
We substitute to find C.
$$4 = \frac{1}{3} \cdot 0^3 + 0^2 - 3 \cdot 0 + C \quad \text{Substituting 4 for } y$$
$$\text{and 0 for } x$$
$$4 = C$$

Thus the solution is
$$y = \frac{1}{3}x^3 + x^2 - 3x + 4.$$

9. Solve $f'(x) = x^{2/3} - x$, given that $f(1) = -6$. First find the general solution.
$$f(x) = \int f'(x) \, dx$$
$$f(x) = \frac{3}{5}x^{5/3} - \frac{1}{2}x^2 + C$$
We substitute to find C.
$$-6 = \frac{3}{5} \cdot 1^{5/3} - \frac{1}{2} \cdot 1^2 + C$$
$$\text{Substituting } -6 \text{ for } f(x)$$
$$\text{and 1 for } x$$
$$-6 = \frac{3}{5} - \frac{1}{2} + C$$
$$-6 - \frac{3}{5} + \frac{1}{2} = C$$
$$-\frac{60}{10} - \frac{6}{10} + \frac{5}{10} = C$$
$$-\frac{61}{10} = C$$

Thus the solution is $f(x) = \dfrac{3}{5}x^{5/3} - \dfrac{1}{2}x^2 - \dfrac{61}{10}$.

11. Show that $y = x \ln x + 3x - 2$ is a solution to $y'' - \dfrac{1}{x} = 0$.

First find y' and y''.
$$y = x \ln x + 3x - 2$$
$$y' = x \cdot \frac{1}{x} + 1 \cdot \ln x + 3$$
$$= \ln x + 4$$
$$y'' = \frac{1}{x}$$

Substitute as follows in the differential equation.

$$
\begin{array}{c|c}
y'' - \dfrac{1}{x} = 0 & \\
\hline
\dfrac{1}{x} - \dfrac{1}{x} & 0 \\
0 &
\end{array}
$$

Thus, $y = x \ln x + 3x - 2$ is a solution of $y'' - \dfrac{1}{x} = 0$.

13. Show that $y = e^x + 3xe^x$ is a solution to $y'' - 2y' + y = 0$.
First find y' and y''.
$$y = e^x + 3xe^x$$
$$y' = e^x + 3(xe^x + e^x)$$
$$= e^x + 3xe^x + 3e^x$$
$$= 4e^x + 3xe^x$$
$$y'' = 4e^x + 3(xe^x + e^x)$$
$$= 4e^x + 3xe^x + 3e^x$$
$$= 7e^x + 3xe^x$$

Substitute as follows in the differential equation.

$$y'' - 2y' + y = 0$$

$(7e^x + 3xe^x) - 2(4e^x + 3xe^x) + (e^x + 3xe^x)$	0
$7e^x + 3xe^x - 8e^x - 6xe^x + e^x + 3xe^x$	
$(7 - 8 + 1)e^x + (3 - 6 + 3)xe^x$	
$0 + 0$	
0	

Thus, $y = e^x + 3xe^x$ is a solution of $y'' - 2y' + y = 0$.

15. Solve $\dfrac{dy}{dx} = 4x^3 y$.

We first separate variables as follows.

$dy = 4x^3 y \, dx$ Multiplying by dx

$\dfrac{dy}{y} = 4x^3 \, dx$ Multiplying by $\dfrac{1}{y}$

We then integrate both sides.

$$\int \dfrac{dy}{y} = \int 4x^3 \, dx$$

$\ln y = x^4 + C$ $(y > 0)$

$y = e^{x^4 + C}$ Definition of logarithms

$y = e^{x^4} \cdot e^C$

Thus

$y = C_1 \, e^{x^4}$, where $C_1 = e^C$.

17. Solve $3y^2 \dfrac{dy}{dx} = 5x$

We first separate variables as follows.

$3y^2 \, dy = 5x \, dx$ Multiplying by dx

We then integrate both sides.

$$\int 3y^2 \, dy = \int 5x \, dx$$

$3 \cdot \dfrac{y^3}{3} = 5 \cdot \dfrac{x^2}{2} + C$

$y^3 = \dfrac{5}{2}x^2 + C$

$y = \sqrt[3]{\dfrac{5}{2}x^2 + C}$

19. Solve $\dfrac{dy}{dx} = \dfrac{2x}{y}$.

We first separate variables as follows.

$dy = \dfrac{2x}{y} \, dx$ Multiplying by dx

$y \, dy = 2x \, dx$ Multiplying by y

We then integrate both sides.

$$\int y \, dy = \int 2x \, dx$$

$\dfrac{y^2}{2} = x^2 + C$

$y^2 = 2x^2 + 2C$

$y^2 = 2x^2 + C_1$ $(C_1 = 2C)$

Thus, $y = \sqrt{2x^2 + C_1}$ and $y = -\sqrt{2x^2 + C_1}$, where $C_1 = 2C$.

21. Solve $\dfrac{dy}{dx} = \dfrac{3}{y}$.

We first separate variables as follows.

$dy = \dfrac{3}{y} \, dx$

$y \, dy = 3 \, dx$

We then integrate both sides.

$$\int y \, dy = \int 3 \, dx$$

$\dfrac{y^2}{2} = 3x + C$

$y^2 = 6x + 2C$

$y^2 = 6x + C_1$ $(C_1 = 2C)$

Thus, $y = \sqrt{6x + C_1}$ and $y = -\sqrt{6x + C_1}$, where $C_1 = 2C$.

23. Solve $y' = 3x + xy$.

We first separate variables as follows.

$\dfrac{dy}{dx} = 3x + xy$ Replacing y' with $\dfrac{dy}{dx}$

$\dfrac{dy}{dx} = x(3 + y)$

$dy = x(3 + y) \, dx$ Multiplying by dx

$\dfrac{dy}{3 + y} = x \, dx$ Multiplying by $\dfrac{1}{3 + y}$

We then integrate both sides.

$$\int \dfrac{dy}{3 + y} = \int x \, dx$$

$\ln(3 + y) = \dfrac{x^2}{2} + C$ $(3 + y > 0)$

$3 + y = e^{x^2/2 + C}$

$3 + y = e^{x^2/2} \cdot e^C$

$y = e^{x^2/2} \cdot e^C - 3$

$y = C_1 \, e^{x^2/2} - 3$ $(C_1 = e^C)$

We substitute to find C_1.

$5 = C_1 \, e^{0^2/2} - 3$ Substituting 0 for x and 5 for y

$8 = C_1 \, e^0$

$8 = C_1 \cdot 1$

$8 = C_1$

Thus the solution is $y = 8 \, e^{x^2/2} - 3$.

25. Solve $y' = 5y^{-2}$.

We first separate variables as follows.

$\dfrac{dy}{dx} = \dfrac{5}{y^2}$ Replacing y' with $\dfrac{dy}{dx}$

$y^2 \dfrac{dy}{dx} = 5$

$y^2 \, dy = 5 \, dx$

We then integrate both sides.

$\int y^2 \, dy = \int 5 \, dx$

$\dfrac{y^3}{3} = 5x + C$

$y^3 = 15x + 3C$

$y^3 = 15x + C_1 \quad (C_1 = 3C)$

We substitute to find C_1.

$3^3 = 15 \cdot 2 + C_1 \quad$ Substituting 3 for y and 2 for x

$27 = 30 + C_1$

$-3 = C_1$

Thus,

$y^3 = 15x - 3$

$y = \sqrt[3]{15x - 3}$

27. Solve $\dfrac{dy}{dx} = 3y$.

We first separate variables as follows.

$dy = 3y \, dx$

$\dfrac{dy}{y} = 3 \, dx$

We then integrate both sides.

$\int \dfrac{dy}{y} = \int 3 \, dx$

$\ln y = 3x + C \quad (y > 0)$

$y = e^{3x+C} \quad$ Definition of logarithms

$y = e^{3x} \cdot e^C$

Thus $y = C_1 \, e^{3x}$, where $C_1 = e^C$.

29. Solve $\dfrac{dP}{dt} = 2P$.

We first separate variables as follows.

$dP = 2P \, dt$

$\dfrac{dP}{P} = 2 \, dt$

We then integrate both sides.

$\int \dfrac{dP}{P} = \int 2 \, dt$

$\ln P = 2t + C \quad (P > 0)$

$P = e^{2t+C}$

$P = e^{2t} \cdot e^C$

Thus $P = C_1 \, e^{2t}$, where $C_1 = e^C$.

31. Solve $f'(x) = \dfrac{1}{x} - 4x + \sqrt{x}$, given that $f(1) = \dfrac{23}{3}$.

$f(x) = \int f'(x) \, dx$

$f(x) = \ln x - 2x^2 + \dfrac{2}{3}x^{3/2} + C$

Substitute to find C.

$\dfrac{23}{3} = \ln 1 - 2 \cdot 1^2 + \dfrac{2}{3} \cdot 1^{3/2} + C$

$\dfrac{23}{3} = 0 - 2 + \dfrac{2}{3} + C$

$\dfrac{23}{3} = -\dfrac{4}{3} + C$

$\dfrac{27}{3} = C$

$9 = C$

Then $f(x) = \ln x - 2x^2 + \dfrac{2}{3}x^{3/2} + 9$.

33. $C'(x) = 2.6 - 0.02x$

$C(x) = \int C'(x) \, dx = 2.6x - 0.01x^2 + K.$

Using K for the constant

We substitute to find K.

$C(0) = 2.6(0) - 0.01(0^2) + K = 120$

$K = 120$

Thus,

$C(x) = 2.6x - 0.01x^2 + 120$ and

$A(x) =$ the average cost of producing x units

$\qquad = \dfrac{C(x)}{x}$

$\qquad = \dfrac{2.6x - 0.01x^2 + 120}{x}$

$\qquad = 2.6 - 0.01x + \dfrac{120}{x}$

35. a) $\quad \dfrac{dP}{dC} = \dfrac{-200}{(C+3)^{3/2}}$

$P(C) = \int \dfrac{-200}{(C+3)^{3/2}} \, dC$

$\qquad = -200 \int (C+3)^{-3/2} \, dC$

$\qquad = -200 \dfrac{(C+3)^{-1/2}}{-1/2} + K$

$\qquad = \dfrac{400}{\sqrt{C+3}} + K \qquad$ Using K for the constant

We substitute to find K.

$P(C) = \dfrac{400}{\sqrt{C+3}} + K$

$10 = \dfrac{400}{\sqrt{61+3}} + K \quad$ Substituting 10 for P and 61 for C

$10 = \dfrac{400}{8} + K$

$10 = 50 + K$

$-40 = K$

Thus the solution is $P(C) = \dfrac{400}{\sqrt{C+3}} - 40.$

b) The firm will break even when $P = 0$.

$$P = \frac{400}{\sqrt{C+3}} - 40$$

$$0 = \frac{400}{\sqrt{C+3}} - 40 \quad \text{Substituting 0 for } P$$

$$40 = \frac{400}{\sqrt{C+3}}$$

$$40\sqrt{C+3} = 400$$

$$\sqrt{C+3} = 10$$

$$C+3 = 100$$

$$C = 97$$

When the total cost is $97, the firm breaks even.

37. a) $\dfrac{dR}{dS} = \dfrac{k}{S+1}$

We first separate variables.

$$dR = \frac{k}{S+1}dS \qquad \text{Multiplying by } dS$$

We then integrate both sides.

$$\int dR = \int \frac{k}{S+1}dS$$

$$R = k\ln(S+1) + C \qquad (S+1) > 0$$

b) Substitute to find C.

$$R = k\ln(S+1) + C$$

$$0 = k\ln(0+1) + C$$

$$0 = k\ln 1 + C$$

$$0 = k \cdot 0 + C$$

$$0 = C$$

Thus, $R = k\ln(S+1)$.

c) $R(0) = 0$ is reasonable because there is no pleasure from 0 units.

39. $E(p) = \dfrac{p}{200-p}; \; x = 190$ when $p = 10$

$$E(p) = -\frac{p}{x} \cdot \frac{dx}{dp}$$

$$-\frac{p}{x} \cdot \frac{dx}{dp} = \frac{p}{200-p}$$

We separate variables.

$$\frac{dx}{x} = \frac{p}{200-p} \cdot \left(-\frac{dp}{p}\right)$$

$$\frac{dx}{x} = -\frac{dp}{200-p}$$

We then integrate both sides.

$$\int \frac{dx}{x} = \int -\frac{dp}{200-p}$$

$$\ln x = \ln(200-p) + C$$

We substitute to find C.

$$\ln 190 = \ln(200-10) + C$$

$$\ln 190 = \ln 190 + C$$

$$0 = C$$

Thus, $\ln x = \ln(200-p)$

$$x = 200 - p$$

The demand function is $x = 200 - p$.

41. $E(p) = n$ for some constant n and all $p > 0$

$$E(p) = -\frac{p}{x} \cdot \frac{dx}{dp}$$

$$-\frac{p}{x} \cdot \frac{dx}{dp} = n$$

We separate variables.

$$\frac{dx}{x} = n\left(-\frac{dp}{p}\right)$$

$$\frac{dx}{x} = -\frac{n}{p}dp$$

We then integrate both sides.

$$\int \frac{dx}{x} = \int -\frac{n}{p}dp$$

$$\ln x = -n\ln p + C$$

$$\ln x = -n\ln p + \ln C_1 \qquad (\ln C_1 = C)$$

$$\ln x = \ln C_1 - \ln p^n \qquad (-n\ln p = -\ln p^n)$$

$$\ln x = \ln\left(\frac{C_1}{p^n}\right)$$

$$x = \frac{C_1}{p^n}$$

43. $\dfrac{dR}{dS} = k \cdot \dfrac{R}{S}$

We first separate variables.

$$\frac{dR}{R} = k \cdot \frac{dS}{S} \qquad \text{Multiplying by } \frac{dS}{R}$$

We then integrate both sides.

$$\int \frac{dR}{R} = \int k \cdot \frac{dS}{S}$$

$$\ln R = k\ln S + C$$

$$\ln R = k\ln S + \ln C_1 \qquad (\ln C_1 = C)$$

$$\ln R = \ln S^k + \ln C_1 \qquad (k\ln S = \ln S^k)$$

$$\ln R = \ln(S^k \cdot C_1)$$

$$R = C_1 \cdot S^k, \text{ where } C_1 = e^C$$

45. $\dfrac{dy}{dx} = \dfrac{5}{y}$

First we separate the variables.

$$y\,dy = 5\,dx$$

We then integrate both sides.

$$\int y\,dy = \int 5\,dx$$

$$\frac{y^2}{2} = 5x + C$$

$$y^2 = 10x + 2C$$

$$y^2 = 10x + C_1 \qquad (C_1 = 2C)$$

Thus, $y = \sqrt{10x + C_1}$ and $y = -\sqrt{10x + C}$, where $C_1 = 2C$.

$$y_1 = \sqrt{10x + C_1}, \quad y_2 = -\sqrt{10x + C_1}$$

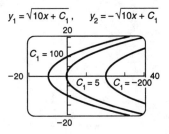

Chapter 7

Functions of Several Variables

1. $f(x,y) = x^2 - 2xy$

$f(0,-2) = 0^2 - 2 \cdot 0 \cdot (-2)$ Substituting 0 for x
and -2 for y

$\quad = 0 - 0$

$\quad = 0$

$f(2,3) = 2^2 - 2 \cdot 2 \cdot 3$ Substituting 2 for x
and 3 for y

$\quad = 4 - 12$

$\quad = -8$

$f(10,-5) = 10^2 - 2 \cdot 10 \cdot (-5)$ Substituting 10 for
x and -5 for y

$\quad = 100 + 100$

$\quad = 200$

3. $f(x,y) = 3^x + 7xy$

$f(0,-2) = 3^0 + 7 \cdot 0 \cdot (-2)$ Substituting 0 for x
and -2 for y

$\quad = 1 + 0$

$\quad = 1$

$f(-2,1) = 3^{-2} + 7 \cdot (-2) \cdot 1$ Substituting -2 for
x and 1 for y

$\quad = \frac{1}{9} - 14$

$\quad = -13\frac{8}{9},\ \text{or}\ -\frac{125}{9}$

$f(2,1) = 3^2 + 7 \cdot 2 \cdot 1$ Substituting 2 for x
and 1 for y

$\quad = 9 + 14$

$\quad = 23$

5. $f(x,y) = \ln x + y^3$

$f(e,2) = \ln e + 2^3$ Substituting e for x
and 2 for y

$\quad = 1 + 8$

$\quad = 9$

$f(e^2,4) = \ln e^2 + 4^3$ Substituting e^2 for x
and 4 for y

$\quad = 2 + 64$

$\quad = 66$

$f(e^3,5) = \ln e^3 + 5^3$ Substituting e^3 for x
and 5 for y

$\quad = 3 + 125$

$\quad = 128$

7. $f(x,y,z) = x^2 - y^2 - z^2$

$f(-1,2,3) = (-1)^2 - 2^2 + 3^2$ Substituting -1 for x,
2 for y, and 3 for z

$\quad = 1 - 4 + 9$

$\quad = 6$

$f(2,-1,3) = 2^2 - (-1)^2 + 3^2$ Substituting 2 for x,
-1 for y, and 3 for z

$\quad = 4 - 1 + 9$

$\quad = 12$

9. $C_2 = \left(\dfrac{V_2}{V_1}\right)^{0.6} C_1$

$C_2 = \left(\dfrac{160,000}{80,000}\right)^{0.6} \cdot 100,000$

Substituting 100,000 for C_1,
80,000 for V_1, and 160,000 for V_2

$\quad = 2^{0.6}(100,000)$

$\quad \approx (1.5157166)(100,000)$

$\quad \approx \$151,571.66$

11. $Y(D,P) = \dfrac{D}{P}$

$Y\left(\$1.20, \$50\frac{7}{16}\right) = \dfrac{\$1.20}{\$50\frac{7}{16}}$ Substituting $\$1.20$ for
D and $\$50\frac{7}{16}$ for P

$\quad = \dfrac{1.20}{50.4375}$

$\quad \approx 0.0237$

$\quad \approx 2.4\%$

13. $S = \dfrac{aV}{0.51d^2}$

$S = \dfrac{0.78(1,600,000)}{0.51(100)^2}$ Substituting 0.78 for a,
1,600,000 for V, and 100
for d

$\quad = \dfrac{1,248,000}{5100}$

$\quad \approx 244.7\ \text{mph}$

15. $\boxed{tw}$

17. $W(v, T) = 91.4 - \dfrac{(10.45 + 6.68\sqrt{v} - 0.447v)(457 - 5T)}{110}$

$W(25, 30)$

$= 91.4 - \dfrac{(10.45 + 6.68\sqrt{25} - 0.447 \cdot 25)(457 - 5 \cdot 30)}{110}$

$= 91.4 - \dfrac{(10.45 + 33.4 - 11.175)(307)}{110}$

$= 91.4 - \dfrac{(32.675)(307)}{110}$

$= 91.4 - \dfrac{10,031.225}{110}$

$= 91.4 - 91.2$

$\approx 0° \text{ F}$ To the nearest degree

19. $W(v, T) = 91.4 - \dfrac{(10.45 + 6.68\sqrt{v} - 0.447v)(457 - 5T)}{110}$

$W(40, 20)$

$= 91.4 - \dfrac{(10.45 + 6.68\sqrt{40} - 0.447 \cdot 40)(457 - 5 \cdot 20)}{110}$

$= 91.4 - \dfrac{(10.45 + 42.248 - 17.88)(357)}{110}$

$= 91.4 - \dfrac{(34.818)(357)}{110}$

$= 91.4 - \dfrac{12,430.026}{110}$

$= 91.4 - 113.0$

$\approx -22° \text{ F}$ To the nearest degree

21. See the answer section in the text.

23. See the answer section in the text.

25. See the answer section in the text.

Exercise Set 7.2

1. $z = 2x - 3xy$

Find $\dfrac{\partial z}{\partial x}$.

$z = 2\underline{x} - 3\underline{x}y$ The variable is underlined; y is treated as a constant.

$\dfrac{\partial z}{\partial x} = 2 - 3y$

Find $\dfrac{\partial z}{\partial y}$.

$z = 2x - 3x\underline{y}$ The variable is underlined; x is treated as a constant.

$\dfrac{\partial z}{\partial y} = -3x$

Find: $\dfrac{\partial z}{\partial x}\Big|_{(-2,-3)}$

$\dfrac{\partial z}{\partial x} = 2 - 3y$

$\dfrac{\partial z}{\partial x}\Big|_{(-2,-3)} = 2 - 3(-3)$ Substituting -3 for y

$= 2 + 9$

$= 11$

Find: $\dfrac{\partial z}{\partial y}\Big|_{(0,-5)}$

$\dfrac{\partial z}{\partial y} = -3x$

$\dfrac{\partial z}{\partial y}\Big|_{(0,-5)} = -3(0)$ Substituting 0 for x

$= 0$

3. $z = 3x^2 - 2xy + y$

Find $\dfrac{\partial z}{\partial x}$.

$z = 3\underline{x}^2 - 2\underline{x}y + y$ The variable is underlined; y is treated as a constant.

$\dfrac{\partial z}{\partial x} = 6x - 2y$

Find $\dfrac{\partial z}{\partial y}$.

$z = 3x^2 - 2x\underline{y} + \underline{y}$ The variable is underlined; x is treated as a constant.

$\dfrac{\partial z}{\partial y} = -2x + 1$

Find: $\dfrac{\partial z}{\partial x}\Big|_{(-2,-3)}$

$\dfrac{\partial z}{\partial x} = 6x - 2y$

$\dfrac{\partial z}{\partial x}\Big|_{(-2,-3)} = 6(-2) - 2(-3)$ Substituting -2 for x and -3 for y

$= -12 + 6$

$= -6$

Find: $\dfrac{\partial z}{\partial y}\Big|_{(0,-5)}$

$\dfrac{\partial z}{\partial y} = -2x + 1$

$\dfrac{\partial z}{\partial y}\Big|_{(0,-5)} = -2 \cdot 0 + 1$ Substituting 0 for x

$= 1$

5. $f(x, y) = 2x - 3y$

Find f_x.

$f(x, y) = 2\underline{x} - 3y$ The variable is underlined; y is treated as a constant.

$f_x = 2$

Find f_y.

$f(x, y) = 2x - 3\underline{y}$ The variable is underlined; x is treated as a constant.

$f_y = -3$

Find $f_x(-2, 4)$.

$f_x = 2$

f_x has the value 2 for any x and y.

$f_x(-2, 4) = 2$

Find $f_y(4, -3)$.

$f_y = -3$

f_y has the value -3 for any x and y.

$f_y(4, -3) = -3$

7. $f(x,y) = \sqrt{x^2 + y^2}$

$\quad\quad = (x^2 + y^2)^{1/2}$

Find f_x.

$\quad f(x,y) = (\underline{x}^2 + y^2)^{1/2}$ The variable is underlined.

$\quad f_x = \frac{1}{2}(x^2 + y^2)^{-1/2} \cdot 2x$

$\quad\quad = x(x^2 + y^2)^{-1/2}$, or $\dfrac{x}{\sqrt{x^2 + y^2}}$

Find f_y.

$\quad f(x,y) = (x^2 + \underline{y}^2)^{1/2}$ The variable is underlined.

$\quad f_y = \frac{1}{2}(x^2 + y^2)^{-1/2} \cdot 2y$

$\quad\quad = y(x^2 + y^2)^{-1/2}$, or $\dfrac{y}{\sqrt{x^2 + y^2}}$

Find $f_x(-2, 1)$.

$\quad f_x = \dfrac{x}{\sqrt{x^2 + y^2}}$

$\quad f_x(-2,1) = \dfrac{-2}{\sqrt{(-2)^2 + 1^2}}$ Substituting -2 for x and 1 for y

$\quad\quad = \dfrac{-2}{\sqrt{5}}$

Find $f_y(-3, -2)$.

$\quad f_y = \dfrac{y}{\sqrt{x^2 + y^2}}$

$\quad f_y(-3,-2) = \dfrac{-2}{\sqrt{(-3)^2 + (-2)^2}}$ Substituting -3 for x and -2 for y

$\quad\quad = \dfrac{-2}{\sqrt{13}}$

9. $f(x,y) = e^{2x+3y}$

Find f_x.

$\quad f(x,y) = e^{2\underline{x}+3y}$ The variable is underlined.

$\quad f_x = 2e^{2x+3y}$

Find f_y.

$\quad f(x,y) = e^{2x+3\underline{y}}$ The variable is underlined.

$\quad f_y = 3e^{2x+3y}$

11. $f(x,y) = e^{xy}$

Find f_x.

$\quad f(x,y) = e^{\underline{x}y}$ The variable is underlined.

$\quad f_x = y\, e^{xy}$

Find f_y.

$\quad f(x,y) = e^{x\underline{y}}$ The variable is underlined.

$\quad f_y = x\, e^{xy}$

13. $f(x,y) = y \ln(x+y)$

Find f_x.

$\quad f(x,y) = y \ln(\underline{x}+y)$ The variable is underlined.

$\quad f_x = y \cdot \dfrac{1}{x+y}$

$\quad\quad = \dfrac{y}{x+y}$

Find f_y.

$\quad f(x,y) = \underline{y} \ln(x+\underline{y})$ The variable is underlined.

$\quad f_y = y \cdot \dfrac{1}{x+y} + 1 \cdot \ln(x+y)$

$\quad\quad = \dfrac{y}{x+y} + \ln(x+y)$

15. $f(x,y) = x \ln(xy)$

Find f_x.

$\quad f(x,y) = \underline{x} \ln(\underline{x}y)$ The variable is underlined.

$\quad f_x = x \cdot \left(y \cdot \dfrac{1}{xy}\right) + 1 \cdot \ln(xy)$

$\quad\quad = 1 + \ln(xy)$

Find f_y.

$\quad f(x,y) = x \ln(x\underline{y})$ The variable is underlined.

$\quad f_y = x \cdot \left(x \cdot \dfrac{1}{xy}\right)$

$\quad\quad = \dfrac{x}{y}$

17. $f(x,y) = \dfrac{x}{y} - \dfrac{y}{x}$

Find f_x.

$\quad f(x,y) = \dfrac{x}{y} - \dfrac{y}{x}$

$\quad\quad = \dfrac{1}{y} \cdot \underline{x} - y \cdot \underline{x}^{-1}$ The variable is underlined.

$\quad f_x = \dfrac{1}{y} - y \cdot (-1) \cdot x^{-2}$

$\quad\quad = \dfrac{1}{y} + \dfrac{y}{x^2}$

Find f_y.

$\quad f(x,y) = \dfrac{x}{y} - \dfrac{y}{x}$

$\quad\quad = x \cdot \underline{y}^{-1} - \dfrac{1}{x} \cdot \underline{y}$ The variable is underlined.

$\quad f_y = x \cdot (-1) \cdot y^{-2} - \dfrac{1}{x}$

$\quad\quad = -\dfrac{x}{y^2} - \dfrac{1}{x}$

19. $f(x,y) = 3(2x+y-5)^2$

Find f_x.

$\quad f(x,y) = 3(2\underline{x}+y-5)^2$ The variable is underlined.

$\quad f_x = 3 \cdot 2(2x+y-5) \cdot 2$

$\quad\quad = 12(2x+y-5)$

Find f_y.

$\quad f(x,y) = 3(2x+\underline{y}-5)^2$ The variable is underlined.

$\quad f_y = 3 \cdot 2(2x+y-5) \cdot 1$

$\quad\quad = 6(2x+y-5)$

21. $f(b, m) = (m + b - 4)^2 + (2m + b - 5)^2 + (3m + b - 6)^2$

Find $\dfrac{\partial f}{\partial b}$.

$f(b, m) = (m + \underline{b} - 4)^2 + (2m + \underline{b} - 5)^2 + (3m + \underline{b} - 6)^2$

                 The variable is underlined.

$\dfrac{\partial f}{\partial b} = 2(m + b - 4) \cdot 1 + 2(2m + b - 5) \cdot 1 + 2(3m + b - 6) \cdot 1$

$\quad = 2m + 2b - 8 + 4m + 2b - 10 + 6m + 2b - 12$

$\quad = 12m + 6b - 30$

Find $\dfrac{\partial f}{\partial m}$.

$f(b, m) = (\underline{m} + b - 4)^2 + (2\underline{m} + b - 5)^2 + (3\underline{m} + b - 6)^2$

                 The variable is underlined.

$\dfrac{\partial f}{\partial m} = 2(m + b - 4) \cdot 1 + 2(2m + b - 5) \cdot 2 + 2(3m + b - 6) \cdot 3$

$\quad = 2m + 2b - 8 + 8m + 4b - 20 + 18m + 6b - 36$

$\quad = 28m + 12b - 64$

23. $f(x, y, \lambda) = 3xy - \lambda(2x + y - 8)$

Find f_x.

$f(x, y, \lambda) = 3\underline{x}y - \lambda(2\underline{x} + y - 8)$ The variable is underlined.

$f_x = 3y - \lambda \cdot 2$

$\quad = 3y - 2\lambda$

Find f_y.

$f(x, y, \lambda) = 3x\underline{y} - \lambda(2x + \underline{y} - 8)$ The variable is underlined.

$f_y = 3x - \lambda \cdot 1$

$\quad = 3x - \lambda$

Find f_λ.

$f(x, y, \lambda) = 3xy - \underline{\lambda}(2x + y - 8)$ The variable is underlined.

$f_\lambda = -(2x + y - 8)$

25. $f(x, y, \lambda) = x^2 + y^2 - \lambda(10x + 2y - 4)$

Find f_x.

$f(x, y, \lambda) = \underline{x}^2 + y^2 - \lambda(10\underline{x} + 2y - 4)$

                 The variable is underlined.

$f_x = 2x - \lambda \cdot 10$

$\quad = 2x - 10\lambda$

Find f_y.

$f(x, y, \lambda) = x^2 + \underline{y}^2 - \lambda(10x + 2\underline{y} - 4)$

                 The variable is underlined.

$f_y = 2y - \lambda \cdot 2$

$\quad = 2y - 2\lambda$

Find f_λ.

$f(x, y, \lambda) = x^2 + y^2 - \underline{\lambda}(10x + 2y - 4)$

                 The variable is underlined.

$f_\lambda = -(10x + 2y - 4)$

27. a) $p(x, y) = 1800 \, x^{0.621} \, y^{0.379}$

$p(2500, 1700) = 1800(2500)^{0.621}(1700)^{0.379}$

           Substituting 2500 for x and 1700 for y

$\quad = 1800(128.860777)(16.762552)$

$\quad \approx 3,888,064 \text{ units}$

b) Find $\dfrac{\partial p}{\partial x}$.

$p(x, y) = 1800 \, \underline{x}^{0.621} \, y^{0.379}$ The variable is underlined.

$\dfrac{\partial p}{\partial x} = 1800(0.621 \, x^{-0.379}) \, y^{0.379}$

$\quad - 1117.8 \, x^{-0.379} \, y^{0.379}$

$\quad = 1117.8 \left(\dfrac{y}{x}\right)^{0.379}$

Find $\dfrac{\partial p}{\partial y}$.

$p(x, y) = 1800 \, x^{0.621} \, \underline{y}^{0.379}$ The variable is underlined.

$\dfrac{\partial p}{\partial y} = 1800 \, x^{0.621} \, (0.379 \, y^{-0.621})$

$\quad = 682.2 \, x^{0.621} \, y^{-0.621}$

$\quad = 682.2 \left(\dfrac{x}{y}\right)^{0.621}$

c) $\boxed{tw}$

d) $\dfrac{\partial p}{\partial x} = 1117.8 \left(\dfrac{y}{x}\right)^{0.379}$

$\left.\dfrac{\partial p}{\partial x}\right|_{(2500, 1700)} = 1117.8 \left(\dfrac{1700}{2500}\right)^{0.379}$ Substituting

$\quad = 1117.8(0.68)^{0.379}$

$\quad = 1117.8(0.864014)$

$\quad \approx 965.8$

$\dfrac{\partial p}{\partial y} = 682.2 \left(\dfrac{x}{y}\right)^{0.621}$

$\left.\dfrac{\partial p}{\partial y}\right|_{(2500, 1700)} = 682.2 \left(\dfrac{2500}{1700}\right)^{0.621}$

                    Substituting

$\quad = 682.2(1.470588)^{0.621}$

$\quad = 682.2(1.270609)$

$\quad \approx 866.8$

29. $T_h = 1.98T - 1.09(1 - H)(T - 58) - 56.9$

When $T = 85°$ and $H = 60\%$, or 0.6:

$T_h = 1.98(85) - 1.09(1 - 0.6)(85 - 58) - 56.9$

$\quad = 168.3 - 1.09(0.4)(27) - 56.9$

$\quad = 168.3 - 11.772 - 56.9$

$\quad = 99.628$

$\quad \approx 99.6°$ To the nearest tenth

31. $T_h = 1.98T - 1.09(1 - H)(T - 58) - 56.9$

When $T = 90°$ and $H = 100\%$, or 1:

$T_h = 1.98(90) - 1.09(1 - 1)(90 - 58) - 56.9$

$\quad = 178.2 - 1.09(0)(32) - 56.9$

$\quad = 178.2 - 0 - 56.9$

$\quad = 121.3°$

33. $\boxed{tw}$

35. $E = 206.835 - 0.846w - 1.015s$

When $w = 146$ and $s = 5$:

$E = 206.835 - 0.846(146) - 1.015(5)$

$\quad = 206.835 - 123.516 - 5.075$

$\quad = 78.244$

37. $E = 206.835 - 0.846w - 1.015s$

$\dfrac{\partial E}{\partial w} = -0.846$

39. $f(x,t) = \dfrac{x^2 + t^2}{x^2 - t^2}$

Find f_x.

$f(x,t) = \dfrac{\underline{x}^2 + t^2}{\underline{x}^2 - t^2}$ The variable is underlined.

$f_x = \dfrac{(x^2 - t^2)(2x) - 2x(x^2 + t^2)}{(x^2 - t^2)^2}$

$\quad = \dfrac{2x^3 - 2xt^2 - 2x^3 - 2xt^2}{(x^2 - t^2)^2}$

$\quad = \dfrac{-4xt^2}{(x^2 - t^2)^2}$

Find f_t.

$f(x,t) = \dfrac{x^2 + \underline{t}^2}{x^2 - \underline{t}^2}$ The variable is underlined.

$f_t = \dfrac{(x^2 - t^2)(2t) - (-2t)(x^2 + t^2)}{(x^2 - t^2)^2}$

$\quad = \dfrac{2x^2t - 2t^3 + 2x^2t + 2t^3}{(x^2 - t^2)^2}$

$\quad = \dfrac{4x^2t}{(x^2 - t^2)^2}$

41. $f(x,t) = \dfrac{2\sqrt{x} - 2\sqrt{t}}{1 + 2\sqrt{t}}$

Find f_x.

$f(x,t) = \dfrac{2\underline{x}^{1/2} - 2t^{1/2}}{1 + 2t^{1/2}}$ The variable is underlined.

$f_x = \dfrac{(1 + 2t^{1/2})\left(2 \cdot \frac{1}{2}x^{-1/2}\right) - 0(2x^{1/2} - 2t^{1/2})}{(1 + 2t^{1/2})^2}$

$\quad = \dfrac{1 + 2\sqrt{t}}{\sqrt{x}(1 + 2\sqrt{t})^2}$

$\quad = \dfrac{1}{\sqrt{x}(1 + 2\sqrt{t})}$

Find f_t.

$f(x,t) = \dfrac{2x^{1/2} - 2\underline{t}^{1/2}}{1 + 2\underline{t}^{1/2}}$ The variable is underlined.

$f_t =$

$\dfrac{(1 + 2t^{1/2})\left(-2 \cdot \frac{1}{2}t^{-1/2}\right) - \left(2 \cdot \frac{1}{2}t^{-1/2}\right)(2x^{1/2} - 2t^{1/2})}{(1 + 2t^{1/2})^2}$

$\quad = \dfrac{(-t^{-1/2} - 2t^0) - (2x^{1/2}t^{-1/2} - 2t^0)}{(1 + 2t^{1/2})^2}$

$\quad = \dfrac{-t^{-1/2} - 2 - 2x^{1/2}t^{-1/2} + 2}{(1 + 2t^{1/2})^2}$

$\quad = \dfrac{t^{-1/2}(-1 - 2x^{1/2})}{(1 + 2t^{1/2})^2}$

$\quad = \dfrac{-1 - 2\sqrt{x}}{\sqrt{t}(1 + 2\sqrt{t})^2}$

43. $f(x,t) = 6x^{2/3} - 8x^{1/4}t^{1/2} - 12x^{-1/2}t^{3/2}$

Find f_x.

$f(x,t) = 6\underline{x}^{2/3} - 8\underline{x}^{1/4}t^{1/2} - 12\underline{x}^{-1/2}t^{3/2}$

   The variable is underlined.

$f_x = 6 \cdot \dfrac{2}{3}x^{-1/3} - 8 \cdot \dfrac{1}{4}x^{-3/4} \cdot t^{1/2} -$

$\qquad\qquad 12 \cdot \left(-\dfrac{1}{2}\right)x^{-3/2} \cdot t^{3/2}$

$\quad = 4x^{-1/3} - 2x^{-3/4}t^{1/2} + 6x^{-3/2}t^{3/2}$

Find f_t.

$f(x,t) = 6x^{2/3} - 8x^{1/4}\underline{t}^{1/2} - 12x^{-1/2}\underline{t}^{3/2}$

   The variable is underlined.

$f_t = -8x^{1/4} \cdot \dfrac{1}{2}t^{-1/2} - 12x^{-1/2} \cdot \dfrac{3}{2}t^{1/2}$

$\quad = -4x^{1/4}t^{-1/2} - 18x^{-1/2}t^{1/2}$

45. $\boxed{tw}$

Exercise Set 7.3

1. $f(x,y) = 3x^2 - xy + y$

Find f_x.

$f(x,y) = 3\underline{x}^2 - \underline{x}y + y$ The variable is underlined.

$f_x = 6x - y$

Find f_{xx}.

$f_x = 6\underline{x} - y$ The variable is underlined.

$f_{xx} = 6$

Find f_{xy}.

$f_x = 6x - \underline{y}$ The variable is underlined.

$f_{xy} = -1$

Find f_y.

$f(x,y) = 3x^2 - x\underline{y} + \underline{y}$ The variable is underlined.

$f_y = -x + 1$

Find f_{yx}.

$f_y = -\underline{x} + 1$ The variable is underlined.

$f_{yx} = -1$

Find f_{yy}.

$f_y = -x + 1$

$f_{yy} = 0$ $-x+1$ is treated as a constant.

3. $f(x,y) = 3xy$

Find f_x.

$f(x,y) = 3\underline{x}y$

$f_x = 3y$

Find f_{xx}.

$f_x = 3y$

$f_{xx} = 0$ $3y$ is treated as a constant.

Find f_{xy}.

$f_x = 3\underline{y}$

$f_{xy} = 3$

Find f_y.

$f(x,y) = 3x\underline{y}$

$f_y = 3x$

Find f_{yx}.

$f_y = 3\underline{x}$

$f_{yx} = 3$

Find f_{yy}.

$f_y = 3x$

$f_{yy} = 0$ $3x$ is treated as a constant.

5. $f(x,y) = x^5y^4 + x^3y^2$

Find f_x.

$f(x,y) = \underline{x}^5y^4 + \underline{x}^3y^2$

$f_x = 5x^4 \cdot y^4 + 3x^2 \cdot y^2$

$= 5x^4y^4 + 3x^2y^2$

Find f_{xx}.

$f_x = 5\underline{x}^4y^4 + 3\underline{x}^2y^2$

$f_{xx} = 5 \cdot 4x^3 \cdot y^4 + 3 \cdot 2x \cdot y^2$

$= 20x^3y^4 + 6xy^2$

Find f_{xy}.

$f_x = 5x^4\underline{y}^4 + 3x^2\underline{y}^2$

$f_{xy} = 5x^4 \cdot 4y^3 + 3x^2 \cdot 2y$

$= 20x^4y^3 + 6x^2y$

Find f_y.

$f(x,y) = x^5\underline{y}^4 + x^3\underline{y}^2$

$f_y = x^5 \cdot 4y^3 + x^3 \cdot 2y$

$= 4x^5y^3 + 2x^3y$

Find f_{yx}.

$f_y = 4\underline{x}^5y^3 + 2\underline{x}^3y$

$f_{yx} = 4 \cdot 5x^4 \cdot y^3 + 2 \cdot 3x^2 \cdot y$

$= 20x^4y^3 + 6x^2y$

Find f_{yy}.

$f_y = 4x^5\underline{y}^3 + 2x^3\underline{y}$

$f_{yy} = 4x^5 \cdot 3y^2 + 2x^3$

$= 12x^5y^2 + 2x^3$

7. $f(x,y) = 2x - 3y$

Find f_x.

$f(x,y) = 2\underline{x} - 3y$

$f_x = 2$

Find f_{xx}.

$f_x = 2$

$f_{xx} = 0$

Find f_{xy}.

$f_x = 2$

$f_{xy} = 0$

Find f_y.

$f(x,y) = 2x - 3\underline{y}$

$f_y = -3$

Find f_{yx}.

$f_y = -3$

$f_{yx} = 0$

Find f_{yy}.

$f_y = -3$

$f_{yy} = 0$

9. $f(x,y) = e^{2xy}$

Find f_x.

$f(x,y) = e^{2\underline{x}y}$

$f_x = 2y\, e^{2xy}$

Find f_{xx}.

$f_x = 2y\, e^{2\underline{x}y}$

$f_{xx} = 2y \cdot 2y\, e^{2xy}$

$= 4y^2\, e^{2xy}$

Find f_{xy}.

$f_x = 2\underline{y}\, e^{2x\underline{y}}$

$f_{xy} = 2(y \cdot 2x\, e^{2xy} + e^{2xy})$

$= 4xy\, e^{2xy} + 2\, e^{2xy}$

Find f_y.

$f(x,y) = e^{2x\underline{y}}$

$f_y = 2x\, e^{2xy}$

Find f_{yx}.

$f_y = 2\underline{x}\, e^{2\underline{x}y}$

$f_{yx} = 2(x \cdot 2y\, e^{2xy} + e^{2xy})$

$= 4xy\, e^{2xy} + 2\, e^{2xy}$

Find f_{yy}.

$f_y = 2x\, e^{2x\underline{y}}$

$f_{yy} = 2x \cdot 2x\, e^{2xy}$

$= 4x^2\, e^{2xy}$

11. $f(x, y) = x + e^y$

Find f_x.
$$f(x, y) = \underline{x} + e^y$$
$$f_x = 1$$

Find f_{xx}.
$$f_x = 1$$
$$f_{xx} = 0$$

Find f_{xy}.
$$f_x = 1$$
$$f_{xy} = 0$$

Find f_y.
$$f(x, y) = x + e^{\underline{y}}$$
$$f_y = e^y$$

Find f_{yx}.
$$f_y = e^y$$
$$f_{yx} = 0 \qquad e^y \text{ is treated as a constant.}$$

Find f_{yy}.
$$f_y = e^{\underline{y}}$$
$$f_{yy} = e^y$$

13. $f(x, y) = y \ln x$

Find f_x.
$$f(x, y) = y \ln \underline{x}$$
$$f_x = y \cdot \frac{1}{x}$$
$$= \frac{y}{x}, \text{ or } yx^{-1}$$

Find f_{xx}.
$$f_x = \frac{y}{\underline{x}}, \text{ or } y\underline{x}^{-1}$$
$$f_{xx} = y \cdot (-1)x^{-2}$$
$$= -\frac{y}{x^2}$$

Find f_{xy}.
$$f_x = \frac{\underline{y}}{x}, \text{ or } \frac{1}{x} \cdot \underline{y}$$
$$f_{xy} = \frac{1}{x}$$

Find f_y.
$$f(x, y) = \underline{y} \ln x$$
$$f_y = \ln x$$

Find f_{yx}.
$$f_y = \ln \underline{x}$$
$$f_{yx} = \frac{1}{x}$$

Find f_{yy}.
$$f_y = \ln x$$
$$f_{yy} = 0 \qquad \ln x \text{ is treated as a constant.}$$

15. $f(x, y) = \dfrac{x}{y^2} - \dfrac{y}{x^2}$

Find f_x.
$$f(x, y) = \underline{x}y^{-2} - \underline{x}^{-2}y$$
$$f_x = y^{-2} - (-2)x^{-3}y$$
$$= y^{-2} + 2x^{-3}y$$

Find f_{xx}.
$$f_x = y^{-2} + 2\underline{x}^{-3}y$$
$$f_{xx} = 2(-3)x^{-4}y$$
$$= -6x^{-4}y$$
$$= \frac{-6y}{x^4}$$

Find f_{xy}.
$$f_x = \underline{y}^{-2} + 2x^{-3}\underline{y}$$
$$f_{xy} = -2y^{-3} + 2x^{-3}$$
$$= -\frac{2}{y^3} + \frac{2}{x^3}$$
$$= \frac{-2x^3 + 2y^3}{y^3 x^3}$$
$$= \frac{2(y^3 - x^3)}{y^3 x^3}$$

Find f_y.
$$f(x, y) = x\underline{y}^{-2} - x^{-2}\underline{y}$$
$$f_y = x(-2)y^{-3} - x^{-2}$$
$$= -2xy^{-3} - x^{-2}$$

Find f_{yx}.
$$f_y = -2\underline{x}y^{-3} - \underline{x}^{-2}$$
$$f_{yx} = -2y^{-3} - (-2)x^{-3}$$
$$= -2y^{-3} + 2x^{-3}$$
$$= -\frac{2}{y^3} + \frac{2}{x^3}$$
$$= \frac{2(y^3 - x^3)}{y^3 x^3} \qquad \text{See the simplification of } f_{xy}.$$

Find f_{yy}.
$$f_y = -2x\underline{y}^{-3} - x^{-2}$$
$$f_{yy} = -2x \cdot (-3)y^{-4}$$
$$= 6xy^{-4}$$
$$= \frac{6x}{y^4}$$

17. $f(x, y) = \ln(x^2 + y^2)$

Find f_x.
$$f(x, y) = \ln(\underline{x}^2 + y^2)$$
$$f_x = \frac{2x}{x^2 + y^2}$$

Find f_{xx}.

$$f_x = \frac{2x}{x^2 + y^2}$$

$$f_{xx} = \frac{(x^2 + y^2)2 - 2x \cdot 2x}{(x^2 + y^2)^2}$$

$$= \frac{2x^2 + 2y^2 - 4x^2}{(x^2 + y^2)^2}$$

$$= \frac{2y^2 - 2x^2}{(x^2 + y^2)^2}$$

Find f_y.

$$f(x, y) = \ln(x^2 + \underline{y}^2)$$

$$f_y = \frac{2y}{x^2 + y^2}$$

Find f_{yy}.

$$f_y = \frac{2y}{x^2 + \underline{y}^2}$$

$$f_{yy} = \frac{(x^2 + y^2)2 - 2y \cdot 2y}{(x^2 + y^2)^2}$$

$$= \frac{2x^2 + 2y^2 - 4y^2}{(x^2 + y^2)^2}$$

$$= \frac{2x^2 - 2y^2}{(x^2 + y^2)^2}$$

$\dfrac{\partial^2 f}{\partial x^2} + \dfrac{\partial^2 f}{\partial y^2}$	0
$\dfrac{2y^2 - 2x^2}{(x^2 + y^2)^2} + \dfrac{2x^2 - 2y^2}{(x^2 + y^2)^2}$	0
$\dfrac{0}{(x^2 + y^2)^2}$	
0	

Thus, f is a solution to $\dfrac{\partial^2 f}{\partial x^2} + \dfrac{\partial^2 f}{\partial y^2} = 0$.

19.

$$f(x, y) = \begin{cases} \dfrac{xy(x^2 - y^2)}{x^2 + y^2}, & \text{for } (x, y) \neq (0, 0), \\[2mm] 0, & \text{for } (x, y) = (0, 0). \end{cases}$$

a) Find $f_x(0, y)$.

$$\lim_{h \to 0} \frac{f(h, y) - f(0, y)}{h}$$

$$= \lim_{h \to 0} \frac{\dfrac{hy(h^2 - y^2)}{h^2 + y^2} - \dfrac{0 \cdot y(0^2 - y^2)}{0^2 + y^2}}{h} \quad \text{Substituting}$$

$$= \lim_{h \to 0} \frac{hy(h^2 - y^2)}{h(h^2 + y^2)}$$

$$= \lim_{h \to 0} \frac{y(h^2 - y^2)}{(h^2 + y^2)}$$

$$= \frac{y(-y^2)}{y^2}$$

$$= -\frac{y^3}{y^2}$$

$$= -y$$

Thus, $f_x(0, y) = -y$.

b) Find $f_y(x, 0)$.

$$\lim_{h \to 0} \frac{f(x, h) - f(x, 0)}{h}$$

$$= \lim_{h \to 0} \frac{\dfrac{xh(x^2 - h^2)}{x^2 + h^2} - \dfrac{x \cdot 0(x^2 - 0^2)}{x^2 + 0^2}}{h}$$

$$\qquad\qquad\qquad\qquad\qquad\qquad \text{Substituting}$$

$$= \lim_{h \to 0} \frac{xh(x^2 - h^2)}{h(x^2 + h^2)}$$

$$= \lim_{h \to 0} \frac{x(x^2 - h^2)}{x^2 + h^2}$$

$$= \frac{x(x^2)}{x^2}$$

$$= x$$

Thus, $f_y(x, 0) = x$.

c) Find $f_{yx}(0, 0)$.

$$\lim_{h \to 0} \frac{f_y(h, 0) - f_y(0, 0)}{h}$$

$$= \lim_{h \to 0} \frac{h - 0}{h} \qquad \text{Substituting; } f_y(x, 0) = x.$$

$$= \lim_{h \to 0} 1$$

$$= 1$$

Find $f_{xy}(0, 0)$.

$$\lim_{h \to 0} \frac{f_x(0, h) - f_x(0, 0)}{h}$$

$$= \lim_{h \to 0} \frac{-h - 0}{h} \qquad [f_x(0, y) = -y]$$

$$= \lim_{h \to 0} -1$$

$$= -1$$

Thus, $f_{yx}(0, 0) \neq f_{xy}(0, 0)$.

Exercise Set 7.4

1. $f(x, y) = x^2 + xy + y^2 - y$

Find f_x.

$$f(x, y) = \underline{x}^2 + \underline{x}y + y^2 - y \quad \text{The variable is}$$
$$\qquad\qquad\qquad\qquad\qquad\qquad \text{underlined.}$$

$$f_x = 2x + y$$

Find f_y.

$$f(x, y) = x^2 + x\underline{y} + \underline{y}^2 - \underline{y} \quad \text{The variable is}$$
$$\qquad\qquad\qquad\qquad\qquad\qquad \text{underlined.}$$

$$f_y = x + 2y - 1$$

Find f_{xx} and f_{xy}.

$$f_x = 2\underline{x} + y \qquad\quad f_x = 2x + \underline{y}$$

$$f_{xx} = 2 \qquad\qquad\quad f_{xy} = 1$$

Find f_{yy}.

$$f_y = x + 2\underline{y} - 1$$

$$f_{yy} = 2$$

Solve the system of equations $f_x = 0$ and $f_y = 0$:

$$2x + y = 0 \quad (1)$$
$$x + 2y - 1 = 0 \quad (2)$$

Solving Eq. (1) for y we get $y = -2x$. Substituting $-2x$ for y in Eq. (2) and solving we get

$$x + 2(-2x) - 1 = 0$$
$$x - 4x - 1 = 0$$
$$-3x = 1$$
$$x = -\frac{1}{3}$$

To find y when $x = -\frac{1}{3}$, we substitute $-\frac{1}{3}$ for x in in either Eq. (1) or Eq. (2). We use Eq. (1).

$$2\left(-\frac{1}{3}\right) + y = 0$$
$$-\frac{2}{3} + y = 0$$
$$y = \frac{2}{3}$$

Thus, $\left(-\frac{1}{3}, \frac{2}{3}\right)$ is our candidate for a maximum or minimum.

We have to check to see if $f\left(-\frac{1}{3}, \frac{2}{3}\right)$ is a maximum or minimum.

$$D = f_{xx}(a, b) \cdot f_{yy}(a, b) - \left[f_{xy}(a, b)\right]^2$$
$$D = f_{xx}\left(-\frac{1}{3}, \frac{2}{3}\right) \cdot f_{yy}\left(-\frac{1}{3}, \frac{2}{3}\right) - \left[f_{xy}\left(-\frac{1}{3}, \frac{2}{3}\right)\right]^2$$
$$D = 2 \cdot 2 - 1^2 \quad \text{For all values of } x \text{ and } y, \, f_{xx} = 2,$$
$$\qquad\qquad f_{yy} = 2, \text{ and } f_{xy} = 1$$
$$= 4 - 1$$
$$= 3$$

Thus $D = 3$ and $f_{xx}\left(-\frac{1}{3}, \frac{2}{3}\right) = 2$. Since $D > 0$ and $f_{xx}\left(-\frac{1}{3}, \frac{2}{3}\right) > 0$, it follows that f has a relative minimum at $\left(-\frac{1}{3}, \frac{2}{3}\right)$ and that the minimum is found as follows:

$$f(x, y) = x^2 + xy + y^2 - y$$
$$f\left(-\frac{1}{3}, \frac{2}{3}\right) = \left(-\frac{1}{3}\right)^2 + \left(-\frac{1}{3}\right) \cdot \frac{2}{3} + \left(\frac{2}{3}\right)^2 - \frac{2}{3}$$
$$= \frac{1}{9} - \frac{2}{9} + \frac{4}{9} - \frac{2}{3}$$
$$= \frac{1}{9} - \frac{2}{9} + \frac{4}{9} - \frac{6}{9}$$
$$= -\frac{3}{9}$$
$$= -\frac{1}{3}$$

The relative minimum value of f is $-\frac{1}{3}$ at $\left(-\frac{1}{3}, \frac{2}{3}\right)$.

3. $f(x, y) = 2xy - x^3 - y^2$.

Find f_x.

$$f(x, y) = 2\underline{x}y - \underline{x}^3 - y^2$$
$$f_x = 2y - 3x^2$$

Find f_y.

$$f(x, y) = 2x\underline{y} - x^3 - \underline{y}^2$$
$$f_y = 2x - 2y$$

Find f_{xx} and f_{xy}.

$$f_x = 2y - 3\underline{x}^2 \qquad f_x = 2\underline{y} - 3x^2$$
$$f_{xx} = -6x \qquad\qquad f_{xy} = 2$$

Find f_{yy}.

$$f_y = 2x - 2\underline{y}$$
$$f_{yy} = -2$$

Solve the system of equations $f_x = 0$ and $f_y = 0$:

$$2y - 3x^2 = 0 \quad (1)$$
$$2x - 2y = 0 \quad (2)$$

Solving Eq. (1) for $2y$ we get $2y = 3x^2$. Substituting $3x^2$ for $2y$ in Eq. (2) and solving we get

$$2x - 3x^2 = 0$$
$$x(2 - 3x) = 0$$
$$x = 0 \text{ or } 2 - 3x = 0 \quad \text{Principle of zero products}$$
$$x = 0 \text{ or } \qquad 2 = 3x$$
$$x = 0 \text{ or } \qquad \frac{2}{3} = x$$

To find y when $x = 0$ we substitute 0 for x in either Eq. (1) or Eq. (2). We use Eq. (2).

$$2 \cdot 0 - 2y = 0$$
$$-2y = 0$$
$$y = 0$$

Thus $(0, 0)$ is one critical value (candidate for a maximum or minimum). To find the other critical value we substitute $\frac{2}{3}$ for x in either Eq. (1) or Eq. (2). We use Eq. (2).

$$2 \cdot \frac{2}{3} - 2y = 0$$
$$\frac{4}{3} - 2y = 0$$
$$-2y = -\frac{4}{3}$$
$$y = \frac{2}{3}$$

Thus $\left(\frac{2}{3}, \frac{2}{3}\right)$ is the other critical point.

We have to check whether $(0, 0)$ and $\left(\frac{2}{3}, \frac{2}{3}\right)$ yield maximum or minimum values.

For $(0, 0)$:

$$D = f_{xx}(0, 0) \cdot f_{yy}(0, 0) - \left[f_{xy}(0, 0)\right]^2$$
$$= 0 \cdot (-2) - 2^2 \quad \begin{bmatrix} f_{xx}(0, 0) = -6 \cdot 0 = 0 \\ f_{yy}(0, 0) = -2 \\ f_{xy}(0, 0) = 2 \end{bmatrix}$$
$$= -4$$

Since $D < 0$, it follows that $f(0, 0)$ is neither a maximum nor a minimum, but a saddle point.

For $\left(\frac{2}{3}, \frac{2}{3}\right)$:

$$D = f_{xx}\left(\frac{2}{3}, \frac{2}{3}\right) \cdot f_{yy}\left(\frac{2}{3}, \frac{2}{3}\right) - \left[f_{xy}\left(\frac{2}{3}, \frac{2}{3}\right)\right]^2$$

$$= -4 \cdot (-2) - 2^2 \qquad \left[\begin{array}{l} f_{xx}\left(\frac{2}{3}, \frac{2}{3}\right) = -6 \cdot \frac{2}{3} = -4 \\[2mm] f_{yy}\left(\frac{2}{3}, \frac{2}{3}\right) = -2 \\[2mm] f_{xy}\left(\frac{2}{3}, \frac{2}{3}\right) = 2 \end{array}\right]$$

$$= 8 - 4$$

$$= 4$$

Thus $D = 4$ and $f_{xx}\left(\frac{2}{3}, \frac{2}{3}\right) = -4$. Since $D > 0$ and $f_{xx}\left(\frac{2}{3}, \frac{2}{3}\right) < 0$, it follows that f has a relative maximum at $\left(\frac{2}{3}, \frac{2}{3}\right)$ and that maximum is found as follows:

$$f(x, y) = 2xy - x^3 - y^2$$

$$f\left(\frac{2}{3}, \frac{2}{3}\right) = 2 \cdot \frac{2}{3} \cdot \frac{2}{3} - \left(\frac{2}{3}\right)^3 - \left(\frac{2}{3}\right)^2$$

$$= \frac{8}{9} - \frac{8}{27} - \frac{4}{9}$$

$$= \frac{4}{9} - \frac{8}{27}$$

$$= \frac{12}{27} - \frac{8}{27}$$

$$= \frac{4}{27}$$

The relative maximum value of f is $\frac{4}{27}$ at $\left(\frac{2}{3}, \frac{2}{3}\right)$.

5. $f(x, y) = x^3 + y^3 - 3xy$

Find f_x.

$$f(x, y) = \underline{x}^3 + y^3 - 3\underline{x}y$$

$$f_x = 3x^2 - 3y$$

Find f_y.

$$f(x, y) = x^3 + \underline{y}^3 - 3x\underline{y}$$

$$f_y = 3y^2 - 3x$$

Find f_{xx} and f_{xy}.

$$f_x = 3\underline{x}^2 - 3y \qquad f_x = 3x^2 - 3\underline{y}$$

$$f_{xx} = 6x \qquad\qquad f_{xy} = -3$$

Find f_{yy}.

$$f_y = 3\underline{y}^2 - 3x$$

$$f_{yy} = 6y$$

Solve the system of equations $f_x = 0$ and $f_y = 0$:

$$3x^2 - 3y = 0$$

$$3y^2 - 3x = 0$$

We multiply each equation by $\frac{1}{3}$.

$$x^2 - y = 0 \qquad (1)$$

$$y^2 - x = 0 \qquad (2)$$

Solving Eq. (1) for y we get $y = x^2$. Substituting x^2 for y in Eq. (2) and solving we get

$$(x^2)^2 - x = 0$$

$$x^4 - x = 0$$

$$x(x^3 - 1) = 0$$

$$x = 0 \text{ or } x^3 - 1 = 0 \quad \text{Principle of zero products}$$

$$x = 0 \text{ or } \qquad x^3 = 1$$

$$x = 0 \text{ or } \qquad x = 1$$

To find y when $x = 0$, substitute 0 for x in either Eq. (1) or Eq. (2). We use Eq. (2).

$$y^2 - 0 = 0$$

$$y^2 = 0$$

$$y = 0$$

Thus, $(0, 0)$ is one critical value. To find the other critical values, we substitute 1 for x in either Eq. (1) or Eq. (2). We use Eq. (2).

$$y^2 - 1 = 0$$

$$y^2 = 1$$

$$y = \pm 1$$

Thus $(1, -1)$ and $(1, 1)$ are also critical points.

We have to check whether $(0, 0)$, $(1, -1)$ and $(1, 1)$ yield maximum or minimum values.

For $(0, 0)$:

$$D = f_{xx}(0, 0) \cdot f_{yy}(0, 0) - \left[f_{xy}(0, 0)\right]^2$$

$$= 0 \cdot 0 - (-3)^2 \qquad \left[\begin{array}{l} f_{xx}(0, 0) = 6 \cdot 0 = 0 \\ f_{yy}(0, 0) = 6 \cdot 0 = 0 \\ f_{xy}(0, 0) = -3 \end{array}\right]$$

$$= 0 - 9$$

$$= -9$$

Since $D < 0$, it follows that $f(0, 0)$ is neither a maximum nor a minimum, but a saddle point.

For $(1, -1)$:

$$D = f_{xx}(1, -1) \cdot f_{yy}(1, -1) - \left[f_{xy}(1, -1)\right]^2$$

$$= 6 \cdot (-6) - (-3)^2 \left[\begin{array}{l} f_{xx}(1, -1) = 6 \cdot 1 = 6 \\ f_{yy}(1, -1) = 6 \cdot (-1) = -6 \\ f_{xy}(1, -1) = -3 \end{array}\right]$$

$$= -36 - 9$$

$$= -45$$

Since $D < 0$, it follows that $f(1, -1)$ is neither a maximum nor a minimum, but a saddle point.

For $(1, 1)$:

$$D = f_{xx}(1, 1) \cdot f_{yy}(1, 1) - \left[f_{xy}(1, 1)\right]^2$$

$$= 6 \cdot 6 - (-3)^2 \quad \left[\begin{array}{l} f_{xx}(1, 1) = 6 \cdot 1 = 6 \\ f_{yy}(1, 1) = 6 \cdot 1 = 6 \\ f_{xy}(1, 1) = -3 \end{array}\right]$$

$$= 36 - 9$$

$$= 27$$

Thus $D = 27$ and $f_{xx}(1, 1) = 6$. Since $D > 0$ and

$f_{xx}(1,1) > 0$, it follows that f has a relative minimum at $(1,1)$ and that minimum is found as follows:

$$f(x,y) = x^3 + y^3 - 3xy$$
$$f(1,1) = 1^3 + 1^3 - 3 \cdot 1 \cdot 1$$
$$= 1 + 1 - 3$$
$$= -1$$

The relative minimum value of f is -1 at $(1,1)$.

7. $f(x,y) = x^2 + y^2 - 2x + 4y - 2$

Find f_x.
$$f(x,y) = \underline{x}^2 + y^2 - 2\underline{x} + 4y - 2$$
$$f_x = 2x - 2$$

Find f_y.
$$f(x,y) = x^2 + \underline{y}^2 - 2x + 4\underline{y} - 2$$
$$f_y = 2y + 4$$

Find f_{xx} and f_{xy}.
$$f_x = 2\underline{x} - 2 \qquad f_x = 2x - 2$$
$$f_{xx} = 2 \qquad\qquad f_{xy} = 0$$

Find f_{yy}.
$$f_y = 2\underline{y} + 4$$
$$f_{yy} = 2$$

Solve the system of equations $f_x = 0$ and $f_y = 0$.
$$2x - 2 = 0 \qquad 2y + 4 = 0$$
$$2x = 2 \qquad\quad 2y = -4$$
$$x = 1 \qquad\qquad y = -2$$

The only critical value is $(1,-2)$. We have to check to see if $f(1,-2)$ is a maximum or a minimum.

$$D = f_{xx}(1,-2) \cdot f_{yy}(1,-2) - \left[f_{xy}(1,-2)\right]^2$$
$$= 2 \cdot 2 - 0^2 \qquad \begin{bmatrix} f_{xx}(1,-2) = 2 \\ f_{yy}(1,-2) = 2 \\ f_{xy}(1,-2) = 0 \end{bmatrix}$$
$$= 4$$

Thus $D = 4$ and $f_{xx}(1,-2) = 2$. Since $D > 0$ and $f_{xx}(1,-2) > 0$, it follows that f has a relative minimum at $(1,-2)$ and that minimum is found as follows:

$$f(x,y) = x^2 + y^2 - 2x + 4y - 2$$
$$f(1,-2) = 1^2 + (-2)^2 - 2 \cdot 1 + 4 \cdot (-2) - 2$$
$$= 1 + 4 - 2 - 8 - 2$$
$$= -7$$

The relative minimum value of f is -7 at $(1,-2)$.

9. $f(x,y) = x^2 + y^2 + 2x - 4y$

Find f_x.
$$f(x,y) = \underline{x}^2 + y^2 + 2\underline{x} - 4y$$
$$f_x = 2x + 2$$

Find f_y.
$$f(x,y) = x^2 + \underline{y}^2 + 2x - 4\underline{y}$$
$$f_y = 2y - 4$$

Find f_{xx} and f_{xy}.
$$f_x = 2\underline{x} + 2 \qquad f_x = 2x + 2$$
$$f_{xx} = 2 \qquad\qquad f_{xy} = 0$$

Find f_{yy}.
$$f_y = 2\underline{y} - 4$$
$$f_{yy} = 2$$

Solve the system of equations $f_x = 0$ and $f_y = 0$.
$$2x + 2 = 0 \qquad 2y - 4 = 0$$
$$2x = -2 \qquad\quad 2y = 4$$
$$x = -1 \qquad\qquad y = 2$$

The only critical value is $(-1,2)$. We have to check to see if $f(-1,2)$ is a maximum or a minimum.

$$D = f_{xx}(-1,2) \cdot f_{yy}(-1,2) - \left[f_{xy}(-1,2)\right]^2$$
$$= 2 \cdot 2 - 0^2 \qquad \begin{bmatrix} f_{xx}(-1,2) = 2 \\ f_{yy}(-1,2) = 2 \\ f_{xy}(-1,2) = 0 \end{bmatrix}$$
$$= 4$$

Thus $D = 4$ and $f_{xx}(-1,2) = 2$. Since $D > 0$ and $f_{xx}(-1,2) > 0$, it follows that f has a relative minimum at $(-1,2)$ and that minimum is found as follows:

$$f(x,y) = x^2 + y^2 + 2x - 4y$$
$$f(-1,2) = (-1)^2 + 2^2 + 2(-1) - 4 \cdot 2$$
$$= 1 + 4 - 2 - 8$$
$$= -5$$

The relative minimum value of f is -5 at $(-1,2)$.

11. $f(x,y) = 4x^2 - y^2$

Find f_x.
$$f(x,y) = 4\underline{x}^2 - y^2$$
$$f_x = 8x$$

Find f_y.
$$f(x,y) = 4x^2 - \underline{y}^2$$
$$f_y = -2y$$

Find f_{xx} and f_{xy}.
$$f_x = 8\underline{x} \qquad f_x = 8x$$
$$f_{xx} = 8 \qquad\quad f_{xy} = 0$$

Find f_{yy}.
$$f_y = -2\underline{y}$$
$$f_{yy} = -2$$

Solve the system of equations $f_x = 0$ and $f_y = 0$.
$$8x = 0 \qquad -2y = 0$$
$$x = 0 \qquad\quad y = 0$$

The only critical value is $(0,0)$. We have to check to see if $f(0,0)$ is a maximum or a minimum.

$$D = f_{xx}(0,0) \cdot f_{yy}(0,0) - \left[f_{xy}(0,0)\right]^2$$
$$= 8 \cdot (-2) - 0^2 \qquad \begin{bmatrix} f_{xx}(0,0) = 8 \\ f_{yy}(0,0) = -2 \\ f_{xy}(0,0) = 0 \end{bmatrix}$$
$$= -16$$

Since $D < 0$, it follows that $(0,0)$ is neither a maximum nor a minimum, but a saddle point.

Thus f has no relative maximum value or relative minimum value.

13. $R(x, y) = 17x + 21y$

$C(x, y) = 4x^2 - 4xy + 2y^2 - 11x + 25y - 3$

Total profit, $P(x, y)$, is given by

$$P(x, y)$$
$$= R(x, y) - C(x, y)$$
$$= (17x + 21y) - (4x^2 - 4xy + 2y^2 - 11x + 25y - 3)$$
$$= 17x + 21y - 4x^2 + 4xy - 2y^2 + 11x - 25y + 3$$
$$= -4x^2 - 2y^2 + 4xy + 28x - 4y + 3$$

Find P_x.
$$P(x, y) = -4\underline{x}^2 - 2y^2 + 4\underline{x}y + 28\underline{x} - 4y + 3$$
$$P_x = -8x + 4y + 28$$

Find P_y.
$$P(x, y) = -4x^2 - 2\underline{y}^2 + 4x\underline{y} + 28x - 4\underline{y} + 3$$
$$P_y = -4y + 4x - 4$$

Find P_{xx} and P_{xy}.
$$P_x = -8\underline{x} + 4y + 28 \qquad P_x = -8x + 4\underline{y} + 28$$
$$P_{xx} = -8 \qquad\qquad P_{xy} = 4$$

Find P_{yy}.
$$P_y = -4\underline{y} + 4x - 4$$
$$P_{yy} = -4$$

Solve the system of equations $P_x = 0$ and $P_y = 0$.
$$-8x + 4y + 28 = 0 \quad (1)$$
$$-4y + 4x - 4 = 0 \quad (2)$$

Adding these equations, we get
$$-4x + 24 = 0.$$
Then
$$-4x = -24$$
$$x = 6.$$

To find y when $x = 6$, we substitute 6 for x in either Eq. (1) or Eq. (2). We use Eq. (1).
$$-8x + 4y + 28 = 0$$
$$-8 \cdot 6 + 4y + 28 = 0 \quad \text{Substituting}$$
$$-48 + 4y + 28 = 0$$
$$4y - 20 = 0$$
$$4y = 20$$
$$y = 5$$

Thus, $(6, 5)$ is our candidate for a maximum or minimum. We have to check to see if $P(6, 5)$ is a maximum.
$$D = P_{xx}(6, 5) \cdot P_{yy}(6, 5) - \left[P_{xy}(6, 5)\right]^2$$
$$= (-8)(-4) - 4^2 \left[\begin{array}{l} P_{xx}(6, 5) = -8 \\ P_{yy}(6, 5) = -4 \\ P_{xy}(6, 5) = 4 \end{array} \right]$$
$$= 32 - 16$$
$$= 16$$

Thus $D = 16$ and $P_{xx} = -8$. Since $D > 0$ and $P_{xx}(6, 5) < 0$, it follows that P has a relative maximum at $(6, 5)$. Thus to maximize profit, the company must produce and sell 6 thousand of the \$17 radio and 5 thousand of the \$21 radio.

15. $P(a, p) = 2ap + 80p - 15p^2 - \dfrac{1}{10}a^2p - 100$

Find P_a.
$$P(a, p) = 2\underline{a}p + 80p - 15p^2 - \frac{1}{10}\underline{a}^2p - 100$$
$$P_a = 2p - \frac{1}{5}ap$$

Find P_p.
$$P(a, p) = 2a\underline{p} + 80\underline{p} - 15\underline{p}^2 - \frac{1}{10}a^2\underline{p} - 100$$
$$P_p = 2a + 80 - 30p - \frac{1}{10}a^2$$

Find P_{aa} and P_{ap}.
$$P_a = 2p - \frac{1}{5}\underline{a}p \qquad P_a = 2\underline{p} - \frac{1}{5}a\underline{p}$$
$$P_{aa} = -\frac{1}{5}p \qquad\qquad P_{ap} = 2 - \frac{1}{5}a$$

Find P_{pp}.
$$P_p = 2a + 80 - 30\underline{p} - \frac{1}{10}a^2$$
$$P_{pp} = -30$$

Solve the system of equations $P_a = 0$ and $P_p = 0$.
$$2p - \frac{1}{5}ap = 0 \qquad\qquad (1)$$
$$2a + 80 - 30p - \frac{1}{10}a^2 = 0 \qquad (2)$$

Solving Eq. (1) for a we get $a = 10$. Substituting 10 for a in Eq. (2) and solving we get
$$2 \cdot 10 + 80 - 30p - \frac{1}{10}(10)^2 = 0 \quad \text{Substituting 10 for } a$$
$$20 + 80 - 10 = 30p$$
$$90 = 30p$$
$$3 = p$$

Thus, $(10, 3)$ is a candidate for a maximum or a minumum. We have to check to see if $P(10, 3)$ is a maximum.
$$D = P_{aa}(10, 3) \cdot P_{pp}(10, 3) - \left[P_{ap}(10, 3)\right]^2$$
$$= \left(-\frac{3}{5}\right)(-30) - 0^2 \left[\begin{array}{l} P_{aa}(10, 3) = -\frac{1}{5} \cdot 3 = -\frac{3}{5} \\ P_{pp}(10, 3) = -30 \\ P_{ap}(10, 3) = 2 - \frac{1}{5} \cdot 10 = 0 \end{array} \right]$$
$$= 18$$

Thus $D = 18$ and $P_{aa} = -\dfrac{3}{5}$. Since $D > 0$ and $P_{aa}(10, 3) < 0$, it follows that P has a relative maximum at $(10, 3)$. Thus to maximize profit, the company must spend 10 million dollars on advertising and charge $3 per item. The maximum profit is found as follows:

$$P(a, p) = 2ap + 80p - 15p^2 - \frac{1}{10}a^2p - 100$$

$$P(10, 3) = 2 \cdot 10 \cdot 3 + 80 \cdot 3 - 15 \cdot 3^2 - \frac{1}{10} \cdot 10^2 \cdot 3 - 100$$

Substituting

$$= 60 + 240 - 135 - 30 - 100$$

$$= 35$$

The maximum profit is $35 million.

17.

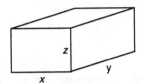

Let x, y, and z represent the dimensions of the container as shown in the drawing.

$$V = xyz$$

$$320 = xyz \quad (V = 320 \text{ ft}^3)$$

$$\frac{320}{xy} = z$$

Now we can express the cost as a function of two variables, x and y. The area of the bottom is xy ft^2, so the cost of the bottom is $5xy$. The area of each of two of the sides is xz, or $x\left(\dfrac{320}{xy}\right) = \dfrac{320}{y}$, so the area of both sides is $2 \cdot \dfrac{320}{y}$, or $\dfrac{640}{y}$. The area of each of the remaining sides is yz, or $y\left(\dfrac{320}{xy}\right) = \dfrac{320}{x}$, so the area of both such sides is $2 \cdot \dfrac{320}{x}$, or $\dfrac{640}{x}$. Then the total area of the four sides is $\dfrac{640}{y} + \dfrac{640}{x}$, and the cost of the four sides is $4\left(\dfrac{640}{y} + \dfrac{640}{x}\right)$, or $\dfrac{2560}{y} + \dfrac{2560}{x}$. Now we can write the cost function.

$$\begin{array}{ccc} \text{Total} \\ \text{cost} \end{array} = \begin{array}{c} \text{Cost of} \\ \text{bottom} \end{array} + \begin{array}{c} \text{Cost of} \\ \text{sides} \end{array}$$

$$C(x, y) = 5xy + \left(\frac{2560}{y} + \frac{2560}{x}\right)$$

We try to find a minimum value for $C(x, y)$.

$$C_x = 5y - \frac{2560}{x^2} \qquad C_y = 5x - \frac{2560}{y^2}$$

$$C_{xx} = \frac{5120}{x^3} \qquad C_{yy} = \frac{5120}{y^3}$$

$$C_{xy} = 5$$

Solve the system of equations $C_x = 0$, $C_y = 0$.

$$5y - \frac{2560}{x^2} = 0 \quad (1)$$

$$5x - \frac{2560}{y^2} = 0 \quad (2)$$

Solve Eq. (1) for y:

$$5y - \frac{2560}{x^2} = 0$$

$$5y = \frac{2560}{x^2}$$

$$y = \frac{512}{x^2}$$

Substitute $\dfrac{512}{x^2}$ for y in Eq. (2) and solve for x:

$$5x - \frac{2560}{\left(\frac{512}{x^2}\right)^2} = 0$$

$$5x - \frac{2560}{\frac{262,144}{x^4}} = 0$$

$$5x - \frac{2560x^4}{262,144} = 0$$

$$5x - \frac{5x^4}{512} = 0 \quad \text{Simplifying}$$

$$2560x - 5x^4 = 0 \quad \text{Multiplying by 512}$$

$$5x(512 - x^3) = 0$$

$$5x = 0 \quad \text{or} \quad 512 - x^3 = 0$$

$$x = 0 \quad \text{or} \quad 512 = x^3$$

$$x = 0 \quad \text{or} \quad 8 = x$$

Since none of the dimensions can be 0, only $x = 8$ has meaning in this application. Substitute 8 for x in Eq. (1) to find y:

$$5y - \frac{2560}{x^2} = 0$$

$$5y - \frac{2560}{8^2} = 0$$

$$5y - \frac{2560}{64} = 0$$

$$5y - 40 = 0$$

$$5y = 40$$

$$y = 8$$

Check to determine if $C(8, 8)$ is a minimum.

$$D = C_{xx}(8, 8) \cdot C_{yy}(8, 8) - [C_{xy}(8, 8)]^2$$

$$= \frac{5120}{8^3} \cdot \frac{5120}{8^3} - 5^2$$

$$= \frac{5120}{512} \cdot \frac{5120}{512} - 25$$

$$= 10 \cdot 10 - 25$$

$$= 100 - 25$$

$$= 75$$

Since $D > 0$ and $C_{xx}(8, 8) = 10 > 0$, C has a minimum at $(8, 8)$. Thus, the dimensions of the bottom are 8 ft by 8 ft. The height is $\dfrac{320}{8 \cdot 8}$, or 5 ft.

19. a) $\quad q_1 = 64 - 4p_1 - 2p_2 \quad$ (1)

$\qquad q_2 = 56 - 2p_1 - 4p_2 \quad$ (2)

$\qquad R = p_1 q_1 + p_2 q_2 \qquad$ Total revenue

$\qquad R = p_1(64 - 4p_1 - 2p_2) + p_2(56 - 2p_1 - 4p_2)$

$\qquad\qquad\qquad\qquad\qquad$ Substituting

$\qquad = 64p_1 - 4p_1^2 - 2p_1p_2 + 56p_2 - 2p_1p_2 - 4p_2^2$

$\qquad = 64p_1 - 4p_1^2 - 4p_1p_2 + 56p_2 - 4p_2^2$

b) We now find the values of p_1 and p_2 to maximize total revenue.

$\quad R_{p_1} = 64 - 8p_1 - 4p_2 \qquad R_{p_2} = -4p_1 + 56 - 8p_2$

$\quad R_{p_1p_1} = -8 \qquad\qquad\qquad R_{p_2p_2} = -8$

$\qquad\qquad R_{p_1p_2} = -4$

Solve the system of equations $Rp_1 = 0$ and $Rp_2 = 0$.

$\qquad 64 - 8p_1 - 4p_2 = 0$

$\qquad -4p_1 + 56 - 8p_2 = 0$

The solution of the system is $p_1 = 6$ and $p_2 = 4$.

We check to determine if $R(6, 4)$ is a maximum.

$\quad D = R_{p_1p_1}(6, 4) \cdot R_{p_2p_2}(6, 4) - \left[R_{p_1p_2}(6, 4)\right]^2$

$\quad D = -8 \cdot (-8) - (-4)^2$

$\qquad = 64 - 16$

$\qquad = 48$

Since $D > 0$ and $R_{p_1p_2}(6, 4) = -8 < 0$, R has a maximum at $(6, 4)$. Then to maximize revenue p_1 must be 6×10, or \$60 and p_2 must be 4×10, or \$40.

c) We substitute 6 for p_1 and 4 for p_2 in Eqs. (1) and (2) to find q_1 and q_2.

$\quad q_1 = 64 - 4p_1 - 2p_2$

$\quad q_1 = 64 - 4 \cdot 6 - 2 \cdot 4$

$\qquad = 64 - 24 - 8$

$\qquad = 32 \quad$ (hundreds)

$\quad q_2 = 56 - 2p_1 - 4p_2$

$\quad q_2 = 56 - 2 \cdot 6 - 4 \cdot 4$

$\qquad = 56 - 12 - 16$

$\qquad = 28 \quad$ (hundreds)

d) To maximize revenue 3200 units of the \$60 calculator and 2800 units of the \$40 calculator must be produced and sold. The maximum revenue is found as follows.

$\quad R = \$60 \cdot 3200 + \$40 \cdot 2800$

$\qquad = \$192,000 + \$112,000$

$\qquad = \$304,000$

21. $f(x, y) = e^x + e^y - e^{x+y}$

We first find f_x, f_y, f_{xx}, f_{yy}, and f_{xy}.

$\quad f_x = e^x - e^{x+y}$

$\quad f_y = e^y - e^{x+y}$

$\quad f_{xx} = e^x - e^{x+y}$

$\quad f_{yy} = e^y - e^{x+y}$

$\quad f_{xy} = -e^{x+y}$

Next we solve the system of equations $f_x = 0$ and $f_y = 0$.

$\quad e^x - e^{x+y} = 0 \quad$ (1)

$\quad e^y - e^{x+y} = 0 \quad$ (2)

We can solve the first equation for y.

$\quad e^x - e^{x+y} = 0$

$\qquad e^x = e^{x+y}$

$\qquad x = x + y$

$\qquad 0 = y$

We can solve the second equation for x.

$\quad e^y - e^{x+y} = 0$

$\qquad e^y = e^{x+y}$

$\qquad y = x + y$

$\qquad 0 = x$

Thus, $(0, 0)$ is a candidate for a maximum or a minimum.

We use the D-test to check $f(0, 0)$.

$\quad D = f_{xx}(0, 0) \cdot f_{yy}(0, 0) - \left[f_{xy}(0, 0)\right]^2$

$\qquad = 0 \cdot 0 - (-1)^2$

$\qquad = -1$

Since $D < 0$, it follows that $f(0, 0)$ is neither a maximum nor a minimum, but a saddle point.

23. $f(x, y) = 2y^2 + x^2 - x^2y$

We first find f_x, f_y, f_{xx}, f_{yy}, and f_{xy}.

$\quad f_x = 2x - 2xy$

$\quad f_y = 4y - x^2$

$\quad f_{xx} = 2 - 2y$

$\quad f_{yy} = 4$

$\quad f_{xy} = -2x$

Solve the system of equations $f_x = 0$ and $f_y = 0$.

$\quad 2x - 2xy = 0 \quad$ (1)

$\quad 4y - x^2 = 0 \quad$ (2)

Solve Eq. (2) for y.

$\quad 4y - x^2 = 0$

$\qquad 4y = x^2$

$\qquad y = \dfrac{x^2}{4}$

Substitute $\dfrac{x^2}{4}$ for y in Eq. (1) and solve for x.

$\quad 2x - 2x \cdot \dfrac{x^2}{4} = 0$

$\qquad 2x - \dfrac{x^3}{2} = 0$

$\qquad 4x - x^3 = 0$

$\qquad x(4 - x^2) = 0$

$\quad x(2 - x)(2 + x) = 0$

$\quad x = 0 \text{ or } 2 - x = 0 \text{ or } 2 + x = 0$

$\quad x = 0 \text{ or } \quad x = 2 \text{ or } \quad x = -2$

When $x = 0$, $y = \dfrac{0^2}{4} = 0$.

When $x = 2$, $y = \dfrac{2^2}{4} = 1$.

When $x = -2$, $y = \dfrac{(-2)^2}{4} = 1$.

The critical values are $(0,0)$, $(2,1)$, and $(-2,1)$.

We check all critical points to see if they yield maximum or minimum values.

For $(0,0)$:

$$D = f_{xx}(0,0) \cdot f_{yy}(0,0) - \left[f_{xy}(0,0)\right]^2$$

$$= 2 \cdot 4 - 0^2 \qquad \begin{bmatrix} f_{xx}(0,0) = 2 \\ f_{yy}(0,0) = 4 \\ f_{xy}(0,0) = 0 \end{bmatrix}$$

$$= 8$$

Since $D > 0$ and $f_{xx} > 0$, f has a relative minimum at $(0,0)$ and that minimum value is found as follows:

$$f(x,y) = 2y^2 + x^2 - x^2 y$$

$$f(0,0) = 2 \cdot 0^2 + 0^2 - 0^2 \cdot 0 = 0$$

The relative minimum of f is 0 at $(0,0)$.

For $(2,1)$:

$$D = f_{xx}(2,1) \cdot f_{yy}(2,1) - \left[f_{xy}(2,1)\right]^2$$

$$= 0 \cdot 4 - (-4)^2 \qquad \begin{bmatrix} f_{xx}(2,1) = 0 \\ f_{yy}(2,1) = 4 \\ f_{xy}(2,1) = -4 \end{bmatrix}$$

$$= -16$$

For $(-2,1)$:

$$D = f_{xx}(-2,1) \cdot f_{yy}(-2,1) - \left[f_{xy}(-2,1)\right]^2$$

$$= 0 \cdot 4 - 4^2 \qquad \begin{bmatrix} f_{xx}(-2,1) = 0 \\ f_{yy}(-2,1) = 4 \\ f_{xy}(-2,1) = 4 \end{bmatrix}$$

$$= -16$$

Since $D < 0$ for both $(2,1)$ and $(-2,1)$, f has neither a maximum nor a minimum at these points. Both $(2,1)$ and $(-2,1)$ are saddle points.

25. $\boxed{tw}$

27. Minimum $= -5$ at $(0,0)$

29. None

Exercise Set 7.5

1. a) The data points are $(0, 5.1)$, $(1, 8.8)$, $(2, 10.0)$, $(3, 11.1)$, and $(4, 12.6)$.

The points on the regression line are $(0, y_1)$, $(1, y_2)$, $(2, y_3)$, $(3, y_4)$, and $(4, y_5)$.

The y-deviations are $y_1 - 5.1$, $y_2 - 8.8$, $y_3 - 10.0$, $y_4 - 11.1$, and $y_5 - 12.6$.

We want to minimize

$$S = (y_1 - 5.1)^2 + (y_2 - 8.8)^2 + (y_3 - 10.0)^2 +$$
$$(y_4 - 11.1)^2 + (y_5 - 12.6)^2$$

where $y_1 = m \cdot 0 + b$,

$\qquad y_2 = m \cdot 1 + b$,

$\qquad y_3 = m \cdot 2 + b$,

$\qquad y_4 = m \cdot 3 + b$,

$\qquad y_5 = m \cdot 4 + b$.

Substituting, we get

$S = (b - 5.1)^2 + (m + b - 8.8)^2 + (2m + b - 10.0)^2 + (3m + b - 11.1)^2 + (4m + b - 12.6)^2$.

In order to minimize S we find the first partial derivatives.

$$\frac{\partial S}{\partial b} = 2(b-5.1) + 2(m+b-8.8) + 2(2m+b-10.0) +$$
$$2(3m + b - 11.1) + 2(4m + b - 12.6)$$

$$= 2b - 10.2 + 2m + 2b - 17.6 + 4m + 2b - 20.0 +$$
$$6m + 2b - 22.2 + 8m + 2b - 25.2$$

$$= 20m + 10b - 95.2$$

$$\frac{\partial S}{\partial m} = 2(m + b - 8.8) + 2(2m + b - 10.0)(2) +$$
$$2(3m + b - 11.1)(3) + 2(4m + b - 12.6)(4)$$

$$= 2m + 2b - 17.6 + 8m + 4b - 40.0 +$$
$$18m + 6b - 66.6 + 32m + 8b - 100.8$$

$$= 60m + 20b - 225$$

We set these derivatives equal to 0 and solve the resulting system of equations.

$$20m + 10b - 95.2 = 0,$$
$$60m + 20b - 225 = 0$$

The solution of the system is $b = 6.06$ and $m = 1.73$. We use the D-test to verify that $(6.06, 1.73)$ is a relative minimum. We first find the second-order partial derivatives.

$$S_{bb} = 10, \quad S_{bm} = 20, \quad S_{mm} = 60$$

$$D = S_{bb}(6.06, 1.73) \cdot S_{mm}(6.06, 1.73) -$$
$$\left[S_{bm}(6.06, 1.73)\right]^2$$

$$= 10 \cdot 60 - 20^2$$

$$= 600 - 400$$

$$= 200$$

Since $D > 0$ and $S_{bb}(6.06, 1.73) = 10 > 0$, S has a relative minimum at $(6.06, 1.73)$. The regression line is $y = 1.73x + 6.06$.

b) In 2005, $x = 2005 - 1993 = 12$.

$y = 1.73(12) + 6.06 = 26.82$ million

In 2010, $x = 2010 - 1993 = 17$.

$y = 1.73(17) + 6.06 = 35.47$ million

3. a) The data points are $(1, 70.9)$, $(2, 73.2)$, $(3, 74.8)$, $(4, 77.5)$, and $(5, 78.6)$.

The points on the regression line are $(1, y_1)$, $(2, y_2)$, $(3, y_3)$, $(4, y_4)$, and $(5, y_5)$.

The y-deviations are $y_1 - 70.9$, $y_2 - 73.2$, $y_3 - 74.8$, $y_4 - 77.5$, and $y_5 = 78.6$.

We want to minimize
$$S = (y_1 - 70.9)^2 + (y_2 - 73.2)^2 + (y_3 - 74.8)^2 + (y_4 - 77.5)^2 + (y_5 - 78.6)^2$$

where $y_1 = m \cdot 1 + b = m + b,$

$\qquad y_2 = m \cdot 2 + b = 2m + b,$

$\qquad y_3 = m \cdot 3 + b = 3m + b,$

$\qquad y_4 = m \cdot 4 + b = 4m + b,$

$\qquad y_5 = m \cdot 5 + b = 5m + b.$

Substituting, we get
$$S = (m + b - 70.9)^2 + (2m + b - 73.2)^2 +$$
$$(3m + b - 74.8)^2 + (4m + b - 77.5)^2 +$$
$$(5m + b - 78.6)^2.$$

In order to minimize S we find the first partial derivatives.
$$\frac{\partial S}{\partial b} = 2(m + b - 70.9) + 2(2m + b - 73.2) +$$
$$2(3m + b - 74.8) + 2(4m + b - 77.5) +$$
$$2(5m + b - 78.6)$$
$$= 2m + 2b - 141.8 + 4m + 2b - 146.4 +$$
$$6m + 2b - 149.6 + 8m + 2b - 155 +$$
$$10m + 2b - 157.2$$
$$= 30m + 10b - 750$$

$$\frac{\partial S}{\partial m} = 2(m + b - 70.9) + 2(2m + b - 73.2)(2) +$$
$$2(3m + b - 74.8)(3) + 2(4m + b - 77.5)(4) +$$
$$2(5m + b - 78.6)(5)$$
$$= 2m + 2b - 141.8 + 8m + 4b - 292.8 +$$
$$18m + 6b - 448.8 + 32m + 8b - 620 +$$
$$50m + 10b - 786$$
$$= 110m + 30b - 2289.4$$

We set these derivatives equal to 0 and solve the resulting system of equations.
$$30m + 10b - 750 = 0,$$
$$110m + 30b - 2289.4 = 0$$

The solution of the system is $b = 69.09$ and $m = 1.97$. We use the D-test to verify that $S(69.09, 1.97)$ is a relative minimum. We first find the second-order partial derivatives.

$$S_{bb} = 10, \ S_{bm} = 30, \ S_{mm} = 110$$
$$D = S_{bb}(69.09, 1.97) \cdot S_{mm}(69.09, 1.97) -$$
$$\qquad [S_{bm}(69.09, 1.97)]^2$$
$$= 10 \cdot 110 - 30^2$$
$$= 1100 - 900$$
$$= 200$$

Since $D > 0$ and $S_{bb}(69.09, 1.97) = 10 > 0$, S has a relative minimum at $(69.09, 1.97)$. The regression line is $y = 1.97x + 69.09$.

b) In 2000, $x = 6$.
$$y = 1.97(6) + 69.09$$
$$= 11.82 + 69.09$$
$$= 80.91$$

In 2010, $x = 7$.
$$y = 1.97(7) + 69.09$$
$$= 13.79 + 69.09$$
$$= 82.88$$

5. a) The data points are $(70, 75)$, $(60, 62)$, and $(85, 89)$.

The points on the regression line are $(70, y_1)$, $(60, y_2)$, and $(85, y_3)$.

The y-deviations are $y_1 - 75$, $y_2 - 62$, and $y_3 - 89$.

We want to minimize
$$S = (y_1 - 75)^2 + (y_2 - 62)^2 + (y_3 - 89)^2$$

where $y_1 = m \cdot 70 + b = 70m + b$

$\qquad y_2 = m \cdot 60 + b = 60m + b$

$\qquad y_3 = m \cdot 85 + b = 85m + b.$

Substituting we get
$$S = (70m + b - 75)^2 + (60m + b - 62)^2 + (85m + b - 89)^2.$$

In order to minimize S we find the first partial derivatives.
$$\frac{\partial S}{\partial b} = 2(70m + b - 75) + 2(60m + b - 62) +$$
$$\qquad 2(85m + b - 89)$$
$$= 140m + 2b - 150 + 120m + 2b - 124 +$$
$$\qquad 170m + 2b - 178$$
$$= 430m + 6b - 452$$

$$\frac{\partial S}{\partial m} = 2(70m + b - 75) \cdot 70 + 2(60m + b - 62) \cdot 60 +$$
$$\qquad 2(85m + b - 89) \cdot 85$$
$$= 9800m + 140b - 10,500 + 7200m + 120b -$$
$$\qquad 7440 + 14,450m + 170b - 15,130$$
$$= 31,450m + 430b - 33,070$$

We set these derivatives equal to 0 and solve the resulting system
$$430m + 6b - 452 = 0 \quad (1)$$
$$31,450m + 430b - 33,070 = 0 \quad (2)$$

The solution of the system is $b = -\dfrac{47}{38}$ and $m = \dfrac{203}{190}$.

We use the D-test to verify that $S\left(-\dfrac{47}{38}, \dfrac{203}{190}\right)$ is a relative minimum. We first find S_{bb}, S_{bm}, and S_{mm}.

$$S_{bb} = 6, \ S_{bm} = 430, \ S_{mm} = 31,450$$

$$D = S_{bb}\left(-\frac{47}{38}, \frac{203}{190}\right) \cdot S_{mm}\left(-\frac{47}{38}, \frac{203}{190}\right) -$$

$$\left[S_{bm}\left(-\frac{47}{38}, \frac{203}{190}\right)\right]^2$$

$$= 6 \cdot (31,450) - 430^2$$

$$= 188,700 - 184,900$$

$$= 3800$$

Since $D > 0$ and $S_{bb}\left(-\frac{47}{38}, \frac{203}{190}\right) > 0$, it follows that S has a relative minimum at $\left(-\frac{47}{38}, \frac{203}{190}\right)$.

The regression line is

$$y = \frac{203}{190}x - \frac{47}{38}$$

or

$$y = 1.068421x - 1.236842$$

b) $y = 1.068421x - 1.236842$

$\quad y = 1.068421(81) - 1.236842$ Substituting 81
                                          for x

$\quad y = 86.542101 - 1.236842$

$\quad y = 85.305259$

$\quad y \approx 85.3$

7. $\boxed{tw}$

9. a) Converting the times to decimal notation and using the STAT package on a grapher, we get $y = -0.005925x + 15.54734$.

   b) We predict that the world record will be about 3.709 min, or about 3:42.54 in 1998. We predict that the record will be about 3.6734 min, or about 3:40.40 in 2004.

   c) Using a grapher, we would have predicted that the record would be about 3.3786 min, or about 3:44.32 in 1993. Using the equation in part (a), we would have predicted that the record would be about 3:44.33.

Exercise Set 7.6

1. Find the maximum value of

   $$f(x,y) = xy$$

   subject to the constraint

   $$2x + y = 8.$$

   We form the new function F given by

   $$F(x,y,\lambda) = xy - \lambda(2x + y - 8).$$
                   Expressing $2x + y = 8$ as
                   $2x + y - 8 = 0$

   We find the first partial derivatives.

   $$F(x,y,\lambda) = \underline{x}y - \lambda(2\underline{x} + y - 8)$$
   $$F_x = y - 2\lambda$$
   $$F(x,y,\lambda) = x\underline{y} - \lambda(2x + \underline{y} - 8)$$
   $$F_y = x - \lambda$$
   $$F(x,y,\lambda) = xy - \underline{\lambda}(2x + y - 8)$$
   $$F_\lambda = -(2x + y - 8)$$

We set these derivatives equal to 0 and solve the resulting system.

$$y - 2\lambda = 0 \quad (1)$$
$$x - \lambda = 0 \quad (2)$$
$$2x + y - 8 = 0 \quad (3) \quad [-(2x + y - 8) = 0 \text{ or}$$
$$ 2x + y - 8 = 0]$$

Solving Eq. (1) for y, we get

$$y - 2\lambda = 0$$
$$y = 2\lambda.$$

Solving Eq. (2) for x, we get

$$x - \lambda = 0$$
$$x = \lambda.$$

Substituting 2λ for y and λ for x in Eq. (3) we get

$$2x + y - 8 = 0$$
$$2\lambda + 2\lambda - 8 = 0$$
$$4\lambda = 8$$
$$\lambda = 2.$$

Then

$$x = \lambda = 2$$

and

$$y = 2\lambda = 2 \cdot 2 = 4.$$

The maximum of f subject to the contraint occurs at $(2,4)$ and is

$$f(2,4) = 2 \cdot 4 = 8. \quad [f(x,y) = xy]$$

3. Find the maximum value of

   $$f(x,y) = 4 - x^2 - y^2$$

   subject to the constraint

   $$x + 2y = 10.$$

   We form the new function F given by

   $$F(x,y,\lambda) = 4 - x^2 - y^2 - \lambda(x + 2y - 10).$$
                   Expressing $x + 2y = 10$ as
                   $x + 2y - 10 = 0$

   We find the first partial derivatives.

   $$F(x,y,\lambda) = 4 - \underline{x}^2 - y^2 - \lambda(\underline{x} + 2y - 10)$$
   $$F_x = -2x - \lambda$$
   $$F(x,y,\lambda) = 4 - x^2 - \underline{y}^2 - \lambda(x + 2\underline{y} - 10)$$
   $$F_y = -2y - 2\lambda$$
   $$F(x,y,\lambda) = 4 - x^2 - y^2 - \underline{\lambda}(x + 2y - 10)$$
   $$F_\lambda = -(x + 2y - 10)$$

We set these derivatives equal to 0 and solve the resulting system.

$$2x + \lambda = 0 \quad (1) \quad (-2x - \lambda = 0 \text{ or}$$
$$ 2x + \lambda = 0)$$
$$y + \lambda = 0 \quad (2) \quad (-2y - 2\lambda = 0 \text{ or}$$
$$ y + \lambda = 0)$$
$$x + 2y - 10 = 0 \quad (3) \quad [-(x + 2y - 10) = 0 \text{ or}$$
$$ x + 2y - 10 = 0]$$

Solving Eq. (1) for x, we get

$$2x + \lambda = 0$$
$$2x = -\lambda.$$
$$x = -\frac{1}{2}\lambda$$

Solving Eq. (2) for y, we get

$$y + \lambda = 0$$
$$y = -\lambda.$$

Substituting $-\frac{1}{2}\lambda$ for x and $-\lambda$ for y in Eq. (3), we get,

$$x + 2y - 10 = 0$$
$$-\frac{1}{2}\lambda + 2(-\lambda) - 10 = 0$$
$$-\frac{5}{2}\lambda - 10 = 0$$
$$-\frac{5}{2}\lambda = 10$$
$$\lambda = -4.$$

Then

$$x = -\frac{1}{2}\lambda = -\frac{1}{2}(-4) = 2$$

and

$$y = -\lambda = -(-4) = 4.$$

The maximum of f subject to the contraint occurs at $(2, 4)$ and is

$$f(2, 4) = 4 - 2^2 - 4^2 \quad [f(x, y) = 4 - x^2 - y^2]$$
$$= 4 - 4 - 16$$
$$= -16.$$

5. Find the minimum value of

$$f(x, y) = x^2 + y^2$$

subject to the constraint

$$2x + y = 10.$$

We form the new function F given by

$$F(x, y, \lambda) = x^2 + y^2 - \lambda(2x + y - 10).$$
$$\text{Expressing } 2x + y = 10 \text{ as}$$
$$2x + y - 10 = 0$$

We find the first partial derivatives.

$$F(x, y, \lambda) = \underline{x}^2 + y^2 - \lambda(2\underline{x} + y - 10)$$
$$F_x = 2x - 2\lambda$$
$$F(x, y, \lambda) = x^2 + \underline{y}^2 - \lambda(2x + \underline{y} - 10)$$
$$F_y = 2y - \lambda$$
$$F(x, y, \lambda) = x^2 + y^2 - \underline{\lambda}(2x + y - 10)$$
$$F_\lambda = -(2x + y - 10)$$

We set these derivatives equal to 0 and solve the resulting system.

$$x - \lambda = 0 \quad (1) \quad (2x - 2\lambda = 0 \text{ or}$$
$$ x - \lambda = 0)$$
$$2y - \lambda = 0 \quad (2)$$
$$2x + y - 10 = 0 \quad (3) \quad [-(2x + y - 10) = 0 \text{ or}$$
$$ 2x + y - 10 = 0]$$

Solving Eq. (1) for x, we get

$$x - \lambda = 0$$
$$x = \lambda.$$

Solving Eq. (2) for y, we get

$$2y - \lambda = 0$$
$$2y = \lambda$$
$$y = \frac{1}{2}\lambda.$$

Substituting λ for x and $\frac{1}{2}\lambda$ for y in Eq. (3), we get,

$$2x + y - 10 = 0$$
$$2\lambda + \frac{1}{2}\lambda - 10 = 0 \quad \text{Substituting}$$
$$\frac{5}{2}\lambda = 10$$
$$\lambda = 4.$$

Then

$$x = \lambda = 4$$

and

$$y = \frac{1}{2}\lambda = \frac{1}{2} \cdot 4 = 2.$$

The minimum of f subject to the contraint occurs at $(4, 2)$ and is

$$f(4, 2) = 4^2 + 2^2 \quad [f(x, y) = x^2 + y^2]$$
$$= 16 + 4$$
$$= 20.$$

7. Find the minimum value of

$$f(x, y) = 2y^2 - 6x^2$$

subject to the constraint

$$2x + y = 4.$$

We form the new function F given by

$$F(x, y, \lambda) = 2y^2 - 6x^2 - \lambda(2x + y - 4).$$
$$\text{Expressing } 2x + y = 4 \text{ as}$$
$$2x + y - 4 = 0$$

We find the first partial derivatives.

$$F(x, y, \lambda) = 2y^2 - 6\underline{x}^2 - \lambda(2\underline{x} + y - 4)$$
$$F_x = -12x - 2\lambda$$
$$F(x, y, \lambda) = 2\underline{y}^2 - 6x^2 - \lambda(2x + \underline{y} - 4)$$
$$F_y = 4y - \lambda$$
$$F(x, y, \lambda) = 2y^2 - 6x^2 - \underline{\lambda}(2x + y - 4)$$
$$F_\lambda = -(2x + y - 4)$$

We set these derivatives equal to 0 and solve the resulting system.

$$12x + 2\lambda = 0 \quad (1) \quad (-12x - 2\lambda = 0 \text{ or}$$
$$ 12x + 2\lambda = 0)$$
$$4y - \lambda = 0 \quad (2)$$
$$2x + y - 4 = 0 \quad (3) \quad [-(2x + y - 4) = 0 \text{ or}$$
$$ 2x + y - 4 = 0]$$

Solving Eq. (1) for x, we get

$$12x + 2\lambda = 0$$
$$12x = -2\lambda$$
$$x = -\frac{\lambda}{6}.$$

Solving Eq. (2) for y, we get

$$4y - \lambda = 0$$
$$4y = \lambda$$
$$y = \frac{\lambda}{4}.$$

Substituting $-\dfrac{\lambda}{6}$ for x and $\dfrac{\lambda}{4}$ for y in Eq. (3), we get,

$$2x + y - 4 = 0$$
$$2\left(-\frac{\lambda}{6}\right) + \frac{\lambda}{4} - 4 = 0$$
$$-\frac{\lambda}{3} + \frac{\lambda}{4} = 4$$
$$-4\lambda + 3\lambda = 48$$
$$-\lambda = 48$$
$$\lambda = -48.$$

Then

$$x = -\frac{\lambda}{6} - \frac{(-48)}{6} = 8$$

and

$$y = \frac{\lambda}{4} = \frac{-48}{4} = -12.$$

The minimum of f subject to the contraint occurs at $(8, -12)$ and is

$$f(8, -12) = 2(-12)^2 - 6 \cdot 8^2 \quad [f(x,y) = 2y^2 - 6x^2]$$
$$= 2 \cdot 144 - 6 \cdot 64$$
$$= 288 - 384$$
$$= -96.$$

9. Find the minimum value of

$$f(x, y) = x^2 + y^2 + z^2$$

subject to the constraint

$$y + 2x - z = 3.$$

We form the new function F given by

$$F(x, y, z, \lambda) = x^2 + y^2 + z^2 - \lambda(y + 2x - z - 3).$$

Expressing $y + 2x - z = 3$ as
$$y + 2x - z - 3 = 0$$

We find the first partial derivatives.

$$F(x, y, z, \lambda) = \underline{x}^2 + y^2 + z^2 - \lambda(y + 2\underline{x} - z - 3)$$
$$F_x = 2x - 2\lambda$$
$$F(x, y, z, \lambda) = x^2 + \underline{y}^2 + z^2 - \lambda(\underline{y} + 2x - z - 3)$$
$$F_y = 2y - \lambda$$
$$F(x, y, z, \lambda) = x^2 + y^2 + \underline{z}^2 - \lambda(y + 2x - \underline{z} - 3)$$
$$F_z = 2z + \lambda$$
$$F(x, y, z, \lambda) = x^2 + y^2 + z^2 - \underline{\lambda}(y + 2x - z - 3)$$
$$F_\lambda = -(y + 2x - z - 3)$$

We set these derivatives equal to 0 and solve the resulting system.

$$2x - 2\lambda = 0 \quad (1)$$
$$2y - \lambda = 0 \quad (2)$$
$$2z + \lambda = 0 \quad (3)$$
$$y + 2x - z - 3 = 0 \quad (4) \quad [-(y + 2x - z - 3) = 0 \text{ or}$$
$$y + 2x - z - 3 = 0]$$

Solving Eq. (1) for x, we get

$$2x - 2\lambda = 0$$
$$2x = 2\lambda$$
$$x = \lambda$$

Solving Eq. (2) for y, we get

$$2y - \lambda = 0$$
$$2y = \lambda$$
$$y = \frac{1}{2}\lambda.$$

Solving Eq. (3) for z, we get

$$2z + \lambda = 0$$
$$2z = -\lambda$$
$$z = -\frac{1}{2}\lambda.$$

Substituting λ for x, $\dfrac{1}{2}\lambda$ for y, and $-\dfrac{1}{2}\lambda$ for z in Eq. (4), we get,

$$y + 2x - z - 3 = 0$$
$$\frac{1}{2}\lambda + 2\lambda - \left(-\frac{1}{2}\lambda\right) - 3 = 0$$
$$3\lambda - 3 = 0$$
$$3\lambda = 3$$
$$\lambda = 1.$$

Then

$$x = \lambda = 1$$
$$y = \frac{1}{2}\lambda = \frac{1}{2} \cdot 1 = \frac{1}{2}$$

and

$$z = -\frac{1}{2}\lambda = -\frac{1}{2} \cdot 1 = -\frac{1}{2}.$$

The minimum of f subject to the contraint occurs at $\left(1, \dfrac{1}{2}, -\dfrac{1}{2}\right)$ and is

$$f\left(1, \frac{1}{2}, -\frac{1}{2}\right) = 1^2 + \left(\frac{1}{2}\right)^2 + \left(-\frac{1}{2}\right)^2$$
$$[f(x, y, z) = x^2 + y^2 + z^2]$$
$$= 1 + \frac{1}{4} + \frac{1}{4}$$
$$= 1\frac{1}{2}, \text{ or } \frac{3}{2}.$$

11. Find the minimum value of

$$f(x, y) = xy \quad \text{(Product is } xy.\text{)}$$

subject to the constraint

$$x + y = 70. \quad \text{(Sum is 70.)}$$

We form the new function F given by
$$F(x, y, \lambda) = xy - \lambda(x + y - 70).$$
$$\text{Expressing } x + y = 70 \text{ as}$$
$$x + y - 70 = 0$$
We find the first partial derivatives.
$$F_x = y - \lambda$$
$$F_y = x - \lambda$$
$$F_\lambda = -(x + y - 70)$$
We set these derivatives equal to 0 and solve the resulting system.
$$y - \lambda = 0 \quad (1)$$
$$x - \lambda = 0 \quad (2)$$
$$x + y - 70 = 0 \quad (3) \quad [-(x + y - 70) = 0 \text{ or}$$
$$x + y - 70 = 0]$$
Solving Eq. (1) for y, we get
$$y - \lambda = 0$$
$$y = \lambda.$$
Solving Eq. (2) for x, we get
$$x - \lambda = 0$$
$$x = \lambda.$$
Substituting λ for x and for y in Eq. (3), we get
$$x + y - 70 = 0$$
$$\lambda + \lambda - 70 = 0$$
$$2\lambda = 70$$
$$\lambda = 35$$
Then $x = y = \lambda = 35$.

The maximum of f subject to the constraint occurs at $(35, 35)$. Thus, the two numbers whose sum is 70 that have the maximum product are 35 and 35.

13. Find the minimum value of
$$f(x, y) = xy \qquad \text{(Product is } xy.)$$
subject to the constraint
$$x - y = 6. \qquad \text{(Difference is 6.)}$$
We form the new function F given by
$$F(x, y, \lambda) = xy - \lambda(x - y - 6).$$
$$\text{Expressing } x - y = 6 \text{ as}$$
$$x - y - 6 = 0$$
We find the first partial derivatives.
$$F_x = y - \lambda$$
$$F_y = x + \lambda$$
$$F_\lambda = -(x - y - 6)$$
We set these derivatives equal to 0 and solve the resulting system.
$$y - \lambda = 0 \quad (1)$$
$$x + \lambda = 0 \quad (2)$$
$$x - y - 6 = 0 \quad (3) \quad [-(x - y - 6) = 0 \text{ or}$$
$$x - y - 6 = 0]$$
Solving Eq. (1) for y, we get
$$y - \lambda = 0$$
$$y = \lambda.$$

Solving Eq. (2) for x, we get
$$x + \lambda = 0$$
$$x = -\lambda.$$
Substituting λ for y and $-\lambda$ for x in Eq. (3), we get
$$x - y - 6 = 0$$
$$-\lambda - \lambda - 6 = 0$$
$$-2\lambda = 6$$
$$\lambda = -3.$$
Then
$$x = -\lambda = -(-3) = 3$$
and
$$y = \lambda = -3.$$
The minimum of f subject to the constraint occurs at $(3, -3)$. Thus, the two numbers whose difference is 6 that have the minimum product are 3 and -3.

15.
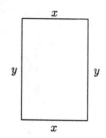

$$A = xy \qquad\qquad \text{Area}$$
$$P = 2x + 2y = 39 \qquad \text{Perimeter}$$
We need to maximize the area A
$$A = xy$$
subject to the constraint
$$2x + 2y = 39.$$
We form the new function F given by
$$F(x, y, \lambda) = xy - \lambda(2x + 2y - 39).$$
$$\text{Expressing } 2x + 2y = 39 \text{ as}$$
$$2x + 2y - 39 = 0$$
We find the first partial derivatives.
$$F_x = y - 2\lambda$$
$$F_y = x - 2\lambda$$
$$F_\lambda = -(2x + 2y - 39)$$
We set these derivatives equal to 0 and solve the resulting system.
$$y - 2\lambda = 0 \quad (1)$$
$$x - 2\lambda = 0 \quad (2)$$
$$2x + 2y - 39 = 0 \quad (3) \quad [-(2x + 2y - 39) = 0 \text{ or}$$
$$2x + 2y - 39 = 0]$$
Solving Eq. (1) for y, we get
$$y - 2\lambda = 0$$
$$y = 2\lambda.$$
Solving Eq. (2) for x, we get
$$x - 2\lambda = 0$$
$$x = 2\lambda.$$

Substituting 2λ for x and for y in Eq. (3), we get

$$2x + 2y - 39 = 0$$
$$2(2\lambda) + 2(2\lambda) - 39 = 0$$
$$8\lambda = 39$$
$$\lambda = \frac{39}{8}$$

Then

$$x = y = 2\lambda = 2 \cdot \frac{39}{8} = \frac{39}{4} \text{ or } 9\frac{3}{4}.$$

The maximum value of A subject to the constraint occurs at $\left(\frac{39}{4}, \frac{39}{4}\right)$. Thus, the dimensions of the paper that will give the most typing area, subject to the perimeter constraint of 39 in. are $9\frac{3}{4}$ in. $\times 9\frac{3}{4}$ in..

The maximum area is $A = \frac{39}{4} \cdot \frac{39}{4} = \frac{1521}{16} = 95\frac{1}{16}$ in^2.

The area of the standard $8\frac{1}{2} \times 11$ paper is $\frac{17}{2} \cdot \frac{11}{1} = \frac{187}{2}$, or $93\frac{1}{2}$ in^2. The perimeter of an $8\frac{1}{2} \times 11$ sheet of paper is 39 in., but its area is less than the area of a sheet whose perimeter is also 39 in. but whose dimensions are $9\frac{3}{4} \times 9\frac{3}{4}$.

17. We want to minimize the function s given by

$$s(h, r) = 2\pi r h + 2\pi r^2$$

subject to the volume constraint

$$\pi r^2 h = 27, \text{ or } \pi r^2 h - 27 = 0.$$

We form the new function S given by

$$S(h, r, \lambda) = 2\pi r h + 2\pi r^2 - \lambda(\pi r^2 h - 27).$$

We find the first partial derivatives.

$$\frac{\partial S}{\partial h} = 2\pi r - \lambda \pi r^2,$$

$$\frac{\partial S}{\partial r} = 2\pi h + 4\pi r - 2\lambda \pi r h,$$

$$\frac{\partial S}{\partial \lambda} = -(\pi r^2 h - 27).$$

We set these derivatives equal to 0 and solve the resulting system.

$$2\pi r - \lambda \pi r^2 = 0 \quad (1)$$
$$2\pi h + 4\pi r - 2\lambda \pi r h = 0 \quad (2)$$
$$\pi r^2 h - 27 = 0 \quad (3) \quad [-(\pi r^2 h - 27) = 0 \text{ or } \pi r^2 h - 27 = 0]$$

Note that we can solve Eq. (1) for r:

$$\pi r(2 - \lambda r) = 0$$
$$\pi r = 0 \text{ or } 2 - \lambda r = 0$$
$$r = 0 \text{ or } \quad r = \frac{2}{\lambda}$$

Note $r = 0$ cannot be a solution to the original problem, so we continue by substituting $2/\lambda$ for r in Eq. (2).

$$2\pi h + 4\pi \cdot \frac{2}{\lambda} - 2\lambda \pi \cdot \frac{2}{\lambda} \cdot h = 0$$

$$2\pi h + \frac{8\pi}{\lambda} - 4\pi h = 0$$

$$\frac{8\pi}{\lambda} - 2\pi h = 0$$

$$-2\pi h = -\frac{8\pi}{\lambda}$$

$$h = \frac{4}{\lambda}$$

Since $h = 4/\lambda$ and $r = 2/\lambda$, it follows that $h = 2r$. Substituting $2r$ for h in Eq. (3) yields

$$\pi r^2 \cdot 2r - 27 = 0$$
$$2\pi r^3 - 27 = 0$$
$$2\pi r^3 = 27$$
$$r^3 = \frac{27}{2\pi}$$
$$r = \sqrt[3]{\frac{27}{2\pi}} \approx 1.6 \text{ ft.}$$

So when $r = 1.6$ ft and $y = 2(1.6) = 3.2$ ft, the surface area is a minimum and is about

$$2\pi(1.6)(3.2) + 2\pi(1.6)^2, \text{ or } 48.3 \text{ ft}^2.$$

(Answers will vary due to rounding differences.)

19. We need to find the maximum value of

$$S(L, M) = ML - L^2$$

subject to the constraint

$$M + L = 80.$$

We form the new function F given by

$$F(L, M, \lambda) = ML - L^2 - \lambda(M + L - 80).$$
$$\text{Expressing } M + L = 80 \text{ as}$$
$$M + L - 80 = 0$$

We find the first partial derivatives.

$$F_L = M - 2L - \lambda$$
$$F_M = L - \lambda$$
$$F_\lambda = -(M + L - 80)$$

We set these derivatives equal to 0 and solve the resulting system.

$$M - 2L - \lambda = 0 \quad (1)$$
$$L - \lambda = 0 \quad (2)$$
$$M + L - 80 = 0 \quad (3) \quad [-(M + L - 80) = 0 \text{ or } M + L - 80 = 0]$$

Solving Eq. (2) for L we get

$$L - \lambda = 0$$
$$L = \lambda.$$

Substituting λ for L in Eq. (1) and solving for M, we get

$$M - 2L - \lambda = 0$$
$$M - 2\lambda - \lambda = 0$$
$$M = 3\lambda$$

Substituting λ for L and 3λ for M in Eq. (3), we get

$$M + L - 80 = 0$$
$$3\lambda + \lambda - 80 = 0$$
$$4\lambda = 80$$
$$\lambda = 20$$

Then

$$L = \lambda = 20$$

and

$$M = 3\lambda = 3 \cdot 20 = 60.$$

The maximum value of S subject to the constraint occurs at $(20, 60)$ and is

$$S(20, 60) = 60 \cdot 20 - 20^2 \qquad [S(L, M) = ML - L^2]$$
$$= 1200 - 400$$
$$= 800.$$

21. a) The area of the floor is xy.

The cost of the floor is $4xy$.

The area of the walls is $2xz + 2yz$.

The cost of the walls is $3(2xz + 2yz)$.

The area of the ceiling is xy.

The cost of the ceiling is $3xy$.

$$C(x, y, z) = 4xy + 3(2xz + 2yz) + 3xy$$
$$= 7xy + 6xz + 6yz$$

b) We need to find the minimum value of

$$C(x, y, z) = 7xy + 6xz + 6yz$$

subject to the constraint

$$xyz = 252,000. \qquad (\text{Volume} = l \cdot w \cdot h)$$

We form the new function N given by

$$N(x, y, z, \lambda) = 7xy + 6xz + 6yz - \lambda(xyz - 252,000)$$

Expressing $xyz = 252,000$ as

$$xyz - 252,000 = 0$$

We find the first partial derivatives.

$$N_x = 7y + 6z - \lambda yz$$
$$N_y = 7x + 6z - \lambda xz$$
$$N_z = 6x + 6y - \lambda xy$$
$$N_\lambda = -(xyz - 252,000)$$

We set these derivatives equal to 0 and solve the resulting system.

$$7y + 6z - \lambda yz = 0 \qquad (1)$$
$$7x + 6z - \lambda xz = 0 \qquad (2)$$
$$6x + 6y - \lambda xy = 0 \qquad (3)$$
$$xyz - 252,000 = 0 \qquad (4)$$

Solving Eq. (2) for x and Eq. (1) for y, we get

$$x = \frac{6z}{\lambda z - 7} \quad \text{and} \quad y = \frac{6z}{\lambda z - 7}.$$

Thus, $x = y$.

Substituting x for y we get the following system.

$$7x + 6z - \lambda xz = 0 \text{ or } 7x + 6z - \lambda xz = 0 \quad (5)$$
$$6x + 6x - \lambda xx = 0 \qquad\qquad 12x - \lambda x^2 = 0 \quad (6)$$
$$xxz - 252,000 = 0 \qquad x^2 z - 252,000 = 0 \quad (7)$$

Solving Eq. (6) for x, we get

$$12x - \lambda x^2 = 0$$
$$x(12 - \lambda x) = 0$$
$$x = 0 \text{ or } 12 - \lambda x = 0$$
$$x = 0 \text{ or } \qquad x = \frac{12}{\lambda}$$

We only consider $x = \frac{12}{\lambda}$ since x cannot be 0 in the original problem. We continue by substituting $\frac{12}{\lambda}$ for x in Eq. (7) and solving for z.

$$\left(\frac{12}{\lambda}\right)^2 z - 252,000 = 0$$
$$\frac{144}{\lambda^2} z = 252,000$$
$$z = \frac{252,000}{144} \lambda^2$$
$$z = 1750\lambda^2$$

Next we substitute $\frac{12}{\lambda}$ for x and $1750\lambda^2$ for z in Eq. (5) and solve for λ.

$$7 \cdot \frac{12}{\lambda} + 6 \cdot 1750\lambda^2 - \lambda \cdot \frac{12}{\lambda} \cdot 1750\lambda^2 = 0$$
$$\frac{84}{\lambda} + 10,500\lambda^2 - 21,000\lambda^2 = 0$$
$$\frac{84}{\lambda} = 10,500\lambda^2$$
$$84 = 10,500\lambda^3$$
$$\frac{84}{10,500} = \lambda^3$$
$$\frac{1}{125} = \lambda^3$$
$$\frac{1}{5} = \lambda$$

Thus,

$$x = \frac{12}{\lambda} = \frac{12}{1/5} = 12 \cdot \frac{5}{1} = 60$$
$$y = \frac{12}{\lambda} = \frac{12}{1/5} = 60$$
$$z = 1750\lambda^2 = 1750\left(\frac{1}{5}\right)^2 = \frac{1750}{25} = 70.$$

The minimum total cost is obtained when the dimensions are 60 ft by 60 ft by 70 ft. The minimum cost is found as follows:

$$C(x, y, z) = 7xy + 6xz + 6yz$$
$$C(60, 60, 70) = 7 \cdot 60 \cdot 60 + 6 \cdot 60 \cdot 70 + 6 \cdot 60 \cdot 70$$
$$= 25,200 + 25,200 + 25,200$$
$$= \$75,600$$

23. We need to find the minimum value of

$$C(x, y) = 10 + \frac{x^2}{6} + 200 + \frac{y^3}{9} \quad [C(x,y) = C(x) + C(y)]$$

$$= \frac{x^2}{6} + \frac{y^3}{9} + 210$$

subject to the constraint

$$x + y = 10,100.$$

We form the new function N given by

$$N(x, y, \lambda) = \frac{x^2}{6} + \frac{y^3}{9} + 210 - \lambda(x + y - 10,100)$$

Expressing $x + y = 10,100$ as
$$x + y - 10,100 = 0$$

We first find the partial derivatives.

$$N_x = \frac{1}{3}x - \lambda$$

$$N_y = \frac{1}{3}y^2 - \lambda$$

$$N_\lambda = -(x + y - 10,100)$$

We set these derivatives equal to 0 and solve the resulting system.

$$\frac{1}{3}x - \lambda = 0 \quad (1)$$

$$\frac{1}{3}y^2 - \lambda = 0 \quad (2)$$

$$x + y - 10,100 = 0 \quad (3) \quad [-(x + y - 10,100) = 0 \text{ or}$$
$$x + y - 10,100 = 0]$$

Solve Eq. (1) for x.

$$\frac{1}{3}x - \lambda = 0$$

$$\frac{1}{3}x = \lambda$$

$$x = 3\lambda$$

Solve Eq. (2) for y^2.

$$\frac{1}{3}y^2 - \lambda = 0$$

$$\frac{1}{3}y^2 = \lambda$$

$$y^2 = 3\lambda$$

Thus, $x = y^2$.

Substitute y^2 for x in Eq. (3) and solve for y.

$$y^2 + y - 10,100 = 0$$

$$(y + 101)(y - 100) = 0$$

$$y + 101 = 0 \quad \text{or} \quad y - 100 = 0$$

$$y = -101 \quad \text{or} \quad y = 100$$

Since y cannot be -101 in the original problem, we only consider $y = 100$. If $y = 100$, $x = 100^2$, or $10,000$. To minimize total costs, 10,000 units on A and 100 units on B should be made.

25. Find the minimum value of

$$f(x, y) = 2x^2 + y^2 + 2xy + 3x + 2y$$

subject to the constraint

$$y^2 = x + 1.$$

We form the new function F given by

$$F(x, y, \lambda) = 2x^2 + y^2 + 2xy + 3x + 2y - \lambda(y^2 - x - 1)$$

Expressing $y^2 = x + 1$ as
$$y^2 - x - 1 = 0$$

We find the first partial derivatives.

$$F_x = 4x + 2y + 3 + \lambda$$

$$F_y = 2y + 2x + 2 - 2\lambda y$$

$$F_\lambda = -(y^2 - x - 1)$$

We set these derivatives equal to 0 and solve the resulting system.

$$4x + 2y + 3 + \lambda = 0 \quad \text{or} \quad 4x + 2y + 3 + \lambda = 0 \quad (1)$$

$$2y + 2x + 2 - 2\lambda y = 0 \qquad\qquad y + x + 1 - \lambda y = 0 \quad (2)$$

$$-(y^2 - x - 1) = 0 \qquad\qquad\qquad y^2 - x - 1 = 0 \quad (3)$$

Multiply Eq. (1) by y and add the result to Eq. (2).

$$4xy + 2y^2 + 3y + \lambda y = 0 \quad (1)$$

$$\underline{y + x + 1 - \lambda y = 0 \quad (2)}$$

$$4y + x + 4xy + 2y^2 + 1 = 0 \quad \text{(Adding)} \quad (4)$$

Solve Eq. (3) for x.

$$y^2 - x - 1 = 0$$

$$y^2 - 1 = x$$

Substitute $y^2 - 1$ for x in Eq. (4) and solve for y.

$$4y + (y^2 - 1) + 4(y^2 - 1)y + 2y^2 + 1 = 0$$

$$4y + y^2 - 1 + 4y^3 - 4y + 2y^2 + 1 = 0$$

$$4y^3 + 3y^2 = 0$$

$$y^2(4y + 3) = 0$$

$$y^2 = 0 \quad \text{or} \quad 4y + 3 = 0$$

$$y = 0 \quad \text{or} \qquad y = -\frac{3}{4}$$

For $y = 0$, $x = 0^2 - 1$, or -1.

For $y = -\frac{3}{4}$, $x = \left(-\frac{3}{4}\right)^2 - 1 = \frac{9}{16} - \frac{16}{16} = -\frac{7}{16}$.

When $x = -1$ and $y = 0$,

$$f(-1, 0) = 2(-1)^2 + (0)^2 + 2(-1)(0) + 3(-1) + 2(0)$$

$$= 2 + 0 + 0 - 3 + 0$$

$$= -1$$

When $x = -\dfrac{7}{16}$ and $y = -\dfrac{3}{4}$.

$$f\left(-\frac{7}{16}, -\frac{3}{4}\right)$$

$$= 2\left(-\frac{7}{16}\right)^2 + \left(-\frac{3}{4}\right)^2 + 2\left(-\frac{7}{16}\right)\left(-\frac{3}{4}\right) + 3\left(-\frac{7}{16}\right) +$$

$$2\left(-\frac{3}{4}\right)$$

$$= \frac{98}{256} + \frac{9}{16} + \frac{42}{64} - \frac{21}{16} - \frac{6}{4}$$

$$= \frac{49}{128} - \frac{12}{16} + \frac{21}{32} - \frac{3}{2}$$

$$= \frac{49 \cdot}{128} - \frac{96}{128} + \frac{84}{128} - \frac{192}{128}$$

$$= -\frac{155}{128} \qquad \text{Minimum}$$

Thus, the minimum value of f is $-\dfrac{155}{128}$ at $\left(-\dfrac{7}{16}, -\dfrac{3}{4}\right)$.

27. Find the maximum value of
$$f(x, y, z) = x^2 y^2 z^2$$
subject to the constraint
$$x^2 + y^2 + z^2 = 1.$$
We form the new function F given by
$$F(x, y, z, \lambda) = x^2 y^2 z^2 - \lambda(x^2 + y^2 + z^2 - 1)$$
$$\text{Expressing } x^2 + y^2 + z^2 = 1 \text{ as}$$
$$x^2 + y^2 + z^2 - 1 = 0$$
We find the first partial derivatives.
$$F_x = 2xy^2 z^2 - 2\lambda x$$
$$F_y = 2x^2 yz^2 - 2\lambda y$$
$$F_z = 2x^2 y^2 z - 2\lambda z$$
$$F_\lambda = -(x^2 + y^2 + z^2 - 1)$$

We set these derivatives equal to 0 and solve the resulting system.

$$2xy^2 z^2 - 2\lambda x = 0 \quad \text{or} \quad x(y^2 z^2 - \lambda) = 0 \quad (1)$$
$$2x^2 yz^2 - 2\lambda y = 0 \quad \text{or} \quad y(x^2 z^2 - \lambda) = 0 \quad (2)$$
$$2x^2 y^2 z - 2\lambda z = 0 \quad \text{or} \quad z(x^2 y^2 - \lambda) = 0 \quad (3)$$
$$-(x^2 + y^2 + z^2 - 1) = 0 \quad \text{or} \quad x^2 + y^2 + z^2 - 1 = 0 \quad (4)$$

Note that for $x = 0$, $y = 0$, or $z = 0$, $f(x, y, z) = 0$. For all values of x, y, and $z \neq 0$, $f(x, y, z) > 0$. Thus the maximum value of f cannot occur when any or all of the variables is 0. Thus, we will only consider nonzero values of x, y, and z.

Using the Principle of Zero Products, we get:

From Eq. (1)   From Eq. (2)   From Eq. (3)
$y^2 z^2 - \lambda = 0$ $x^2 y^2 - \lambda = 0$ $x^2 y^2 - \lambda = 0$
$\quad y^2 z^2 = \lambda$ $\quad x^2 z^2 = \lambda$ $\quad x^2 y^2 = \lambda$

Thus, $y^2 z^2 = x^2 z^2 = x^2 y^2$ and $x^2 = y^2 = z^2$.

Substitute x^2 for y^2 and z^2 in Eq. (4) and solve for x.

$$x^2 + y^2 + z^2 - 1 = 0$$
$$x^2 + x^2 + x^2 - 1 = 0$$
$$3x^2 = 1$$
$$x^2 = \frac{1}{3}$$
$$x = \pm\sqrt{\frac{1}{3}}, \text{ or } \pm\frac{1}{\sqrt{3}}$$

Now $x^2 = y^2 = z^2$ so we can find y and z.

$$y^2 = \frac{1}{3} \qquad\qquad z^2 = \frac{1}{3}$$
$$y = \pm\frac{1}{\sqrt{3}} \qquad z = \pm\frac{1}{\sqrt{3}}$$

For $\left(\pm\dfrac{1}{\sqrt{3}}, \pm\dfrac{1}{\sqrt{3}}, \pm\dfrac{1}{\sqrt{3}}\right)$:

$$f(x, y, z) = \left(\pm\frac{1}{\sqrt{3}}\right)^2 \cdot \left(\pm\frac{1}{\sqrt{3}}\right)^2 \cdot \left(\pm\frac{1}{\sqrt{3}}\right)^2$$
$$= \frac{1}{3} \cdot \frac{1}{3} \cdot \frac{1}{3} = \frac{1}{27}.$$

Thus f has a maximum value of $\dfrac{1}{27}$ at $\left(\pm\dfrac{1}{\sqrt{3}}, \pm\dfrac{1}{\sqrt{3}}, \pm\dfrac{1}{\sqrt{3}}\right)$.

29. Find the maximum value of
$$f(x, y, z, t) = x + y + z + t$$
subject to the constraint
$$x^2 + y^2 + z^2 + t^2 = 1.$$
We form the new function F given by
$$F(x, y, z, t, \lambda) = x + y + z + t - \lambda(x^2 + y^2 + z^2 + t^2 - 1)$$
$$\text{Expressing } x^2 + y^2 + z^2 + t^2 = 1 \text{ as}$$
$$x^2 + y^2 + z^2 + t^2 - 1 = 0$$
We find the first partial derivatives.
$$F_x = 1 - 2\lambda x$$
$$F_y = 1 - 2\lambda y$$
$$F_z = 1 - 2\lambda z$$
$$F_t = 1 - 2\lambda t$$
$$F_\lambda = -(x^2 + y^2 + z^2 + t^2 - 1)$$

We set these derivatives equal to 0 and solve the resulting system.

$$1 - 2\lambda x = 0 \quad (1)$$
$$1 - 2\lambda y = 0 \quad (2)$$
$$1 - 2\lambda z = 0 \quad (3)$$
$$1 - 2\lambda t = 0 \quad (4)$$
$$x^2 + y^2 + z^2 + t^2 - 1 = 0 \quad (5)$$
$$[-(x^2 + y^2 + z^2 + t^2 - 1) = 0$$
$$\text{or } x^2 + y^2 + z^2 + t^2 - 1 = 0]$$

Solving the first four equations for x, y, z, and t respectively, we get

$$x = \frac{1}{2\lambda}$$

$$y = \frac{1}{2\lambda}$$

$$z = \frac{1}{2\lambda}$$

$$t = \frac{1}{2\lambda}.$$

Thus, $x = y = z = t$.

Next we substitute $\frac{1}{2\lambda}$ for x, y, z, and t in Eq. (5) and solve for λ.

$$x^2 + y^2 + z^2 + t^2 - 1 = 0 \qquad (5)$$

$$\left(\frac{1}{2\lambda}\right)^2 + \left(\frac{1}{2\lambda}\right)^2 + \left(\frac{1}{2\lambda}\right)^2 + \left(\frac{1}{2\lambda}\right)^2 - 1 = 0$$

Substituting

$$4\left(\frac{1}{4\lambda^2}\right) - 1 = 0$$

$$\frac{1}{\lambda^2} = 1$$

$$1 = \lambda^2$$

$$\pm 1 = \lambda$$

When $\lambda = 1$, $x = y = z = t = \frac{1}{2}$ and

$$f\left(\frac{1}{2}, \frac{1}{2}, \frac{1}{2}, \frac{1}{2}\right) = \frac{1}{2} + \frac{1}{2} + \frac{1}{2} + \frac{1}{2} = 2.$$

When $\lambda = -1$, $x = y = z = t = -\frac{1}{2}$ and

$$f\left(-\frac{1}{2}, -\frac{1}{2}, -\frac{1}{2}, -\frac{1}{2}\right) =$$

$$-\frac{1}{2} + \left(-\frac{1}{2}\right) + \left(-\frac{1}{2}\right) + \left(-\frac{1}{2}\right) = -2.$$

Thus f has a maximum value of 2 at $\left(\frac{1}{2}, \frac{1}{2}, \frac{1}{2}, \frac{1}{2}\right)$.

31. $P(x, y, \lambda) = p(x, y) - \lambda(c_1 x + c_2 y - B)$

Expressing $c_1 x + c_2 y = B$ as

$c_1 x + c_2 y - B = 0$

$P_x = p_x - \lambda c_1$

$P_y = p_y - \lambda c_2$

We set these derivatives equal to 0 and solve for λ.

$$p_x - \lambda c_1 = 0 \qquad\qquad p_y - \lambda c_2 = 0$$

$$p_x = \lambda c_1 \qquad\qquad p_y = \lambda c_2$$

$$\frac{p_x}{c_1} = \lambda \qquad\qquad \frac{p_y}{c_2} = \lambda$$

Thus, $\lambda = \frac{p_x}{c_1} = \frac{p_y}{c_2}$.

33. $\boxed{tw}$

35. - 41.

Exercise Set 7.7

1. $\int_0^1 \int_0^1 2y \, dx \, dy$

$= \int_0^1 \left(\int_0^1 2y \, dx\right) dy$

We first evaluate the inside integral.

$\int_0^1 2y \, dx$

$= 2y \int_0^1 dx$ \qquad $2y$ is a constant

$= 2y \left[x\right]_0^1$ \qquad Integrating with respect to x

$= 2y(1 - 0)$

$= 2y$

Then we evaluate the outside integral.

$\int_0^1 \left(\int_0^1 2y \, dx\right) dy$

$= \int_0^1 2y \, dy$ \qquad $\left(\int_0^1 2y \, dx = 2y\right)$

$= \left[y^2\right]_0^1$ \qquad Integrating with respect to y

$= 1^2 - 0^2$

$= 1$

3. $\int_{-1}^1 \int_x^1 xy \, dy \, dx$

$= \int_{-1}^1 \left(\int_x^1 xy \, dy\right) dx$

We first evaluate the inside integral.

$\int_x^1 xy \, dy$

$= x \int_x^1 y \, dy$ \qquad x is a constant

$= x \left[\frac{y^2}{2}\right]_x^1$ \qquad Integrating with respect to y

$= x\left(\frac{1^2}{2} - \frac{x^2}{2}\right)$

$= \frac{x}{2} - \frac{x^3}{2}$

$= \frac{1}{2}(x - x^3)$

Then we evaluate the outside integral.

$$\int_{-1}^1 \left(\int_x^1 xy \, dy\right) dx$$

$$= \int_{-1}^1 \frac{1}{2}(x - x^3) \, dx \qquad \left[\int_x^1 xy \, dy = \frac{1}{2}(x - x^3)\right]$$

$$= \frac{1}{2}\int_{-1}^{1}(x - x^3)\,dx$$

$$= \frac{1}{2}\left[\frac{x^2}{2} - \frac{x^4}{4}\right]_{-1}^{1} \qquad \text{Integrating with respect to } x$$

$$= \frac{1}{2}\left[\left(\frac{1^2}{2} - \frac{1^4}{4}\right) - \left(\frac{(-1)^2}{2} - \frac{(-1)^4}{4}\right)\right]$$

$$= \frac{1}{2}\left[\left(\frac{1}{2} - \frac{1}{4}\right) - \left(\frac{1}{2} - \frac{1}{4}\right)\right]$$

$$= \frac{1}{2}\left[\frac{1}{2} - \frac{1}{4} - \frac{1}{2} + \frac{1}{4}\right]$$

$$= \frac{1}{2}\cdot 0$$

$$= 0$$

5. $\int_{0}^{1}\int_{-1}^{3}(x + y)\,dy\,dx$

$$= \int_{0}^{1}\left(\int_{-1}^{3}(x + y)\,dy\right)dx$$

We first evaluate the inside integral.

$$\int_{-1}^{3}(x + y)\,dy$$

$$= \left[xy + \frac{y^2}{2}\right]_{-1}^{3} \qquad \text{Integrating with respect to } y;\ x \text{ is a constant}$$

$$= \left(3x + \frac{3^2}{2}\right) - \left(-x + \frac{(-1)^2}{2}\right)$$

$$= 3x + \frac{9}{2} + x - \frac{1}{2}$$

$$= 4x + 4$$

Then we evaluate the outside integral.

$$\int_{0}^{1}\left(\int_{-1}^{3}(x + y)\,dy\right)dx$$

$$= \int_{0}^{1}(4x + 4)\,dx \qquad \left[\int_{-1}^{3}(x + y)\,dy = 4x + 4\right]$$

$$= 4\int_{0}^{1}(x + 1)\,dx$$

$$= 4\left[\frac{x^2}{2} + x\right]_{0}^{1} \qquad \text{Integrating with respect to } x$$

$$= 4\left[\left(\frac{1^2}{2} + 1\right) - \left(\frac{0^2}{2} + 0\right)\right]$$

$$= 4\left(\frac{1}{2} + 1\right)$$

$$= 6$$

7. $\int_{0}^{1}\int_{x^2}^{x}(x + y)\,dy\,dx$

$$= \int_{0}^{1}\left(\int_{x^2}^{x}(x + y)\,dy\right)dx$$

We first evaluate the inside integral.

$$\int_{x^2}^{x}(x + y)\,dy$$

$$= \left[xy + \frac{y^2}{2}\right]_{x^2}^{x} \qquad \text{Integrating with respect to } y;\ x \text{ is a constant}$$

$$= \left[\left(x\cdot x + \frac{x^2}{2}\right) - \left(x\cdot x^2 + \frac{(x^2)^2}{2}\right)\right]$$

$$= \left[x^2 + \frac{x^2}{2} - x^3 - \frac{x^4}{2}\right]$$

$$= \frac{3}{2}x^2 - x^3 - \frac{1}{2}x^4$$

Then we evaluate the outside integral.

$$\int_{0}^{1}\left(\int_{x^2}^{x}(x + y)\,dy\right)dx$$

$$= \int_{0}^{1}\left(\frac{3}{2}x^2 - x^3 - \frac{1}{2}x^4\right)dx$$

$$\left[\int_{x^2}^{x}(x + y)\,dy = \frac{3}{2}x^2 - x^3 - \frac{1}{2}x^4\right]$$

$$= \left[\frac{x^3}{2} - \frac{x^4}{4} - \frac{x^5}{10}\right]_{0}^{1} \qquad \text{Integrating with respect to } x$$

$$= \left(\frac{1^3}{2} - \frac{1^4}{4} - \frac{1^5}{10}\right) - \left(\frac{0^3}{2} - \frac{0^4}{4} - \frac{0^5}{10}\right)$$

$$= \left(\frac{1}{2} - \frac{1}{4} - \frac{1}{10}\right) - 0$$

$$= \frac{10}{20} - \frac{5}{20} - \frac{2}{20}$$

$$= \frac{3}{20}$$

9. $\int_{0}^{2}\int_{0}^{x}(x + y^2)\,dy\,dx$

$$= \int_{0}^{2}\left(\int_{0}^{x}(x + y^2)\,dy\right)dx$$

We first evaluate the inside integral.

$$\int_{0}^{x}(x + y^2)\,dy$$

$$= \left[xy + \frac{y^3}{3}\right]_{0}^{x} \qquad \text{Integrating with respect to } y;\ x \text{ is a constant}$$

$$= \left(x\cdot x + \frac{x^3}{3}\right) - \left(x\cdot 0 + \frac{0^3}{3}\right)$$

$$= x^2 + \frac{x^3}{3}$$

Then we evaluate the outside integral.

$$\int_0^2 \left(\int_0^x (x+y^2)\, dy \right) dx$$

$$= \int_0^2 \left(x^2 + \frac{x^3}{3} \right) dx \qquad \left[\int_0^x (x+y^2)\, dy = x^2 + \frac{x^3}{3} \right]$$

$$= \left[\frac{x^3}{3} + \frac{x^4}{12} \right]_0^2 \qquad \text{Integrating with respect to } x$$

$$= \left(\frac{2^3}{3} + \frac{2^4}{12} \right) - \left(\frac{0^3}{3} + \frac{0^4}{12} \right)$$

$$= \frac{8}{3} + \frac{16}{12}$$

$$= \frac{8}{3} + \frac{4}{3}$$

$$= \frac{12}{3}, \text{ or } 4$$

11. $\int_0^1 \int_0^{1-x^2} (1-y-x^2)\, dy\, dx$

$$= \int_0^1 \left(\int_0^{1-x^2} (1-y-x^2)\, dy \right) dx$$

We first evaluate the inside integral.

$\int_0^{1-x^2} (1-y-x^2)\, dy$

$$= \left[y - \frac{y^2}{2} - x^2 y \right]_0^{1-x^2} \qquad \begin{array}{l}\text{Integrating with respect} \\ \text{to } y; \ x \text{ is a constant}\end{array}$$

$$= \left[(1-x^2) - \frac{(1-x^2)^2}{2} - x^2(1-x^2) \right] - \left[0 - \frac{0^2}{2} - x^2 \cdot 0 \right]$$

$$= 1 - x^2 - \frac{1 - 2x^2 + x^4}{2} - x^2 + x^4$$

$$= 1 - x^2 - \frac{1}{2} + x^2 - \frac{1}{2}x^4 - x^2 + x^4$$

$$= \frac{1}{2}x^4 - x^2 + \frac{1}{2}$$

Then we evaluate the outside integral.

$$\int_0^1 \left(\int_0^{1-x^2} (1-y-x^2)\, dy \right) dx$$

$$= \int_0^1 \left(\frac{1}{2}x^4 - x^2 + \frac{1}{2} \right) dx$$

$$= \left[\frac{x^5}{10} - \frac{x^3}{3} + \frac{1}{2}x \right]_0^1 \qquad \text{Integrating with respect to } x$$

$$= \left(\frac{1^5}{10} - \frac{1^3}{3} + \frac{1}{2} \cdot 1 \right) - \left(\frac{0^5}{10} - \frac{0^3}{3} + \frac{1}{2} \cdot 0 \right)$$

$$= \frac{1}{10} - \frac{1}{3} + \frac{1}{2}$$

$$= \frac{3}{30} - \frac{10}{30} + \frac{15}{30}$$

$$= \frac{8}{30}$$

$$= \frac{4}{15}$$

13. $f(x,y) = x^2 + \frac{1}{3}xy$

$0 \le x \le 1$

$0 \le y \le 2$

Find

$\int_0^2 \int_0^1 f(x,y)\, dx\, dy$

$$= \int_0^2 \left(\int_0^1 \left(x^2 + \frac{1}{3}xy \right) dx \right) dy$$

We first evaluate the inside integral.

$\int_0^1 \left(x^2 + \frac{1}{3}xy \right) dx$

$$= \left[\frac{x^3}{3} + \frac{1}{3}y \cdot \frac{x^2}{2} \right]_0^1 \qquad \begin{array}{l}\text{Integrating with respect} \\ \text{to } x; \ y \text{ is a constant}\end{array}$$

$$= \left(\frac{1^3}{3} + \frac{1}{3}y \cdot \frac{1^2}{2} \right) - \left(\frac{0^3}{3} + \frac{1}{3}y \cdot \frac{0^2}{2} \right)$$

$$= \frac{1}{3} + \frac{1}{6}y$$

Then we evaluate the outside integral.

$$\int_0^2 \left(\int_0^1 \left(x^2 + \frac{1}{3}xy \right) dx \right) dy$$

$$= \int_0^2 \left(\frac{1}{3} + \frac{1}{6}y \right) dy$$

$$= \left[\frac{1}{3}y + \frac{1}{6} \cdot \frac{y^2}{2} \right]_0^2 \qquad \text{Integrating with respect to } y$$

$$= \left[\frac{1}{3}y + \frac{1}{12}y^2 \right]_0^2$$

$$= \left(\frac{1}{3} \cdot 2 + \frac{1}{12} \cdot 2^2 \right) - \left(\frac{1}{3} \cdot 0 + \frac{1}{12} \cdot 0^2 \right)$$

$$= \frac{2}{3} + \frac{1}{3}$$

$$= 1$$

15. $\int_0^1 \int_1^3 \int_{-1}^2 (2x + 3y - z)\, dx\, dy\, dz$

We first evaluate the x-integral.

$\int_{-1}^2 (2x + 3y - z)\, dx$

$$= \left[x^2 + 3xy - xz \right]_{-1}^2 \qquad \begin{array}{l}\text{Integrating with respect} \\ \text{to } x; \ y \text{ and } z \text{ are constants}\end{array}$$

$$= [2^2 + 3 \cdot 2 \cdot y - 2 \cdot z] - [(-1)^2 + 3 \cdot (-1) \cdot y - (-1)z]$$

$$= 4 + 6y - 2z - 1 + 3y - z$$

$$= 9y - 3z + 3$$

Then we evaluate the y-integral.

$$\int_1^3 \left(\int_{-1}^2 (2x + 3y - z)\, dx \right) dy$$

$$= \int_1^3 (9y - 3z + 3)\, dy$$

$$= \left[\frac{9}{2}y^2 - 3zy + 3y \right]_1^3 \qquad \begin{array}{l}\text{Integrating with respect} \\ \text{to } y; \ z \text{ is a constant}\end{array}$$

$$= \left(\frac{9}{2} \cdot 3^2 - 3z \cdot 3 + 3 \cdot 3 \right) - \left(\frac{9}{2} \cdot 1^2 - 3z \cdot 1 + 3 \cdot 1 \right)$$

$$= \frac{81}{2} - 9z + 9 - \frac{9}{2} + 3z - 3$$

$$= 42 - 6z$$

Next we evaluate the z-integral.

$$\int_0^1 \left(\int_1^3 \int_{-1}^2 (2x + 3y - z) \, dx \, dy \right) dz$$

$$= \int_0^1 (42 - 6z) \, dz$$

$$= 6 \int_0^1 (7 - z) \, dz$$

$$= 6 \left[7z - \frac{z^2}{2} \right]_0^1 \quad \text{Integrating with respect to } z$$

$$= 6 \left[\left(7 \cdot 1 - \frac{1^2}{2} \right) - \left(7 \cdot 0 - \frac{0^2}{2} \right) \right]$$

$$= 6 \left(7 - \frac{1}{2} \right)$$

$$= 6 \cdot \frac{13}{2}$$

$$= 39$$

17. $\int_0^1 \int_0^{1-x} \int_0^{2-x} xyz \, dz \, dy \, dx$

We first evaluate the z-integral.

$$\int_0^{2-x} xyz \, dz$$

$$= xy \int_0^{2-x} z \, dz \qquad x \text{ and } y \text{ are constants}$$

$$= xy \left[\frac{z^2}{2} \right]_0^{2-x} \qquad \text{Integrating with respect to } z$$

$$= xy \left[\frac{(2-x)^2}{2} - \frac{0^2}{2} \right]$$

$$= xy \cdot \frac{4 - 4x + x^2}{2}$$

$$= 2xy - 2x^2 y + \frac{1}{2} x^3 y$$

Then we evaluate the y-integral.

$$\int_0^{1-x} \left(\int_0^{2-x} xyz \, dz \right) dy$$

$$= \int_0^{1-x} \left(2xy - 2x^2 y + \frac{1}{2} x^3 y \right) dy$$

$$= \left[xy^2 - x^2 y^2 + \frac{x^3 y^2}{4} \right]_0^{1-x} \qquad \begin{array}{l} \text{Integrating with} \\ \text{respect to } y; x \text{ is} \\ \text{a constant} \end{array}$$

$$= \left[x(1-x)^2 - x^2(1-x)^2 + \frac{x^3(1-x)^2}{4} \right] -$$

$$\qquad \left[x \cdot 0^2 - x^2 \cdot 0^2 + \frac{x^3 \cdot 0^2}{4} \right]$$

$$= x(1 - 2x + x^2) - x^2(1 - 2x + x^2) +$$

$$\qquad \frac{x^3(1 - 2x + x^2)}{4}$$

$$= x - 2x^2 + x^3 - x^2 + 2x^3 - x^4 + \frac{1}{4} x^3 - \frac{1}{2} x^4 + \frac{1}{4} x^5$$

$$= \frac{1}{4} x^5 - \frac{3}{2} x^4 + \frac{13}{4} x^3 - 3x^2 + x$$

Then we evaluate the x-integral.

$$\int_0^1 \left(\int_0^{1-x} \int_0^{2-x} xyz \, dz \, dy \right) dx$$

$$= \int_0^1 \left(\frac{1}{4} x^5 - \frac{3}{2} x^4 + \frac{13}{4} x^3 - 3x^2 + x \right) dx$$

$$= \left[\frac{x^6}{24} - \frac{3x^5}{10} + \frac{13x^4}{16} - x^3 + \frac{x^2}{2} \right]_0^1 \quad \begin{array}{l} \text{Integrating with} \\ \text{respect to } x \end{array}$$

$$= \left(\frac{1}{24} - \frac{3}{10} + \frac{13}{16} - 1 + \frac{1}{2} \right) - 0$$

$$= \frac{10}{240} - \frac{72}{240} + \frac{195}{240} - \frac{240}{240} + \frac{120}{240}$$

$$= \frac{13}{240}$$

19. $\boxed{tw}$

Children's Literature